Multidisciplinary Approaches to Service-Oriented Engineering

Mehdi Khosrow-Pour
Information Resources Management Association, USA

A volume in the Advances in Computer and Electrical Engineering (ACEE) Book Series

IGI Global
DISSEMINATOR OF KNOWLEDGE

Published in the United States of America by
IGI Global
Engineering Science Reference (an imprint of IGI Global)
701 E. Chocolate Avenue
Hershey PA, USA 17033
Tel: 717-533-8845
Fax: 717-533-8661
E-mail: cust@igi-global.com
Web site: http://www.igi-global.com

Library of Congress Cataloging-in-Publication Data

Names: Khosrow-Pour, Mehdi, 1951- editor.
Title: Multidisciplinary approaches to service-oriented engineering / Mehdi Khosrow-Pour, editor.
Description: Hershey, PA : Engineering Science Reference, [2018] | Includes bibliographical references and index.
Identifiers: LCCN 2017058070| ISBN 9781522559511 (hardcover) | ISBN 9781522559528 (ebook)
Subjects: LCSH: Service-oriented architecture (Computer science)
Classification: LCC TK5105.5828 .M85 2018 | DDC 004.6/54--dc23 LC record available at https://lccn.loc.gov/2017058070

This book is published in the IGI Global book series Advances in Computer and Electrical Engineering (ACEE) (ISSN: 2327-039X; eISSN: 2327-0403)

British Cataloguing in Publication Data
A Cataloguing in Publication record for this book is available from the British Library.

Advances in Computer and Electrical Engineering (ACEE) Book Series

Srikanta Patnaik
SOA University, India

ISSN:2327-039X
EISSN:2327-0403

Mission

The fields of computer engineering and electrical engineering encompass a broad range of interdisciplinary topics allowing for expansive research developments across multiple fields. Research in these areas continues to develop and become increasingly important as computer and electrical systems have become an integral part of everyday life.

The **Advances in Computer and Electrical Engineering (ACEE) Book Series** aims to publish research on diverse topics pertaining to computer engineering and electrical engineering. **ACEE** encourages scholarly discourse on the latest applications, tools, and methodologies being implemented in the field for the design and development of computer and electrical systems.

Coverage

- Analog Electronics
- Sensor Technologies
- VLSI Design
- Electrical Power Conversion
- Programming
- VLSI Fabrication
- Power Electronics
- Computer science
- Algorithms
- Digital Electronics

IGI Global is currently accepting manuscripts for publication within this series. To submit a proposal for a volume in this series, please contact our Acquisition Editors at Acquisitions@igi-global.com or visit: http://www.igi-global.com/publish/.

The Advances in Computer and Electrical Engineering (ACEE) Book Series (ISSN 2327-039X) is published by IGI Global, 701 E. Chocolate Avenue, Hershey, PA 17033-1240, USA, www.igi-global.com. This series is composed of titles available for purchase individually; each title is edited to be contextually exclusive from any other title within the series. For pricing and ordering information please visit http://www.igi-global.com/book-series/advances-computer-electrical-engineering/73675. Postmaster: Send all address changes to above address.

Titles in this Series

For a list of additional titles in this series, please visit: www.igi-global.com/book-series

Soft-Computing-Based Nonlinear Control Systems Design
Uday Pratap Singh (Madhav Institute of Technology and Science, India) Akhilesh Tiwari (Madhav Institute of Technology and Science, India) and Rajeev Kumar Singh (Madhav Institute of Technology and Science, India)
Engineering Science Reference • copyright 2018 • 388pp • H/C (ISBN: 9781522535317) • US $245.00 (our price)

EHT Transmission Performance Evaluation Emerging Research and Opportunities
K. Srinivas (Transmission Corporation of Andhra Pradesh Limited, India) and R.V.S. Satyanarayana (Sri Venkateswara University College of Engineering, India)
Engineering Science Reference • copyright 2018 • 160pp • H/C (ISBN: 9781522549413) • US $145.00 (our price)

Fuzzy Logic Dynamics and Machine Prediction for Failure Analysis
Tawanda Mushiri (University of Johannesburg, South Africa) and Charles Mbowhwa (University of Johannesburg, South Africa)
Engineering Science Reference • copyright 2018 • 301pp • H/C (ISBN: 9781522532446) • US $225.00 (our price)

Creativity in Load-Balance Schemes for Multi/Many-Core Heterogeneous Graph Computing Emerging Research and Opportunities
Alberto Garcia-Robledo (Center for Research and Advanced Studies of the National Polytechnic Institute (Cinvestav-Tamaulipas), Mexico) Arturo Diaz-Perez (Center for Research and Advanced Studies of the National Polytechnic Institute (Cinvestav-Tamaulipas), Mexico) and Guillermo Morales-Luna (Center for Research and Advanced Studies of the National Polytechnic Institute (Cinvestav-IPN), Mexico)
Engineering Science Reference • copyright 2018 • 217pp • H/C (ISBN: 9781522537991) • US $155.00 (our price)

Free and Open Source Software in Modern Data Science and Business Intelligence Emerging Research and Opportunities
K.G. Srinivasa (CBP Government Engineering College, India) Ganesh Chandra Deka (M. S. Ramaiah Institute of Technology, India) and Krishnaraj P.M. (M. S. Ramaiah Institute of Technology, India)
Engineering Science Reference • copyright 2018 • 189pp • H/C (ISBN: 9781522537076) • US $190.00 (our price)

Design Parameters of Electrical Network Grounding Systems
Osama El-Sayed Gouda (Cairo University, Egypt)
Engineering Science Reference • copyright 2018 • 316pp • H/C (ISBN: 9781522538530) • US $235.00 (our price)

IGI Global
DISSEMINATOR OF KNOWLEDGE

701 East Chocolate Avenue, Hershey, PA 17033, USA
Tel: 717-533-8845 x100 • Fax: 717-533-8661
E-Mail: cust@igi-global.com • www.igi-global.com

Table of Contents

Detailed Table of Contents

Cloud computing is an emerging technology that attracts the attention of an increasing number of organizations that are willing to virtualize their infrastructure. Towards this direction, enterprises can deploy their applications on different public virtual machines (VMs), in order to reduce their operational costs, as well as optimize the performance of the offered applications. In such heterogeneous multi-cloud environments, the need for monitoring the quality of (a) the acquired resources, (b) the services offered to the final users, and (c) the offered service-based processes, and adapting them has come into force. Current research approaches addressing these areas are limited and most of them target a subset of these offerings. This chapter aims at proposing a novel event-based approach for cross-layer monitoring and adaptation of service-based applications (SBAs) deployed on multiple clouds. The proposed approach is empirically evaluated based on an invoice management application.

The purpose of this chapter is to clarify a three-part entrepreneurial strategy in a Japanese accounting cloud service with two-sided markets and freemium. The service provider established its platform as one of industry platform in Fintech and made its initial public offering in five years since its entrepreneurship. The authors studied the service provider and obtained three findings. First, the service provider dares to adopt its cloud platform with single architecture. Second, it builds up its service platform with two business strategies, two marketing strategies, and two management of technology. Third, the integration of these strategies becomes a three-part entrepreneurial strategy, which realizes economies of scale and scope cyclically. Finally, it realizes agglomeration economies in the service platform. Therefore, start-up with two-sided markets and freemium not only dares to design its service platform with single architecture but also should build up the three-part entrepreneurial strategy to formulate service platform as market before building up its business ecosystem.

The aim of the Cloud "as-a-service" concept is provisioning of software services to facilitate access to resources to a range of different users in different locations. Service adaptation through localization in this context can facilitate the internationalization and localization of services by allowing their adaption to different locales. Three localization dimensions are investigated: (1) lingual localization by providing service-level language translation techniques to adopt services to different languages; (2) regulatory localization by providing standards-based mappings to achieve regulatory compliance with regionally varying laws, standards, and regulations; and (3) social localization by taking into account preferences and customs for individuals and the groups or communities in which they participate. The objective is to support and implement an explicit modelling of aspects that are relevant to localization and runtime support consisting of tools and middleware services to automating the deployment based on models of locales, driven by the two localization dimensions.

With the increasing maturity of cloud technologies and the growing demand from customers, the cloud computing ecosystem has been expanding continuously with both incumbents and new entrants, whereby it has become more distributed and less transparent. For cloud service providers previously focusing on growth strategies, it is now necessary to shift the attention to providing service efficiently, as well as profitably. Based on 14 explorative interviews with cloud service experts, the relationship between cloud service provider profitability and value facilitation, which stands for the capability to build up resources in advance of future customer engagements, is investigated. The results indicate a positive relationship between cloud service profitability and value facilitation and deliver valuable insights for both researchers and practitioners. In particular, guidelines on how to design profitable cloud service offerings are discussed.

Researchers have done tremendous works for data security, but a robust security mechanism is not available yet. Researchers are doing continuous work to build robust SaaS mechanism. SaaS has several security vulnerabilities. Data security is still the most important challenge to researcher and they can constantly do research to protect the data over the network but they are facing numerous technical challenges to completely secure the cloud network and cloud storage. This work would be helpful to understand data security and privacy problems. Researchers can find the new way to understand SaaS security vulnerabilities and currently available solutions.

Change is an integral part of any software system. Predicting the impact of making changes through techniques of change impact analysis helps engineers identify and analyze those parts of the system that will be potentially affected by requested change(s). This chapter presents a semi-automated approach to (a) compute likelihood of impacted functions in a system through identification and analysis of functional

to check their compatibility by invoking the model checker NuSMV. Furthermore, the authors extended D&A4WSC so that it permits a web services choreography described in a WS-CDL specification. For this they developed a translation from WS-CDL to a composition of oWF-nets, so that one can verify this choreography by the aforementioned approach.

The periodic interactions represented as a dynamic network possess two aspects: structure and weight. This chapter introduces and explores the use of a third aspect that is associated with the periodic interactions, namely the directional aspect. Moreover, the authors have showcased how some applications require mining of patterns on both aspects—1) on direction and 2) on weight of directed interactions—for a better understanding of their behavior. With the aim of overcoming the limitation of existing frameworks, which only mine periodic patterns individually, a framework is proposed to mine periodic patterns on both the aspects together. Further, the patterns are analyzed to develop a better understanding of the dynamic network. A set of six parameters, defined later in the chapter, are used to conduct a microscopic study on the behavior of interactions. The framework is tested on real world and synthetic datasets. The results highlight its practical scalability and prove its efficiency.

Hybrid regression testing approaches involve the combinations of test suite selections, prioritizations, and minimizations. The hybrid approaches must reduce size of test suite to minimal level and enhance fault detection rate. The chapter proposes a new hybrid regression testing approach that reduces the number of test cases by reducing the paths of source code on the basis of the dependency between the statements and the changes. The proposed technique is evaluated to be better than the existing hybrid approach in terms of percentage savings in test cases and fault detection rate.

Software quality is imperative for industrial strength software. This quality will be often determined by a few components present in the software which decides the entire functionality. If any of these components are not rigorously tested, the quality will be highly affected. Without knowing which of these components are really critical, it will not be possible to perform high level testing. Hence, to predict such fault-prone or critical components from the software prior to testing and prioritizing them during the testing process, an agent-based approach is proposed in this chapter. The framework developed as part of this work will

Preface

Multidisciplinary Approaches to Service-Oriented Engineering is a vital reference source that explores some of the challenges associated with bringing system theories, model-driven software engineering techniques, and various ontologies into a systematic method for engineering service-oriented systems. Through discussions centered on topics such as: service-oriented architectures, semantic web, cloud computing, and regression testing, theoreticians, educators, developers, IT consultants, researchers, practitioners, and professionals, can further enrich their knowledge and stimulate future innovations within their respective industries and fields.

This reference source is organized into 15 chapters contributed by global experts, drawing on their experiences, observations, and research findings. A brief description of each of the chapters can be found in the following paragraphs.

In Chapter 1, the authors identify how and why supply chain resources and skills are deployed by franchisors to retain their franchisees thanks to a high level of service quality and therefore avoid the termination of contractual relationships. Also, based on a case study driven within a European franchise network, it is possible to conclude that the supply chain resources and competences of a franchisor play an important role in the duration of the franchisor-franchisee relationship when customized solutions are proposed thanks to information technologies.

In Chapter 2, the authors explore diabetes systems within service-oriented architectures with a decision aspect (SOAda) and provides clinical management and advice for patients. They aim to identify the current research state in the area, assist in deriving the key problems and features with the SOAda, and use these key features so that they could serve as a guide in the development of a new tool to address the decision problems in SOAda.

In Chapter 3, the authors address how enterprises can deploy their applications on different public virtual machines to reduce their operational costs, as well as optimize the performance of the offered applications. They also propose a novel event-based approach for cross-layer monitoring and adaptation of service-based applications deployed on multiple clouds. The proposed approach is empirically evaluated based on an invoice management application.

In Chapter 4, the authors clarify a three-part entrepreneurial strategy in a Japanese accounting cloud service with two-sided markets and freemium. They studied how the service provider established its platform as one of the industry platforms in Fintech.

In Chapter 5, the authors explore service adaptation through localization. Three localization dimensions are investigated: lingual localization by providing service-level language translation techniques to adapt services to different languages; regulatory localization by providing standards-based mappings to achieve regulatory compliance with regionally varying laws, standards, and regulations; and social localization by taking into account preferences and customs for individuals and the groups or communities in which they participate.

In Chapter 6, the authors investigate the relationship between cloud service provider profitability and value facilitation, which stands for the capability to build up resources in advance of future customer engagements. The results indicate a positive relationship between cloud service profitability and value facilitation.

In Chapter 7, the authors present research on how to protect data over networks while facing the challenges to completely secure the cloud network and cloud storage. They also explore new ways to understand Software as a Service (SaaS) security vulnerabilities and available solutions.

In Chapter 8, the authors present a semi-automated approach to compute the likelihood of impacted functions in a system through the identification and analysis of functional dependencies between them to assist software engineers in selective regression testing. This technique first classifies the impact set data into two categories based on their type of impact propagation. Next it performs a prioritization on the classified data to rank functions into higher and lower levels according to the degree of impact they will make.

In Chapter 9, the author gives a hybrid approach for the prioritization of software requirements by considering the regression testing efforts as one of the criteria. The author gives software engineers a technique for prioritizing both the requirements and aspects.

In Chapter 10, the authors explore the existing work done in the area of applying intelligent approaches to web personalization and highlighting the application of intelligent agents in this domain for solving many existing issues like personalized content management, user profile learning, modeling, and adaptive interactions with users.

In Chapter 11, the author provides a formal framework for ensuring the model checking-based verification of the composition of Web services. The author proposes to model the Web services composition by an interaction of open workflow nets: a special class of Petri nets.

In Chapter 12, the authors showcase how some applications require the mining of patterns on both the aspects of the direction and weight of the directed interactions, for a better understanding of their behavior. With the aim of overcoming the limitation of existing frameworks, which only mine periodic patterns individually, a framework is proposed to mine periodic patterns on both the aspects together.

In Chapter 13, the author proposes a new hybrid regression testing approach that reduces the number of test cases by reducing the paths of source code on the basis of the dependency between the statements and the changes. The proposed technique is evaluated to be better than the existing hybrid approach in terms of percentage savings in test cases and fault detection rates.

In Chapter 14, the author addresses how software quality is imperative for industrial strength software. This quality will be often determined by a few components present in the software which decides the entire functionality. If any of these components are not rigorously tested, the quality will be highly affected. The framework developed as part of this work, will certainly reduce the field failures and thus will improve the software quality.

In Chapter 15, the authors explore a new approach for the exploitation of Fuzzy Relational Databases (FRDB) described by the model GEFRED. This approach consists of a new technique for extracting summary fuzzy data, Fuzzy SAINTETIQ, based on the classification of fuzzy data and Formal Concepts analysis; the assessment of flexible queries in the context of FDB based on the set of fuzzy summaries generated by the Fuzzy SAINTETIQ system; the repair and substitution of unanswered queries.

The comprehensive coverage this publication offers is sure to enhance its reader's understanding of the latest topics and discoveries pertaining to service-oriented engineering. Furthermore, the contributions included within this publication will be instrumental to the expansion of knowledge offerings in diverse fields and will inspire future research and innovations.

Chapter 1
Exploring the Franchisor–Franchisee Relationship:
A Logistical Service-Oriented Perspective

Thierry Allègre
Aix-Marseille University (AMU), France

François Fulconis
Avignon University (UAPV), France

Gilles Paché
Aix-Marseille University (AMU), France

ABSTRACT

In the past 40 years, the franchising system has undergone a remarkable expansion, increasingly retaining the attention of economy and service management researchers. A recurring question relates to the sources of competitive advantage that a franchise network may have. This chapter intends to contribute to the debate based on the fundamentals of the resource/competence-based view, and applying the reflection to the case of the logistical service. The aim is to identify how and why supply chain resources and skills are deployed by franchisors to retain their franchisees thanks to a high level of service quality and therefore avoid the termination of contractual relationships. Based on a case study driven within a European franchise network, it is possible to conclude that the supply chain resources and competences of a franchisor play an important role in the duration of the franchisor-franchisee relationship when customized solutions are proposed thanks to information technologies.

INTRODUCTION

The study of franchise networks, and in particular of their governance, has given rise to many works putting forward important points: the development of marketing know-how from the franchisor to retain franchisees, the setting up of a central purchasing unit to enable a better bargaining with the suppliers, the financial support for struggling franchisees, etc. (Blair & Lafontaine, 2010; Webber, 2013; Meiklejohn,

DOI: 10.4018/978-1-5225-5951-1.ch001

2014; Beere, 2017). On the other hand, not much is said about the logistical support as a key resource to improving the performance of the franchising system. This is particularly surprising given that the large retailing industry has developed supply chain strategies since the 1980s in order to place the logistical performance within the centre of a policy of cost leadership, but also of differentiation (Fernie & Sparks, 2014: Ayers & Odegaard, 2017). Walmart's success, for example, is largely due to these reasons, with the early introduction of new technologies (satellite radio guidance of trucks, RFID tracking and tracing, etc.). It would be surprising that the franchising system, based on an important network of stores, does not also integrate a reflection regarding the supply chain to improve the franchisees' level of perceived satisfaction.

In a conventional way, the franchising system is similar to a partnership contract between two interdependent entities: the franchisor and franchisee. If the franchisor is at the root of the original concept, it does not have the sufficient financial resources to develop it on a large scale. It then turns to franchisees, which embrace the concept but will have to pay royalties on sales to benefit from the franchisor's support in terms of training, management tools, and access to the central purchasing unit, or the supply system. The aim of this chapter is to study the franchising system by questioning the role of logistical service as a key resource to increase the franchisees' level of perceived satisfaction. The research led by Grace, Frazer, Weaven and Dant (2016) regarding the factors of a franchisee's trust in their franchisor emphasised the importance of organisational factors such as the franchisee's commitment in the franchising system, the franchisee's perception of a strong team culture, and the franchisee's perception of the franchisor's character. However, the authors notice two other factors, more operational in nature: the franchisee's perception of the franchisor's competence and the franchisee's trust in the franchising system. In the two latter cases, it is interesting to know if the logistical expertise of the franchisor and more broadly, the way it manages its franchisees supply system, is part of the critical factors. A positive answer would mean that the logistical service plays an essential role in the quality of the contractual relationship between a franchisor and a franchisee, and therefore, it contributes to a sustainable process of value creation.

The conceptual model used to study the role of logistical service in the franchisor-franchisee relationship is the resource/competence-based view, later called RCBV. This model appears as one of the most appropriate as it examines the different resources at a firm's disposal in order to develop a sustainable competitive advantage. The resources in question articulate around competences allowing to deploy itself in the most efficient manner possible. To the extent that a franchise network relies on a group project that unites several partners, the franchisor's capacity to manage good relationships with its franchisees will constitute a central element to its dynamism. These are singular relational competences, poorly researched in the context of franchising system, and that deserves a particular attention. More precisely, it will be important to know in what way an efficient logistical service, capable of bringing a high level of service quality and reactivity at an acceptable cost for the franchisees, shows the characteristics of a key resource in the meaning of the RCBV. Even though Dupuis (1986) highlighted, over thirty years ago, that the franchisor's failures regarding logistics generated bad feelings by the French franchisees, no research program has been launched since then to investigate in this direction. The aim here is to suggest paths to fill in the gap identified from Dupuis' analysis (1986).

The results of a case study regarding a franchise network of the commercial catering industry in France will be presented and analysed in this chapter. A participant observation method was used for this case study. According to Bogdan, Taylor and DeVault (2015), the major advantage of this method is to result in an extensive period of social interaction between the researcher and the people being investigated in their own environment. Using a longitudinal research, one of its authors has followed the evolution of logistical patterns as well as strategies lead by the franchisor. While being involved in the management of interfaces between the franchisees stores, he had access to the indicators measuring the perceived level of efficiency of the logistical service. The methodology retained is of a qualitative and exploratory nature. It is based on the synthesis of direct observations made during strategic committee and Board meetings, as well as observations made on the ground, to the extent that, on an informal level, any visit of a place can be an excuse to a spontaneous gathering of data (Gagnon, 2012). The present case study, which comes close to the work done by Fristedt *et al.* (2012), does not intend to generalize. Its target is only to base itself on already identifiable outlines in order to understand, and with time, improve the governance of franchise networks.

THEORETICAL FOUNDATIONS

Franchise is a contractual model of development through a network, based on a close cooperation between legally and financially independent firms. The franchisor has developed an original commercial concept, and has marketing, purchasing and logistics (supply chain) know-how allowing to ensure the selling of a set of products and/or services. The latter is entrusted with franchisees who accept to take the risk of operating the concept developed, in compensation of a sufficient income. In this way, the franchisor and the franchisees participate in the construction of a franchise network intended to be durable, provided that it can generate sufficient shared profits. In the end, the franchising system can resemble the shape of collective entrepreneurship of which the franchisor is certainly the initiator, but for which the strength of innovation depends directly on the implication and the commitment of the franchisees in the common project. Implication and commitment are all the more difficult to obtain as the franchisees remain in an asymmetric situation *vis-à-vis* the franchisor, in terms of information, education or even contractual clauses (Buchan, 2013). To what extent the logistical service really constitutes a key resource for a franchise network? To answer this central question, we will use the RCBV, which origins are found in Penrose's seminal work (1959/2013).

Resource/Competence-Based View

At first approximation, the RCBV tries to explain the constitution of a competitive advantage based on distinct resources and competences held by a firm (Prahalad & Hamel, 1990; Peteraf, 1993). The central idea is that the companies are indeed capable of accumulating resources and competences that will transform into a sustainable advantage regarding the competition if they are rare, value creating, non-substitutable and difficult to imitate (Wernerfelt, 1984; Barney, 1991). This type of approach originally looked at resources and competences of a firm in particular, before developing the study within inter-organisational networks (including franchising system), to eventually consider the network as being the appropriate level to capture the localization of resources and competences (Afuah, 2000). The RCBV fills in the conceptual gap partly left by Porter's (1980) studies, in which are recognised the importance of managerial decisions and the strategies established for the firm's success: the latter is no longer perceived as an entity selected by its environment, but as an actor controlling its own future. The strategic intent is, *de facto*, one of the founding principles of this approach that refers to a more proactive approach than the corporate strategy. In other words, only the firms that know-how to make good use of certain critical resources within an environment can truly develop a sustainable competitive advantage.

Preliminary Definitions

Although it is disputed by certain academics (for a synthesis, see Kraaijenbrink, Spender & Groen, 2010), Penrose's (1959/2013) contribution is recognised as:

one of the more influential books of the second half of the twentieth century, bridging strategic management and organisational economics (Kor & Mahoney, 2000).

In order to provide a better knowledge of the creation of a competitive advantage, Penrose (1959/2013) suggests an explanatory logic to establish causal links between resources, opportunities and competitive advantage, which contribute to the creation of the competitive advantage itself. Extended by contributions aiming at better understanding the methods of support of the competitive advantage, Penrose's (1959/2013) work appears at the heart of the current development of modern RCBV within strategic management. The RCBV attributes to the company a triple function of resources allocation, creation of new resources and promotion of the income (Quélin, 1996). It perceives the company as a *collection of productive resources* (Penrose, 1959/2013), and difference in performance between companies and the heterogeneity of business organisation types are explained by the fact that they do not all possess the same set of resources and competences.

One of the main problems with the RCBV resides in the difficulty to define precisely the concepts used and make them operational for field research (Priem & Butler, 2001; Armstrong & Shimizu, 2007; Almarri & Gardiner, 2014). Sanchez, Heene and Thomas (1996) have brought necessary clarifications in a work of reference. *Resources* are defined as assets exchangeable on the market. It can be physical assets (equipment, premises, raw materials, etc.), financial resources, human capital, organisational resources (coordination mechanisms, planning, etc.), intangible assets (patents, brands, property rights) and information in the broad sense. *Competences* designate, on the other hand, the firm's capability to match, coordinate and deploy the resources to accomplish tasks (see Figure 1). The competences are not exchangeable on the market. They are "woven" internally by the teams interacting with one another, along with their environment. Therefore, it is important to avoid fragmented thinking by underlying, likewise Williamson (1999) that any firm is at the same time a bundle of related resources and competences, a bundle of routines, and a bundle of transactions (contracts).

Tangible and Intangible Resources

More recently, the resources are defined by Hunt and Davis (2008) as

the tangible and intangible entities available to the organisation that enable it to produce efficiently and/or effectively a market offering that has value for some market segment.

It appears from this definition that a resource can be tangible or intangible, that is to refer to a material or non-material existence, and owned or controlled by companies (Wernerfelt, 1984; Grant, 1991): the tangible resources affect all the elements of a firm having a physical existence; the intangible resources are based on information and knowledge. If we refer to the supply chains, a warehouse is a tangible resource whereas a share of information between supply chain members is an intangible resource, and the importance relies in knowing how tangible and intangible resources are bundled together (Rungtusan-

Figure 1. Typology of competences
Source: Adapted from Sanchez, Heene and Thomas (1996).

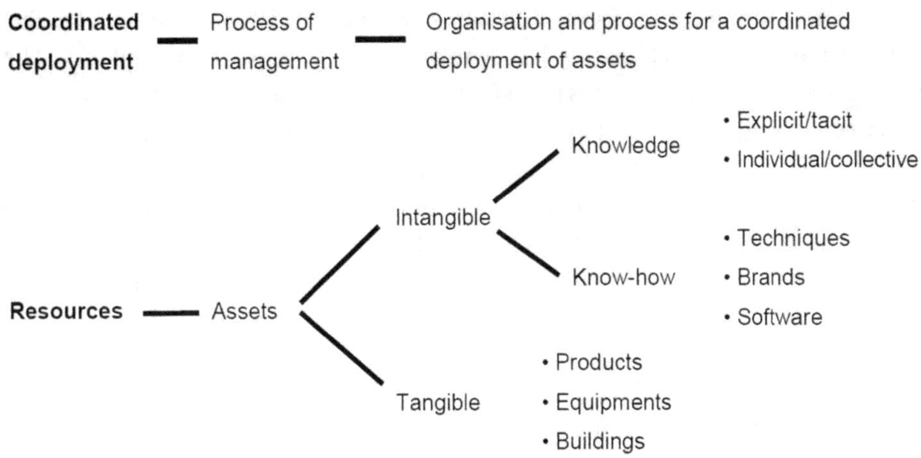

atham, Salvador, Forza & Choi, 2003; Brandon-Jones, Squire, Autry & Petersen, 2014). The immaterial resources can be ordered in two groups:

- The organisational knowledge constitutes a set of intra-organisational resources, mainly developed through learning; they correspond to accumulated collective knowledge that exceeds the sum of individual knowledge, and they can be shared to make the organisation more effective.
- The intangible "marginal" resources stem from the connection between the firm and its environment, such as the customer's loyalty or the knowledge deriving from the partnerships, the intangible marginal resources go beyond the organisation's border, as opposed to the organisational knowledge.

Sanchez, Heene and Thomas (1996) offer to distinguish the resources and assets on the one hand and the function of coordinated deployment of these same resources on the other hand. The management consists precisely in managing organisational processes that allow the firm to act by implementing the existing resources as well as providing itself with new resources and building new processes. Here, the competence essentially refers to the management process and to the intangible organisational elements built around the knowledge and the know-how, collective and individual; a "key competence" enables to achieve a sustainable competitive advantage (Grant, 1991), whether it is based on a differentiation strategy, a focus strategy or a cost leadership strategy. Sanchez (2000) underlines that the integration of issues regarding the inter-organisational relationships is one of the development paths of the competence approach:

By combining their resources in networks, firms that cooperate can benefit together from advantages such as asset pools, interconnections of assets and fixed durations to which an independent firm could not have access (Sanchez, 2000).

In the end, when resources are managed in such a way that their outcomes cannot be imitated, this ultimately creates a competitive barrier to entry (Mahoney & Pandian, 1992; Lavie, 2006).

Nowadays, the transaction cost economy, following Williamson (1975), is admitted as the most complete theoretical framework for addressing issues related to the company's boundaries. However, if it suggests that all governance structure aims at saving up transaction and production costs, on the contrary, the RCBV is based on the growth of created value and insists on the difficulty of accumulating resources internally. Often tacit in nature, non-codified and linked to employees, the resources enable to ensure the durability of a competitive position (Polanyi, 1967/2009). To the extent that it often constitutes the driving force of its advantage, management's role consists then in researching resources and competences that are difficult to imitate. The issue of the management of inter-organisational relationships is central here as it represents –or would represent– a means of access, acquisition or exchange of knowledge and competences enabling partner companies to increase their knowledge base and maintain their development (Kogut & Zander, 1993; Quélin, 1996; Nielsen, 2005; Costa & Monteiro, 2016). Inter-organisational relationships, more specifically the cooperation strategies between companies, rivals sometimes, aim at reinforcing the competitive potential of these companies by satisfying both efficiency and effectiveness requirements. The issue is then the relationship between the RCBV and the transaction cost theory. This relationship appears especially through the concept of asset specificity. If generic assets can be obtained thanks to market mechanisms with low transaction costs, the specific (or idiosyncratic) assets call for the use of a governance structure as a firm or an inter-organisational cooperation.

Individual and Collective Levels

Beyond the identification of competences, appears the question of knowing what prevails in their construction. In other words, are the key competences at an individual or a collective level? Regarding this subject, the researchers are unanimous: in organisations, the competences are systemic by nature. Like so, Black and Boal (1994) admit that the connections woven between individuals of an organisation account for as much, if not more, than the capabilities of these same individuals in isolation. Synergy effects create and develop, making inimitable the network of relationships at the origin of the competences; the organisations are the place of creation of the collective competences that are not transferable as such. Persais (2004) defines the relational competences as *the capability of a firm to weave and maintain a positive and durable link with key actors of the environment*. For this author, the concept is wider than that of the network resources to the extent that it embraces all the actors around *relational competences* developed through continuous interactions that link them together. For many years, strategic procurement management, which seeks to capture and stabilise the external resources to increase the performance of the firm, naturally puts the accent on relational competences (Cox, 1996; Paulraj, Lado & Chen, 2008; Whipple, Wiedmer & Boyer, 2015).

As an inter-organisational system, is a supply chain based, in the same way, on the establishment of relational competences that allow the development of a sustainable competitive advantage? And to achieve that, is the naturally collective functioning of a supply chain involving, for example a manufacturer, a large retailer and a logistics service provider (LSP), based on inter-organisational resources that can explain the performance gaps between different supply chains? For Hunt and Davis (2012), the answer is clearly positive. A supply chain is characterised by singular capacities to learn, innovate or adapt to environmental transformations. This will only be the case if the supply chain members interact efficiently between each other. Relational resources, allowing to resolve conflicts and to create trust, participate in the creation of relational competences, whose impact is positive in terms of improved coordination of logistical operations. This improved coordination betters value creation across the supply chain, in particular by avoiding wastes of means and a useless recurrence of capacity (of transport, of storage, etc.).

Franchising System

All franchising systems rely on a set of complex interactions between a franchisor and a variety of organisational entities. Chen (2018) identifies four main entities in the framework of a franchisor's learning process, of which Figure 2 represents the main characteristics: (1) *business units and customers*, in the first line of which are the franchisees; (2) *prospective franchisees*, that need to be captured to develop the franchise network in the following years; (3) *suppliers*, bringing resources and competences needed for the functioning of a franchise network; and (4) the *governmental institutions*, in charge of the regulation of contractual relationships, both on legal and financial levels, in particular to resolve any litigation between franchisors and franchisees. The aim of the four entities identified by Chen (2018) is to perform

Figure 2. Complex interactions in the franchising system
Source: Chen (2018).

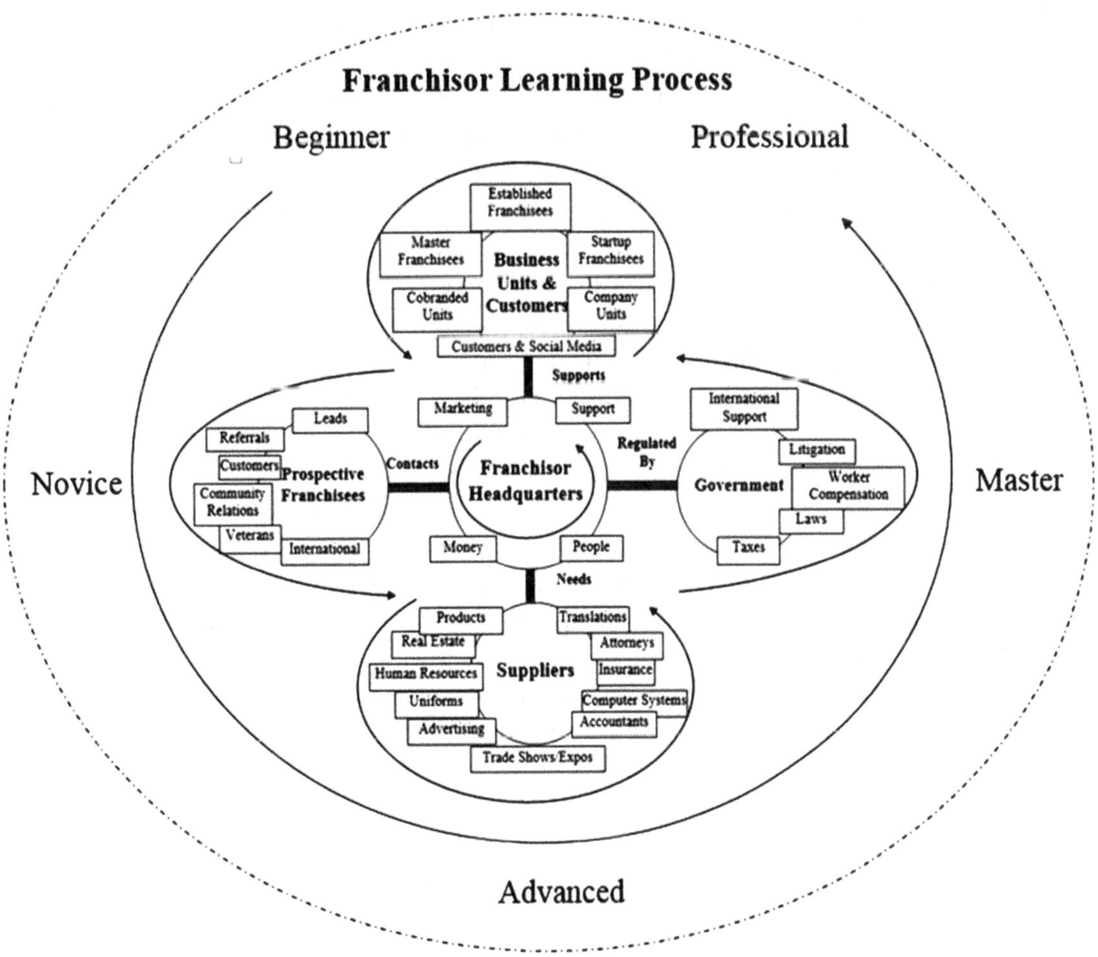

a certain number of marketing and supply activities, in particular in the sales of products and services, in the creation of powerful brand names essential to attract new franchisees, and in the efficient management of activities of logistical support, including information systems.

De Castro, Mota and Marnoto (2009) question the pertinence of the two usual reasons retained to explain the success of the franchising system: a way to face the scarcity of resources and/or a mechanism leading the franchisor and the franchisee to align on the same incentive mechanism. According to the authors, such a perspective minimises the learning processes in franchise networks and the construction of a type of relational capital between the partners. The latter is due to the idiosyncratic nature of the relationships that influence the manner of mobilising the resources, of arranging them according to a common project, beyond the simple replication and utilisation of a "good recipe". If the capability to associate, coordinate and deploy resources to accomplish tasks of different nature is indisputable at the levels of purchasing, marketing, logistics or the layout of the stores, the relational competences to govern the network constitutes an approach of utmost importance on a strategic level, and demands singular organisational know-how. Unsurprisingly, the analysis led by Chabowski, Hult and Mena (2011) confirms that the channel relationship management constitutes an important theme of research on the franchising system in the years 2000-2010.

As Perdreau, Le Nadant and Cliquet (2015) noted in reference to critical resource perspective and to the agency theory, a franchise-type development helps to rapidly mobilise the financial resources, as well as the managerial resources to materialise a common project. The inputs are both those of the franchisor and the franchisee. The franchisor's inputs stem from the specific know-how of the network (for example, the selection of the best locations) and of the storefront branding (for example, the advertising investments); López-Fernández and López-Bayón (2018), for that matter, underline that some early terminations in franchising are accountable to a lack of experience of the franchisor, perceived by the franchisees, in terms of both marketing and operations management. On the other side, the franchisee contributes to the common project with inputs in terms of knowledge or specific know-how of the local market. Along with Perdreau, Le Nadant and Cliquet (2015) precision, the resource is difficult to imitate (except by franchisees of a rival network on the same catchment area), without a real substitute (except through managers employed in stores), relatively rare and with qualities in terms of understanding and reactivity of the market which confer it an intrinsic value. In reference to the contributions of Wernerfelt (1984) and Barney (1991), here are the dimensions which give the resources the opportunity to bring to a firm a sustainable competitive advantage.

The main strength of the franchisor is also the ability to transform the tacit knowledge issued from the franchisees in formalised knowledge that can be spread out in the entire franchise network (Lindblom & Tikkanen, 2010), as well as the efficient transfer of intellectual capital between franchisor and franchisees (Sanfelix & Puig, 2018). More generally, the sharing of information between franchisors and franchisees to improve the performance of a franchise network constitutes a major research theme, in particular at the level of functioning of the supply chain, organised by the franchisor, and more generally the value co-creation process (Paswan, D'Souza & Rajamma, 2014). Indeed, as Yan and Wang (2012) underline, the demand forecasting constitutes a critical point at two levels for all franchise networks: (1) it allows to adapt the inventory level of the market demand more rigorously; (2) it leads to strongly reducing the cost of inventory shortage. The authors conclude that the sharing of information between franchisors and franchisees generates optimal shared profits for the partners, which is verified with the help of a Bayesian forecasting methodology in high-tech industries, where the volatility of the market demand is very important. In other words, an efficient management of marketing and logistics relationships in the franchise network directly impacts the performance of value creation.

The work led in France by El Akremi, Herrbach, Mignonac and Perrigot (2009) on the know-how of the franchise networks places it perfectly within this issue. It relies on the RCBV to identify and rank the know-how at the root of competitive advantage in franchise networks. Two main types of know-how have been highlighted. On the one hand, the business know-how, that are specific and difficultly imitable, but nevertheless *codified, taught and transferred to the franchisees* (El Akremi, Herrbach, Mignonac & Perrigot, 2009), they unite all that derives from the positioning of the franchise, both in terms of the characteristics of its offer as well as its demand. On the other hand, the organisational know-how, that are defined as the franchisor's capability to franchise with, on one side, all *the engineering related to the creation and to the transfer of business know-how to its entire pool of franchisees*, on the other side, *the management and the governance of network allowing to minimise the transaction and control costs* (El Akremi, Herrbach, Mignonac & Perrigot, 2009). The authors inscribe inter-organisational know-how in the relational theory where mutual trust and efficient management of conflicts are central concepts; unlike precedent know-how, they are not passed on to franchisees.

Methodologically, El Akremi *et al.* (2009) identify and measure the characteristics inherent to the different know-how of franchise networks, compare the monitoring of these know-how depending on the networks' characteristics (age of the network, size, degree of gender mix, etc.), analyse and explain the links between, on the one hand, the nature of the know-how and their organisation, on the other hand, the competitive advantage within the franchise and, *in fine*, the performance of franchise networks. A factor analysis allows to reveal eight main dimensions of the know-how: (1) codification, transmission and replication; (2) support to human resource management; (3) monitoring and control of units; (4) external communication; (5) support and internal cohesion; (6) organisational agility; (7) purchasing and logistics; and (8) access to finance. The main results indicate that the degree of variance accounted for by the organisational know-how is superior to the degree of variance accounted for by the business know-how. It is, then, the former which contributes a great deal to the efficiency of the networks, illustrating the pertinence of a relational approach: the competitive advantage, the success and the longevity of a network depend more and more on the franchisor's capability to manage relationships with and between franchisees, to mobilise and federate them around a common project.

Logistics for Service-Oriented System

Does the logistical support that a franchisor can bring to its franchisees constitute an inter-organisational know-how of utmost importance in order to ensure the durability of the franchise network? The review of competitive strategies driven in the retailing industry since the 1980s clearly shows that the supply chain performance of a large retailer allows to increase its market shares, to reduce the stores' supply costs and very rapidly adjust the product assortment to unexpected variations of the demand (Fernie & Sparks, 2014). We can therefore imagine that the franchising system, as a component of the retailing industry, can also be sensitive to the supply chain performance, in particular by strongly improving the service delivering process, which Kaner and Karni (2010) highlighted its central character in the competitiveness of modern organisations. The research led by Chancey *et al.* (2016) on a network of restaurant franchise underline, for that matter, that the franchisor was able to carry out a new organisation between the distribution centres and the demand points (franchisees) which made possible an improved management of operational constraints of the customer. This tends to show that, by being able to offer a high level of service in stores, reducing their level of out-of-stock, the franchisor probably increases the level of satisfaction of consumers, and consequently of the franchisee. In other words, it is a question of whether the logistical resources mobilised by a company can enable it, or not, to build a competitive advantage based on the efficient flow management (Panayides, 2007; Wong & Karia, 2010; Christopher, 2016).

Logistical Resources and Competences

By the late 1950s, Selznick's (1957/1984) introduced the concept of distinctive competence in order to represent the singular success of an organisation, based on the activities in which it is better than its competitors. The wise deployment of resources, Selznick (1957/1984) underlined, enables to achieve a unique competitive position, and these resources can be internal or external. Here, the competence embraces the resources, as well as the different types of knowledge and management practices controlled by the company; for that matter, the knowledge traditionally plays an essential role in the explanation of the performance of the franchising system (Jeon, Dant & Baker, 2016). The reasoning is applicable to the supply chain –including the supply chain built by a franchisor– as an organisation that associates many members to provide products to the client under optimal costs, services and responsive conditions. Logistical resources and competences refer to a set of tangible and intangible assets, as well as organisational attributes in the hands of several partners who are committed in sharing them to improve logistical processes. Table 1, adapted from Mentzer, Min and Bobbitt (2004), suggests a classification of logistical resources and competences that influence the long-term profitability of the firm.

Hult, Ketchen and Nichols (2003) underline how the durable commitment of partners in the collective functioning of a supply chain is essential to increase its performance, in particular in terms of risk management, achievement of economies of scale, or acquisition of new technologies. In a general manner, the commitment made by a firm to its partners constitutes an important source of innovation, to the extent that the innovation is less and less the result of the contribution of a single firm, but the result of a continuous interaction process between different partners (Soosay, Hyland & Ferrer, 2008; Oke, Prajogo & Jayaram, 2013; Ralston, Richey & Grawe, 2017). The mutual commitment of partners in a relationship, by developing new logistical resources and competences, consequently constitutes a significant source of creation of competitive advantage. In the context of supply chains, conducting joint investments in technology (transport, handling, etc.), establishing integrated procedures of flow man-

Table 1. Classification of logistical resources and competences

Logistical Resources	Logistical Competences
Tangible Resources	**Demand Management Interface**
• Plants	• Customer service
• Equipments	• Logistics quality
• Raw materials	**Supply Management Interface**
• Distribution centres	• Low cost distribution
• Logistics network	• Low cost supply
Intangible Resources	**Information Management**
• Relationships	• Information sharing
• Management skills	• Connectivity
• Knowledge	**Coordination**
• Logistical expertise	• Internal coordination
• Logistical service	• External coordination

Source: Adapted from Mentzer, Min and Bobbitt (2004).

agement, exchanging or lending staff leads then to a sharing of resources, competences and knowledge between partners (Dyer & Nobeoka, 2000; Min et al., 2005; Blome, Schoenherr & Eckstein, 2014; Tan, Wong & Chung, 2016).

The company Jardiland is an excellent example of the essential role that is played by the logistical investment of the franchisor to improve the quality of service perceived by the franchisees. Jardiland is one of the most powerful networks of garden centres in Europe (Perrigot & Martinez-Ribes, 2008), with over 300 stores commercialising approximately 300,000 references. The management of logistical operations was assigned to DHL Supply Chain, supplying franchisees from two warehouses and guaranteeing availability rates between 95% and 98%. The monitoring of flows is centralised towards a logistics division of Jardiland, which interacts in real-time with DHL Supply Chain to ensure an automatic restocking of franchisees. At the operational level, DHL Supply Chain manages the picking of common products and activities of cross docking from pallets already prepared by the different suppliers. Given that Jardiland organises approximately fifty promotional operations every year to support the sales of its franchisees, DHL Supply Chain is also in charge of the promotional picking by managing the deliveries by the unit or by the parcel in order to optimise the assembly of pallets in warehouses. In total, the logistical investment agreed upon by Jardiland represented 4.5 million Euros, but it enabled the franchisees to go from an average sales turnover of 750 euros to 1,000 euros per square meter of retail space thanks to a centralised and automated restocking system. The resounding success of Jardiland regarding supply chain organisation probably explains the interest aroused by the company Gamm Vert that took control of it in October 2017.

Among the elements that reinforce the partners' will to engage in an exchange relationship, special attention was paid to the communication and, more broadly, the exchange of information within the supply chains (Paulraj, Lado & Chen, 2008; Sezen, 2008; Ramayah & Omar, 2010; Wu, Chuang & Hsu, 2014; Singh & Teng, 2016). Many studies show the central role that is played by inter-organisational communication, as a shared logistical resource, in the maintaining and strengthening of commitment (Modi & Mabert, 2007; Paulraj, Lado & Chen, 2008; Ren, Hu, Ngai & Zhou, 2015). Indeed, efficient and rapidly flowing communication contributes to reducing the risks of failure and conflicts, while increasing the trust they have for one another (Leuschner, Rogers & Charvet, 2013). It participates in the development of common knowledge to supply chain members, by facilitating mutual adjustments in case of problems, or again by participating in the creation of more responsive supply chains (Wu, Yeniyurt, Kim & Cavusgil, 2006; Baihaqi & Sohal, 2013; Tao, Lee & Chew, 2016). The performance of information systems is found at several complementary levels: a decrease of the error rate, a greater speed and an improved reliability in data transfers, a global tracking and tracing of products, a decrease of delivery times, and a control of stock levels. Paulraj, Lado and Chen (2008) have, for that matter, underlined the importance of inter-organisational communication as a *relational competence* in the development of collaborative relationships between supply chain members.

Logistical Expertise and Legitimacy

Through the mastery of a supply chain approach, the franchisor disposes of a power of expertise and legitimacy *vis-à-vis* its franchisees. It mainly relies on the franchisor's ability to control logistical operations, which is part of an overall strategy of standardisation of the elements of its corporate policy, including marketing and human resource management (Kellner, 2017). For example, the establishment of tracking and tracing tools enables an efficient management of flows helping the franchisee to know in real time the status of restocking in progress; moreover, this perfectly corresponds to the qualitative control mechanisms identified by Cordobés Madueño and Soldevila García (2015) in the monitoring of franchise networks. As Croonen and Brand (2013) noted, efficient logistics is one of the main determinants of trustworthiness associated with the franchisor's operational management. The franchisor can directly involve the franchisees in the organisation (or the reorganisation) of the supply chain, it can escalate the logistical issues encountered by the franchisees to modify the structure of the supply chain, or it can also establish from the start a strict bill of specifications to prevent the franchisees from having problems of a logistical nature.

The example of Great Harvest Bread Company in the USA –a franchise network offering an unusual autonomy to its 200 franchisees for franchise operations until the early 2010s– clearly shows that the franchisor only managed to restore the competitiveness of the franchise network because it unilaterally decided to centralise and streamline the entire supply chain. This strategy was led both at the level of deliveries towards bakeries and at the level of procurement of different ingredients from a variety of sources, by strongly increasing the satisfaction of franchisees who had nevertheless lost a significant part of their autonomy (Davis & Serna Molina, 2011). Now, paradoxically, Combs, Ketchen Jr., Shook and Short (2011) indicate that the logistical support brought to the franchisees has not yet been investigated in depth, and it always constitutes a stimulating path of research. Two main issues are to be addressed according to the authors:

How can franchisors create value for franchisees via supply chain? Under what conditions do these value-added supply chains make franchising more attractive than creating one's own business? (Combs, Ketchen Jr., Shook & Short, 2011).

Obviously, these issues refer to the manner in which the quality of support services offered by the franchisor participate or not in the emergence of relational governance, in which franchisee satisfaction has a central place.

From this perspective, the research led by Fristedt *et al.* (2012) on McDonald's Sweden is very instructive regarding the results. The authors' idea was to compare the fundamentals of the supply chain management (SCM) with the firm's practices, in order to know if it had encouraged a genuine supply chain approach, as said in numerous corporate communication documents. According to Fristedt *et al.* (2012), we can confirm that McDonald's Sweden respects the fundamentals of SCM, in particular in terms of coordination and joint planning between parties. Additionally, the supply chain of McDonald's Sweden has come a long way in reducing redundant inventories and speed of operations, which shows as a perfect synchronisation of flows and a significant improvement of inventory management. In a more general way, McDonald's Sweden has developed a recognised logistical expertise, and the trust the franchisees have had in it for numerous years indisputably reassures its power of influence:

The long term relationship that is common for the supply chain of McDonald's and having a mutual trust of each party in the supply chain is a fundamental stage in order to have long relationships that are fruitful for all parties and also benefits the entire supply chain (Fristedt et al., 2012).

This example underlines, if needed, the interest of a field study to evaluate how the logistical service can be a key resource for a franchise network.

CASE STUDY

This part of the chapter will give a concrete dimension to the importance that is necessary to allocate to logistics as a key resource for the efficient functioning of a franchise network. As indicated in the introduction, the participant observation method was privileged. It looks at a worldwide franchise network, largely implanted in France (more than 1,300 stores in 2017), and whose specificity is to lean on the same LSP from its beginning. For confidentiality reasons, we will call it Omega in this chapter. Firstly, we will deal with the logistics dedicated to this franchise network, in a general manner. Secondly, we will highlight a few specificities linked to its management: the management of physical flows and information systems, the organisational architecture and the interdependences that can exist between Omega, its LSP and its franchisees.

Main Characteristics of the Logistical Service

A constant element of the franchising system as a contractual model is to methodologically duplicate the good practices that lead to the success of the brand *vis-à-vis* its clients (Buchan, 2013). This is true within a national network, and above all if the franchise has a global reputation, as the aim will be to reproduce the key learning of the organisation at a larger scale. Beyond the business model and the quality of franchisee recruitment, the quality and the durability of the relationship between the franchisor and its suppliers within an efficient supply chain are essential elements to guarantee the success of the franchise network. The case studied shows the particularity to have developed in the USA in the first place in the 1950s before becoming international at the end of the 1960s. After a growth of thousands of stores, it establishes in France in the early 1980s, where it has had an exceptional growth since then.

The distinct feature of Omega is that it has systematically leaned on its products and service suppliers to develop, as well as and mostly to duplicate and adapt its concept to local constraints, for example in terms of offer, by integrating the logistics issue in its "*glocal*" strategic approach. Omega wishes to bring to its franchisees, logistics is considered as a key factor of success, as much as marketing, to increase its market shares and perceived value of its offer *vis-à-vis* the competitors. Four main characteristics can define the logistics of Omega's network:

- **The Logistics Leans on Historical LSP:** When it comes to France, the LSP that is in charge of the logistical service for franchisees has been part of the distribution system since 1956. In 2017, it was Omega's main LSP worldwide, in charge of over half of the tonnage distributed in the 37,000 stores.
- **Logistics is of an Exclusive Character:** The operating model expected by Omega in France excludes pooling with other clients, just as it excludes the sharing of its logistics with different LSPs; the logistical organisation of the LSP is unique and totally dedicated to the client.
- **Logistics is of a Contributive Nature:** In the operating mode of the dedicated contract, the LSP must think in terms of absorption of fixed costs (in coherence with the growth of the number of stores) and in terms of innovation, in order (1) to enable the decrease of logistical costs; and (2) to bring higher level of distinction with the rest of the market.
- **Logistics Leans on an Equalisation Process:** No matter the localisation of the store, whether it is a few hundred meters from the distribution centre or hundreds kilometres away, the price of the logistical service will be the same for each franchisee, uniquely based on the tonnage delivered.

The quality of the relationship of the LSP with Omega is essential. It is the keystone of a productive and, above all, durable approach. This approach must cover multiple aspects that bring an advantage to the end user: an optimal quality service; the respect and follow-up of seasons; an economically competitive offer; an efficient trading relationship; a constant research for innovation to the benefit of an always more efficient logistical service. This range of parameters contributes to the upholding of the LSP's performance, and in the long term, to the durability of the relationship with Omega. The notion of quality of the logistical service is, here, an essential measurement dimension of performance for the franchisor. Two indicators are retained for all the countries where the client is established: (1) the punctuality of deliveries (any late delivery or delivery made 30 minutes over of the schedule established with the franchisee is considered as "not punctual"); (2) the reliability of the order delivered (any delivery on which a product is missing, broken or in excess is considered as "unreliable"). These indicators allow the measurement, at the level of a country's franchise network, of the rate of guaranteed services to the stores, as well as the comparison of the performance of Omega's LSP worldwide.

The logistical organisation must contribute to the global performance of the franchisor through a control of costs that enables the guarantee of an acceptable selling price for the consumer, as well as the support on an economic level of the growth and the opening of new stores. Without underestimating the profitability of the LSP, whose services are paid for at the right price, Omega has opted for a business model that puts forward the importance of factors of construction of a relational climate based on the commitment with its partner. The presence of a dedicated contract implies a permanent re-actualisation of the cost of the logistical service that takes into account external events (such as the green tax) and the necessary research of an increase in productivity. Ultimately, the LSP is an interface actor placed between the franchisor and the franchisee, and whose relational management competences serves two types of partners: (1) Omega is the global client of the LSP, it decides on the franchise network's strategy, negotiates the logistical sourcing conditions with the franchisees and suppliers, and seeks the best cost/service settlement; (2) the franchisees are evaluating clients of the logistical service, they are considered in a way as tactical purchasing advisors of the implemented logistics.

Logistical Features

The elements presented here above put in perspective the singular character of the relationship that unites the LSP to Omega. It is important to go into further details of the features of the logistical service, in particular compared to the current market's standards of the logistics industry in France (Fulconis, Paché & Roveillo, 2011). The originality of the case studied is the fact that it is based on a dedicated contract. This means that the 550,000 tons delivered each year are done through a single logistical organisation for more than 1,300 stores. For that reason, Omega uses seven distribution centres that perform more than 220,000 deliveries of fresh, frozen and dry products every year that is an average of three deliveries per week and per store. Figure 3 shows the place held by each distribution centre in Omega's supply chain, between the suppliers and the stores. Beyond the storing, order preparation and the transport activities, additional services are offered to complete the logistical offer, and the target of which is to bring added value to the franchise network served.

Figure 3. The place of the distribution centre in the supply chain

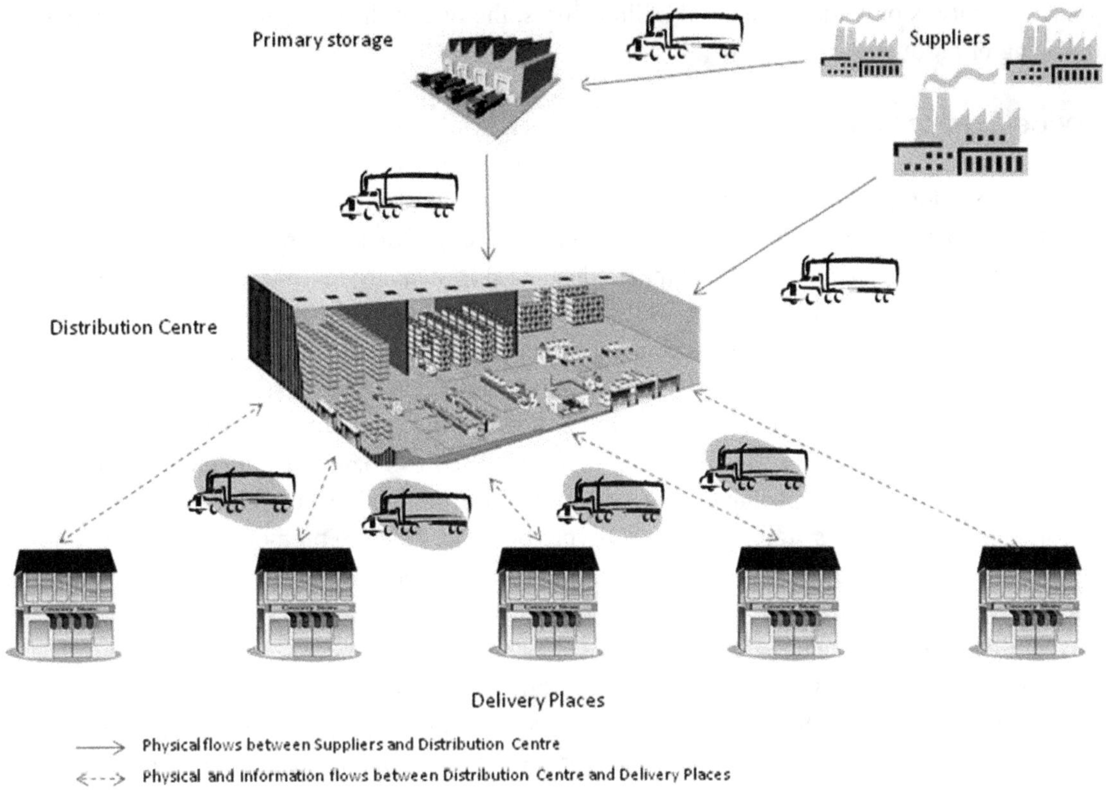

Between Logistics and Procurement

Based on a business model established in the 1970s, Omega's headquarters are in charge, in each country, of the sourcing of the products, i.e. the identification of the needs, the characteristics and the quantities of the products that will be consumed, in particular when it comes to promotional products of which the selling lifespan is limited, as well as the price of the product. Once these elements are put under contract with the supplier of the product, the LSP is then in charge of the management of the supplies towards the distribution centres by regulating the flows to guarantee that each level of the logistical organisation has a sufficient quantity of products and aims at decreasing the out-of-stocks in store. The products become the ownership of the LSP as soon as they arrive at the distribution centre. They are then invoiced to the store when delivered. From this point of view, the LSP is closer to the traditional concept of wholesaling intermediation in the distribution channels.

The relationship between Omega and its LSP shows an important feature concerning the transfer of property. Unlike the most traditional case in logistics industry, where the LSP is not the owner of the product but only the flow manager on the behalf of a shipper, including additional customer services like shelf display or customer billing (Fulconis, Paché & Roveillo, 2011), here it is a trade activity, such as buying and selling products, directly fitted in the logistics activity. However, the trade activity generates no margin for the LSP. The logistical costs resulting from the supplying of stores are simply integrated in the global logistical costs invoiced to the store on the basis of the kilogramme delivered. When the sales turnover of Omega's French LSP is more than one billion euros in 2017, the share linked to its logistics activity is only more than 100 million euros, the rest of the sales turnover corresponds to the purchasing of products.

Service Customisation

In addition to its logistical offer, ten years ago, Omega's LSP had also established an information technology with the objective of supporting the stores in the process of making order. The tools enable a daily collection of scanning data and recreate the products sold in standard logistics units. This simplifies the stores' task in terms of making orders, while reducing stock in the distribution centres. The store can therefore develop a fine tuning of its stock management by taking into account the product mix specific to its catchment area. The aim is to parameter a safety stock level specific to each type of product, as well as measuring possible gaps in stock or productivity. At the level of the franchise network, the data history and the abundance of events in the past ten years enable to benefit from reproducible and quantifiable models as to their impact in a given situation. For example, it is possible to evaluate the impact of a competing franchisor's store, or a BOGO-type (buy one, get one) promotional campaign done locally, and also to bring an efficient support to the opening of one of Omega's new store by identifying the level of expected activity.

In parallel, a complementary computer tool was managed by the LSP to physically organise the storage space in each store in order to optimise the reception capacity. The idea is to consider that the stock management competence is held by the LSP and that sharing the latter with the stores can only positively contribute to a better utilisation of spaces and storage capacity. The combination of data regarding the perspectives in terms of sales turnover (held by the store), of the average consumption per product mix (stemming from the order proposal tool) and of the optimisation of stocks leads to more rational deliveries on a just-in-time model. Indeed, the conjunction of these elements translate into delivery rates adapted to the needs of each store per type of temperature, which size is optimised, and thus, lead to a better global command of logistical costs. Here, we join Streed and Cliquet's (2017) suggestion which considers that the performance of a franchise network rests on the capacity to face a growing individualisation of demand thanks to an advanced customisation of the service offer.

The Distribution Centres in the Supply Organisation

In general, information technologies have undergone an important development in the logistics industry over the past decade, in particular to improve the flow management in the warehouses (Fulconis, Paché & Roveillo, 2011; Ruchton, Croucher & Baker, 2014; Christopher, 2016). The LSP working with Omega is no exception to the rule. It has therefore implemented a warehouse management system (WMS) that enables to track in real time any movement of products from its arrival on the platforms of its distribution centres up to its disposal in the stores. The unit of work is the pallet, and the specification needs that contractually link Omega to its LSP indicates that every product must be located within the three hours that follow the request. From an operational point of view, the pallet is flashed as soon as it arrives in the distribution centre, during its placing in storage, then during the picking phase. The order preparation is done by personnel assigned to this activity using the technology of *picking by voice* (the order clerks are equipped with headphones and the WMS indicates the location of the parcel, as well as the quantities to take).

Concerning the transport, different information systems also support the optimisation of the activity realised within the distribution centres. On the one hand, the LSP uses a delivery round optimisation software, the assignment of which is double: from a strategic point of view, it allows to simulate numerous store supplying plans; from a tactical point of view, given the actual volumes sent, it suggests a daily delivery round. On the other hand, the LSP has equipped delivery vehicles with a telematics box that passes on information linked to the driving (energetic consumption, accumulated braking distance, etc.), as well as the geo-localisation of the products during the journey, which contributes to a more rigorous management of round delivery costs. Finally, the LSP appeals to a transport management system (TMS) that, in the same way as the WMS in the warehouse, aims at structuring the organisation and the daily planning of the delivery activity.

Evidently, taken individually, the elements of the logistical offer brought by Omega through its LSP can without a doubt end up with other LSPs having developed, for example, a sophisticated approach of the management of interfaces between manufacturers and large retailers. However, it is exceptional to find such a set of tools to help a unique client, with such customisation *finesse* for each of the franchisees depending on their mix of products and promotional offers organised locally. Instead of forcing each franchisee in the general logistical model conceived by the LSP depending on its own profitability targets, the logistical service is calibrated as tightly as possible depending on the structuring of the franchise network, its size in number of stores, of its volumetry in terms of tonnage delivered and its marketing complexity. In other words, the LSP has developed a type of "Swiss army knife" of perfectly adapted logistical solutions, that strengthens the durable relationship built between the franchisor and its franchisees, and that, undoubtedly, largely explains the loyalty of the franchisees to the franchise network.

DISCUSSION AND CONCLUSION

The Omega case pertinently shows some of the points addressed in the academic studies dedicated to the management of franchising system. Here, the target is not to adopt a deductive perspective in order to validate or invalidate the research hypothesis, but to underline in what way the choice made by Omega in its relationship with its franchisees correspond to a strategic intent to improve the performance of the franchise network. As we have underlined, the logistical service has a central place for Omega according to a double dimension: a technical dimension linked to the provision of tools allowing to facilitate the management of flows; an inter-organisational dimension linked to the coordination process of exchanges between franchisor and franchisees. These two dimensions are inextricably associated to a global approach of SCM as presented by Fristedt *et al.* (2012). Two aspects draw particular attention in reference to the literature analysed in the chapter: the importance of mobilised competences; the central nature of information sharing. This enables making recommendations for action for the attention of members of a franchising system.

Conceptual and Managerial Implications

The idea of competences, as retained here, refers to tangible elements as well as intangible elements built around the collective knowledge and collaborative learning (Chen, 2018). The relational competences are based more specifically on the quality of the interactions maintained between the actors in the long term, leading to the creation of a positive dynamics in the capability of a franchise network to adapt to its environment. The Omega case clearly shows a strong relational competence based on the performance of the supply chain, it constitutes an organisational know-how in the meaning of El Akremi *et al.* (2009). Indeed, the franchisor suggests that its franchisees adapt its stock management and its delivery rounds to the need of each store, in particular depending on its promotional operations (with the BOGO skill). Even more, the logistical service adapts to the product mix and to the variations of its day-to-day volume, as the supplying system is based on a high frequency. If Omega is capable of such a performance, it means that it is capable of adding value to the accumulated knowledge in the relationship forged with its franchises over twenty-five years. Therefore, Omega benefits from the increasing know-how in its supply chain experience, allowing it to facilitate the opening of new stores and anticipating the flow to its destination. For the franchisee, this is the guarantee of a low level of out-of-stock as from the very first hours of the store's opening.

The relational competence that is established between Omega and its franchisees is based on the sharing of information, facilitating the process of management of physical flows in an unstable environment. This aspect is strongly put forward in the studies of Wu *et al.* (2006), Paulraj, Lado, and Chen (2008), Yan and Wang (2012) and Huo, Han and Prajogo (2016). Without a doubt, it is another aspect of the organisational know-how that Omega develops, and the impact of which is significant in terms of network governance. The WMS, implemented by the franchisor, leads to a tracking and a tracing of products from their arrival on the distribution centres until their availability in stores. Omega's essential strength remains however the ability to consolidate the precise information on the perspective of sales turnover, store by store, and on the average consumption per product mix. The information is passed from the stores in real time and leads to an instantaneous adjustment of the logistical capacities of Omega's LSP to face the demand variations from one day to another. Thus, each disruptive event, just as the real impact of promotional operations on the sales level, is immediately integrated in the process of flow management in the service of the stores.

Even if the Omega case cannot be generalised, it is however possible to progress in the direction suggested by Combs *et al.* (2011) concerning the necessary depth of knowledge of the SCM mechanisms in the franchising system: how can franchisors create value for franchisees via supply chain? Thus, we can observe that if the tangible resources in the matter of logistics hold an important role (warehouses, trucks, etc.), the intangible resources linked to the organisational and managerial knowledge must not be underestimated. The issue of the governance of a franchise network refers to the coordination mechanisms between the actors, and especially in the way with which their continuous interactions favour the performance of the supply chain. In the Omega case, these interactions concern the franchisor and its franchisees, as well as the franchisor and its LSP, that have managed to establish technological and organisational resources facilitating the collective process of value creation. From this point of view, we recognise here the essence of the franchising system the success of which is in function with the commitment of all the actors in the common projects that greatly exceeds the legal barriers of each actor.

Recommendations for Action

To summarise, the success of Omega resides in the existence of dedicated contracts, the genuine understanding of the client's stakes by the LSP and a real will to accompany the management and the monitoring of logistical activities, on an operational and strategic level. Logistics become the "armed wing" of the franchise network and a differentiation criterion for the benefit of the entire network members. The logistical service as a key resource for a franchise network in terms of SCM would have to be further discussed, based on the organisation of transactions and the RCBV, as well as on the technologies. Indeed, if we admit that the relationships correspond to a coherent chain of transactions, considering the logistical service through the only market-hierarchy option –with the transaction as sole unit of analysis– could be reconsidered. This is a major challenge for the contractual approach even though, on the one hand, some studies highlight that the transaction can be the pertinent unit of analysis to understand the acquisition mechanisms of collective knowledge and competences (Zhou et al., 2014; Loebbecke, van Fenema & Powell, 2016) and, on the other hand, some research insists on the complementarity of the approach through both transactions and competences (Williamson, 1999; Jacobides & Winter, 2005; Kim, 2017). From a managerial point of view, the Omega case enables, in a complementary manner, to make three important recommendations for action for the attention of franchising systems.

Firstly, it is necessary that the franchisor maintains a durable and privileged relationship with its LSP for two main reasons: (1) the LSP must immerse itself totally in its client's specific culture; (2) the LSP must develop highly specific investments to customise the service offer for franchisees. Here, a durable relationship means the presence of long term contracts and that are dedicated to the franchisor's needs. Indeed, if undifferentiated logistics can lead an LSP to make significant economies of scale, by managing logistical infrastructures common to several franchise networks, this could be at the cost of a deterioration of the service quality. Indeed, the LSP working in a same warehouse on behalf of several franchisors will seek satisfactory solutions common to all, which can result sometimes in a service quality evaluated as mediocre by one of the franchisors, while being evaluated as satisfactory by one or many other franchisors. However, it is obvious that a poor management of a franchisor's supply chain, in terms of quality of delivery and respect of deadlines, will directly impact the franchisee's level of sales. Moreover, the presence of long-term contracts and that are dedicated to the franchisor's needs makes it easier to introduce marketing innovations in order to meet the very special customisation needs that can issue from franchisees of a same franchise network. As noticed in the Omega case, this customisation of service offer tends to facilitate the loyalty process of franchisors during the renewal of franchise contracts.

Secondly, the franchisor must be capable of building with its suppliers a powerful business ecosystem in the meaning of Moore (1996), that is to say an economic community based on the interaction of companies to produce goods and services by bringing value to consumers. Rather than being considered as a client among others, the franchisor will stimulate its community of suppliers to achieve, from collaborative practices, objectives of costs reduction, synergy of resources and technology innovations. The LSP in charge of the supply of franchisees is naturally involved in this process, for example, by partnering with other suppliers for the purchase of commodities (furniture, consumable goods, etc.). In other words, the franchisor should not work in silo with its suppliers, by looking for a global optimisation of its upstream supply chain, and by involving all the members of its business ecosystem in collective strategies. Therefore, it is easy to imagine that the franchisor's LSP initiates a new way of organising the filling of palettes in the warehouses, by working in common with suppliers, to reduce supply costs borne by the franchisees, without deteriorating the service quality. This approach would be perceived in a very positive way by the franchisees.

Thirdly, it is desirable that the franchisees be greatly involved in the choice of organisation of the supply chain, which they depend on, including by being associated with the final decisions taken by the franchisor. We could therefore imagine the systematising of logistical joint committees consisting of the franchisors' and the franchisees' representatives in order to ratify supply chain choices in terms of redefinition of logistical service costs, budgetary commitment, evaluation of the LSP or again monitoring promotional operations at the level of stock management. The franchisees' representatives will receive a mandate from the entire franchisee community to ensure that the franchisor takes into account the opinion and the recommendations issuing from the field so as to modify the supply chain, if necessary. As underlined by Argyres and Bercovitz (2015), a franchise community can dispose of a strong bargaining power, and the franchisor must take it into account to avoid placing itself in a situation of dependence. The franchisees' representatives, associated with the franchisor's supply chain experts, will speak on behalf of a community, and will not defend their own individual interests. A particular effort must therefore be led by the franchisee community to choose the best representatives, which interest for the supply chain theme should be verified *ex ante*. It is also the franchisor's task to successfully demonstrate that the performance of the supply chain demands a global vision of the franchise network, independently of the local situation experienced by this or that franchisee.

REFERENCES

Afuah, A. (2000). How much do your "co-opetitors" capabilities matter in the face of technological change? *Strategic Management Journal, 21*(3), 387–404. doi:

Almarri, K., & Gardiner, P. (2014). Application of resource-based view to project management research: Supporters and opponents. *Procedia: Social and Behavioral Sciences, 119*, 437–445. doi:10.1016/j.sbspro.2014.03.049

Argyres, N., & Bercovitz, J. (2015). Franchisee associations as sources of bargaining power? Some evidence. *Journal of Economics & Management Strategy, 24*(4), 811–832. doi:10.1111/jems.12111

Armstrong, C., & Shimizu, K. (2007). A review of approaches to empirical research on the resource-based view of the firm. *Journal of Management, 33*(6), 959–986. doi:10.1177/0149206307307645

Ayers, J., & Odegaard, M.-A. (2017). *Retail supply chain management* (2nd ed.). Boca Raton, FL: CRC Press.

Baihaqi, I., & Sohal, A. (2013). The impact of information sharing in supply chains on organisational performance: An empirical study. *Production Planning and Control, 24*(8-9), 743–758. doi:10.1080/09537287.2012.666865

Barney, J. (1991). Firm resources and sustained competitive advantage. *Journal of Management, 17*(1), 99–120. doi:10.1177/014920639101700108

Beere, R. (2017). *The role offranchising on industry evolution: assessing the emergence offranchising and its impact on structural change*. Basingstoke, UK: Palgrave Macmillan. doi:10.1007/978-3-319-49064-9

Black, J., & Boal, K. (1994). Strategic resources: traits, configurations and paths to sustainable competitive advantage. *Strategic Management Journal, 15*(SI), 131-148. doi:10.1002/smj.4250151009

Blair, R., & Lafontaine, F. (2010). *The economics offranchising* (2nd ed.). New York: Cambridge University Press.

Blome, C., Schoenherr, T., & Eckstein, D. (2014). The impact of knowledge transfer and complexity on supply chain flexibility: A knowledge-based view. *International Journal of Production Economics, 147*, 307–316. doi:10.1016/j.ijpe.2013.02.028

Bogdan, R., Taylor, S., & DeVault, M. (2015). *Introduction to qualitative research methods: a guidebook and resource* (4th ed.). New York: John Wiley & Sons.

Brandon-Jones, E., Squire, B., Autry, C., & Petersen, K. (2014). A contingent resource-based perspective of supply chain resilience and robustness. *The Journal of Supply Chain Management, 50*(3), 55–73. doi:10.1111/jscm.12050

Buchan, J. (2013). *Franchisees as consumers: benchmarks, perspectives and consequences*. New York: Springer. doi:10.1007/978-1-4614-5614-8

Chabowski, B., Hult, T., & Mena, J. (2011). The retailing literature as a basis for franchising research: Using intellectual structure to advance theory. *Journal of Retailing, 87*(3), 269–284. doi:10.1016/j.jretai.2011.04.001

Chancey, E., Flores, M., Luis, J., Báez Palma, M., Badillo Valenzuela, O., Cabrera, S., & Arlethe, M. (2016). Redesign of the supply chain of a restaurant franchise in the food industry. *Global Journal of Business Research, 10*(2), 103–111.

Chen, Y.-S. (2018). Utilizing information science and technology in franchise organizations. In M. Khosrow-Pour (Ed.), *Encyclopedia of information science and technology* (4th ed.; pp. 4822–4835). Hershey, PA: IGI Global. doi:10.4018/978-1-5225-2255-3.ch418

Christopher, M. (2016). *Logistics and supply chain management* (5th ed.). Harlow: Pearson Education.

Combs, J., Ketchen, D. Jr, Shook, C., & Short, J. (2011). Antecedents and consequences of franchising: Past accomplishments and futures challenges. *Journal of Management, 37*(1), 99–126. doi:10.1177/0149206310386963

Cordobés Madueño, M., & Soldevila García, P. (2015). Management control in inter-organizational relationships: The case of franchises. *Rivista Innovar, 25*(58), 23–36. doi:10.15446/innovar.v25n58.52357

Costa, V., & Monteiro, S. (2016). Key knowledge management processes for innovation: A systematic literature review. *Journal of Information & Knowledge Management Systems, 46*(3), 386–410. doi:10.1108/VJIKMS-02-2015-0017

Cox, A. (1996). Relational competence and strategic procurement management: Towards an entrepreneurial and contractual theory of the firm. *European Journal of Purchasing & Supply Management, 2*(1), 57–70. doi:10.1016/0969-7012(95)00019-4

Croonen, E., & Brand, M. (2013). What makes franchisees trust their franchisors? In T. Ehrmannn, J. Windsperger, G. Cliquet, & G. Hendrikse (Eds.), *Network governance: alliances, cooperatives and franchise chains* (pp. 207–229). Heidelberg, Germany: Physica-Verlag. doi:10.1007/978-3-7908-2867-2_12

Davis, E., & Serna Molina, J. (2011). *Great Harvest Bread Company: leveraging the supply chain for competitive advantage (Darden Case No. UVA-OM-1437)*. Charlottesville, VA: Darden Graduate School of Business Administration, University of Virginia.

De Castro, L., Mota, J., & Marnoto, S. (2009). Toward a relational perspective of franchising chains. *Service Business, 3*(1), 15–30. doi:10.100711628-008-0050-6

Dupuis, M. (1986). *Distribution, la nouvelle donne*. Paris: Les Editions d'Organisation.

Dyer, J., & Nobeoka, K. (2000). Creating and managing a high-performance knowledge-sharing network: The Toyota case. *Strategic Management Journal, 21*(3), 345–367. doi:

El Akremi, A., Herrbach, O., Mignonac, K., & Perrigot, R. (2009). *Quels sont les savoir-faire sources de l'avantage concurrentiel en franchise?* Paris: Fédération Française de la Franchise.

Fernie, J., & Sparks, L. (Eds.). (2014). *Logistics and retail management: emerging issues and new challenges in the retail supply chain* (4th ed.). London: Kogan Page.

Fristedt, M., Hansson, A., Huge-Brodin, M., Rehme, J., & Sandberg, E. (2012). *Supply chain management in practice. A case study of McDonald's Sweden* (Report LIU-IEI-R-12/0006). Linköping: Department of Management and Engineering, Linköping University.

Fulconis, F., Paché, G., & Roveillo, G. (2011). *La prestation logistique: origines, enjeux et perspectives*. Caen: Editions Management & Société.

Gagnon, Y.-C. (2012). *L'étude de cas comme méthode de recherche* (2nd ed.). Québec: Presses de l'Université du Québec.

Grace, A., Frazer, L., Weaven, S., & Dant, R. (2016). Building franchisee trust in their franchisor: Insights from the franchise sector. *Qualitative Market Research, 19*(1), 65–83. doi:10.1108/QMR-09-2014-0085

Grant, R. (1991). The resource-based theory of competitive advantage: Implications for strategy formulation. *California Management Review, 33*(3), 114–135. doi:10.2307/41166664

Hult, G., Ketchen, D. Jr, & Nichols, E. Jr. (2003). Organizational learning as a strategic resource in supply management. *Journal of Operations Management, 21*(5), 541–556. doi:10.1016/j.jom.2003.02.001

Hunt, S., & Davis, D. (2008). Grounding supply chain management in resource-advantage theory. *The Journal of Supply Chain Management, 44*(1), 10–21. doi:10.1111/j.1745-493X.2008.00042.x

Hunt, S., & Davis, D. (2012). Grounding supply chain management in resource-advantage theory: In defense of a resource-based of the firm. *The Journal of Supply Chain Management, 48*(2), 14–20. doi:10.1111/j.1745-493X.2012.03266.x

Huo, B., Han, Z., & Prajogo, D. (2016). Antecedents and consequences of supply chain information integration: A resource-based view. *Supply Chain Management, 21*(6), 661–677. doi:10.1108/SCM-08-2015-0336

Jacobides, M., & Winter, S. (2005). The co-evolution of capabilities and transaction costs: Explaining the institutional structure of production. *Strategic Management Journal, 26*(5), 395–413. doi:10.1002mj.460

Jeon, H.-J., Dant, R., & Baker, B. (2016). A knowledge-based explanation of franchise system resources and performance. *Journal of Marketing Channels, 23*(3), 97–113. doi:10.1080/1046669X.2016.1186470

Kaner, M., & Karni, R. (2010). An ideation framework for service process improvement. *International Journal of Service Science, Management, Engineering, and Technology, 1*(3), 17–49. doi:10.4018/jssmet.2010070102

Kellner, A. (2017). Human resource management standardisation and adaptation in franchises. *Service Industries Journal, 37*(9-10), 545–566. doi:10.1080/02642069.2017.1339794

Kim, H.-J. (2017). The complementary effects of transaction cost economics and resource-based view: A technological alliance perspective. *International Journal of Business Excellence, 13*(3), 355–376. doi:10.1504/IJBEX.2017.087230

Kogut, B., & Zander, U. (1993). Knowledge of the firm and the evolutionary theory of the multinational corporation. *Journal of International Business Studies, 24*(4), 625–645. doi:10.1057/palgrave.jibs.8490248

Kor, Y., & Mahoney, J. (2000). Penrose resource-based approach: The process and product of research creativity. *Journal of Management Studies, 37*(1), 109–139. doi:10.1111/1467-6486.00174

Kraaijenbrink, J., Spender, J., & Groen, A. (2010). The resource-based view: A review and assessment of its critiques. *Journal of Management, 36*(1), 349–372. doi:10.1177/0149206309350775

Lavie, D. (2006). The competitive advantage of interconnected firms: An extension of the resource-based view. *Academy of Management Review, 31*(3), 638–658. doi:10.5465/AMR.2006.21318922

Leuschner, R., Rogers, D., & Charvet, F. (2013). A meta-analysis of supply chain integration and firm performance. *The Journal of Supply Chain Management, 49*(2), 34–57. doi:10.1111/jscm.12013

Lindblom, A., & Tikkanen, H. (2010). Knowledge creation and business format franchising. *Management Decision, 48*(2), 179–188. doi:10.1108/00251741011022563

Loebbecke, C., van Fenema, P., & Powell, P. (2016). Managing inter-organizational knowledge sharing. *The Journal of Strategic Information Systems, 25*(1), 4–14. doi:10.1016/j.jsis.2015.12.002

López-Fernández, B., & López-Bayón, S. (2018). (forthcoming). Antecedents of early terminations in franchising: Franchisor versus franchisee cancelations. *Small Business Economics*. doi:10.10071118

Mahoney, J., & Pandian, J. (1992). The resource-based view within the conversation of strategic management. *Strategic Management Journal, 13*(5), 363–380. doi:10.1002mj.4250130505

Meiklejohn, A. (Ed.). (2014). *Franchising: cases, materials, and problems*. Chicago, IL: American Bar Association.

Mentzer, J., Min, S., & Bobbitt, M. (2004). Toward a unified theory of logistics. *International Journal of Physical Distribution & Logistics Management, 34*(8), 606–627. doi:10.1108/09600030410557758

Min, S., Roath, A., Daugherty, P., Genchev, S., Chen, H., Arndt, A., & Richey, G. (2005). Supply chain collaboration: What's happening? *International Journal of Logistics Management, 16*(2), 237–256. doi:10.1108/09574090510634539

Modi, B., & Mabert, A. (2007). Supplier development: Improving supplier performance through knowledge transfer. *Journal of Operations Management, 25*(1), 42–64. doi:10.1016/j.jom.2006.02.001

Moore, J. (1996). *The death of competition: leadership and strategy in the age of business ecosystems*. New York: HarperBusiness.

Nielsen, B. (2005). The role of knowledge embeddedness in the creation of synergies in strategic alliances. *Journal of Business Research*, *58*(9), 1194–1204. doi:10.1016/j.jbusres.2004.05.001

Oke, A., Prajogo, D., & Jayaram, J. (2013). Strengthening the innovation chain: The role of internal innovation climate and strategic relationships with supply chain partners. *The Journal of Supply Chain Management*, *49*(4), 43–58. doi:10.1111/jscm.12031

Olavarrieta, S., & Ellinger, A. (1997). Resource-based theory and strategic logistics research. *International Journal of Physical Distribution & Logistics Management*, *27*(9-10), 559–587. doi:10.1108/09600039710188594

Panayides, M. (2007). Effects of organizational learning in third-party logistics. *Journal of Business Logistics*, *28*(2), 133–159. doi:10.1002/j.2158-1592.2007.tb00061.x

Paswan, A., D'Souza, D., & Rajamma, R. (2014). Value co-creation through knowledge exchange in franchising. *Journal of Services Marketing*, *28*(2), 116–125. doi:10.1108/JSM-09-2013-0254

Paulraj, A., Lado, A., & Chen, I. (2008). Inter-organizational communication as a relational competency: Antecedents and performance outcomes in collaborative buyer-supplier relationships. *Journal of Operations Management*, *26*(1), 45–64. doi:10.1016/j.jom.2007.04.001

Penrose, E. (2013). *The theory of the growth of the firm*. Eastford, CT: Martino Fine Books. (Original work published 1959)

Perdreau, F., Le Nadant, A.-L., & Cliquet, G. (2015). Human capital intangibles and performance of franchise networks: A complementary view between agency and critical resource perspectives. *Managerial and Decision Economics*, *36*(2), 121–138. doi:10.1002/mde.2656

Perrigot, R., & Martinez-Ribes, L. (2008). Innovation in franchising: some illustrations from the French franchising chains. *Proceedings of the 7th International Congress on Marketing Trends*, 1-24.

Persais, E. (2004). Les compétences relationnelles peuvent-elles s'avérer stratégiques? *Revue Française de Gestion*, *158*(148), 119–145. doi:10.3166/rfg.148.119-145

Peteraf, M. (1993). The cornerstones of competitive advantage: A resource-based view. *Strategic Management Journal*, *14*(3), 179–191. doi:10.1002mj.4250140303

Polanyi, M. (1967/2009). *The tacit dimension*. Chicago, IL: University of Chicago Press.

Porter, M. (1980). *Competitive advantage: creating and sustaining superior performance*. New York: The Free Press.

Prahalad, C.-K., & Hamel, G. (1990). The core competence of the corporation. *Harvard Business Review*, *68*(3), 79–91.

Priem, R., & Butler, J. (2001). Is the resource-based theory a useful perspective for strategic management research? *Academy of Management Review*, *26*(1), 22–40. doi:10.2307/259392

Quélin, B. (1996). Appropriability and the creation of new capabilities through strategic alliances. In R. Sanchez & A. Heene (Eds.), *Strategic learning and knowledge management* (pp. 139–160). New York: John Wiley & Sons.

Ralston, P., Richey, R., & Grawe, S. (2017). The past and future of supply chain collaboration: A literature synthesis and call for research. *International Journal of Logistics Management*, *28*(2), 508–530. doi:10.1108/IJLM-09-2015-0175

Ramayah, T., & Omar, R. (2010). Information exchange and supply chain performance. *International Journal of Information Technology & Decision Making*, *9*(1), 35–52. doi:10.1142/S0219622010003658

Ren, S., Hu, C., Ngai, E., & Zhou, M. (2015). An empirical analysis of inter-organisational value co-creation in a supply chain: A process perspective. *Production Planning and Control*, *26*(12), 969–980. doi:10.1080/09537287.2014.1002022

Ruchton, A., Croucher, P., & Baker, P. (Eds.). (2014). *The handbook of logistics and distribution management* (5th ed.). London: Kogan Page.

Rungtusanatham, M., Salvador, F., Forza, C., & Choi, T.-Y. (2003). Supply chain linkages and operational performance: A resource-based view perspective. *International Journal of Operations & Production Management*, *23*(9), 1084–1099. doi:10.1108/01443570310491783

Sanchez, R. (2000). Une comparaison des approches de la ressource, des capacités dynamiques, et de la compétence: une contribution à la théorie du management stratégique. In B. Quélin & J.-L. Arrègle (Eds.), *Le management stratégique des compétences* (pp. 55–81). Paris: Ellipses.

Sanchez, R., Heene, A., & Thomas, H. (1996). Towards the theory and practice of competence-based competition. In R. Sanchez, A. Heene, & H. Thomas (Eds.), *Dynamics of competence-based competition: theory and practice in the new strategic management* (pp. 1–35). New York: John Wiley & Sons.

Sanfelix, G., & Puig, F. (2018). New challenges in franchisor-franchisee relationship: An analysis from agency theory perspective. *Cuadernos de Gestión*, *18*(1), 85–102. doi:10.5295/cdg.150610gn

Selznick, P. (1957/1984). *Leadership in administration: a sociological interpretation*. Berkeley, CA: University of California Press.

Sezen, B. (2008). Relative effects of design, integration and information sharing on supply chain performance. *Supply Chain Management*, *13*(3), 233–240. doi:10.1108/13598540810871271

Singh, A., & Teng, J. (2016). Enhancing supply chain outcomes through information technology and trust. *Computers in Human Behavior*, *54*, 290–300. doi:10.1016/j.chb.2015.07.051

Soosay, C., Hyland, P., & Ferrer, M. (2008). Supply chain collaboration: Capabilities for continuous innovation. *Supply Chain Management*, *13*(2), 160–169. doi:10.1108/13598540810860994

Streed, O., & Cliquet, G. (2017). Autonomy in franchising. In F. Hoy, R. Perrigot, & A. Terry (Eds.), *Handbook of research on franchising* (pp. 50–64). Cheltenham, UK: Edward Elgar. doi:10.4337/9781785364181.00011

Tan, K.-H., Wong, W.-P., & Chung, L. (2016). Information and knowledge leakage in supply chain. *Information Systems Frontiers*, *18*(3), 621–638. doi:10.100710796-015-9553-6

Tao, Y., Lee, L.-H., & Chew, E.-P. (2016). Quantifying the effect of sharing information in a supply chain facing supply disruptions. *Asia-Pacific Journal of Operational Research, 33*(4), 1–28. doi:10.1142/S0217595916500299

Webber, R. (2013). *An introduction to franchising*. New York: Palgrave Macmillan. doi:10.1007/978-1-137-29610-8

Wernerfelt, B. (1984). A resource-based view of the firm. *Strategic Management Journal, 5*(2), 171–180. doi:10.1002mj.4250050207

Whipple, J., Wiedmer, R., & Boyer, K. (2015). A dyadic investigation of collaborative competence, social capital, and performance in buyer-supplier relationships. *The Journal of Supply Chain Management, 51*(2), 3–21. doi:10.1111/jscm.12071

Williamson, O. (1975). *Markets and hierarchies: analysis and antitrust implications. A study in the economics of internal organization*. New York: The Free Press.

Williamson, O. (1999). Strategy research: Governance and competence perspectives. *Strategic Management Journal, 20*(12), 1087–1108. doi:

Wong, Y.-C., & Karia, N. (2010). Explaining the competitive advantage of logistics service providers: A resource-based view approach. *International Journal of Production Economics, 128*(1), 51–67. doi:10.1016/j.ijpe.2009.08.026

Wu, F., Yeniyurt, S., Kim, D., & Cavusgil, T. (2006). The impact of information technology on supply chain capabilities and firm performance: A resource-based view. *Industrial Marketing Management, 35*(4), 493–504. doi:10.1016/j.indmarman.2005.05.003

Wu, L., Chuang, C.-H., & Hsu, C.-H. (2014). Information sharing and collaborative behaviors in enabling supply chain performance: A social exchange perspective. *International Journal of Production Economics, 148*, 122–132. doi:10.1016/j.ijpe.2013.09.016

Yan, R., & Wang, K.-Y. (2012). Franchisor-franchisee supply chain cooperation: Sharing of demand forecast information in high-tech industries. *Industrial Marketing Management, 41*(7), 1164–1173. doi:10.1016/j.indmarman.2012.06.013

Zhou, K., Zhang, Q., Sheng, S., Xie, E., & Bao, Y. (2014). Are relational ties always good for knowledge acquisition? Buyer-supplier exchanges in China. *Journal of Operations Management, 32*(3), 88–98. doi:10.1016/j.jom.2014.01.001

Chapter 2
Migration of Diabetes Diagnosis System to Service-Oriented Architectures:
Using Decision Model and Notation

Boumahdi Fatima
Blida University, Algeria

Rezoug Nachida
Blida University, Algeria

Madani Amina
Blida University, Algeria

Chalal Rachid
High School of Computer Science, Algeria

ABSTRACT

Studies in diabetic systems are vital for both diabetes patients and the medical industry because it is possible to use these systems to improve healthcare quality in several ways. These systems can help doctors to treat patients with all types of conditions. This research selected service oriented architecture with a decision aspect (SOAda) as the core architecture for the reason that it functions well when the domain decision is not adequately clear. This chapter demonstrates the study of the diabetes systems within service-oriented architecture and provides clinical management and advices for patients. Its aim is to (1) identify the current research state in the area, (2) assist in deriving the key problems and features with the SOAda, and (3) use these key features so that they could serve as a guide in the development of a new tool to address the decision problems in SOAda.

INTRODUCTION

Method engineering techniques albeit proven effective for paradigms such as function, object and component, are pertinent to limited aspects of service orientation (Baghdadi, 2015). SOA is considered as the current and new generation of architectural style that can be utilized in computer systems. It is also

DOI: 10.4018/978-1-5225-5951-1.ch002

considered a layered architectural style that is constructed around automated services. One of SOA's foundations is the service inventory, and the SOA quality is affected by how the organization of the inventory is done. Within the business services layer, the services are business-oriented, while those found in other layers are seen as technically oriented. Business people play a vital role in defining business-oriented services. This includes defining business process models, corresponding services, and decision models.

Furthermore, there are service roles within SOA as well. As such, it helps to consider a service role as representative of a separation of concern. Majority of common service roles are considered to be task services and data services or information. The recognition of decision services is less common, as it represents the important business logic separation within SOA. The decision model and business decisions offer a unique way of respectfully separating business logic, possibly at a granular level. One can then roll them up into suitable decision services given all of the features of a good-quality service.

The intelligent systems that are utilized in diabetic are vital in the medical industry because they enable nurses and doctors to collect information quickly and process it in different ways to help in making diagnosis and treatment decisions.

Diabetes can be classified into (Mishra, Singh, Rana, & Shailendra, 2014):

1. **Type 1 Diabetes:** Also known as insulin-dependent diabetes mellitus, which describes an autoimmune disease that stops the production of insulin, therefore requiring treatment by insulin injections, as well as the development of a meal plan and regular exercise.
2. **Type 2 Diabetes:** Also known as non-insulin-dependent diabetes mellitus, which happens when tissues stop responding to insulin, and has associations with heredity and obesity and could be controllable by diet; it also makes up 90 percent of all cases, most of which are not diagnosed for years.
3. **Gestational Diabetes:** Happens in pregnancy, usually during the 24th week. A gestational diabetes diagnosis does not indicate that diabetes was there before conception, or that the diabetes will stay after giving birth. However, following the doctor's advice about blood glucose (blood sugar) levels is important while pregnancy is being planned to ensure the health of both the mother and the baby.

Problem

Service Oriented Architecture with a Decision Aspect (SOAda) (Boumahdi & Chalal, 2014) could help in the diagnosis of diabetes patients by modelling and classifying their diabetes and providing diabetic patients with advice on how to manage their condition. This research involves the development of an "SOA" architecture design that can be used to classify diabetes with improved accuracy. Several techniques have been used in the classification of diabetes risk, and they will be described in this chapter. However, these techniques do not have the ability to clearly classify diabetes. The majority of the techniques that are given in literature offer binary classification of whether a patient is diabetic or non-diabetic. In this chapter, the computational challenge is classifying the diabetes with more accuracy, such as demonstrating the decision level from the SOAda (Boumahdi & Chalal, 2014). Instead of just classifying whether a person is diabetic or not, showing their diabetes risk levels will have more use to people as it could help diabetes patients manage their health conditions by living a proper life style based on their risks. In this chapter, the SOA architecture design reported is focused on studying diabetes classification for people that have diabetes. However, one can also apply the model to identify the risk levels for non-diabetics.

Objectives

This chapter's main objective of this chapter is to examine a new, hybrid architecture for classifying diabetes and assess its accuracy and efficiency. The suggested architecture combines SOAda and DMN. Following are the objectives of this chapter:

1. To formulate a decision SOA architecture to diagnose diabetes mellitus and classify the status of the patient as either prediabetic, normal, or diabetic.
2. To implement the SOAda architecture in the application for the diagnosis of diabetes mellitus.
3. To assess the viability of the utilization of a previous patient, as well as using their solutions to serve as a basis for decision-making when encountering similar cases. It also aims to determine if one can use an SOA architecture for support in clinical decisions regarding the treatment and management of diabetes.
4. To create an architecture that can be used to apply a Diabetes SOA system that is also usable for diabetic diagnosis.

Methodology

One can use a build research methodology when a new system design is suggested, which is a combined decision aspect and SAO architecture and for diabetes classification. One can also use the SOAda architecture to extract the various services from a diabetes problem. Furthermore, the extracted services can be utilized in the SOA^{+d} (Boumahdi, Chalal, Guendouz, & Gasmia, 2016) method's rule base to classify and reason the diabetes level.

This research's main focus is to classify the risk levels in type 2 diabetes with the use of the DMN notation to obtain information from the proven cases in type 2 diabetes within the DRD diagram. Furthermore, one can use an experimental methodology to determine the proposed system's efficiency in risk classification.

Organization

Section 1 introduces the design's general and brief concepts, the main problems that the study aims to solve, the solution being proposed, the literature survey of the thesis, and the main objectives that will be met during this work. The rest of the chapter will be organized in the following manner:

Section 2 provides a brief description of the background of the basic concepts that were utilized, such as SOAda architecture, SOA^{+d} method, and DMN notation. Section 3 provides more details about the proposed architecture design for the classification of risk as well as the proposed model's performance. Section 4 provides an analysis of the main related work that is associated with current methodologies for diabetes diagnosis. The last part concludes the chapter and provides some recommendations for future work.

BACKGROUND

The following section discusses DMN, SOAda architecture and SOA+d method. This study has been encouraged through these concepts that focus on the new diabetes diagnosis architecture.

SOAda Architecture

A discrete unit of business functionality is the service that is offered via a service contract (Rivera, 2016). An SOA architecture that covers decision aspect is the SOAda, which requires three levels to be analyzed for building SOA (Boumahdi & Chalal, 2014):

- **SOAda's Business Level:** On a trade level, a brick that can be used again is the business service. It refers to the business functionalities and employs functional services. These fall under services of average granularity.
- **The SOAda's Second Part Refers to the Information Level:** The functionalities offered by the computerized part of the information system are collected via information services, which are employed by the functional services mentioned earlier. We differentiated three categories, i.e., user interface, entities and technical services.
- **The Third Part of the SOAda Can Be Associated with the Decisional Level**: The decisional part of the business process provides functionalities via decision services, which are employed by the company services mentioned earlier. In this level, three services can be associated with SOAda: design, intelligent and choice.

SOA^{+d} Approach

A new approach is the SOA^{+d} that supports the SOAda architecture. A new approach has been put forward by SOA^{+d} to develop SOA with the decision part (Boumahdi, F., Chalal, R., Guendouz, A., & Gasmia, K., 2016). This is based on DMN and SoaML standards that follow a downward approach for identifying the services associated with SOAda. As presented in Figure 1, SOA^{+d} can be described with four phases: identification and categorize service, analysis, realization phase and services modelling.

- **Phase 1: The Analysis Phase:** This involves three steps (Boumahdi, F., Chalal, R., Guendouz, A., & Gasmia, K., 2016):
    - **Activity 1: The Information System Analysis:** A brief study of the different operations associated with a system and their relationships outside and within the system. A use case diagram of UML language is employed to model this activity.
    - **Activity 2: The Business Analysis:** A model that is employed to describe and better understand various business processes associated with the environment to build a system that can be used later. To define this activity, we employ the BPMN 2.0 language.
    - **Activity 3: The Decision Analysis:** Here, we discovered three steps associated with decision analysis: Mention the decision requirements, recognize the decisions and mention the decision logic level. The new notation DMN is employed to model this activity.
- **Phase 2: Identify and Categorize Services:** This is according to the cartographies already proposed by (Boumahdi & Chalal, 2014) to recognize SOAda services.

Figure 1. SOA^{+d} approach

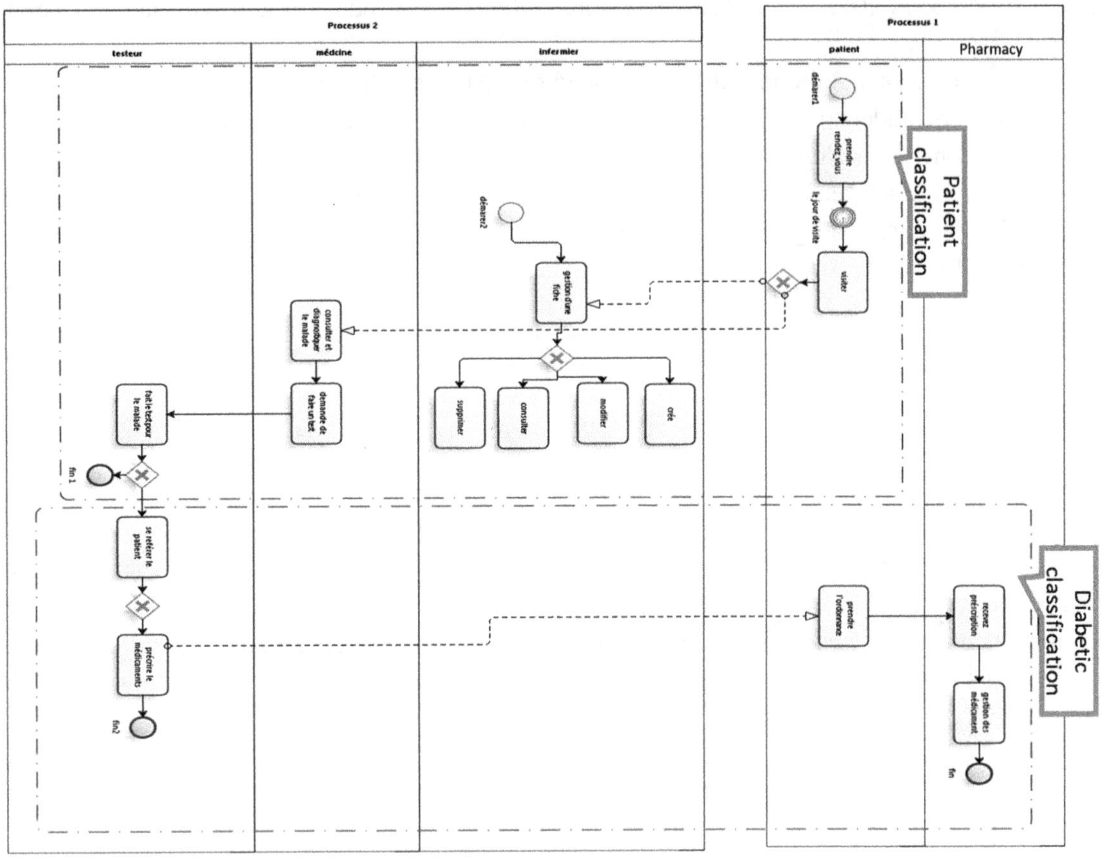

- **Phase 3: Services Modelling:** In this phase, to model services with formalism, we implement a specification at the base of the SoaML (Object Management Group Inc., 2012) that provides a high-level abstraction.
- **Phase 4: The Realization:** The aim of this level is to develop and deploy services when needed. In this phase, there is a technical choice to be carried out.

Decision Model and Notation (DMN)

The Decision Model and Notation's (DMN) (Object Management Group, 2015) primary aim is to offer a common notation for decision logic, which can be easily understood by business analysts, business users and technical developers. DMN offers the constructs to design model decision rules as well as the whole decision-making process itself. There are two levels in the DMN decision model: the decision logic and the decision requirements graph (DRG). The latter shows from where the needed information is originating from, which can be presented with one or more decision requirement diagrams (DRDs) (Thierry Biard, Mauff, Bigand, & Bourey, 2015). The former presents the logic used in making the decision, which is shown in Decision Tables (Mertens, Gailly, & Poels, 2015). The decision table's upper half presents the possible combinations of conditions that may result in certain actions, while the bottom half includes the actions that need to be taken (i.e. outcomes) (Batoulis, K., Meyer, A., Bazhenova, E., Decker, G., Weske, M., & 1., 2015). Thus, for this reason, DMN is employed keeping this capability in mind:

- Thus, a common notation for that is provided by DMN, which:
  - Can be understood by all business users – right from requirements through maintenance.
  - Technical developers can process, who are responsible for automation.
  - Can be managed by business people that are involved in monitoring and managing decisions.
- Build a standardized bridge to close the gap between implementation and the decision design.
- Designed in a way so that it can be employed together with the standard BPMN business process notation.

PROPOSED ARCHITECTURE

In this chapter, SOAda is used in the field of medicine to diagnose diabetes; a family of metabolic disease condition in which increased blood glucose is associated with the patient.

This section discusses the development of the Diabetes Detection System (DDS) that can identify potential diabetes with the aid of the SOAda architecture. Specifically, DDS allows the user to choose the symptoms that they may have by waiving off the need to see a doctor regarding early screening. Based on these symptoms, DDS can also check if the user may have potential risk for diabetes. DDS can sense three possible outcomes: Diabetic Type 1, Healthy and Diabetic Type 2. DMN notation is employed to conceive the solution. DDS follows a SOA architecture that has decision parts. This research vehicle can be represented by DDS to examine if the SOAda architecture is applicable for symptomatic diseases. SOA^{+d} refers to the designing of the Diabetes Disease Diagnosis level for determining and describing the suitability of SOAda architecture for diagnosing diabetes.

Phase 1: The Analysis

Activity 1: The Business Analysis: We start keeping in mind the business aspect, where two sub-processes are defined. We also need to note that this ranking is carried out only after a few exams. The two sub-processes in the Diabetes Diagnosis process are:

Sub-Process 1: Patient Classification

When a patient comes to visit the doctor or make an appointment, the nurse asks for his/her information: If the patient is already registered, the nurse will verify the information, if further deletion or modification is required. If need be, the nurse will then update the form and proceed with the appointment, else the appointment can also be given directly. If the patient is new, registration will be needed, after which a new form needs to be filled with the help of the nurse for direct appointment.

The patient needs to be present on the day fixed by the nurse for appointment. When the turn finally comes, the patient meets the doctor who will examine the information mentioned by the nurse in the form and starts with the process of consultation as well as the relevant diagnosis. During this, the doctor tries to find symptoms of insipidus diabetes: examines if the patient has polydipsia and polyuria. If the case is positive, a clinical examination is followed to determine if the patient is in good health or if the test results indicate the patient is normal. In this case, the doctor will reconfirm if the patient really has insipidus diabetes, and the doctor will not convey any information to the patient until a few additional examinations and tests are performed. After performing these tests, the doctor will examine for signs of decreased blood pressure, increased pulse, decreased weight and increased NA+ blood, after which the doctor can confirm and convey it to the patient that he/she has insipidus diabetes.

Sub-Process 2: Diabetic Classification

We move on to this sub-process after insipidus diabetes is confirmed for the patient. The form is then updated by the doctor while also forwarding a copy to the tester via a computer. The patient is then informed about the type of insipidus diabetes, and which treatment type can give high precision can be determined via some external tests, including the water restriction test and the DHA test.

When the patient goes to the tester for testing, the tester, after performing the task, enters the results on the patient's form and forwards it back to the doctor. When the patient sees the doctor for prescription of medications, the drugs are prescribed by the doctor based on the type of insipid diabetes taken from the analysis results and then a copy of this form is sent to the pharmacist, which will also allow the patient to buy drugs.

The Figure 2 shows the Diabetes Disease Diagnosis process with the language BPMN (Business Process Management Notation).

Activity 2: Information Level: SOA^{+d} recommends employing a use case diagram for modelling the aspect of information system. Figure 3 demonstrates the use case diagram.

Activity 3: Decision Level: We have two decision problems based on the use case as well as the BPMN: the patients' case and the insipidus diabetes's type precision..

Figure 2. Diabetes Disease Diagnosis process

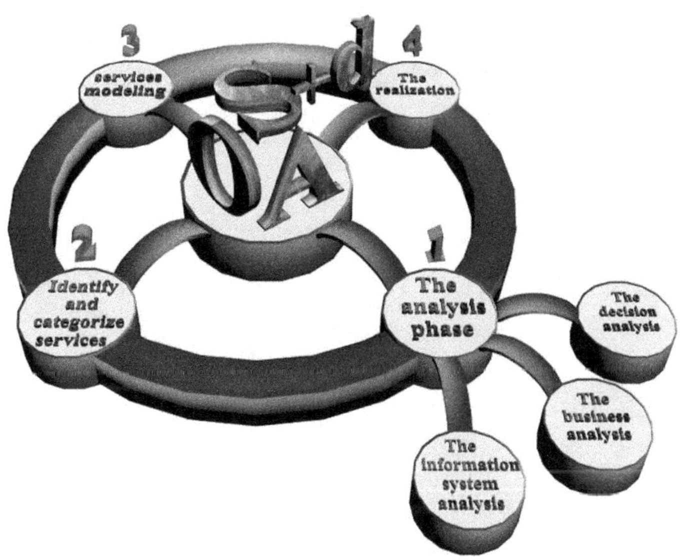

Figure 3. The use case diagram

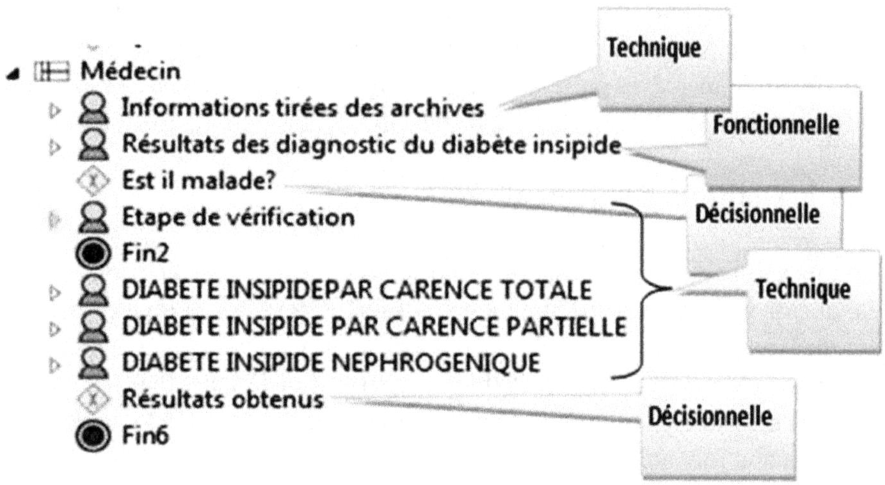

The Case of Patients

Two classes are associated with the patient case: The diabetic patients' class (insipidus diabetes): this categorization is done after confirming that the patient was in good general state during the clinical examination and has shown signs of certain symptoms (polydipsia and polyuria... etc.). The non-diabetic patients class (insipidus diabetes): this segmentation is done if at least one of the symptoms of the previous classification is absent. Figure 4 presents the DMN involved in the 'patient classification' decision.

The Type of Diabetes: This decision is modeled by the DMN diagram, it is presented by the Figure 5.

Figure 4. DMN diagram of patient classification decision

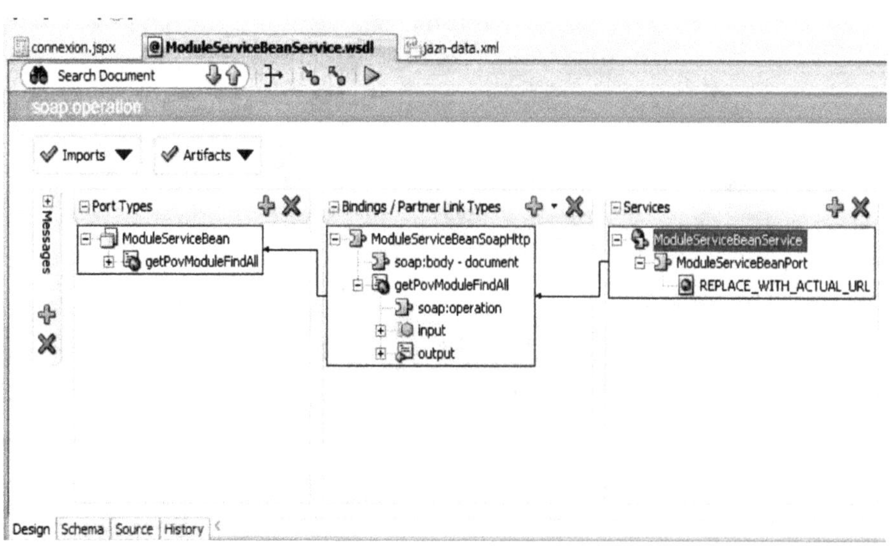

Figure 5. The DMN diagram of diabetes type decision

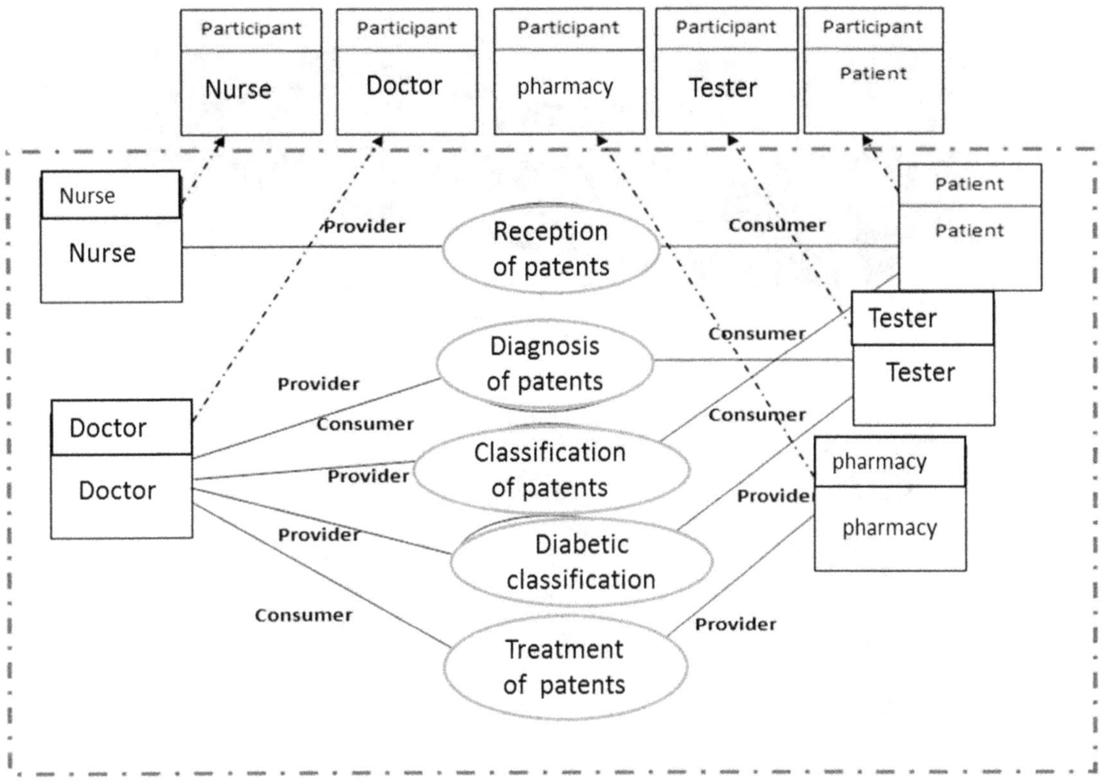

Phase 2: Services Identification

After the use of the analysis defined through the SOA^{+d} approach to Diabetes Disease Diagnosis we found the services of business level list shown in Figure 6.

Phase 3: Services Modeling

After applying the transformation rules as described by (Boumahdi, F., Chalal, R., Guendouz, A., & Gasmia, K., 2016), Figure 7 demonstrates the services modelling fir the Diabetes Disease Diagnosis Process.

Phase 4: Service Realization

The Figure 8 summarizes the Services Integration in Diabetes Disease Diagnosis Process.

As shows the Figure 9 the implementation of Diabetes Disease Diagnosis system in Service Oriented Architecture under SOAda architecture.

Figure 6. Services of Diabetes Disease Diagnosis

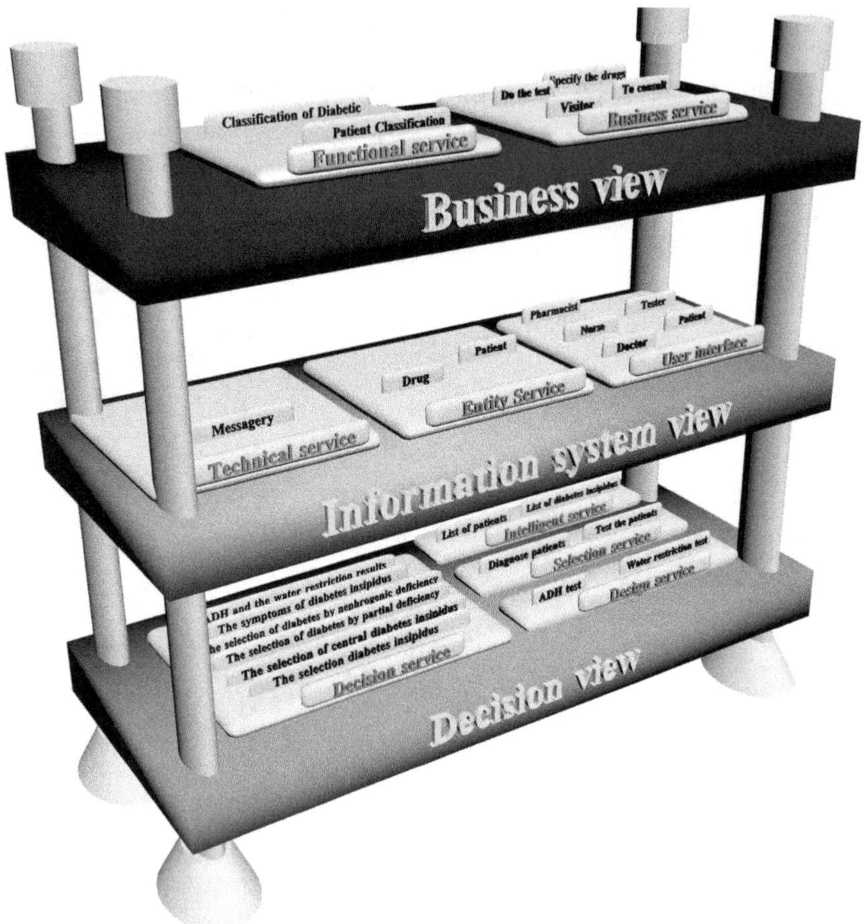

RELATED WORKS

In this section, the primary emphasis is on the work corresponding to the research related to diabetes diagnosis. In the literature, two implementation classes can be identified for the problem 'diabetic diagnosis', which relates to expert systems and case based reasoning (CBR).

Diabetes Diagnosis Using CBR

This section concentrates on employing case-based reasoning (CBR) to manage and treat diabetes. CBR falls under the field of artificial intelligence, where one can make use of previous cases in the past to derive suitable resolution for similar types of problems. The concept follows the dynamic memory theory, in which human beings recall encountered cases of the past to solve problems.

In the medical field, the application of the CBR approach has gained popularity as the core of the reasoning process displays a strong similarity to that of the clinical reasoning. Indeed, the doctors often

Figure 7. Services SoaML Modeling

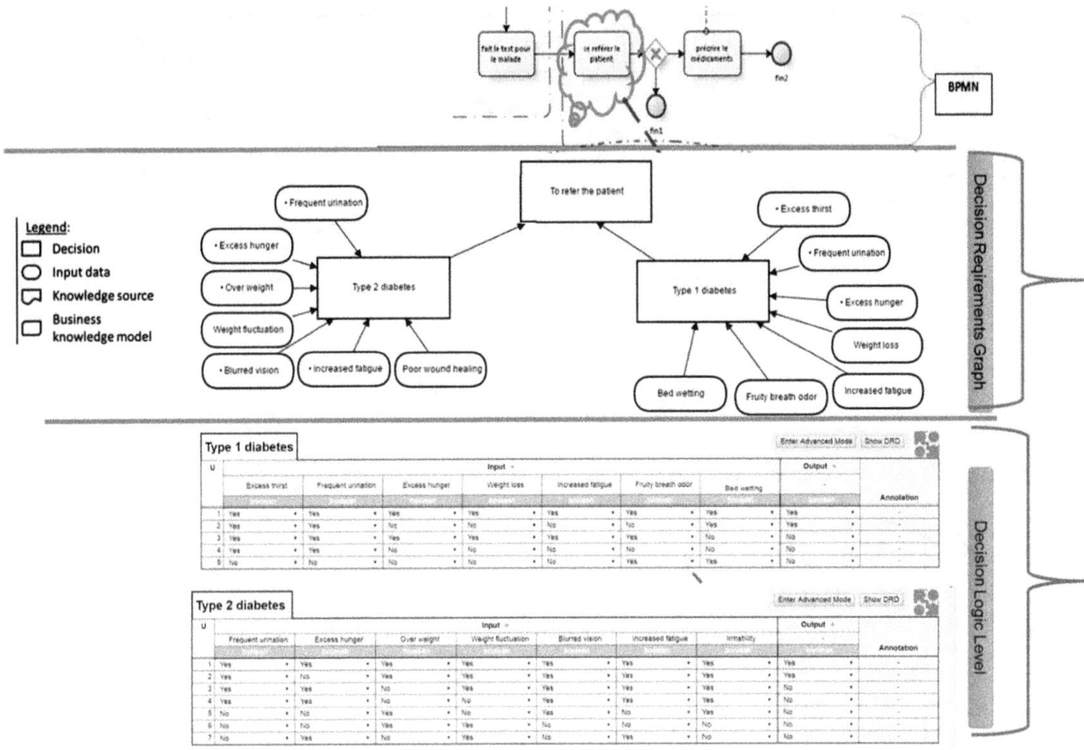

Figure 8. Services Integration

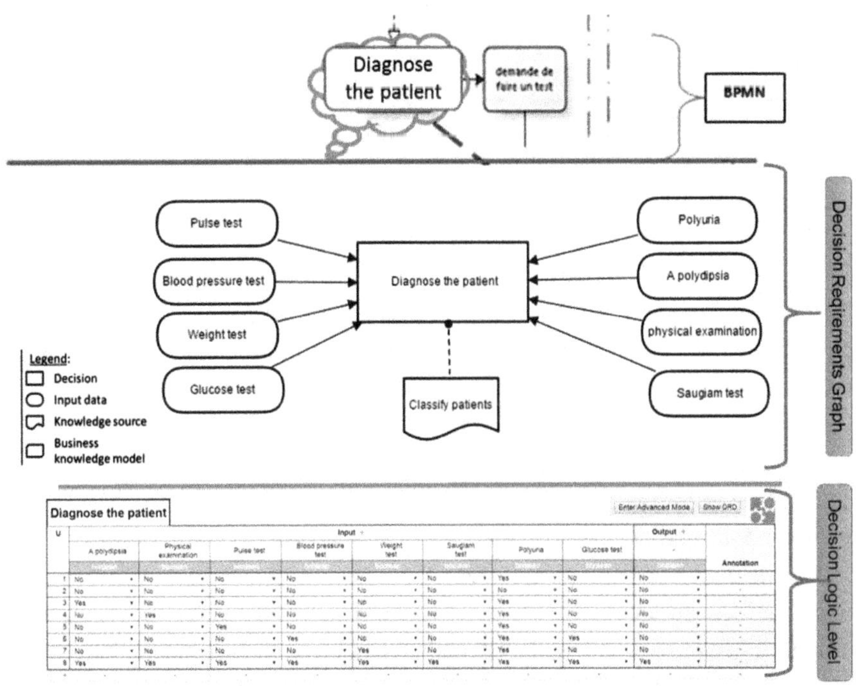

Figure 9. Services implementation

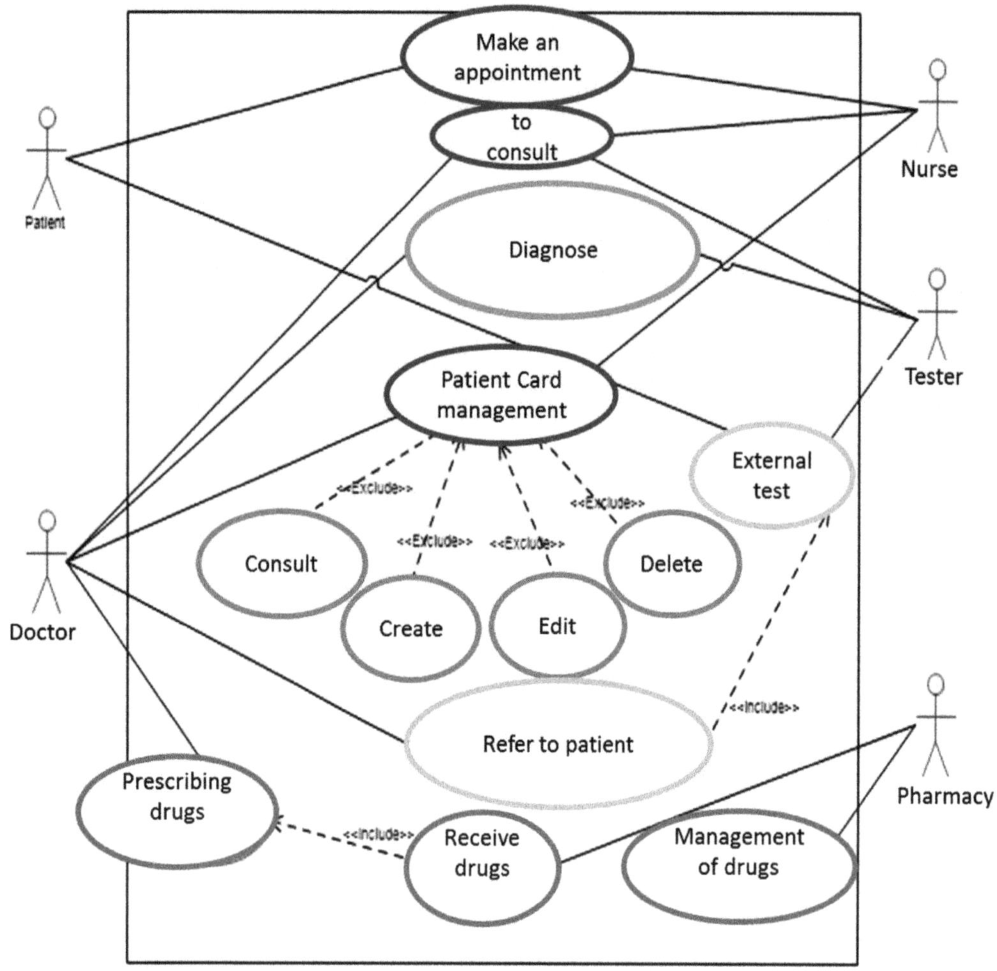

make an attempt to connect the dots between cases and those which they have already experienced in their practice, which is the main principle of this method.

There have been many works conducted related to the CBR of medical decision support systems (David; B & Leake, 1996), an approach was provided by (Cindy Marling & Schwartz, 2008) for decision support, which followed the CBR to manage type 1 diabetes patients. Also, developments of systems like 'PROTOS' for cardiac diagnosis (Porter & Bareiss, 1986) for breast cancer are in place. This list is far to be called exhaustive but it also displays the diversification of the CBR application scope.

When compared with the epidemic of asthma, there have been many works that have allowed understanding this disease in particular, for example, attempting to collect feedback periodically from the recorded data on general medical consultations regarding asthma (Fox, J., Johns, N., Lyons, C., Rahmanzadeh, A., Thomson, R., & Wilson, P., 1997).

Other works were focused on decision support to manage this disease (Szajnar & Setlak, 2011; Austin, Iliffe, Leaning, & Modell, 1996), and systems have been designed specifically for the diagnosis of asthma, such as Proforma (Fox, J., Johns, N., Lyons, C., Rahmanzadeh, A., Thomson, R., & Wilson, P., 1997) and Adema (Sefion, Ennaji, Gailhardou, & Canu, 2003).

Diabetes Diagnosis Using Expert System

Szajnar and Setlak, (2011) put forward the concept of creating an intelligence system to support diabetes diagnostics, in which they employed the start-of-art method designed around artificial intelligence to build a tool for modelling by investigating knowledge gathered from multiple sources. The initial aim of their system was to execute responsibilities of a medical expert for diagnosing diabetes and substituting doctor during the first phase of illness. In the process of diagnostics, the sequence that they follow is: (1) collecting information from patient regarding symptoms; (2) completing basic medical examination in brief; (3) based on the previous data, the system identifies and confirms if the patient has diabetes and then again it decides if it is type1 or type2. A decision tree is employed as a model by the systems for classification.

A natural therapy system was developed by Kumar Jha and Singh (2012) for healing diabetes. It is designed for improving people's health and wellness, and should not cost much. Their main aim was to gather all the natural treatment information regarding diabetes and keep it in one place by employing Expert System Shell for Text Animation (ESTA) as knowledge-based system. ESTA possesses all facilities to write the rules, which also help in making up a knowledge base. Further, ESTA includes an inference engine that can make use of rules in the knowledge base to help select which advice is best for the user. Their system starts with consultation, where users are asked to select the disease (Diabetes) for which they require different type of natural treatment solution, after which the diabetes as a disease as well as their symptoms are described. Then, the natural care (herbal /proper nutrition) treatment solution for diagnosing diabetes as a disease is explained.

Devamalar, Bai, and Srivatsa, (2007) put forward a system that would give accessibility for diabetic expert system from any corner of the world. Relevant knowledge and service information can also be collected, organized, and distributed to the individuals. The dot net framework was employed to design and program project. The system provides the availability to allow detecting and also offer early diagnosis for three types of diabetes, namely type 1, 2, and gestational diabetes in both children and adult.

The proposed (Mishra, P., Singh, D. B. V., Rana, N., & Shailendra, S., 2014) method tries to design and develop a diagnosis system, employing fuzzy logic rule base. A simple set of symptoms is used to examine the developed system, which is then added to clinical data to determine diabetes.

Earlier research studies have shown the development process for Expert System for Diabetics Treatment Advices (Tabibi, Zaki, & Ataeepoor, 2013). To get advices for the treatment of diabetes, an expert system was built. The necessary knowledge was obtained by interviewing medical specialists and nurses in the diabetics department directly, after which derivation of the related rules was carried out. The design of the system was in a rule-based fashion, which was coded by VP-Expert Shell.

CONTRIBUTION

This study evaluated the SOAda architecture in the healthcare industry by employing the elements of the SOA+d method and performing system testing on diabetes diagnosis. Key findings in this section are described below.

The main advantage of this research study is that it allows users to save time as well as minimize costs when visiting a doctor to check if they have diabetes mellitus at early stage. Early detection via symptom selection allows users to control as well as take early prevention steps to get back healthy body free from diabetes.

For both the medical industry and diabetes patients, the research in diabetic systems is crucial. There is a high urgent need for an efficient tool to help with diagnosing and treating diabetes for adding both patients and specialist doctors. This research was inspired by the requirement of such an efficient tool, and therefore this chapter offers a survey on intelligent system designed for diabetes, which is the first step towards building such a tool.

The research study will allow leveraging the benefits through the development of decision systems in medicine. Therefore, it will be possible to develop a web service that will help in combining the decision process across various domains of medicine. Also, this research will offer the basis for building a SOA architecture shell to accelerate development of decision systems in medicine. It will also help new doctors in diagnosing mellitus, as well as improve the availability and the number of resources and activities for people suffering from diabetes, their families and other interested parties. This research have achieved many contributions:

- **SOAda and SOA^{+d} Help in Decision Integration in SOA:** This important role is distinct in literature (example: (Boumahdi, Boulefrakh, & Chalal, 2015)), which is also confirmed through this research. In terms of healthcare organisations' context, it seems that the SOAda architecture can be applied to all fields.
- As a decision-supporting tool, the SOAda framework can be employed for SOA with decision aspect for a healthcare setting or even for other settings with suitable adjustments.
- The building of SOA decision system has confirmed that it is possible to diagnose diabetes mellitus that was only manually diagnosed earlier.
- Decrease the time needed to arrive at a decision, especially when there is an emergency.

CONCLUSION

Since the past decade, SOA has been developed to create some of the most flexible applications that are in use today. Now, SOA is quickly gaining popularity in terms of architecture as a choice for custom application development as well as for commercial software products.

This work developed a diabetes management system that can gather medical readings from patients with diabetes. Data collected are examined by following a proposed medical schema from the healthcare provider. The patient's condition is notified to both patient and healthcare provider, after gathering and evaluating the data. However, there is a need to test various techniques to select the most appropriate techniques appropriate for the proposed model. At this stage, the proposed model is sufficient and can be used for diagnosing and monitoring patients with specific region of interest.

To enhance this research, some suggestions and recommendations that should be followed are as follow:

- Visit concerned people that have expertise in the medical domain regarding diabetes mellitus to determine the most significant attributes employed during diagnosis of diabetes mellitus.
- Build a special application for gestational women for the diagnosis of diabetes mellitus.
- Employ the SOAda architecture in other medical fields like cancer.

REFERENCES

Austin, T., Iliffe, S., Leaning, M., & Modell, M. (1996). A prototype computer decision support system for the management of asthma. *Journal of Medical Systems*, *20*(1), 45–55. doi:10.1007/BF02260873 PMID:8708491

Baghdadi, Y. (2015). Service-Oriented Software Engineering. *International Journal of Systems and Service-Oriented Engineering*, *5*(2), 1–19. doi:10.4018/IJSSOE.2015040101

Batoulis, K., Meyer, A., Bazhenova, E., Decker, G., Weske, M., & 1. (2015). Extracting Decision Logic from Process Models. *Lecture Notes in Computer Science, 9097*, 520–521. 10.1007/978-3-319-19069-3

Biard, T., Mauff, A. L., Bigand, M., & Bourey, J.-P. (2015). Separation of Decision Modeling from Business Process Modeling Using New "Decision Model and Notation" (DMN) for Automating Operational Decision-Making. In IFIP Advances in Information and Communication Technology (Vol. 463, pp. 489–496). Academic Press. doi:10.1007/978-3-319-24141-8

Boumahdi, F., Boulefrakh, H. E., & Chalal, R. (2015). Decision Making and Service Oriented Architecture for Recruitment Process. Using the New Standard Decision Model and Notation (DMN). *ICSEA 2015: The Tenth International Conference on Software Engineering Advances Decision,* 489–493.

Boumahdi, F., & Chalal, R. (2014). SOAda: A New Architecture to Enrich SOA with a Decisional Aspect. *International Journal of Systems and Service-Oriented Engineering*, *4*(2), 13–27. doi:10.4018/ijssoe.2014040102

Boumahdi, F., Chalal, R., Guendouz, A., & Gasmia, K. (2016). SOA (Formula presented.) : A new way to design the decision in SOA—based on the new standard Decision Model and Notation (DMN). *Service Oriented Computing and Applications*, *10*(1), 35–53. doi:10.100711761-014-0162-x

Cindy Marling, J. S., & Schwartz, F. (2008). Case-Based Decision Support for Patients with Type 1 Diabetes on Insulin Pump Therapy. *Advances in Case-Based Reasoning: 9th European Conference, ECCBR 2008,* 325. 10.1007/978-3-540-85502-6_22

David, B., & Leake. (1996). *Case-Based Reasoning: Experiences, lessons and future directions* (AAAI Press). MIT Press. Retrieved from https://dl.acm.org/citation.cfm?id=524680

Devamalar, P. M. B., Bai, V. T., & Srivatsa, S. (2007). An Architecture for a Fully Automated RealTime Web-Centric Expert System. *World Academy of Science, Engineering and Technology, 13*. Retrieved from https://static.aminer.org/pdf/PDF/000/262/917/a_framework_for_the_delivery_of_web_centric_knowledge_management.pdf

Fox, J., Johns, N., Lyons, C., Rahmanzadeh, A., Thomson, R., & Wilson, P. (1997). *PROforma: A general technology for clinical decision support systems, 54 Computer Methods and Programs in Biomedicine*. Elsevier. doi:10.1016/S0169-2607(97)00034-5

Kumar Jha, S., & Singh, D. K. (2012). Development of knowledge Base Expert System for Natural treatment of Diabetes disease. *International Journal of Advanced Computer Science and Applications, 3*(3). Retrieved from www.ijacsa.thesai.org

Mertens, S., Gailly, F., & Poels, G. (2015). Enhancing Declarative Process Models with DMN Decision Logic Steven. In Lecture Notes in Business Information Processing (Vol. 214, pp. 151–165). Springer. doi:10.1007/978-3-319-19237-6

Mishra, P., Singh, D. B. V., Rana, N., & Shailendra, S. (2014). Clinical Decision Support System for Diabetes Disease Diagnosis. *International Conference On Emerging Trends in Mechanical and Electrical Engineering (ICETMEE- 13th-14th March 2014)*, 105–110.

Object Management Group. (2015). *Decision Model and Notation, 1.0(September), 172*. Retrieved from http://www.omg.org/spec/DMN/

Object Management Group Inc. (2012). Service oriented architecture Modeling Language (SoaML) Specification. *OMG*,. Retrieved from http://www.omg.org/spec/SoaML/

Porter, B. W., & Bareiss, E. R. (1986). *PROTOS: An Experiment in Knowledge Acquisition for Heuristic Classification Tasks*. University of Texas at Austin. Retrieved from https://dl.acm.org/citation.cfm?id=899623

Rivera, C. (2016). *The Decision. Mouth* (Vol. 1). 10.5281/ZENODO.154280

Sefion, I., Ennaji, A., Gailhardou, M., & Canu, S. (2003). ADEMA: A decision support system for asthma health care. *Studies in Health Technology and Informatics*, *95*, 623–628. Retrieved from http://www.ncbi.nlm.nih.gov/pubmed/14664057 PMID:14664057

Szajnar, W., & Setlak, G. (2011). A concept of design process of intelligent system supporting diabetes diagnostics. *Methods and Instruments of Artificial Intelligence*, 168–178. Retrieved from http://www.foibg.com/ibs_isc/ibs-20/ibs-20-p21.pdf

Tabibi, S. T., Zaki, T. S., & Ataeepoor, Y. (2013). Developing an Expert System for Diabetics Treatment Advices. *International Journal of Hospital Research*, *2*(3), 155–162.

Chapter 3
SLO-Driven Monitoring and Adaptation of Multi-Cloud Service-Based Applications

Chrysostomos Zeginis
ICS-FORTH, Greece

Kyriakos Kritikos
ICS-FORTH, Greece

Dimitris Plexousakis
ICS-FORTH, Greece

ABSTRACT

Cloud computing is an emerging technology that attracts the attention of an increasing number of organizations that are willing to virtualize their infrastructure. Towards this direction, enterprises can deploy their applications on different public virtual machines (VMs), in order to reduce their operational costs, as well as optimize the performance of the offered applications. In such heterogeneous multi-cloud environments, the need for monitoring the quality of (a) the acquired resources, (b) the services offered to the final users, and (c) the offered service-based processes, and adapting them has come into force. Current research approaches addressing these areas are limited and most of them target a subset of these offerings. This chapter aims at proposing a novel event-based approach for cross-layer monitoring and adaptation of service-based applications (SBAs) deployed on multiple clouds. The proposed approach is empirically evaluated based on an invoice management application.

INTRODUCTION

Service-Oriented Architectures (SOAs) concept emerged in the late 90s, introducing Web services as a new means for delivering software over a network. This architecture paradigm has been adopted, as it provides many opportunities for businesses to automate their processes, by providing services to either end-users or to other services distributed in a network, via published and discoverable interfaces. The

DOI: 10.4018/978-1-5225-5951-1.ch003

evolution of Web services indicates that they are moving beyond the simple exchange of information to the concept of combining existing or new functionality in the form of a service into a complex service-based application (SBA). Consequently, businesses will be able to create composite SBAs, also integrating back-end and older technology systems underlying local or remote applications by wrapping them in the form of services.

With the advent of the new century, Cloud computing has emerged as a new paradigm for delivering "as-a-service" offerings to end users: (1) Software-as-a-Service (SaaS) is the software delivery model adopted in Cloud computing, where web-based applications are made available to users on demand via the Internet from a Cloud provider; (2) Platform-as-a-Service (PaaS) provides the platform to the application owners to develop and deploy their applications on; and (3) Infrastructure-as-a-Service (IaaS) is a resource provisioning model allowing Cloud providers to outsource the resources, i.e., Virtual Machines (VMs), needed by SaaS services.

Furthermore, Cloud computing enables flexibility and elasticity via the flexible resource management that enables to acquire additional resources on demand, as needed by the current situation. In addition, it provides opportunities for optimising the quality and cost of the SBA by selecting the best possible offerings (at different abstraction levels) which might be supplied by different cloud providers. The main benefits of this kind of multi-Cloud deployment (Baryannis, G.; Garefalakis, P.; Kritikos, K.; Magoutis, K.; Papaioannou, A.; Plexousakis, D.; Zeginis, C., 2013) are the following: (1) vendor lock-in is avoided; (2) different application requirements are better satisfied by different Cloud providers, offering special features (e.g., SSD drives, dynamically configurable VMs, better security services); (3) the deployment cost is optimized exploiting the Cloud providers' variant cost policies; (4) ability to decide about the optimal placement of application components (e.g., cross-talk components can be deployed on the same geographical zone or even VM to minimize communication overhead but independent components can be deployed on the most suitable cloud).

While the deployment of applications over multiple, heterogeneous Cloud providers offers more flexibility, certainly not all applications are expected to benefit from it: For example, applications whose different resources exhibit significant cross-talk could suffer from excessive Internet charges as well as performance issues (high network latency across different Clouds). On the other hand, decomposing complex applications at increasingly coarser boundaries may eventually reach the service interfaces between organizations. Reasoning about effective Multi-Cloud deployment at this level requires capturing component dependencies in application models. It is imperative that such distributed hosting environments, where multiple Clouds produce many events across functional layers, exhibit efficient cross-layer monitoring and adaptation mechanisms combining such events and mapping them to suitable adaptation strategies. In (Zeginis, C.; Kritikos, K.; Garefalakis, P.; Konsolaki, K.; Magoutis, K.; Plexousakis, D., 2012) we have thoroughly investigated the need for cross-layer adaptation. Current techniques are mainly fragmented by considering a single SBA layer, while the few cross-layer ones (Popescu, Staikopoulos, Liu, Brogi, & Clarke, 2010) (Zengin, Marconi, & Pistore, 2011) do not consider multi-Cloud aspects. Thus, the need for a cross-layer and multi-Cloud monitoring and adaptation framework emerges.

In such heterogeneous virtualized environments, where multiple Cloud providers are involved, an important concern for a provider of any Cloud-based SBA is to maintain a desired Quality of Service (QoS) level along its entire life cycle. In addition, during application execution, various events are produced by several layers (Cloud and SOA specific) (Zeginis, Konsolaki, Kritikos, & Plexousakis, 2011), which might lead or indicate Service level Objective (SLO) violations. Just as in any distributed application hosting environment, extensive monitoring mechanisms should be realized by the application provider or even a middleware or broker to aid in controlling application performance and functionality and to adapt to variations that might occur at any layer. In this respect, an effective monitoring mechanism must cover the entire range of Cloud SBA layers. Leading Cloud providers (e.g., Amazon AWS, Microsoft Azure, Google Cloud Platform, Rackspace, etc.) are just starting to roll out monitoring solutions, while the rest of the industry is still lagging behind.

In this chapter, a monitoring and adaptation framework is proposed that can have a substantial impact to SOA stakeholders, especially to those who resort in multi-Cloud deployments of their SBAs. This framework enables them to have a clear view of how their applications perform in such distributed virtualized environments. Moreover, the incorporated SLO evaluation and pattern discovery mechanisms assist them in performing a root cause analysis over the monitoring events stream and identifying the main source of malfunction in this multi-cloud environment.

Concerning monitoring, in (Zeginis, C.; Kritikos, K.; Garefalakis, P.; Konsolaki, K.; Magoutis, K.; Plexousakis, D., 2012) a multi-Cloud SBA framework was introduced and evaluated. This work goes a step further supporting multi-Cloud SLO evaluation and SBA adaptation by focusing on an efficient method for processing the huge number of events collected by monitoring mechanisms across different Cloud providers. These events are processed, correlated with each other in order to shape event patterns that are suspicious for causing specific SLO violations. Towards this direction, a novel logic-based pattern discovery algorithm is introduced, as well as an efficient event-based SLO evaluation framework.

The rest of this chapter is structured as follows. Section 2 analyses a motivating example that is used to exemplify the application of the proposed solution. Section 3 describes the proposed multi-Cloud monitoring and adaptation framework, while Section 4 elaborates on the incorporated SLO evaluation functionality. Section 5 proposes an event pattern discovery algorithm that detects event patterns causing specific SLO violations, while Section 6 evaluates the algorithm's accuracy and performance. Finally, Section 7 reviews the related work, while Section 8 concludes this chapter and supplies some future work directions.

USE CASE

An invoice management SBA deployed across Clouds motivates our approach. This SBA is adopted by the CloudSocket[1] EU project, which introduces the notion of Business Process-as-a-Service (BPaaS) as a means of aligning business processes with Cloud deployed workflows. This BPaaS comprises an external CRM SaaS for managing the customer relationships and an internal InvoiceNinja[2] SaaS for creating, editing and sending invoices. These two SaaS services are deployed on different clouds. The CRM SaaS is available over a certain private cloud of the organisation that offers it, while the internal SaaS service needs to be deployed in a public cloud according to the respective SBA requirements. Figure 1 depicts the topology of these two models specified in CAMEL[3], the core modelling and execution language developed in the PaaSage[4] project that enables the specification of multiple aspects of cross-cloud applications.

This BPaaS is characterized by strong interdependencies between layers, and raises various events during its lifecycle (Zeginis, Konsolaki, Kritikos, & Plexousakis, 2011). The InvoiceNinja internal SaaS requires rather high computation and storage capabilities as well as high throughput since information exchange between clients happens frequently. In this respect, it has been deployed on the most suitable VM that instantiates the Amazon EC2 "m1.medium" flavour. The selection of the Amazon VM offering

Figure 1. Topology of invoicing SBA

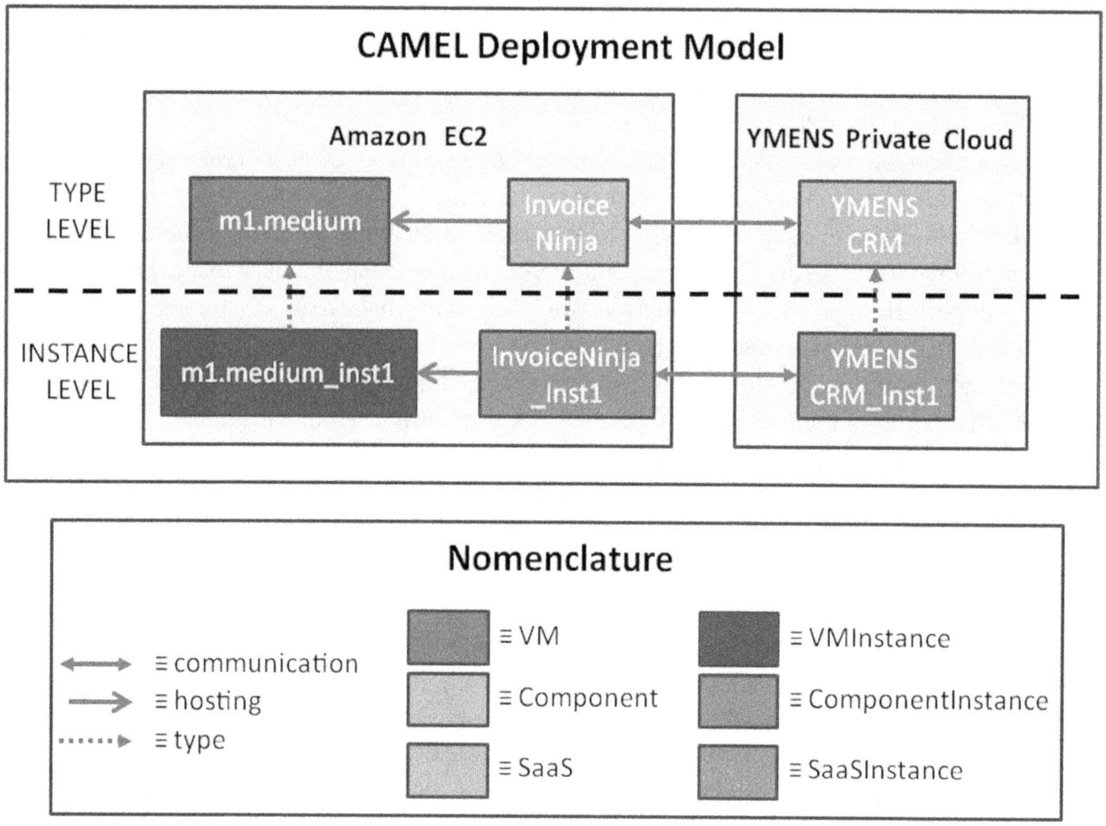

flavour and the external CRM SaaS has been based on the offerings of all Cloud providers with the main goal to optimize the SBA allocation, in terms of quality and cost. Each task in the Send Invoice BPaaS maps to the execution of methods of the respective SaaS services exploited.

The infrastructure supporting this multi Cloud SBA includes the next event sources with particular metric information in parenthesis which are ordered according to the respective layers involved (IaaS, PaaS & SaaS): (1) VM resources (CPU, memory, & disk utilisation); (2) PaaS service used for SaaS hosting; and (3) the hosted or external and the whole application (SaaS response time, availability, etc.). A BP model drives the SBA execution describing the application tasks, involved roles (i.e., who performs the tasks), and control and data flow (Figure 2). Control flow determines to the order of the individual tasks and sub-processes within the BP, while the data flow the flow of information between tasks/services. Thus, BPM events (Key Performance Indicator (KPI) metrics) and SCC events (service response time, throughput) may be captured at the SaaS layer. The main non-functional application goal is to capture warning (identified by we) or critical events (identified by ce) and interrelate them to discover event patterns leading to SLO violations (see sample events below). Warning events refer to measurements that violate the warning threshold of the corresponding SLO and thus indicate a possible malfunction in the SBA execution, while the critical events to violations of the critical SLO's thresholds which indicate a certain malfunction. Some sample (warning and critical) events are the following:

Figure 2. Invoice application workflow

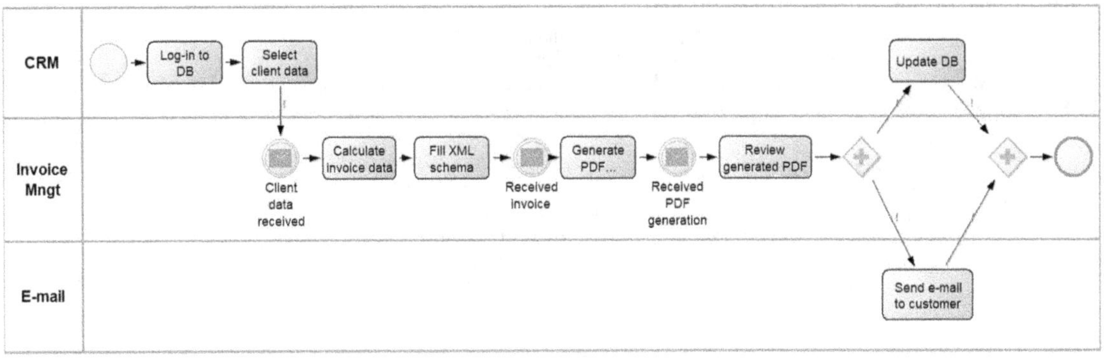

1. ce_1 → Raw Amount of free memory of VM hosting the YMENS CRM SaaS is below the violation threshold.
2. we_2 → Raw Execution time of the YMENS CRM SaaS is above the warning threshold.
3. ce_3 → Duration of the whole workflow (CRM + InvoiceNinja) is above the violation threshold.
4. we_4 → Raw Response time of the Invoice SaaS is above the warning threshold.
5. ce$_{-5}$ → Mean Execution time of the Invoice SaaS is above the violation threshold.
6. we_6 → Mean Availability of the YMENS CRM SaaS is below the warning threshold.
7. we$_{-7}$ → Mean Availability of the Invoice SaaS is below the warning threshold.
8. we_8 → Mean Throughput of the YMENS CRM SaaS is below the warning threshold.
9. ce_9 → Mean CPU load of VM hosting the Invoice SaaS is above the violation threshold.
10. ce_10 → Mean Availability of the whole BPaaS is below the violation threshold.

MONITORING AND ADAPTATION FRAMEWORK

This section analyzes an approach for cross-layer multi-Cloud SBA monitoring and proactive adaptation. The architecture presented in (Zeginis, C.; Kritikos, K.; Garefalakis, P.; Konsolaki, K.; Magoutis, K.; Plexousakis, D., 2012) for cross-layer multi-Cloud SBA monitoring is realized and extended with specific techniques and algorithms to support proactive adaptation. The new architecture (Figure 3) comprises: (a) a *Monitoring Engine*, collecting cross-layer events from various individual layer-specific monitoring components during SBA execution and storing them in a time-series database (TSDB), while also reporting detected violations to *Adaptation Engine* instances; (b) an *Adaptation Engine*, performing cross-layer adaptation actions based on specific adaptation rules; (c) a Model Repository that mediates between these engines. This repository maintains the required models for supporting the cross-cloud application monitoring and adaptation: (a) event models (Zeginis, C.; Kritikos, K.; Garefalakis, P.; Konsolaki, K.; Magoutis, K.; Plexousakis, D., 2012); (b) component models (Zeginis, Konsolaki, Kritikos, & Plexousakis, 2011); (c) adaptation action models describing component-specific adaptation actions (Zeginis, Konsolaki, Kritikos, & Plexousakis, 2011); and (d) the metrics/SLA models. It is a live repository where models are modified based on changes on system or user requirements. The shadowed components indicate where the main contribution of this paper is focused.

Figure 3. Monitoring and adaptation framework for multi-Cloud SBAs

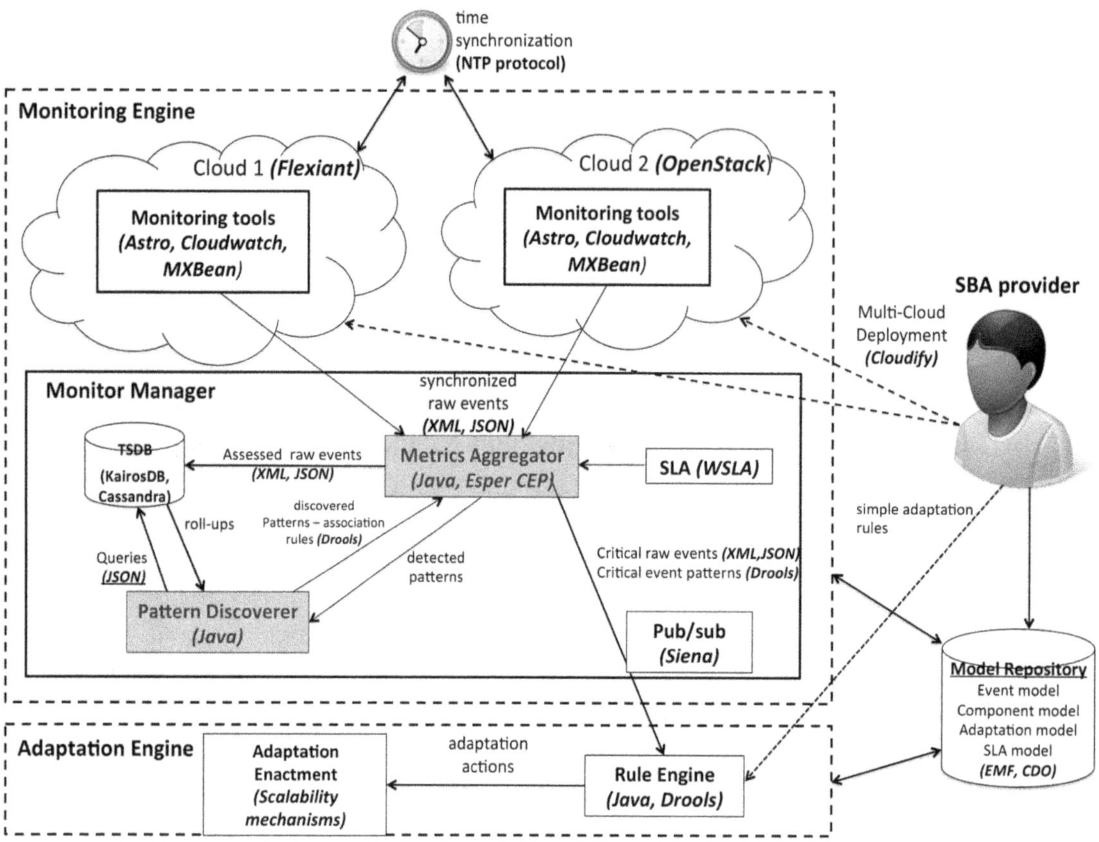

In this multi-Cloud framework, an instance of a complex event processing (CEP) engine is deployed in each of these Clouds, collecting and assessing produced events immediately after their occurrence. The Clouds' monitoring components directly interact with the *Monitor Manager*, which resides on a separate private infrastructure. The *Metrics Aggregator* component performs the assessment by parsing the SLA document, defined in the CAMEL[5] and stored in the *Model Repository*, and then storing the assessed events to the TSDB. The *Pattern Discoverer* component periodically queries the *TSDB* to get the assessed raw events for a specific time interval and identifies raw event patterns leading to SLO violations (mapping to specific aggregate metrics). The *TSDB*, capable of providing roll-ups for specific time intervals, provides the aggregated metric values for specific time intervals, necessary for pattern discovery. The discovered patterns are sent directly to the *Metrics Aggregator* to detect them at runtime. Upon pattern detection, the *Metrics Aggregator* urges the *Rule Engine* of the *Adaptation Engine* to fire the respective adaptation rule (i.e., proactive adaptation) dictating the application of the corresponding adaptation strategy, realized by the *Adaptation Enactment* component. Events are also passed to the *Rule Engine* by the *Metric Aggregator* to perform reactive adaptation. we have proactive adaptation rules and reactive ones. The proactive rely on the composition of warning events while reactive ones on the composition of violation events. Through the passing of the respective event pattern, the Rule Engine can detect which from the proactive or reactive adaptation rules can fire

Figure 3 introduces the main implementation tools for the framework components. NTP[6] realizes time synchronization of monitoring events, designed especially to synchronize time on a machines network, in order to smoothly handle leap seconds with no disruptive events. KairosDB realizes the *TSDB*[7] notion. It is very efficient as it exploits the Cassandra[8] data store, being able to have a very satisfactory capacity level (e.g., store three weeks data in a single row). The algorithm presented in Section 5 realizes the *Pattern Discoverer* by using the Drools Rule Engine[9] to store the discovered patterns as association rules (Agrawal & Srikant, 1994) and Esper Complex Event Processing (CEP) engine[10] to efficiently process the event stream. It also directly interacts with KairosDB API, modified to allow for more comprehensive JSON support. Esper's engine realizes the *Metrics Aggregator* as it exhibits abilities that facilitate the assessment of raw events and the detection of event patterns at runtime. *Adaptation Enactment* is part of ongoing work; However, some adaptation actions have already been realized, especially those mapping to mechanisms provided by Cloud providers (horizontal and vertical scaling).

SLO EVALUATION

Within this framework the SLO evaluation feature plays a critical role as it is the bond between the monitoring and adaptation engines. The proposed SLO Evaluation Engine that lies within the monitoring engine of the Monitoring and Adaptation framework analyzed in the previous section, relies on the architecture depicted in Figure 4. This architecture comprises three main levels: (a) interface; (b) core logic; (c) database (DB). At the interface level, three main actions (add, update, delete) can be performed over an event pattern (EP): add, update and delete. This functionality has been wrapped into the form of a REST service, called *EP Service*, able to parse CAMEL/SRL fragments mapping to the specification of these patterns. Each action, when called, then has an impact over the core logic level of the framework.

Figure 4. SLO evaluation framework

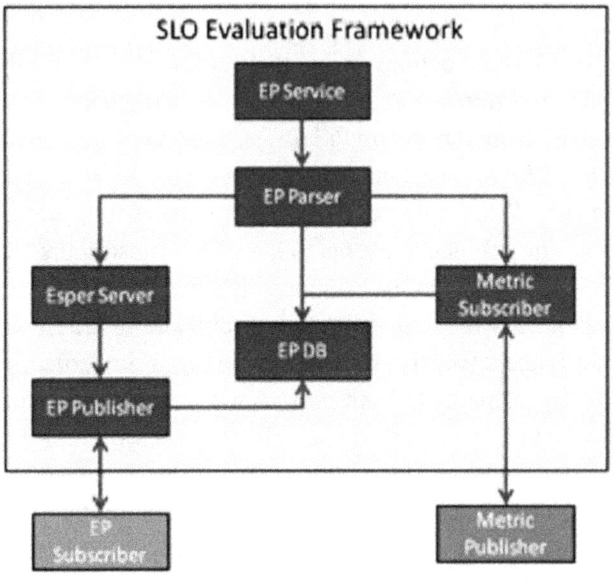

At this second level, there is a main component, called *EP Parser*, which is responsible for processing the EPs obtained from the *EP Service*. Depending then on the action requested, different interactions take place at this level. In the following, the main logic behind each supported action is analyzed:

- **EP Addition:** In case of adding a new EP, the *EP Parser* transforms it into an EP, specified in the Esper EP language[11] of the CEP framework, which is then registered in the *CEP Server*, so that it can be immediately detected. The names of metrics referenced by the EP, i.e., directly involved in the conditions of the EP's events, are also sent to the *Metric Subscriber* which not only informs its local metric list but also registers for subscribing to such metrics, when they are new, in the *Metric Publisher*. In parallel to this registration, the updated metric list of the *Metric Subscriber* is stored in the *EP DB* for fault-tolerance and rapid recovery reasons. The *Metric Publisher* is responsible for publishing the values of metrics monitored to potential subscribers. So, the *EP Parser* stores the new EP in the *EP DB* not only for recovery reasons but also to gather statistics about EPs, while being detected by the CEP Server. The *EP DB* has been realised in the form of a model repository able to store, query and manipulate models of CAMEL, especially EPs, along with their statistics.
- **EP Deletion:** In case of EP deletion, the EP is first fetched from the *EP DB*. Then, in parallel, the *EP Parser* informs both the *CEP Server* and the *Metric Subscriber* to update their structures and take further actions. The *CEP Server* just deregisters the EP's EPL specification. On the other hand, after checking that the EP metrics to be removed are not exploited in other EPs, the *Metric Subscriber* is informed to unsubscribe to these metrics to reduce the system load.
- **EP Update:** In case of EP update, the produced EPL statement by the *EP Parser* is used to update the previous one. In addition, the *Metric Subscriber* is informed for adding or removing metrics which are or not needed any more (by any EP), respectively. While the above actions can take place through the interaction of an external agent/user with the proposed SLO Evaluation framework, we highlight that, in principle, the same interactions could be differently achieved, e.g., via a publish-subscribe mechanism. As the respective functionality has been realized, we could easily switch from one to another mechanism or have both available at the same time.

As there are internally performed actions inside the framework, while it is running, these are now explicated in detail below. As the *Metric Subscriber* subscribes to metrics, it can asynchronously receive measurements for such metrics from the *Metric Publisher*. Such measurements are then transformed into events which are fed into the *CEP Server*. Once all suitable events are received by the latter component, it can detect one or more EPs. When this occurs, this component will inform the *EP Publisher*. The *EP Publisher* is responsible for publishing events to interested subscribers, named as *EP Subscribers*. Such subscribers could be adaptation engines responsible for executing the respective adaptation rule triggered, as, e.g., specified in CAMEL. Apart from this publication, the *EP Publisher* also updates the entries in the *EP DB* to modify the respective statistics of the EP(s) concerned. The proposed architecture exploits publish-subscribe mechanisms to both receive some events/measurements and publish other kinds of events (e.g., EPs). In this respect, it can actually interact with multiple components that might be willing to obtain information from or feed information to this framework. For instance, adaptation responsibility for an SBA management system could be split into multiple instances of an *Adaptation Engine* to balance the respective load. All these instances could then subscribe to the *EP Publisher* to manage their own part of the adaptation space, i.e., only those EPs that concern them.

The presented architecture is logical. This means that it can be flexibly distributed at the physical level. For instance, we could have multiple instances of the framework part that involves the *CEP Server* and the *Event Publisher* to load balance the event workload entering the framework. Alternatively, we could scale out the whole framework into parts that focus on different EP partitions.

PATTERN DISCOVERY

One of the main functionalities of the proposed framework is the discovery of critical event patterns leading to specific SLO violations. This section presents an offline algorithm (Algorithm 1) for discovering event patterns leading to violations of a specific SLO (involving a certain metric). This algorithm is based on propositional logic (Sim, Idrawan, Zutshi, & Srinivasan, 2010) and considers both multi-Cloud SBA component dependencies and aggregate metrics extracted by the TSDB. In particular, the algorithm requires as input the monitoring event stream (returned as JSON file by the TSDB), the concerned application and metric, the metrics classification (i.e., raw or composite), the SLO, the SBA's component model, and sets A, B and D (i.e. the rules that have been detected by the algorithm). Set A is the set of ambiguous association rules (λ=0) (see analysis below) discovered in previous algorithm's iterations. Set B is the set of discarded association rules with low λ' scores identified in previous algorithm's iterations, while set D is the set of already discovered association rules.

The algorithm exploits contingency tables displaying frequency distributions of: (a) candidate patterns and their negations as antecedents as well as (b) the specified metric event and its negation as consequences. These table entries (see Table 1) are utilized to determine the association rules. Only other warning and critical events are assumed to lead to the required critical events (aggregate metric violation). This means that any monitoring event that does not violate the corresponding warning or critical SLO's thresholds are not considered. Therefore, the event stream (line 3) is filtered before the actual discovery process to remove events coming from different Cloud providers than those used by the SBA, as well events that are not related to each other according to the corresponding component dependencies.. Then, the event stream is split based on the aggregate metric's time interval (line 4). The Esper CEP engine performs this by defining an Event Processing Language (EPL) statement for collecting monitoring events and then assessing and storing them in TSDB. Next, intervals are characterized as critical, if there is an aggregate metric violation, or non-critical, otherwise (line 5). For critical intervals, the temporal ordered sets of the raw events subset's power set are calculated, from the first interval event to the one before the last critical raw measurement (e.g., see last interval of Figure 8), as candidate patterns, and then the power set tree is updated. The power set' sets of all non-critical intervals are also calculated as candidate patterns not leading to aggregate metric violations. Sets with events having source components not interrelated with each other are not further processed (lines 6–11). The power sets tree stores only unique power sets.

Algorithm 1. Event pattern discovery algorithm

```
1: Input: event stream, application, metric, interval, SLOs, component model,
set A, set B, set D.
2: Output: pattern graph, pattern ranking, coherent rules, ambiguous rules
3: Filter(raw events);
```

```
4: Divide(event stream);
5: Calculate_average_metric_value();
6: while not end of event stream do
7:          A = events before the last critical raw event in this interval;
8:          I_A = P(A)-> temporal ordered power sets of set A;
9:          filter_power_sets();
10:          update_power_sets_tree();
11: end while
12: for i <- 1; treelevels do
13:          for j <- 1; treebranches do
14:          B_i,j = current branch;
15:          if B_i,j ∉ (D U B) then
16:            while not end of event stream do
17:                 A = events before the last critical raw event in this
interval
18:                 C = critical raw event for the specified metric
19:                 compute S(Bi,j, C) in A;
20:                 compute S(Bi,j,  ¬C) in A;
21:                 compute S(¬Bi,j, C) in A;
22:                 compute S(¬Bi,j, ¬C) in A;
23:                 update contingency table
24:            end while
25:            if (S(Bi,j, C)+ S(¬Bi,j, ¬C))>(S(¬Bi,j, C)+ S(Bi,j, ¬C)) then
26:                 create association rule (B_i,j -> C)
27:                 compute_lambda();
28:                 store (B_i,j -> C) in association rules Knowledge Base;
29:           else if (S(B_i,j,C) + S(¬B_i,j, ¬C) = (S(¬B_i,j,C) + S(B_i,j, ¬C)) then
30:                 Store (B_i,j -> C) as ambiguous rule
31:           else
32:                 compute_lambda'();
33:            end if
34:    end if
35:    end for
36: end for
37: traverse_tree();
38: rank_patterns();
39: filter_discarded_sets();
```

As all candidate patterns are stored in a tree-based structure (see Figure 7), a level- and a branch-based tree traversal is then performed (lines 12–36) to calculate and store the frequencies in each contingency table entry (see below), where each considered pattern reserves its own contingency. Table 1 shows such a contingency table of the event stream depicted in Figure 5. This event stream is a real monitoring event stream captured from the execution of the Invoice BPaaS and shows the frequencies of occurrences of the

Figure 5. Power set tree

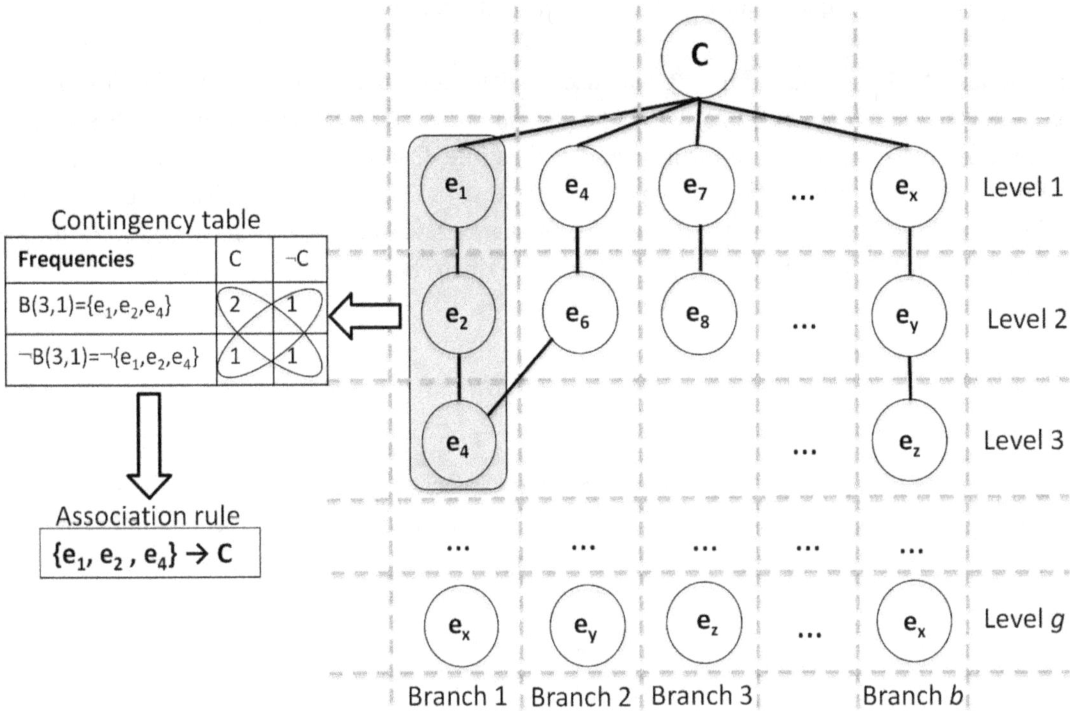

Table 1. Sample Contingency table for the candidate pattern $\{e_1, e_2, e_4\}$

Frequencies	*Y = E3*	*¬Y = ¬E₃*
X = {e1, e2, e4}	2	1
¬X = ¬{e1, e2, e4}	1	1

considered pattern $\{e_1$, e2, e4$\}$ and its negation in critical and in non-critical intervals. $B_{i,j}$ is the concerned sub-branch (i indicates the tree level and j the branch counter), C represents the critical aggregate metric violation (see Figure 6). A pattern's negation means that either of the included events does not appear. For instance, for pattern $\{e_1$, e2, e4$\}$, the next equivalence must hold: $\neg\{e1, e2, e4\} \equiv \neg e_1 \vee \neg e2 \vee \neg e4$

- $S(B_{i,j},C)$ is the frequency of $B_{i,j}$ set in critical intervals
- $S(\neg B_{i,j},\neg C)$ is the frequency of $\neg B_{i,j}$ set in non-critical intervals

Figure 6. Event stream split for pattern discovery algorithm

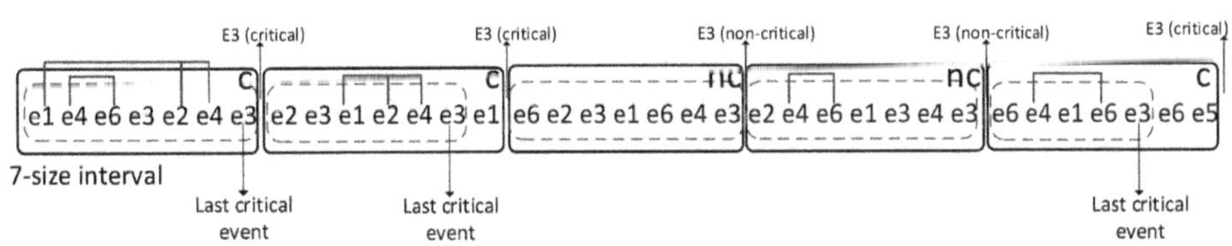

- $S(\neg B_{i,j},C)$ is the frequency of $\neg B_{i,j}$ set in critical intervals
- $S(B_{i,j},\neg C)$ is the frequency of $B_{i,j}$ set in non-critical intervals

Frequencies $S(B_{i,j}, C)$ and $S(\neg B_{i,j}, \neg C)$ are used to determine an association rule. The first proves the association rule under consideration, while the second also further validates it, as the absence of both the pattern and the violation event further validates the rule. Consequently, to determine an association rule (Eq. 1) must hold (lines 25–28), which dictates that the sum of frequencies invigorating the association rule should be greater than the sum of frequencies weakening it so as to accept it. In case of equality (lines 29-30), an ambiguous rule is identified which should be reconsidered in the forthcoming algorithm executions with an updated sample dataset.

$$\left(S\left(Bi,j\ ,C\right)+S\left(\neg Bi,j\ ,\neg C\right)\right)>\left(S\left(\neg Bi,j\ ,C\right)+S\left(Bi,j\ ,\neg C\right)\right) \tag{1}$$

The confidence level λ (line 27) of the determined association rules, indicating the strength of the rule, is calculated by Equation 2. The positive nominator indicates that the rule should be valid while the division with the denominator normalises it. A higher λ score signifies a stronger association rule.

$$\lambda = \frac{\left(\left(\mathrm{S}\left(\mathrm{Bi,j,C}\right)+\mathrm{S}\left(\neg\mathrm{Bi,j,}\neg\mathrm{C}\right)\right)-\left(\left(\mathrm{S}\left(\neg\mathrm{Bi,j,C}\right)+\mathrm{S}\left(\mathrm{Bi,j,}\neg\mathrm{C}\right)\right)\right.}{\mathrm{S}\left(\mathrm{Bi,j,C}\right)+\mathrm{S}\left(\neg\mathrm{Bi,j,}\neg\mathrm{C}\right)+\mathrm{S}\left(\neg\mathrm{Bi,j,C}\right)+\mathrm{S}\left(\mathrm{Bi,j,}\neg\mathrm{C}\right)} \tag{2}$$

Similarly, Equation 3 calculates a discarded set's (violating Equation 1) weakness level λ′ (line 32), standing for a Proportional Reduction of Error measurement and taking positive values. A higher λ′ score signifies a more irrelevant pattern:

$$\lambda' = \left|\frac{\left(\left(\mathrm{S}\left(\mathrm{Bi,j,C}\right)+\mathrm{S}\left(\neg\mathrm{Bi,j,}\neg\mathrm{C}\right)\right)-\left(\left(\mathrm{S}\left(\neg\mathrm{Bi,j,C}\right)+\mathrm{S}\left(\mathrm{Bi,j,}\neg\mathrm{C}\right)\right)\right.}{\mathrm{S}\left(\mathrm{Bi,j,C}\right)+\mathrm{S}\left(\neg\mathrm{Bi,j,}\neg\mathrm{C}\right)+\mathrm{S}\left(\neg\mathrm{Bi,j,C}\right)+\mathrm{S}\left(\mathrm{Bi,j,}\neg\mathrm{C}\right)}\right| \tag{3}$$

The discovered association rules are ranked (line 38) in order to be addressed by the *Adaptation Engine*. Similarly, discarded power sets are ranked (line 39) based on their λ' scores. Sets with high λ' scores (a threshold is applied) are stored as irrelevant to be filtered in next algorithm executions, while sets with low λ' scores are reconsidered in next executions, as there is a tendency that the application behaviour and context changes which might make a weak rule a stronger one.

Figure 6 clarifies the event stream splitting process and the way intervals are processed to identify patterns. Suppose that patterns leading to average Send Invoice SaaS execution time (i.e., violation of E_3 event) violations are searched. The respective event stream comprises 35 events (after event filtering) of Section 2's 6 events. The power set' sets of each interval (connected events in Figure 6) are processed to determine association rules. Thus, the event stream is split into 5 intervals: 3 critical (c mark on interval's upper right corner) and 2 non-critical (nc mark on interval's upper right corner). For critical intervals, the subset before the last critical Send Invoice SaaS execution time event is considered, while for non-critical ones the whole interval. The algorithm discovers two patterns and extracts the association rules: (1) $\{e_1, e_2, e_4 \rightarrow E3\}$ and; (2) $\{e_4, e_6 \rightarrow E3\}$.

The algorithm's time complexity depends on the event stream size n, the average tree level size g and the tree branches b. Thus, as the algorithm requires g*b (i.e., approximately the cardinality of the power sets) iterations on the event stream, its complexity is $O(n * b * g)$. This means that the power set tree is extensively traversed to discover the association rules.

Processing of Event Patterns

This section elaborates on the processing of the discovered patterns both on logical and temporal basis. The discovered patterns can be directly processed by mapping them to a specific adaptation strategy which can be enacted by the *Adaptation Engine*. However, in order to restrict the patterns considered by this engine, we apply a post-processing of the discovered patterns, aggregating and enhancing them, using logical and temporal operators.

The Event meta-model[12] includes all the appropriate properties to express the logical and temporal relationships among the events comprising an event pattern. The former ones are defined by the *logicalOp* attribute, while the latter ones are defined by the *interval* attribute, dictating the maximum time interval between the occurrences of two consecutive events in the discovered event pattern. We give priority to proactive rules with respect to reactive ones. This means that if the second event of an event pattern does not appear within this time interval, then we disregard the considered pattern and either wait for other patterns with the same initial events, if there are any, or otherwise trigger corresponding adaptation rules for any critical events included in these considered initial events. For instance, we assume that the pattern discovery process has identified the following association rules: (i) {ce1, we2, we4 → E3}; and (ii) {ce1, we2, ce5 → E3} with maximum waiting time for the occurrence of the we4 and ce5 events, 5 and 7 seconds respectively (i.e. the mean interval time between the events across the whole monitoring log). If no event appears after 5 seconds of the e2 occurrence, then we wait another 2 seconds to check if the second pattern is detected; otherwise, we trigger the adaptation rule corresponding to the critical ce1 event (i.e., reactive adaptation). We assume that no adaptation rule is triggered for warning events. We have to stress that the reactive rule needs to be executed in any case, unless it conflicts with the proactive one. In this latter case, these two rules are alternative/conflicting, so only one of them can be executed.

As far as the logical interrelationships between the event patterns are concerned, there is support for the logical operators AND, OR and NOT for their post-processing, in order to make them more compact. For instance, we consider the following discovered patterns:

(1) $\{ce_1, we_2, we_4, we_6 \rightarrow E_3\}$
(2) $\{ce_1, we_2, ce_5, we_6 \rightarrow E_3\}$
(3) $\{e_4, e_1, e_5 \rightarrow E_3\}$
(4) $\{e_4, e_2, e_5 \rightarrow E_3\}$
(5) $\{e_4, e_4, e_5 \rightarrow E_3\}$
(6) $\{e_4, e_5, e_5 \rightarrow E_3\}$
(7) $\{e_4, e_6, e_5 \rightarrow E_3\}$
(8) $\{ce_1, we_2, we_4, ce_5, we_7 \rightarrow E_3\}$
(9) $\{ce_1, we_2, ce_5, we_4, ce_1 \rightarrow E_3\}$

The post-processing of the above discovered patterns results in three composite association rules to be addressed by the Adaptation Engine, responsible for mapping them to suitable adaptation strategies. In particular, association rules (1) and (2) result in the composite association rule (a), i.e. either of the events ce4 or we5 may appear between event we2 and we6. Rules (3)–(7) are aggregated in the composite association rule; (b) dictating that any event other than e7 may appear in the second position of the pattern (we assume that events e1–e7 appear in the event stream and event e3 is disregarded when searching for patterns leading to violations of the E3 aggregate metric). In addition, while rules (8) and (9) result in the composite association rule; (c), combining the AND operator for events ce4 and we5 that may occur in either order and OR operator for we7 and we1 events.

(a) $\{ce_1, we_2, (ce_4 \vee we_5), we_6 \rightarrow E_3\}$
(b) $\{e_4, \neg e_7, e_5 \rightarrow E_3\}$
(c) $\{ce_1, we_2, (ce_4 \wedge we_5), (we_7 \vee we_1) \rightarrow E_3\}$

So far, we have identified and processed domain specific event patterns (i.e., patterns for the specific deployment of the SBA) comprising events that are uniquely identified by their ID. For instance, a critical event of the metric e1 indicates a violation of the available free memory of the VM hosting the Send Invoice SaaS. Thus, this type of event patterns is domain specific and is valid only for all the deployments of the considered SBA. However, in our approach, apart from the logical and temporal pattern aggregation described above, further processing of the patterns is performed, in order to identify domain independent patterns, thus leveraging the various contexts in which the SBA is performed. This allows us to derive both cloud-/deployment-specific and cloud-independent event pattern rules. This is achieved by aggregating the already discovered patterns, based on the type of metric (part of the event definition) and on the unique identifier of the raw events. For instance, two events indicating a violation of the CPU utilization of two different VMs can be aggregated to refer to the same abstract metric violation. In this way, we avoid long event patterns containing enough "noise", i.e. many useless events.

EVALUATION

This section describes an experimental evaluation of the proposed pattern discovery algorithm. The aim of this evaluation is to measure: (1) the algorithm's accuracy, in order to optimize the definition of the aggregate metric (i.e., its optimal interval); and (2) the algorithm's performance. An event dataset is used, comprising 100k events, including events from Section 2's 6 metrics for the invoice management multi-Cloud SBA, collected in a 7-minute time period. The events are provided by: a service-level middleware based on the Astro monitoring tool (service availability, execution time, and throughput); the Amazon Cloudwatch PaaS (CPU utilization, data transfer, and disk usage metrics for underlying VMs) and a Cloud resource monitor (CPU load, swap and free memory metrics), for monitoring the Amazon VM hosting the Invoice Ninja SaaS. The event timestamp is defined in millisecond accuracy; during a pre-processing of the event stream we identify five (two 2-size patterns, one 3-size pattern and two 4-size patterns) periodic patterns in it. The main task is to discover patterns causing Invoice Ninja SaaS execution time violations (i.e., violations of E_3). The experiments were performed on a machine with quad-core CPU 2.6GhZ, 8GB RAM memory and Mac OS X Maverick operating system.

The first experiment evaluates the algorithm's raw relevant and absolute accuracy in discovering all incorporated and already-known periodic patterns of the 100k-size dataset. The former considers only the five known patterns, while the latter additionally considers their sub-patterns, as they can also drive proactive adaptation. For instance, pattern $\{e_1, e_2\}$ (i.e., sub-pattern of known $\{e_1, e_2, e_4\}$ pattern) is a relevant pattern as, upon its detection, we can trigger a suitable adaptation strategy. The random patterns in the event stream are ignored, as experiments have shown that they map to perfect accuracy results. Thus, the algorithm's precision and recall is measured while fluctuating the interval size from 4 to 20 events. Figure 7 shows that relevant precision is 1 for small intervals and falls while increasing the interval size, while absolute precision fluctuates similarly at lower levels (as more irrelevant sub-patterns are discovered). This decrease is normal; as for higher intervals more irrelevant patterns (i.e., larger patterns than already-known ones) are retrieved compared to the total discovered patterns. We must note that the precision starts to fall over 8-size intervals, i.e., above the double of the maximum pattern (4 events). Figure 8 shows that the algorithm's absolute recall is 1 for all considered interval sizes, as the predefined event patterns are always discovered, except for 4-size and 6-size intervals where the algorithm fails to discover two and one 4-size patterns respectively, due to interval overlapping. Moreover, relevant recall is always 1, as the discovered sub-patterns compensate the "lost" relevant patterns (for 4- and 6-size intervals), as they also map to adaptation strategies addressing the whole pattern. Considering these accuracy results, metric E_3's optimal definition is to measure it in intervals containing in average 8 events, as this interval size enables the most accurate pattern discovery results. Figure 9 depicts the F-measure results, i.e. an accuracy measurement that considers both precision and recall.

The second experiment evaluates the algorithm's execution time based on (1) the dataset size and; (2) time interval. The former sub-experiment evaluates the algorithm's performance on different dataset sizes for 10-, 20- and 30-size intervals, while the latter is applied on 5, 20 and 50k-size datasets with an interval fluctuation from 5 to 30 events (upper limit to compute power sets). Figures 10 presents

Figure 7. Algorithm's precision

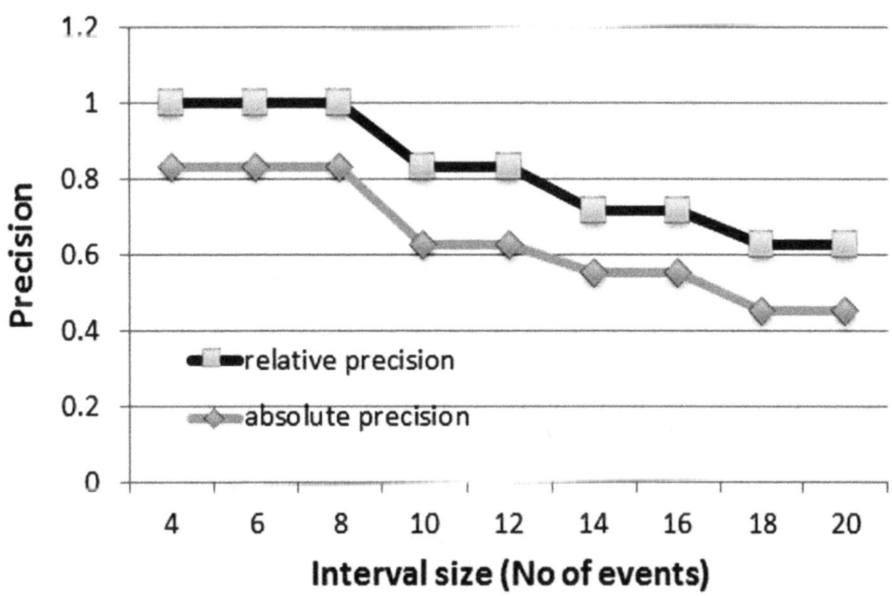

Figure 8. Algorithm's recall

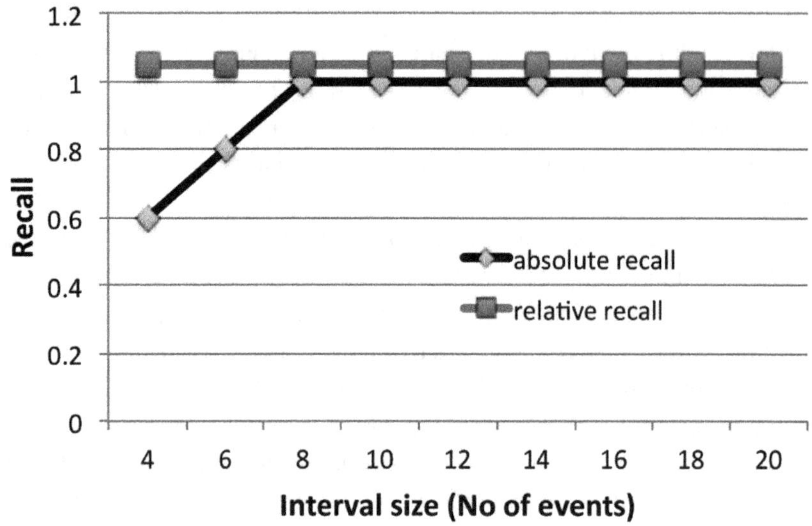

Figure 9. Algorithm's F-measure

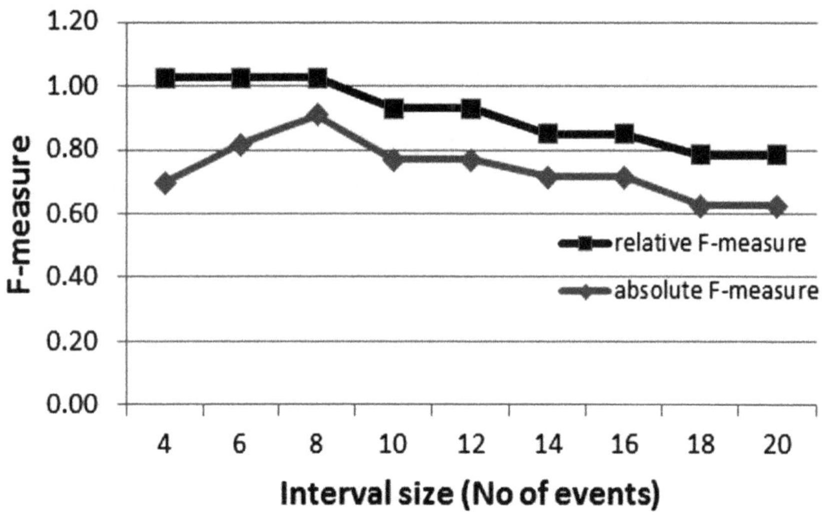

the respective results showing that the algorithm's execution time linearly increases with an increasing dataset size as expected. Larger intervals seems to hurt more the algorithm's performance, due to higher b*g products. However, such execution time is acceptable, as this is an offline algorithm not affecting the overall monitoring and adaptation framework performance. The execution time can scale well according to our needs and the actual pragmatic situations that might occur within a SBA management framework (e.g., the size of a dataset that needs to be evaluated per application, if we consider that the context of this application can be regarded as the filtering criterion over the whole measurement space). Furthermore, the results in Figure 11 reveal a changing relation between execution time and interval

Figure 10. Algorithm's performance while increasing dataset size

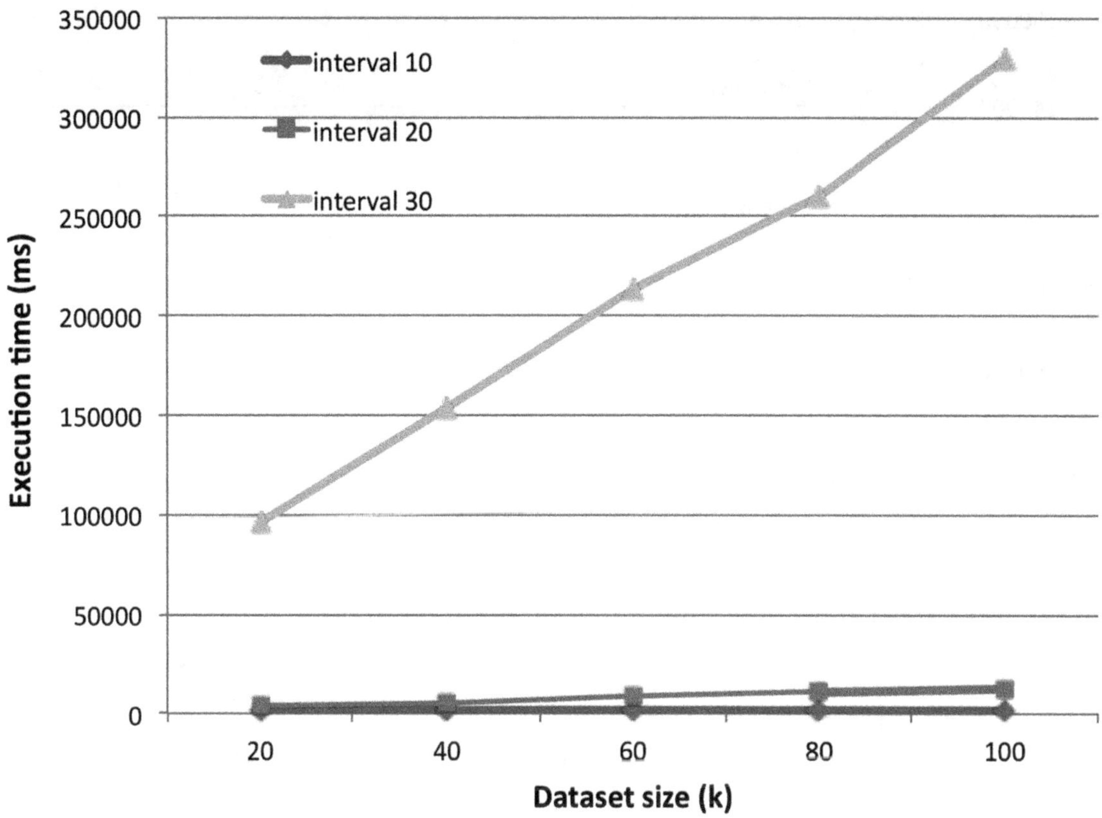

size; for larger intervals, it increases with a burst over 20-size intervals, due to rapid increase of b*g (740 (b=148, g=5) for 30-size interval compared to 92 (b=23, g=4) for the 25-size and 48 (b=12, g=4) for the 20-size intervals), posed by the considered unique sets' high increase. In fact, execution time for the largest interval (30) is more than doubled with respect to 20-size intervals.

8. RELATED WORK

Pattern Discovery Approaches

Various approaches have been proposed in complex event processing and event pattern detection. Most rely on CEP engines that detect complex events continuously and build correlations and relationships between them, such as causal and timing ones. The detection of complex patterns is based on various techniques applied either over event streams (Wu, Diao, & Rizvi, 2006) (Wang, Rundensteiner, & Ellison, 2011) or in an offline manner (Magnusson, 2000) (Hellerstein, Ma, & Perng, 2002). Many research approaches dealing with CEP and event pattern detection have been proposed in the literature and can be categorized in three main directions: (1) mathematical and statistical; (2) temporal and; (3) logic-based approaches.

Figure 11. Algorithm's performance while increasing the interval size

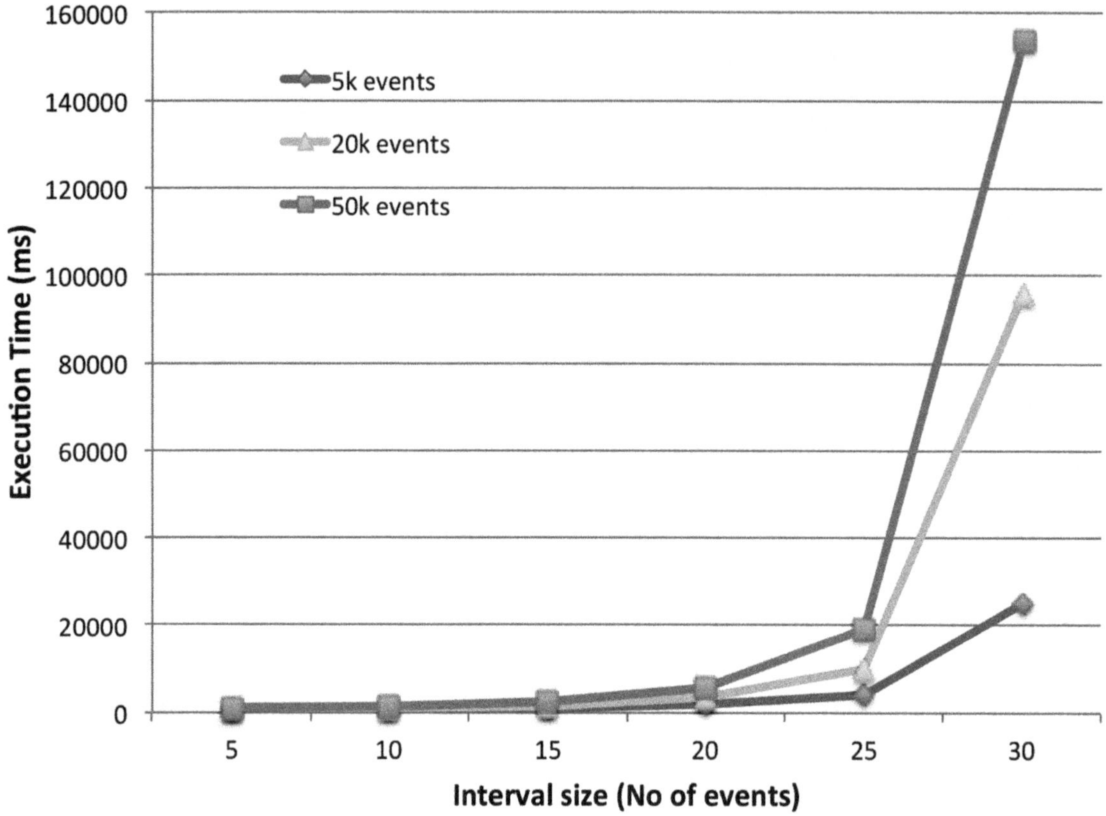

Statistical event detection approaches mainly exploit a user-defined minimum frequency or support (minsup). The springboard of all these approaches is the Apriori algorithm (Agrawal & Srikant, 1994). This algorithm produces the set of all significant association rules (rules relating a set of variables) between items in a large transactions database with a minsup. Similarly, the Lossy Counting algorithm (Manku & Motwani, 2002) computes approximate element frequency counts in a data stream and involves grouping row items into blocks or chunks, and counting within each chunk. It performs better than apriori, as there is no candidate generation phase. Other approaches propose variations of these algorithms, focusing on performance (Bettini, Wang, Jajodia, & Lin, 1998), (Li, Lee, & Shan) and accuracy (Hellerstein, Ma, & Perng, 2002), (Zaki & Hsiao) optimization. All such approaches suffer from many issues (Artikis, A.; Sergot, J. M.; Paliouras, G.), (Sim, Idrawan, Zutshi, & Srinivasan, 2010) spanning: (1) an ideal minsup for every dataset cannot be easily found (Webb & Zhang, 2005); (2) fluctuations in minsup may result in different patterns, even for the same dataset. Thus, the association rule discovery accuracy is subject to the accuracy in determining the minsup;. So, in case of high minsup, a pattern may not be discovered, while a low minsup may lead to the discovery of too many meaningless patterns.

Temporal event processing approaches exploit the temporal relations among an input stream's events. Such approaches can be very useful for deriving implicit information for the temporal ordering of raw data and predicting the future behavior of the monitored application. In (Bettini, Wang, Jajodia, & Lin,

1998) the authors introduce a formal framework for expressing data mining tasks involving time granularities, as well as algorithms for performing these tasks. Time constraints are injected into the system to bound the distance between an event pair in terms of time granularity. For instance, event e2 must happen within two minutes after the occurrence of event e1 so as to consider the combination of e1 and e2 as an event pattern. In (Patnaik, Ramakrishnan, Laxman, & Chandramouli, 2012) a temporal data mining approach is presented for data that cannot fit in memory or are processed at a faster rate than the generation one. The proposed sliding window model slides forward in hops of batches, while only a single batch is available for processing.

Logic-based approaches exploit inferencing to discover patterns defining respective association rules. In (Sim, Idrawan, Zutshi, & Srinivasan, 2010) a pattern discovery approach is proposed mapping logical equivalences based on propositional logic. In particular, a rule mining framework is introduced, generating coherent application domain independent rules for a given dataset that do not require setting an arbitrary minsup. (Artikis, A.; Sergot, J. M.; Paliouras, G.) proposes an event calculus (EC) dialect for efficient run-time recognition that is scalable to large data streams and exploits main EC predicates to discover specific activities. These approaches, which do not suffer from the above limitations, have some extra advantages: enabling ranking of the discovered patterns and discovering potential patterns to be considered for future executions.

SLO Evaluation Approaches

Many research approaches have been proposed targeting SLO evaluation in dynamic environments like the Cloud. In (Ludwig, Heiko; Stamou, Katerina; Mohamed, Mohamed; Mandagere, NagaPramod; Langston, Bryan; Alatorre, Gabriel; Nakamura, Hiroaki; Anya, Obinna; Keller, Alexander, 2015) the rSLA service and language are proposed enabling SLA management. This management services takes as input rSLA specifications and manages their lifecycle. rSLA reuses existing concepts, such as metrics and SLOs, and introduces the concept of measurement directives to enable the measurement of raw metrics. rSLA has been validated using IaaS-related SLOs.

An optimized SLA classification model for cloud computing is proposed in (Wu, Zhu, & Pan, 2013) that satisfies both the requirements of the provider and the consumer. This model distinguishes SLA parameters for each cloud service model, i.e. IaaS, PaaS and IaaS, while also catering for defining responsibilities and obligations of each party. An SLA evaluation method is also introduced, able to produce a final score for a cloud service by considering the preferences and requirements of both the service provider and consumer.

An approach for the performance evaluation of complex multi-tier cloud applications is proposed in (Bahga & Madisetti, 2013). This approach captures the application's workload via three different models: benchmark, workload, and architecture model. Moreover, based on the architecture model, a specific methodology is proposed to facilitate selecting the best and most effective deployment that meets the requirements of the multi-tier application. A comparison of the different deployment architectures is performed to detect system bottlenecks and thus lead to right design choices.

The DeSVi architecture (Emeakaroha, Vincent; Netto, Marco; Calheiros, Rodrigo; Brandic, Ivona; Rajkumar, Buyya; De Rose, Caesar, 2012) supports detecting SLA violations in the Cloud. Upon user request, DeSVi allocates computing resources for a requested service and arranges its deployment on a virtualised environment. Then, application deployment takes place and the LoM2His framework (Emeakaroha, Brandic, Maurer, & Dustdar, 2010) is employed for application monitoring and the translation of low-level metrics into high-level SLAs. DeSVi can only monitor a single data center. It was evaluated using two use cases with heterogeneous workloads for investigating the optimal monitoring interval of SLA parameters.

Other work has focused on SLA-based cloud security monitoring. In (Petcu, 2014) a preliminary research was conducted on the main challenges and barriers in designing a SLA-based security monitoring system. An introduction to the models and methods that can be used to realize such a system were also presented. In (Teshome, Rilling, & Morin, 2017), a research approach to include security monitoring terms in IaaS cloud SLAs was proposed involving a three-step incremental strategy based on the lifecycle of a cloud SLA. The approach was validated by using a network intrusion detection system.

Very few approaches deal with CEP in the cloud. The author in (Higashino, 2016) proposes CEP as a Service (CEPaaS) in order to give to users the ability to use CEP with the advantages offered by a service model. Another CEP-based event correlation approach is proposed in (Leitner, Inzinger, Hummer, Satzger, & Dustdar, 2012), where a hierarchy of predefined events is introduced. Finally, (Meister, Breß, & Saake, 2014) investigates the problem of cost-aware optimization in cloud CEP systems. The factors that should be taken into account to address this problem are presented in specific cost models and the results are validated using three real application scenarios.

Compared to the above related work, the main benefits of the proposed approach are the following. We introduce a novel cross-layer monitoring and adaptation framework that comprises two main functionalities analyzed in this chapter. First, we propose a logic-based algorithm for discovering event patterns leading to critical events within a multi-Cloud SBA's monitoring event stream, enabling injecting proactiveness in an SBA system via mapping event patterns to respective adaptation strategies. This logic-based approach provides a reliable and straightforward way for discovering meaningful critical event patterns within monitoring event streams. Second, an event-based SLO evaluation framework is introduced, that is scalable and distributable across the whole SBA architecture. In fact, this framework relies on a modular architecture with different realization options at the physical level. This framework also encompasses the well-known Esper CEP engine enabling the scalable processing of complex event patterns even in a centralized setting.

CONCLUSION AND FUTURE WORK

This paper has presented a cross-layer monitoring and adaptation framework of multi-cloud SBAs based on event patterns. This approach is part of the framework presented in (Zeginis, C.; Kritikos, K.; Garefalakis, P.; Konsolaki, K.; Magoutis, K.; Plexousakis, D., 2012), extended in a multi-Cloud scope and is validated on invoicing example. They key contributions of this approach are (1) a novel event-based SLO evaluation framework and; (2) a pattern discovery solution for detecting detrimental event patterns causing critical violations during multi-cloud SBA execution that can be further exploited by any adaptation engine to trigger suitable adaptation actions. The evaluation of the latter reveals that its performance and accuracy is optimal when suitably configured by appropriately selecting the dataset and interval (window) size.

As for future work, we plan to optimize the pattern discovery algorithm's accuracy and performance and develop a scalability rule derivation mechanism that could semi-automatically map discovered event patterns to suitable adaptation strategies. Then, upon triggering the respective rule, the framework's adaptation engine will enact the most suitable adaptation strategy, thus preventing the expected SLO violation.

ACKNOWLEDGMENT

This work is supported by CloudSocket project that has been funded within the European Commission's H2020 Program under contract number 644690.

REFERENCES

Agrawal, R., & Srikant, R. (1994). *Fast algorithms for mining association rules in large databases*. VLDB.

Artikis, A., Sergot, J. M., & Paliouras, G. (n.d.). Run-time composite event recognition. In *DEBS* (pp. 69-80). ACM.

Bahga, A., & Madisetti, V. (2013). Performance evaluation approach for multi-tier cloud applications. *Journal of Software Engineering and Applications*.

Baryannis, G., Garefalakis, P., Kritikos, K., Magoutis, K., Papaioannou, A., Plexousakis, D., & Zeginis, C. (2013). *Lifecycle Management of Service-based Applicaitons on Multi-Clouds: A research roadmap. In Multi-Cloud* (pp. 13–20). ACM.

Bettini, C., Wang, X. S., Jajodia, S., & Lin, J. L. (1998, March/April). Discovering frequent event patterns with multiple granularities in time sequencies. *Transactions on Knowledge and Data Engineering*, *10*(2), 222–237. doi:10.1109/69.683754

Blair, G. S., Bencomo, N., & France, R. B. (2009). Models@Runtime. *IEEE Computer*, *10*(10), 22–27. doi:10.1109/MC.2009.326

Emeakaroha, V., Brandic, I., Maurer, M., & Dustdar, S. (2010). *Low level Metrics to High level SLAs - LoM2HiS framework: Bridging the gap between monitored metrics and SLA parameters in cloud environments. HPCS*. IEEE.

Emeakaroha, V., Netto, M., Calheiros, R., Brandic, I., Rajkumar, B., & De Rose, C. (2012). Towards autonomic detection of SLA violations in Cloud infrastructures. *Future Generation Computer Systems*, *28*(7), 1017–1029. doi:10.1016/j.future.2011.08.018

Hellerstein, J. L., Ma, S., & Perng, C. S. (2002). Discovering actionable patterns in event data. *IBM Systems Journal*, *41*(3), 475–493. doi:10.1147j.413.0475

Higashino, W. (2016). *Complex Event Processing as a Service in MultiCloud Environments*. The University of Western Ontario.

Inzinger, C., Satzger, B., Leitner, P., Hummer, W., & Dustdar, S. (2013). *Model-based adaptation of cloud computing applications*. MODELSWARD.

Keller, A., & Kar, G. (2001). Determining service dependencies in distributed systems. *ICC*, *7*, 2084–2088.

Keller, A., & Ludwig, H. (2002). *The WSLA Framework: Specifying and monitoring Service Level Agreements for Web Services*. IBM.

Leitner, P., Inzinger, C., Hummer, W., Satzger, B., & Dustdar, S. (2012). *Application-level performance monitoring of cloud services based on the complex event processing paradigm. In SOCA* (pp. 1–8). IEEE.

Li, H. F., Lee, S. Y., & Shan, M. K. (n.d.). An efficient algorithm for mining frequent itemsets over the entire history of data streams. In *ECML/PKDD* (p. 2004). Springer.

Ludwig, H., Stamou, K., Mohamed, M., Mandagere, N., Langston, B., Alatorre, G., . . . Keller, A. (2015). rSLA: Monitoring SLAs in Dynamic Service Environments. In ICSOC (pp. 139-153). Springer.

Magnusson, M. (2000). Discovering hidden time patterns in behavior: T-patterns and their detection. *Behavior Research Methods, Instruments, & Computers*, 93-110.

Magoutis, K., Devarakonda, M. V., Joukov, N., & Vogl, N. G. (2008). Galapagos: Model-driven discovery of end-to-end application - storage relationships in distributed systems. *Journal of Research and Development (Srinagar)*, *52*(4-5), 367–378. doi:10.1147/rd.524.0367

Manku, G. S., & Motwani, R. (2002). *Approximate frequency counts over data streams. In VLDB* (pp. 346–357). ACM.

Meister, A., Breß, S., & Saake, G. (2014). *Cost-Aware Query Optimization during Cloud-Based Complex Event Processing*. GI-Jahrestagung.

Patnaik, D., Ramakrishnan, N., Laxman, S., & Chandramouli, B. (2012). *Streaming Algorithms for Pattern Discovery over Dynamically Changing Event Sequences.* CoRR, abs/1205.4477

Petcu, D. (2014). *SLA-Based Cloud Security Monitoring: Challenges, Barriers, Models and Methods. In Euro-Par Workshops* (pp. 359–370). Springer.

Popescu, R., Staikopoulos, A., Liu, P., Brogi, A., & Clarke, S. (2010). *Taxonomy-driven adaptation of multi-tier applications using templates. In SASO* (pp. 213–222). IEEE.

Romer, K. (2006). Distributed Mining of Spatio-Temporal Event Patterns in Sensor Networks. *EAWMS Workshop at DCOSS*, 103-116.

Sim, A. T., Idrawan, M., Zutshi, S., & Srinivasan, B. (2010). Logic-based pattern discovery. *Transasctions on Knowledge Data Engineering*, *22*(6), 798–811. doi:10.1109/TKDE.2010.49

Teshome, A., Rilling, L., & Morin, C. (2017). *Verification for Security Monitoring SLAs in IaaS Clouds: the Example of a Network IDS.* Inria Rennes Bretagne Atlantique.

Wang, D., Rundensteiner, E., & Ellison, R. (2011). Active Complex Event Processing over Event Streams. *PVLDB*, *4*(10), 634–645.

Webb, G. I., & Zhang, S. (2005). K-optimal rule discovery. *Data Mining and Knowledge Discovery, 10*(1), 39–79. doi:10.100710618-005-0255-4

Wu, C., Zhu, Y., & Pan, S. (2013). *The SLA Evaluation Model for Cloud Computing. ICCNCE.* Atlantis Press.

Wu, E., Diao, Y., & Rizvi, S. (2006). High-performance complex event processing over streams. In *SIGMOD Conference* (pp. 407-418). ACM.

Zaki, M. J., & Hsiao, C. J. (n.d.). *CHARM: An efficient algorithm for closed itemset mining. SDM.* SIAM.

Zeginis, C., Konsolaki, K., Kritikos, K., & Plexousakis, D. (2011). *Towards practive cross-layer service adaptation. In WISE* (pp. 704–711). Springer.

Zeginis, C., Kritikos, K., Garefalakis, P., Konsolaki, K., Magoutis, K., & Plexousakis, D. (2012). *Towards cross-layer monitoring of multi-Cloud service-based applications. In ESOCC* (pp. 188–195). Springer.

Zengin, A., Marconi, A., & Pistore, M. (2011). *CLAM: Cross-layer adaptation manager for service-based applications. In QASBA* (pp. 21–27). ACM. doi:10.1145/2031746.2031754

ENDNOTES

1 https://site.cloudsocket.eu/
2 https://www.invoiceninja.com/
3 http://camel-dsl.org/
4 https://paasage.ercim.eu/
5 camel-dsl.org/
6 www.ntp.org
7 code.google.com/p/kairosdb
8 cassandra.apache.org/
9 www.jboss.org/drools/
10 esper.codehaus.org/
11 http://www.espertech.com/esper/
12 http://www.didaktorika.gr/eadd/handle/10442/35066?locale=en

Chapter 4
A Case Study of a Three-Part Entrepreneurial Strategy in a Japanese Accounting Cloud Service

Yutaka Mizuno
Nagoya Institute of Technology, Japan

Nobutaka Odake
Nagoya Institute of Technology, Japan

ABSTRACT

The purpose of this chapter is to clarify a three-part entrepreneurial strategy in a Japanese accounting cloud service with two-sided markets and freemium. The service provider established its platform as one of industry platform in Fintech and made its initial public offering in five years since its entrepreneurship. The authors studied the service provider and obtained three findings. First, the service provider dares to adopt its cloud platform with single architecture. Second, it builds up its service platform with two business strategies, two marketing strategies, and two management of technology. Third, the integration of these strategies becomes a three-part entrepreneurial strategy, which realizes economies of scale and scope cyclically. Finally, it realizes agglomeration economies in the service platform. Therefore, start-up with two-sided markets and freemium not only dares to design its service platform with single architecture but also should build up the three-part entrepreneurial strategy to formulate service platform as market before building up its business ecosystem.

INTRODUCTION

A cloud service is an information processing platform. It consists of end-users, players to supply complementary resources, and players to intermediate its services in a platform. Thus, a cloud service builds a business ecosystem and one of the important social infrastructures. High-tech start-ups "unicorn,"

DOI: 10.4018/978-1-5225-5951-1.ch004

whose corporate value break over $ 1 billion, has been developing a new type of industrial platforms utilizing cloud computing.

In the global market, start-ups create new financial cloud services in the financial sector. A term "Fintech" combines two terms of "Financial" and "Technology." Fintech is the digital (r)evolution in the financial sector and algorithm-based banking with the human touch (Dapp, 2014). Some pieces of financial services are provided by high-tech start-ups. These high-tech start-ups have been developing niche and riche financial services rapidly on a service platform utilizing financial engineering and ICT technologies.

In Japan, a new type of accounting cloud services had started in 2013. These accounting cloud services provide an accounting book service for individuals and enterprises. The customers of these services can manage their own daily income and expenditure on the accounting cloud service. In October 2017, one of Japanese accounting cloud service made an initial public offering in only five years since its entrepreneurship.

The objective of this study clarifies a three-part entrepreneurial strategy in a Japanese accounting cloud service with two-sided markets structure and freemium business model in Fintech service industry in Japan.

RELATED WORK

In this section, this paper fills key research gaps in the existing literature: service science, innovation, social network, platform strategy, and network economy.

Service Science

Grönroos (1994) indicated five service management approaches. Vargo and Lusch (2004) developed a vision of service-dominant logic and proposed eight basic propositions.

IBM developed a vision of a multidisciplinary approach called SSME (Services sciences, Management and Engineering), defining it as the application of scientific, management and engineering disciplines to create knowledge and develop solutions for service problems (Stauss, B., Engelmann, K., Kremer, A., & Luhn, A., 2007). Research of SSME needs to integrate the perspectives of management and marketing on the one, as well as IT and engineering sciences along with social sciences on the other (Stauss, B., Engelmann, K., Kremer, A., & Luhn, A., 2007). It also needs to consider other academic disciplines: economics and law, operations research, industrial engineering, computer science, MBA and management consulting, management information systems and knowledge management, organizational studies and organizational learning, urban planning, ecosystem services and nature's services, complexity science and complex adaptive systems for social systems research (Spohrer, 2007). More often than not, service research is confronted with a lack of data and short time series, which hampers the development of rigorous econometric analysis and thus also weakens the explanatory power of the results (Stare & Rubalcaba, 2007).

Innovation

Rogers (2003) had been studying a diffusion of product innovation and a mechanism of the diffusion process, by which an innovation was communicated through certain channels over time among the members of a social system. Bass (1969) proposed a mathematical diffusion model of innovation. Chesbrough (2003) developed a concept of "Open Innovation." Chesbrough (2010) proposed the four key elements of open service innovation: think of business as a service, co-creation with customers, open innovation, and transformed business models. A vision of open & close strategy is a method of Management of Technology (MOT), which combines internal "Closed" innovation and external "Open" innovation (Ogawa, 2014). At the open & close strategy, the term "Open" refers to a mechanism to create the strategic technologies and products of the company. The term "Close" refers to a mechanism to make for a pre-determined technical area to protect as a source of value and not to propagate the source of value out of the company. The open & close strategy combines these two mechanisms into one.

Duncan (1976) discussed that the ambidextrous organization designed dual structures for innovation. He broke down the innovation process into two general stages: initiation and implementation. March (1991) defined that exploration includes things captured by terms such as search, variation, risk taking, experimentation, play, flexibility, discovery, innovation; and that exploitation includes such things as refinement, choice, production, efficiency, selection, implementation, execution. The exploration is in a trade-off relationship with exploitation and has a vulnerability in comparison with the exploitation (March, 1991). An ambidexterity, maintaining an appropriate balance between exploration and exploitation, is a primary factor in system survival and prosperity (March, 1991). An organizational ambidexterity is defined as the capacity to simultaneously achieve alignment and adaptability at a business-unit level (Gibson & Birkinshaw, 2004).

Social Network

Granovetter (1983) indicated the strength of weak ties in interpersonal ties. Coleman (1988) defined that social capital by its function has a variety of entities with two elements in common. One is that they all consist of some aspect of social structure. Another is that they facilitate certain actions of actors -whether persons or corporate actors- within the structure. There are two arguments of network structures that create social capital: closure of social networks, and structural holes of social networks. Closure of social networks as social structure facilitates social capital (Coleman, 1988). Structural holes are an opportunity to broker the flow of information between people, and control the projects that bring together people from opposite sides of the hole (Burt, 2000). Burt (2001) discussed that structural holes and network closure can be brought together in a productive way. Burt (2001) concluded,

while brokerage across structural holes is the source of added value, closure can be critical to realizing the value buried in the structural holes.

Kim and Aldrich (2005) mentioned that entrepreneurs can apparently reap substantial returns on their social capital and boost their chances of commercial success. They offered a simple scenario that entrepreneurs can reach out within two steps to one million potential resource providers by leveraging their direct ties.

Platform Strategy

Gawer and Cusumano (2002) defined that:

Platforms are industry building blocks which attract other firms' investment and innovation on add-on products or services analyzed the competitive strategy of major enterprises...

in the viewpoint of the platform formation and leadership. Platforms are products, services, or technologies that serve as foundations upon which other parties can build complementary products, services, or technologies (Gawer & Cusumano, 2002). A cloud service exchanges various services on an ICT (Information and Communication Technology) platform. Platform strategies are concerned with how to manage and integrate the internal resources of an enterprise. Moore (1993) defined a business ecosystem that crosses a variety of industries. He pointed out that every business ecosystem develops in four distinct stages: birth, expansion, leadership, and self-renewal, or, if not self-renewal, death. Iansiti and Levien (2004) broke strategy of a platform firm of a business ecosystem down into three patterns; keystone, dominator, and niche player. The business ecosystem involves managing and integrating the external resources of enterprises. Gawer and Cusumano (2008) explored the issue of platform leadership in information technology industries and pointed out two strategies as coring and tipping. Gawer (2014) broke platforms into two types; platforms as markets and platforms as technological architectures. He also broke technological platforms into three categories; internal platform, supply-chain platform, and industry platform. He presented a corresponding coherent set of attributes (organizational form, interfaces, set of accessible capabilities, governance) for each type of technological platform and represented an organizational continuum of technological platforms in his empirical studies.

Network Economy

Arthur (1994) established a theory of network economy and explained that the network effect increased returns to scale and had characteristics of path-dependence. The network effect is that the growth of users of a platform increases the growth of profits derived from the platform. Rochet and Tirole (2003) established a theory of two-sided markets structure. A two-sided (or more generally multi-sided) markets structure is defined as multiple markets in which one or several platforms enables interactions between end-users, and tries to get the two (or multiple) sides "on board" by appropriately charging each side (Rochet & Tirole, 2003). Eisenmann et al. (2006) stated that *the platform value to any given user largely depends on the number of users on the network's other side with two-sided network effects* and that *value grows as the platform matches demand from both sides.* The two-sided network effect is that the growth of one side users on a platform increases the growth of profits of other side users (Eisenmann, T., Parker, G., & Van Alstyne, M. W., 2006). Multi-sided markets structure contains several two-sided markets in its platform. Hagiu (2007) defined that a multi-sided platform is both a platform and a market intermediary. Hagiu (2009) mentioned that the most successful multi-sided platforms were constantly evolving, increasing their depth and/or reach and in the process redefining their boundaries and those of entire industries.

One business model of a two-sided market structure is "Freemium." The term combines two other terms, "Free" and "Premium". Freemium is a business model by which a product or a service (typically a digital offering such as software, media, games or web services) is provided free from charge, but a premium is charged for advanced features, functionality, or virtual goods (2009). A freemium market strategy combines free-of-charge users and paid users onto one platform. Kumar (2014) stated that free features were a potent marketing tool and that the model allowed a new venture to scale up and attract a user base without expending resources on costly ad campaigns or a traditional sale force. Mizuno and Odake (2015; 2016a; 2016b; 2017) have surveyed the progressive formation of the best accounting cloud service with a two-sided markets structure and a freemium business model in Japan.

METHODS AND MATERIALS

In this section, this paper elucidates research setting, research design, and data collection.

Research Setting

Research of SSME needs to integrate the perspectives of management and marketing on the one, as well as IT and engineering sciences along with social sciences on the other (Stauss, B., Engelmann, K., Kremer, A., & Luhn, A., 2007). It also needs to consider other academic disciplines (Spohrer, 2007). More often than not, service research is confronted with a lack of data and short time series, which hampers the development of rigorous econometric analysis and thus also weakens the explanatory power of the results (Stare & Rubalcaba, 2007).

Since business strategy, marketing strategy, and MOT (management of technology) interact each other, it is necessary to consider a three-part strategy. In this article, the authors adopt the integrated perspectives of these three strategies on the one. The authors have been studying one specific accounting cloud service with the integrated perspective from its establishment to its initial public offering. The goal of this article clarifies the three-part entrepreneurial strategy in a Japanese accounting cloud with two-sided markets structure and freemium business model in "Fintech" service industry in Japan.

Research Design

The authors employ an exploratory case study research strategy, applying qualitative methods (Eisenhardt, 1989; Eisenhardt & Graebner, 2007) to build explanations of the phenomena. Figure 2(a) indicates our integrated research perspective.

The technological architecture of a cloud service is not only modular design but also core and periphery. The core is cloud services in the cloud platform. The periphery faces two sides: a platform as market and a platform as technological architecture. Complimentary resources in periphery constitute of technological elements and market elements. Technological elements include technological improvements, technological partnerships, original service developing (cloud service OEM), and so on. Market elements include associated cloud services, business partnerships, capital alliances, and others. Figure 2(b) indicates our analytical structure.

First, this paper studies current status of Japanese accounting cloud services with two-sided markets structure. As a result, this paper chooses the best accounting cloud service as a case study and estimates service adopters of the cloud service. Second, this paper divides three stages of diffusion of the cloud service. Then, this paper examines platform development and business ecosystem formation of the accounting cloud service in each stage. Third, this paper discusses business strategies, marketing strategies, and management of technology in the accounting cloud service. Finally, this paper concludes the results of the discussion.

Data Collection

This study relies on public archives including press releases, financial reports, industry databases, and publications.

AN ACCOUNTING CLOUD SERVICE IN JAPAN

In this section, this paper elucidates current status of Japanese accounting cloud services with two-sided markets structure, selects a case company, and introduces the case company.

Accounting Cloud Service in Japan

In the global market, start-ups create new financial cloud services in the financial sector. Some pieces of financial services are provided by high-tech start-ups. Fintech includes social lending, crowdfunding, payment service, block-chain, accounting service, housekeeping book service, and financial information security service. Yano Research Institute, a famous Japanese private marketing research company, forecasts the domestic market scale of Fintech venture sales volume in Japan. The report said that the financial services provided by high-tech start-ups became 9.9 Billion JPY in 2016. Table 1 shows the domestic Fintech market forecast.

An accounting cloud service provides an accounting book service for individuals and enterprises. There are two conventional models of accounting cloud services. One is accounting cloud service for individuals provided by Internet service providers. Another is service for companies provided by system integrators. There are new accounting cloud services for both of individuals and companies with the two-sided markets and freemium. Table 2 shows the list of the new accounting cloud services. These service providers have been growing rapidly.

Table 1. Domestic Fintech market forecast

Fiscal year	2015	2016	2017	2018	2019	2020	2021
Financial market (Billion JPY)	4.89	9.91	19.00	31.90	46.15	61.80	80.80

(Source: Yano Research Institute Ltd. (2017))

Table 2. List of accounting cloud services in Japan

Markets structure	Index	Service name	Establishment	Service in	IPO
Two-sided markets and freemium	A	Freee	2012/7	2013/3	-
	B	Money Forward	2012/5	2012/12	2017/10

(Source: Authors)

Figure 3 shows comparison of search volumes of two accounting cloud services on Google Trends. Both accounting cloud services had started their service almost at the same time. The accounting cloud service B has gathered most search volume. It made an initial public offering in October 2017. Thus, this paper chose the accounting cloud service B as a case study.

Business Overview of the Accounting Cloud Service B

The service provider of the accounting cloud service B established in May 2012. It had started the accounting cloud service in December 2012. Table 3 shows the company's overview in November 2017. Table 4 indicates sales volume and profit of the service provider.

Forecast of Adopters of the Accounting Cloud Service B

The accounting cloud service helps their customers to keep accounting books in a smart way. It aggregates transaction data from Internet banking accounts once a day. In addition, it associated with the other cloud services. Customers of the service can manage daily income and expenditure on the accounting cloud service automatically. The service provider combined a cloud service with smartphone applications into one cloud service platform. Once users had registered the accounting cloud service, they used to access the service via a smartphone application or a bookmark on Internet Browser, they did not search the service on search engines (e.g., Google). The search volume of the service name can estimate the adopters of the accounting cloud service. This paper employs the Bass diffusion model (Bass, 1969) and estimated a long-range forecast with accumulated search volume on the Google Trends. Figure 4 indicates the estimated adopters as of August, 2015. Since the service provider had started the service, the customers of the service have achieved one million in nineteen months and two million in thirty months. This paper divides three stages of diffusion of the accounting cloud service and analyzes business activities of each stage. These three stages are the entrepreneurial stage, progressive stage, and jump-up stage.

Table 3. Company's overview

Company	Money Forward, Inc.
Established	2012/05
Capital	2.3 billion JPY
Employees	212 persons
Address	Minato Ward, Tokyo
Market capitalization	56.0 billion JPY

(Source: Company's home page, (Retrieved in November 2017)

Table 4. Sales volume and profit

(Million JPY)	2013/11	2014/11	2015/11	2016/11
Sales volume	3	76	441	1,542
Profit	-107	-549	-1,142	-888

(Source: Nikkei Financial Report)

BUSINESS DEVELOPMENTS UNTIL INITIAL PUBLIC OFFERING

In this section, this paper examines platform development and business ecosystem formation of the accounting cloud service in the entrepreneurial, progressive, and jump-up stage.

Entrepreneurial Stage

The innovations of the accounting cloud service are a daily aggregation of transaction data in Internet banking and other associated cloud services. Then, business drivers of the accounting cloud service are the number of aggregation accounts and associated services. Figure 5 indicates business drivers in the entrepreneurial stage.

The service provider had released a trial version of the accounting cloud service for individuals in December 2012. It had released several smart-phone applications in March 2013. Then, it had distributed its cloud service in the market rapidly. In July 2013, it had released the official version of service for individuals. When the updated service for individuals had released, the search volume has increased. After the official service for companies had released, the search volume has increased more than the volume of its service for individuals. The accounting cloud service has grown as the two-sided markets structure. The service provider could increase its capital at each stage when it had released its new services of trial services and official services for individuals and companies.

The core services are (1) trial service for individuals, (6) CSV (Comma-Separated Values) data download service, (8) budget management service, (10) official service for individuals, (11) CSV data upload service, (13) trial service for business, (15) official service for business, (16) Final tax return service, (24) original service developing, (25) original service developing, and (31) billing service. Technological elements are (2) iPhone Apps, (3) Facebook ID, (4) Android Apps, (5) Google ID, (7) third party Apps, (9) iPad Apps, (12) widget, (17) receipts recognize Apps, and (21) e-tax service. Market elements are (14) tax accountant partnership, (18) Internet-shop, (19) net-shop, (20) crowdsourcing, (22) net-shop, (23) payment, (26) payment, (27) affiliate, (28) affiliate, (29) tax accountant partnership, (30) business partnership, (32) crowdsourcing, and (33) payment.

An auto aggregation service is one of core service of the accounting cloud service. The auto aggregation service aggregates transaction data from associated financial cloud services into the accounting cloud service once a day. These financial services include bank accounts, credit card accounts, security accounts, pension accounts, electronic money accounts, point card accounts, mail order site accounts, and so on.

Free users and paid users constitute a two-sided markets structure. A boarding strategy had connected paid users with free users. It helped to migrate from free users to paid users. The experimentation of creating free users is an exploration process (March, 1991) to explore paid users. The two-sided network effects of free users had accelerated an exploitation process (March, 1991) for paid users.

The service provider employed a specific associating strategy to connect with the same categorized cloud services. First, this strategy avoided single point failure. Second, it avoided technological lock-in by other cloud service at the same. Third, it realized structural holes (Burt, 2000) and values buried in networks. Figure 6 shows this associating strategy.

The layered business ecosystem had been expanding into both horizontal and vertical directions. Figure 7 shows the layered business ecosystem in the entrepreneurial stage. Technological elements of the accounting cloud service had expanded environmental-independence and neutrality of network path. Market elements had expanded path-dependence of its service process. For example, the accounting cloud service could aggregate sales data from net-shop and submit a bill to its associated services.

The service provider adopted the two-sided markets structure and freemium business model to realize an effective go-to-market strategy in the entrepreneurial stage. First, it had expanded user's convenience with environmental-independence and neutrality of network path to increase its user lead. Second, it had charged its service for free to gather its customers as value co-creators. Third, it had expanded the number of aggregation process for migrating from free users to premium users. Fourth, it had expanded its accounting and incubating services for migrating from premium customers to cooperate customers with path-dependence of associated services.

Progressive Stage

In the progressive stage, the number of the aggregation account has not been increasing a lot. Instead of the number of aggregation account, the service provider has been increasing the number of the associated service. Figure 8 indicates business drivers in the progressive stage. The service provider has been releasing core services for companies, which include an official release of billing (Index 46), automatic receipt recognition service (trial 57), payroll accounting (trial 58, official 64), and negation service of monetary claim and debt (trial 60, official 70).

There are eleven technological elements, which include five data exchange methods and five original service developments. The five data exchange methods include a data import method from other accounting cloud services' file formats, and two APIs (Application Program Interfaces) from other cloud services. There are twenty-two associated services, which include three sales management services, seven accounting services, five payment services, and seven business alliances. The service provider had been focusing on small and midsize enterprises in the progressive stage.

The service provider has gathered one million customers in May 2014 and two million customers in April 2015. Its customers have increased from one million to two million in a year. Thus, it has succeeded in the management of its entrepreneurial process. In July 2015, the search volume has increased rapidly, because the service provider has carried out a joint campaign with the TV program. The premium users of the accounting cloud service could watch business on-demand video contents that TV program provides for free for three months. In addition, the Japanese government has planned to reform its social security and tax number in October 2015, which is called "My Number." The service provider has released a trial version of social security and tax number management service. Then, the search volume has increased rapidly.

The service provider has employed a specific associating strategy to connect with the different categorized customer groups in the progressive stage. Figure 9 shows this associating strategy.

The customer groups are divided by affinity for the cloud service and smartphone. The customer group 1 in the entrepreneurial stage is individuals and solopreneurs. The customer group 2 is middle ages working in SMEs. The customer group 1 and the group 2 constitute a new two-sided markets structure. The marketing strategy has connected the customer group 1 with the customer group 2. The experimentation of creating the customer group 1 is the exploration process to explore the customer group 2. The two-sided network effect of the customer group 1 has accelerated the exploitation process for the customer group 2.

In addition, the structure of associated service groups is the same with the structure of customer groups. Associated services are categorized into two categories. One is pre-accounting service that proceeds before the process of accounting cloud service. The pre-service includes account aggregation service, automatic receipt recognition service, payment service, and so on. Another is post-accounting service that proceeds after the process of accounting cloud service. The post-service includes billing service, payroll management, and so on.

This associating strategy in the progressive stage is summarized as next. First, the service provider expanded economies of scope. Second, it constituted a two-sided markets structure with not only customer groups but also associated service groups. Third, it realized two structural holes: customer groups and associated services. Fourth, it realized values buried in these two structural holes.

The layered ecosystem of the accounting cloud service has been expanding its services into both horizontal and vertical directions steadily in the progressive stage. Figure 10 shows the layered business ecosystem in the progressive stage.

In the progressive stage, the platform as technological architecture has expanded online radio and MVNO (Mobile Virtual Network Operator). The platform as market has expanded customer group of SMEs and business alliances. Theses business alliances include credit firms, major banks, local banks, and TV station. The partners of business alliance have rich social capitals. The service provider has adopted several APIs to realize tight connections with associated services of business alliances.

The service provider has been realizing additional effective go-to-market strategy in the progressive stage successively. First, it has expanded a new customer group in the progressive stage, which constituted a two-sided markets structure with the previous customer group in the entrepreneurial stage. Second, it has expanded not only platform as technological architecture, but also platform as market. Third, it has expanded its platform as technological architecture with environmental-independence and neutrality of network path to increase its user lead. Fourth, it has expanded its platform as market with path-dependence of its services by the reason of association with business alliances such as banks, credit firms, and so on. Fifth, it had begun to connect its associated cloud services with tight coupling methods such as APIs.

Jump-Up Stage

Figure 11 indicates business drivers in the jump-up stage. In the jump-up stage, the service provider announced that the accounting cloud service has finished to establishing all of aggregation account of domestic banks in August 2015. This means that the developing team of the aggregation function could become to focus on the service developing. Instead of the number of aggregation account, the accounting cloud service has been increasing the number of the associated service rapidly. In July 2015, the service provider announced that the accounting cloud service would provide new core service for company, which was a trial release of social security and tax number management service in August 2015. This announcement caused the rapid growth of search volume of Google trends. In October 2015, the service provider released the official version of social security and tax number management service. In October 2016, the service provider was stung by a lawsuit brought by its competitor concerning about auto journal process with artificial intelligent technology in accounting (Index B) and won the case in July 2017. It gathered three million customers in November 2015, four million customers in September 2016, and five million customers in April 2017. In the jump-up stage, the service provider formed a capital alliance with two local banks, raised funds from three venture capitals and eight local banks. Thus, the accounting cloud service succeeded in the management of its progressive process in the jump-up stage.

In this stage, the service provider developed sixty-four technological elements and associated fifty-four market elements. Figure 12 indicates the summary of these elements. The service provider developed nineteen data exchange methods that include three APIs developed by the service provider. It provided twenty-one original service developments for associated partners that include specific industries (e.g. for private doctors, private dentist, and so on) and small financial agency (e.g. local banks and a stock company). It associated thirty services with the social networks that include nine business alliances (seven capital alliances) with local banks, four business alliances with retail industry, four business partnerships with local government and chamber of commerce, and two business alliances with electric companies.

The service provider employed a new specific associating strategy to connect with the different categorized customer groups in the jump-up stage. Figure 13 shows the new associating strategy. The new customer groups are business partner's customers. The service provider connected its ecosystem with business partner's ecosystem with original service developing methods. The original service developing method looks like an original equipment developing in product industry. The service provider customized its accounting cloud service for business partners. Then, it provided its service with the name of its business partner. The accounting data of the original service developing may be provided for the business partner for service improvement and innovation under the business partnership.

The service provider's platform and its business partner's platforms constitute a new two-sided markets structure. The marketing strategy has connected the service provider's platform with business partner's platforms. The experimentation of creating the platform is the exploration process to explore business partner's platforms.

This associating strategy in the jump-up stage can be considered as indirect ties (Kim & Aldrich, 2005). The service provider has been associating with specific industries, banks, local governments, and chambers of commerce. These organizations have rich social capitals. The new associating strategy can reap substantial returns on the shared social capital of the service provider and its business partners. Thus, the service provider can boost their chances of commercial success.

In fact, the search volume of the accounting cloud service has been boosting in the jump-up stage than before. Figure 14 indicates the estimated adopters as of April, 2017. Indexes indicate the number of million customers of the accounting cloud service. The maximum estimated adopters increased from 5 million to 6.7 million (about 1.34 times). The network economies such as the accounting cloud service can apply the economies of increasing returns (Arthur, 1994). The profit of the accounting cloud service will become more than 1.34 times in future. Thus, the service provider can establish more sustainable business than before. In October 2017, the service provider listed on the Mothers of Tokyo Stock Exchange.

The layered ecosystem of the accounting cloud service was expanding its services into both horizontal and vertical directions steadily in the jump-up stage. Figure 15 shows the layered ecosystem with platform as market and as technological architecture. In this stage, the platform as technological architecture expanded wearable devices and web TV. The platform as market expanded customer groups of specific industries and banks employing original service developing methods with co-developing APIs. In addition, the platform as market was also associating with local governments and chambers of commerce to re-vitalize SMEs with improving their accounting process.

The new associating strategy in the jump-up stage is summarized as next. First, the service provider established business partnership with specific industries, small financial agency, local government, and chamber of commerce. Second, it connected its business ecosystem and business partner's ecosystem with the method of the original service developing. Third, it constituted a two-sided markets structure with its services and business partner's customers via their indirect ties. Fourth, the shared social capital of the associated ecosystem could boost their chances of commercial success. Fifth, the new associating strategy constituted structural holes between the service provider's platform and business partner's platforms.

Summary of Service Formation

In the entrepreneurial stage, the service provider had started the accounting cloud service. The business chances of this stage were the wide-spread of Internet banking and new small business promotion act. The service provider did not have rich managemental resources. It had focused on its platform as technological architecture. First, it had built up its single core technological platform, which had dared to adopt one double-entry bookkeeping system of accounting cloud service not only for individuals but also for enterprises. Second, it had been expanding core services and technological elements such as smartphone applications and the number of aggregation account simultaneously. Third, it had started expanding market elements of associated services such as sales, accounting, and payment. Finally, it had succeeded in establishing its platform as technological architecture.

In the progressive stage, the service provider did not have a rich number of its customers. The business chances of this stage were the promotion of social security and tax number and promotion of cultivating Fintech. The service provider has focused on its platform as market. First, it has finished expanding the number of aggregation account as the main business driver in the previous stage, which could decrease the number of its developing team of aggregation account and increase the number of its developing team of service association. Second, it has been expanding core services and market elements such as sales, accounting, and payment. Third, the service provider has started to expand its business alliances such as credit firms, major banks, local banks, and TV station. These business alliances have contributed its name recognition of the accounting cloud service. Fourth, the service provider has started to expand its support programs and to deliver its closed services to its lead customers with branching. Finally, the service provider has succeeded in establishing its platform as market. The success of business development in this stage could gather venture capital from financial sectors. Thus, the service provider could store management resources.

In the jump-up stage, the service provider did not have enough number of its customers and rich social capital. The business chances of this stage were the rapid growth of Fintech. The service provider was powering up not only platform as technological architecture but also platform as market. It formed alliances with technological partners (e.g. system integrators) and market partners (e.g. financial organization). First, it co-developed APIs with the technological partners. It built up abilities of tight linkage between its platform and partner's platforms. Second, it formed tight alliances with its market partners. It tied up the capital alliance with financial groups. Third, it has been co-developing original service developing with its market partners. This strategy could utilize indirect ties with its market partners. Fourth, the shared social capital of the business ecosystem has been boosting its commercial success.

Table 5 shows summary of service formation of the accounting cloud service.

DISCUSSION

In this section, this paper discusses management of technology, marketing strategies, and business strategies in the accounting cloud service, and summarizes these strategies as a three-part entrepreneurial strategy.

Management of Technology

The service provider has two specific management of technology: layer management of platform and an open & close strategy.

Table 5. Summary of service formation

Stage	Entrepreneurial stage	Progressive stage	Jump-up stage
Period	2012/05 - 2014/5	2014/05 - 2015/08	2015/08 - 2017/04
User lead	Yuppie ages	Middle ages	Customers of business partners
Business driver	Numbers of aggregation account	Numbers of associated service	Numbers of business alliance of banks and industry groups
Innovations	·Auto-aggregation of various transactions ·Association of various cloud services	·Business alliance with banks and credit firms ·Business partnership with broadcasting medias	·Capital alliance with financial group ·Original service developing (Cloud service OEM)
Communication channel	·Face to face communication between customers ·Application store of smart-phone	·Mass media (Broadcasting and newspaper) ·Sales Branch	·Sales channel of financial group ·Branching and service depots
Social system	·Diffusion of Internet-banking ·New small business promotion act	·Promotion of social security and tax number ·Promotion of cultivating Fintech	·Rapid growth of Fintech ·Association with chambers of commerce and local governments
Cultivation of trust	·Disclosure of official information ·Observation of service level agreement	·Enriches of support programs ·Fairness and neutrality of service	·Information Security (ISO 27000) ·Alliance of information security

(Source: Authors, based on Mizuno & Odake, 2017)

Layer Management of Platform

The service provider manages its cloud platform in a hierarchical way. The platform as technological architecture has three layers: application environment, computer device, and computer network. The platform as market has four layers: customer group, system interface, social networks, and business on the ground. Figure 16 shows the layer management of platform.

The service provider develops its platform as technological architecture with expanding path-independence. It pursues not only environmental-independence in the application environment and computer device layer, but also neutrality of network path in the computer network layer.

The service provider develops its platform as market with expanding path-dependence. It invites a new customer group step by step in the customer group layer. It exploits market opportunities from solopreneurs to customers of specific institution via SMEs. It synchronizes its new service group with its new customer group. It tightens up on its social ties with complimentary resources from business partnership to capital alliance via business alliance in the social network layer. It manages its system interfaces from loose coupling to tight coupling in the system interface layer.

A pair of the customer group and the service group constitutes a two-sided markets structure. Customer groups constitute a multi-sided markets structure. Service groups also constitute a multi-sided markets structure. Thus, the service platform constitutes not only a multiplex markets structure but also a multi-dimension service structure. The service provider capsules the multiplex and multi-dimension structure in its cloud platform with closure of network. The closure of network can be critical to realizing the value buried in the structural holes of associated services in its business ecosystem.

Open and Close Strategy

The service provider is start-up. Start-up has fewer accumulations of intellectual assets inside than the conventional companies. It is necessary for start-up to develop software for its service alone or purchase from other company. Because start-up don't have enough management resources, it cannot develop the full-set services or purchase software. Then, it must concentrate its core area of its business domain and expand its service association with other companies. This strategy realizes that the service provider as start-up can accelerate its service development and release its service earlier.

The closed area is in the service platform. Competitors cannot perform pressures for the independent cloud service directly. The service provider can lead its service innovation in the core area alone and keep intellectual assets within its service platform. In the closed area, the service provider realizes successive service formation. The key of formulating the successive service formation is system interface technology. The service provider needs to realize sophisticate services. To realize sophisticate services, it sophisticates its system interfaces with its associated services by tight coupling such as API connection. It co-develops these APIs under the business alliance with business partners. The tight coupling can realize its service orchestration with its associated services. To sophisticate its system interfaces, the service provider standardizes and manages technological specifications of its system interfaces. The standardization and management of technological specifications cannot only tighten up on added values distributed by the service platform, but also capsule the added values within its closed platform. This strategy can utilize the path-dependence of the successive service formation in the accounting cloud service.

The open area faces the business ecosystem of the accounting cloud service. In the open area, it is risky for start-up to associate with other companies. If the associated companies are influential in the market or have strong technical advantages, start-up may receive pressures to adopt only the associated company's service and cannot keep its independence. The service provider creates the environmental-independent platform, associates with duplicated services as a redundant configuration, and realizes neutrality of the network path. Then, it can escape from single-point node failure and technical lock-in by other companies. Therefore, it can manage its sustainable growth and keep independence of its business activities. In the open area, to realize successive service formation, the service provider has started to associate its service with partner's services under the business collaboration. Then, it tightened up with business alliances with business partners. Finally, it forms capital alliances. The capital alliances contribute to its initial public offering. Table 6 shows the open & close strategy. Figure 17 indicates the added value distribution utilizing the open & close strategy.

Table 6. Open and close strategy

Condition	Closed Area (Platform development)	Open Area (Ecosystem formation)
Service provider has a strong technology	·Realization of successive services ·Realization of services orchestration with associated services ·Standardization of system interfaces ·Utilization of path-dependence in market	·Environmental-independence ·Neutrality of network path ·Utilization of path-independence of technological architecture
Competitors have strong technologies	·Association of partner's services ·Standardization of services ·Collaborative development of system interfaces	·Business partnership ·Business alliance ·Capital alliance

(Source: Authors)

Marketing Strategies

The service provider operates two specific marketing strategies: a boarding strategy and a spiral-structured marketing strategy.

Boarding Strategy

The service provider adopts two-sided markets and freemium. Two-sided markets structure is a structural hole. It consists of free users, premium users, and paid users. The premium and paid users defray all its service cost. Then, it is necessary for the service provider to establish a substantial increase in paid users. A boarding strategy is a way to migrate from free users to paid users. Figure 18 shows the boarding strategy.

Step 1: The service provider expands user's convenience with environmental-independence of its local system and neutrality of network path to increase its user lead.
Step 2: It charges its service for free to gather its users as value co-creators.
Step 3: It expands the number of its interface of aggregation process for migrating from free users to premium users.
Step 4: It expands its associated services for migrating from premium users to paid users.
Step 5: It invites a new customer group. Then, it starts to repeat new boarding process for the new customer group.

Because the service provider is start-up, it has limited management resources and knowledge. It is necessary for the service provider to co-create its services with its users. First, the service provider starts to co-create its service for free users because of unnecessary of high specialties in accounting. Step by step, it co-creates its services with its users who need more specialties. This boarding strategy migrates from user lead to paid users, and establishes its two-sided markets structure in each customer group.

Spiral-Structured Marketing Strategy

The service provider establishes a specific marketing strategy. Figure 19 indicates the spiral-structured market strategy. As the result that the service provider forced its users of service for individuals to use double-entry bookkeeping on the platform of the accounting cloud service, it establishes single system architecture. The single system architecture realizes a structural hole between free users and paid users. It is not logical in the sight of user's serviceability to adopt double-entry bookkeeping in the services for individuals. But, it offers sophisticated functions to deduce a corresponding and opposite entry automatically in the account aggregation process and the data entry process in the smart-phone applications. As the result that the single system architecture realizes successive developments of its services, the service provider can optimize total cost of system developments in a minimum.

The service provider needs to realize low cost structure of the accounting cloud service. As the result that it provided its service as path-independent and charged its service for free, it can gather more user lead and free users. Because of the economies of scale, it can realize low price services and gather a large scale of its customers.

The service provider needs to realize high quality accounting services. As the result that it adopted Open Service Innovation, it can accelerate and deepen services innovation. It develops high functional services such as account aggregation process and associated services with other cloud services onto its single platform. Because of the economics of scope, it can realize high quality services and increase paid users. Thus, it can match demands from both free users and paid users. On single cloud platform, free users and paid users can be attracted to each other. The two-sided network effects between free users and paid users increase the value of the accounting cloud service. Thus, the service provider can grow its service sustainably and cyclically.

The service provider needs to gather more customers, because the two-sided markets structure and freemium business model has a weakness in its profitability. The service provider invites new customer groups who have more user lead and requests more specialty of accounting process. The new customer groups also become two-sided markets structure, which establishes multi-sided markets structure. The service provider can expand economies of scale and scope step by step. Finally, it realizes agglomeration economies within its service platform. Thus, it can accelerate and deepen its business activities sustainably.

Business Strategies

The service provider had designed and has been operating two specific business strategies: an organizational continuum and organizational ambidexterity.

Organizational Continuum

In the entrepreneurial stage, the service provider had invited individuals and solopreneurs as its first customer group. It had built up its platform with system linkages of B to B connections using loose coupling interfaces within net-firms. Its service platform had become the internal platform. The main system interface of B to B connections is a screen scrapping technology. The screen scrapping technology is a type of method that gathers transaction data from a screen of associated service on the internet browser. The service provider can't control the layout of the screen of associated service, because the layout is an intellectual property of associated service. Thus, the method might be a weak tie (Granovetter, 1983).

In the progressive stage, the service provider has invited SMEs as its second customer group improving its system interfaces from loose coupling to tight coupling within business alliances. Its service platform has become the supply-chain platform and formed a cloud ecosystem. In this stage, the service provider has started to adopt APIs provided by associated services. These interfaces could control data exchanges in detail. The technological specification of the interface is stably-maintained. Thus, the adoption of these interfaces could contribute to establishing closure of social networks (Coleman, 1988).

In the jump-up stage, the service provider invited partner's customers within industrial networks as its third customer group. Its service platform becomes one of industry platform in Japanese Fintech industry and forms a ground-ecosystem (business on the ground utilizing cloud service). In this stage, the service provider was developing not only its own APIs for associated services, but also industrial standard APIs with its alliance partners (e.g. financial organizations and system integrators). These APIs are a tight coupling. Because, these interfaces can manage the service orchestration between the accounting cloud service and its service partners. The service orchestration can sophisticate the service processes between the service provider and partners' service processes. The service provider established indirect ties (Kim & Aldrich, 2005) between its own customer group and its partners' customers.

The growth of the service platform realizes an organization continuum (Gawer, 2014). Table 7 shows the organizational continuum in the accounting cloud service.

The main driver to form the cloud service platform widely within one specific industry is the strength of system interfaces between its platform and complimentary resources. The service provider has been forming the system interfaces from loose coupling to tight coupling utilizing weak ties, closure of networks, and indirect ties.

Organizational Ambidexterity

The nature of the two-sided markets structure and freemium is not only a combination of exploration and exploitation, but also a combination of an initiation and implementation. The service provider designs dual structures of {exploration, exploitation} and/or {initiation, implementation} for innovation: {free users, paid users}, {customer group(i), customer group(i+1)}, {own platform, partner's platform}, {customer group, service group}, {trial release, official release}, {aggregation process, association process}, {pre-accounting service, post accounting service}, {path-independence, path-dependence}, and {loose coupling, tight coupling}.

The service provider adopts a two-phase development and turned two cycles of its system development process. It has been adding two-sided markets structures onto its service platform step by step. Moreover, it connects these additional two-sided markets structures with a relationship of exploration and exploitation. The relationship of exploration and exploitation realizes organizational ambidexterity, which can be aligned and efficient in its management of innovation and as well as being adaptive to changes in the environment at the same time. In addition, the unit of the additional two-sided markets structure grows from users to a customer group, and finally becomes a platform. Thus, the organizational ambidexterity of the service provider realizes the organizational continuum in the accounting cloud service.

Table 7. Organizational continuum

Category	Internal platforms	Supply-chain platforms	Industry platforms
Stage	Entrepreneurial	Progressive	Jump-up
Organizational form	B to B connections	Cloud-ecosystem	Ground-ecosystem
Interface	Open	Somewhat open	Close
Accessible capabilities	Within net-firms	Within business alliances	Within industry networks
Governance	Business partnerships	Business alliances	Capital alliances
User lead	Individuals Solopreneurs	SMEs	Partner's customers
Business form	B to C	B to B	B to B to C
Unit of two-side markets	Users	Customer group	Platform
Social networks	Weak ties	Closure of network	Indirect ties
System interfaces	Loose coupling	Loose to tight coupling	Tight coupling

(Source: Authors, based on Gawer, 2014)

Figure 1. Two-sided market structure and freemium business model (Source: Authors)

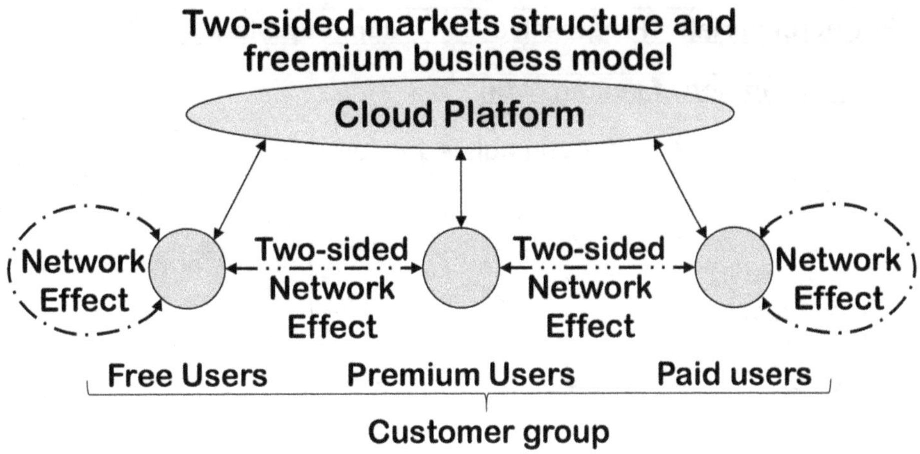

Figure 2. Research perspective and analytical structure (Source: Authors)

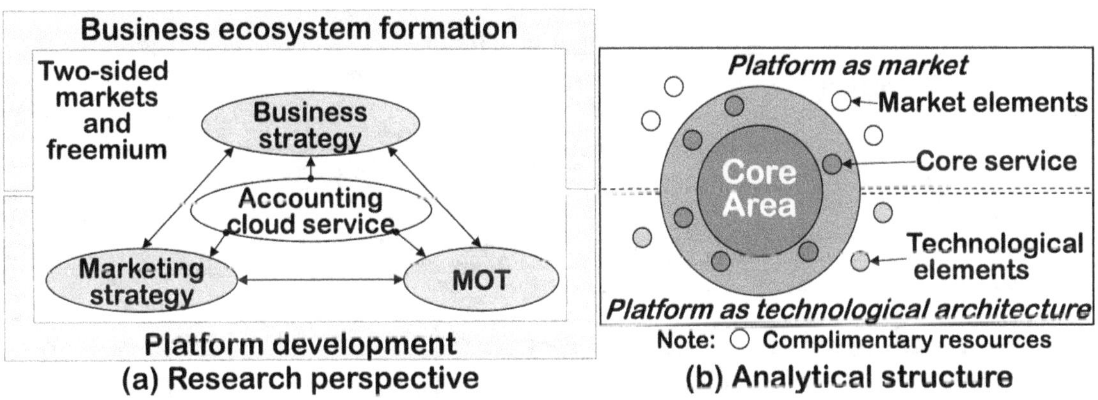

Figure 3. Comparison of search volumes (Source: Google Trends)

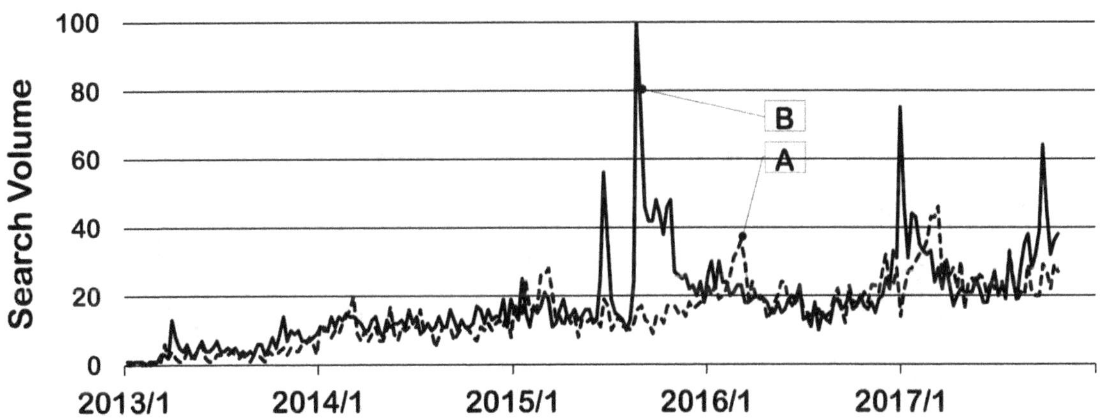

Figure 4. Forecast of adopters of the accounting cloud service (Source: Authors)

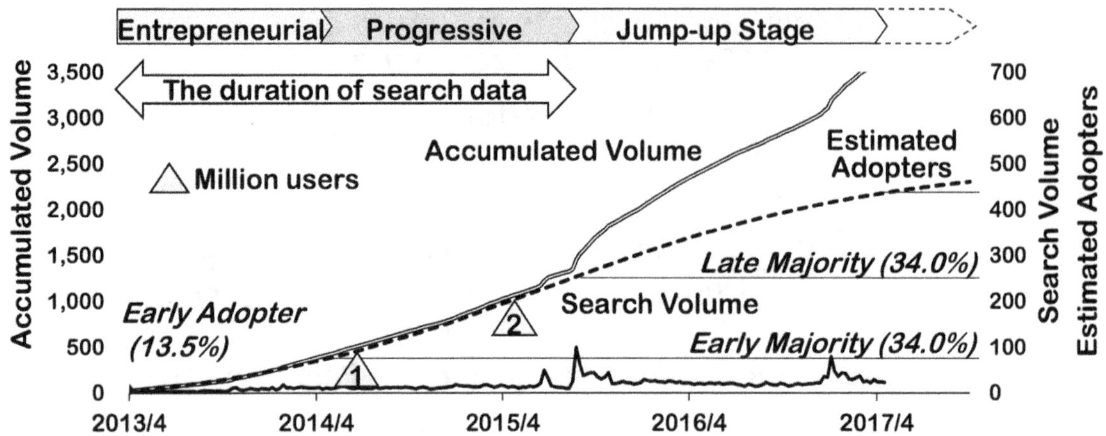

Figure 5. Business drivers in the entrepreneurial stage (Source: Authors, based on Mizuno & Odake (2015))

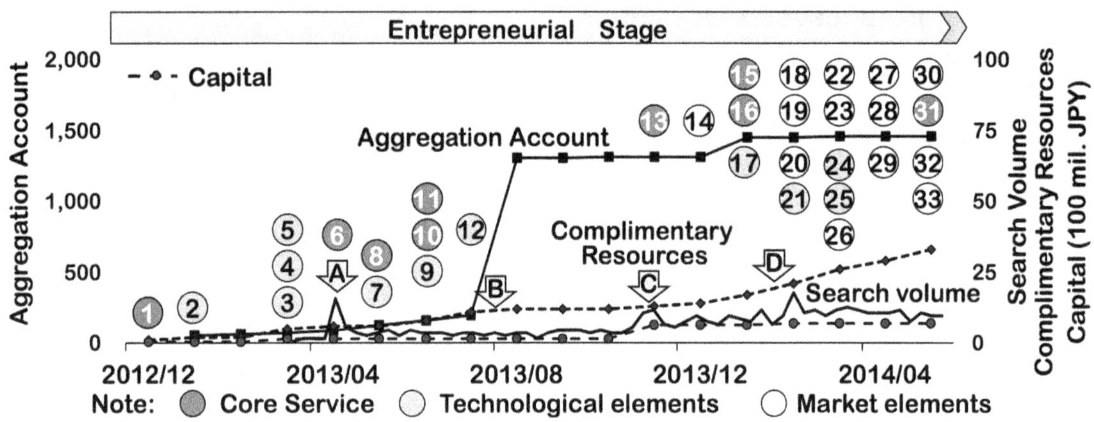

Figure 6. Structural holes in entrepreneurial stage (Source: Authors)

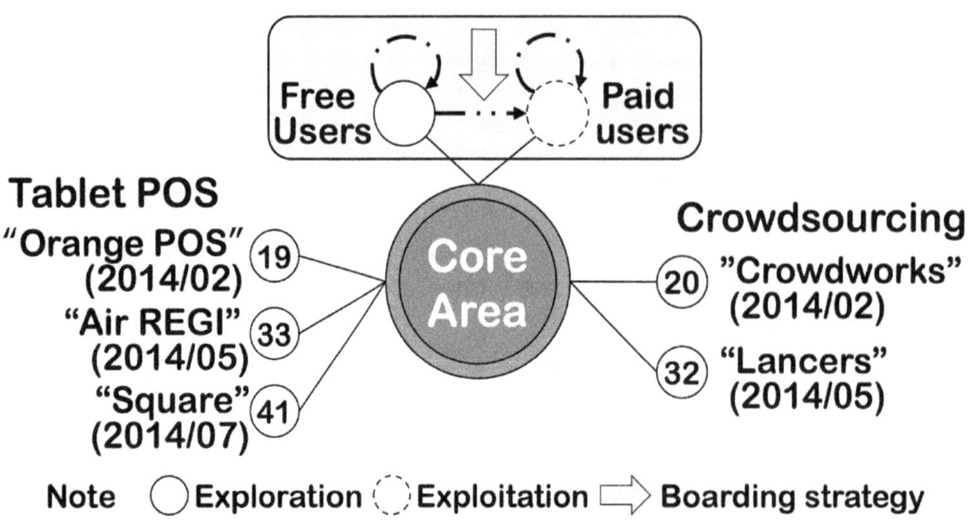

Figure 7. Layered ecosystem in the entrepreneurial stage (Source: Authors, based on Mizuno & Odake (2015))

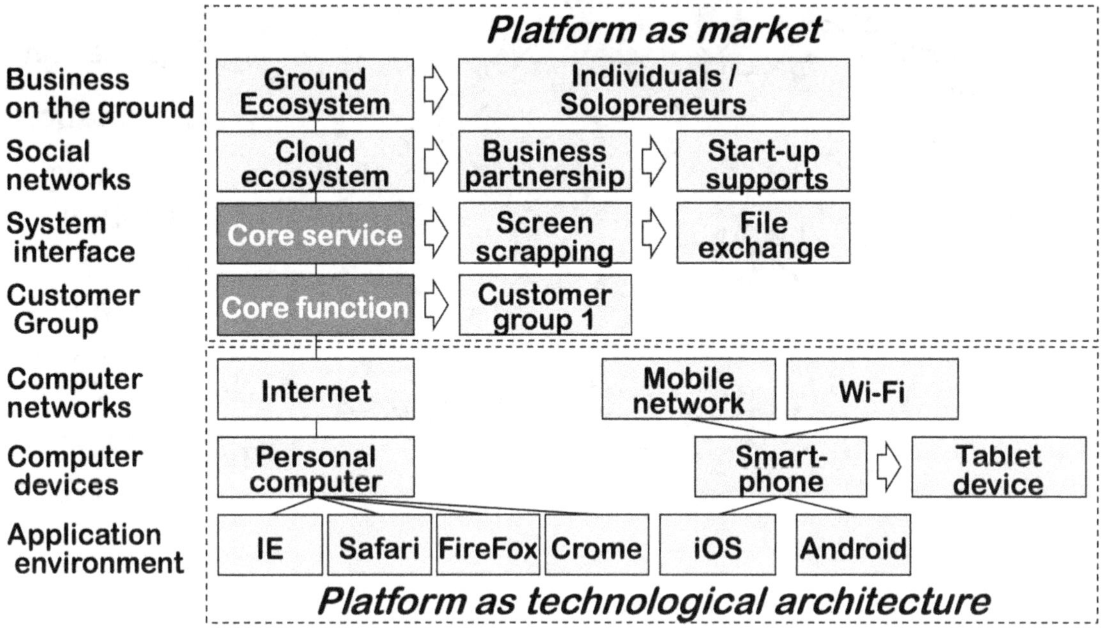

Figure 8. Business drivers in the progressive stage (Source: Authors, based on Mizuno & Odake (2016a))

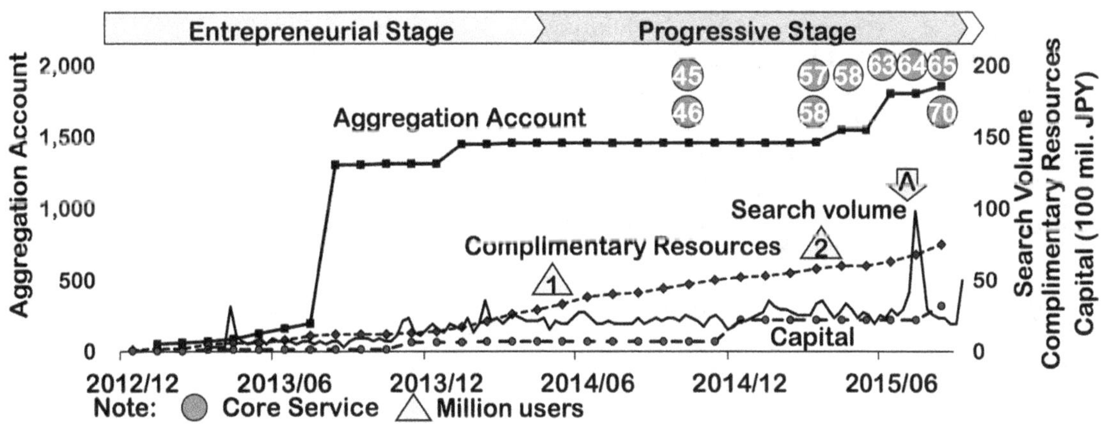

Three-Part Entrepreneurial Strategy

The service provider designs and operates a three-part entrepreneurial strategy, which integrates two management of technology, two marketing strategies, and two business strategies. Figure 20 illustrates the entrepreneurial three-part strategy.

First, the service provider adopts the core architecture as double-entry bookkeeping. It develops a cloud platform with single architecture as core and periphery structure. It manages its service platform

Figure 9. Structural holes in the progressive stage (Source: Authors, based on Mizuno & Odake (2016a))

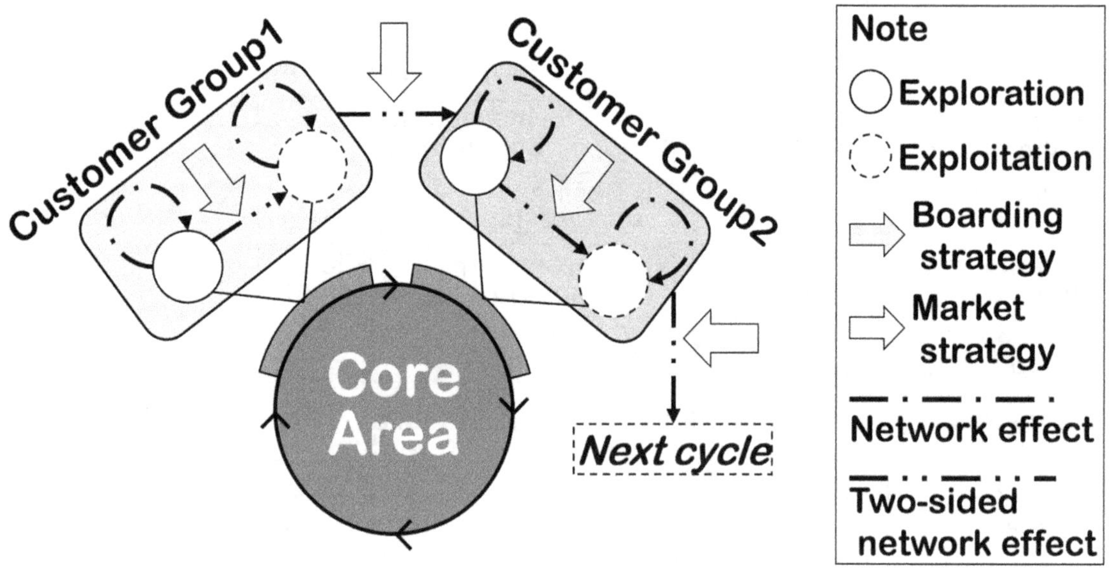

Figure 10. Layered ecosystem in the progressive stage (Source: Authors, based on Mizuno & Odake (2016a))

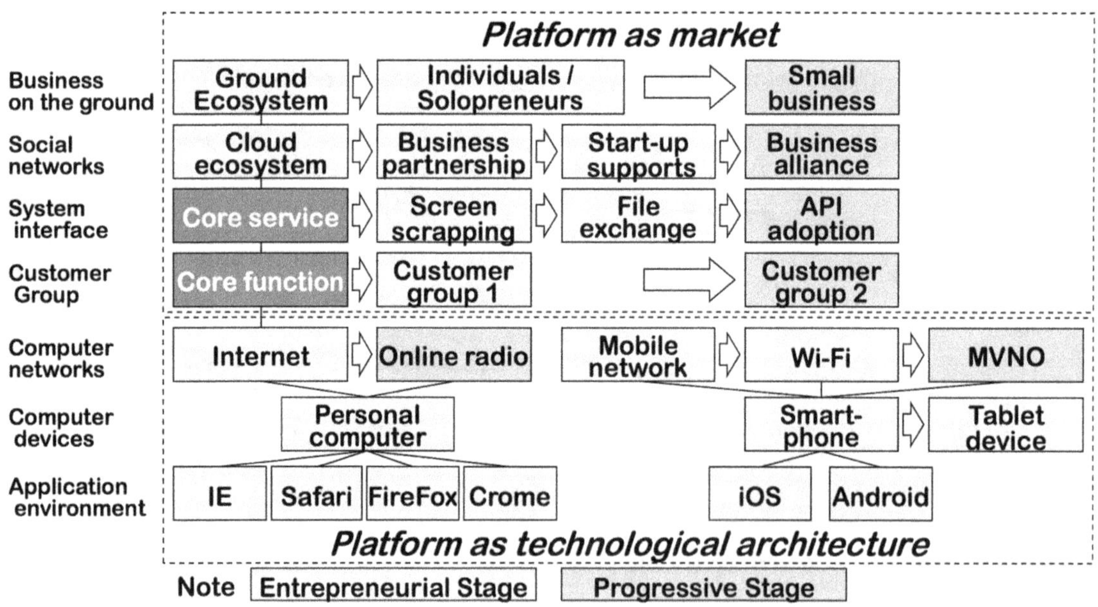

in a hierarchical way as layer management of platform. The layer management of platform can realize the open & close strategy and capsule its value within its service platform.

Second, the service provider gathers free users. It gives user's convenience for free. It develops its platform as path-independence of technological architecture. The path-independence can gather free

Figure 11. Business drivers in the jump-up stage (Source: Authors)

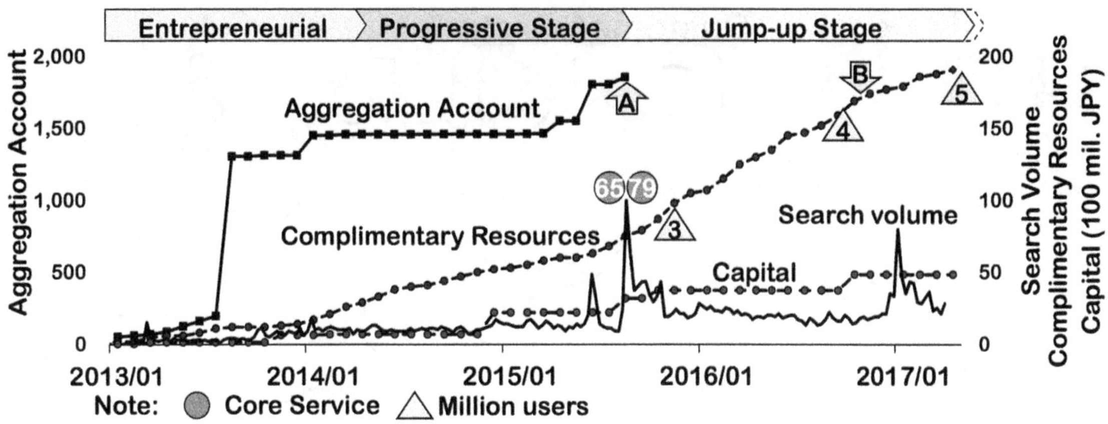

Figure 12. Core and peripheries in the jump-up stage (Source: Authors)

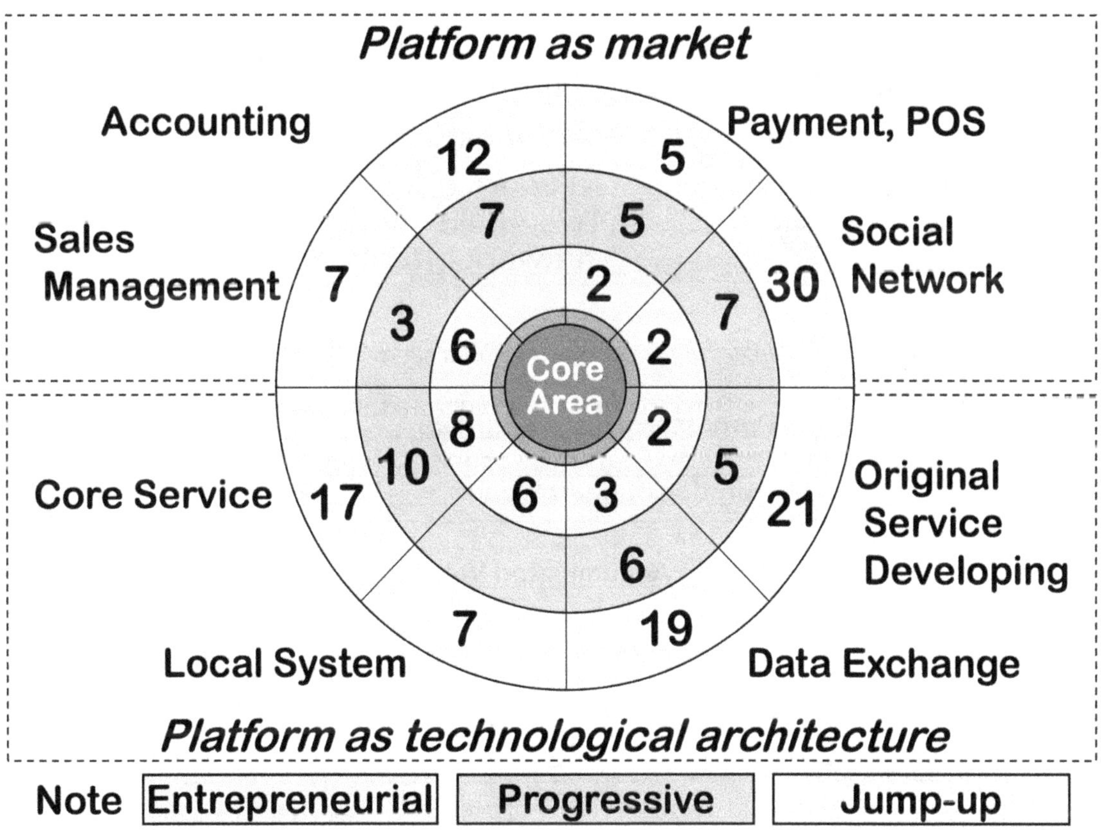

Figure 13. Structural holes and indirect ties in the jump-up stage (Source: Authors)

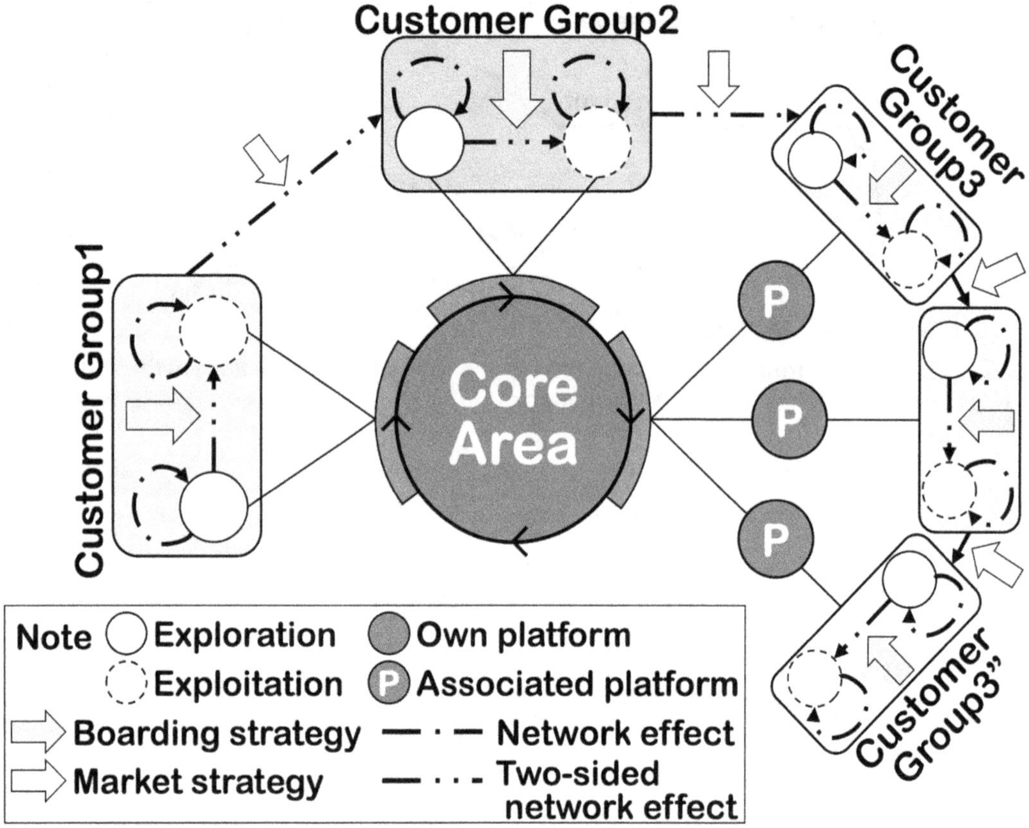

Figure 14. Re-prediction of adopters of the accounting cloud service (Source: Authors)

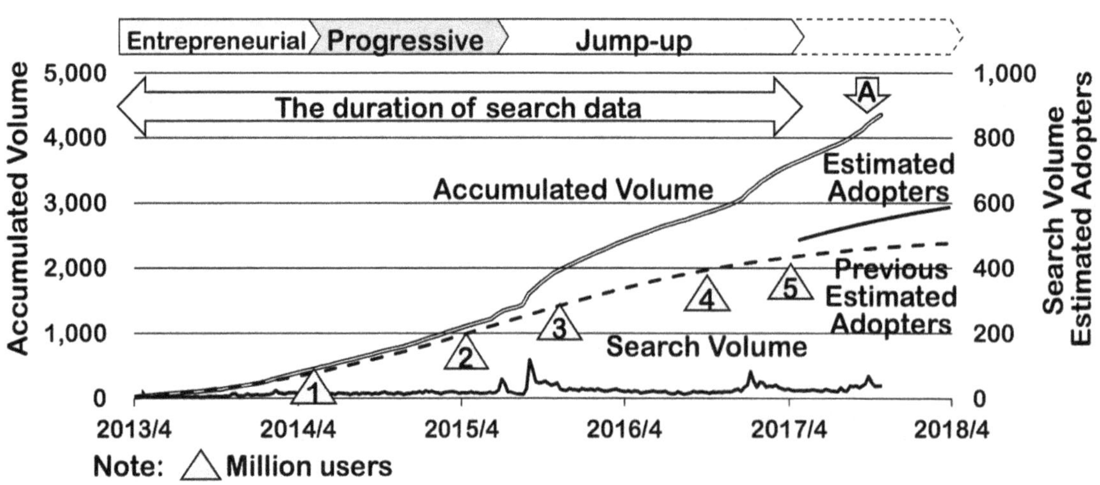

Figure 15. Layered ecosystem in the jump-up stage (Source: Authors)

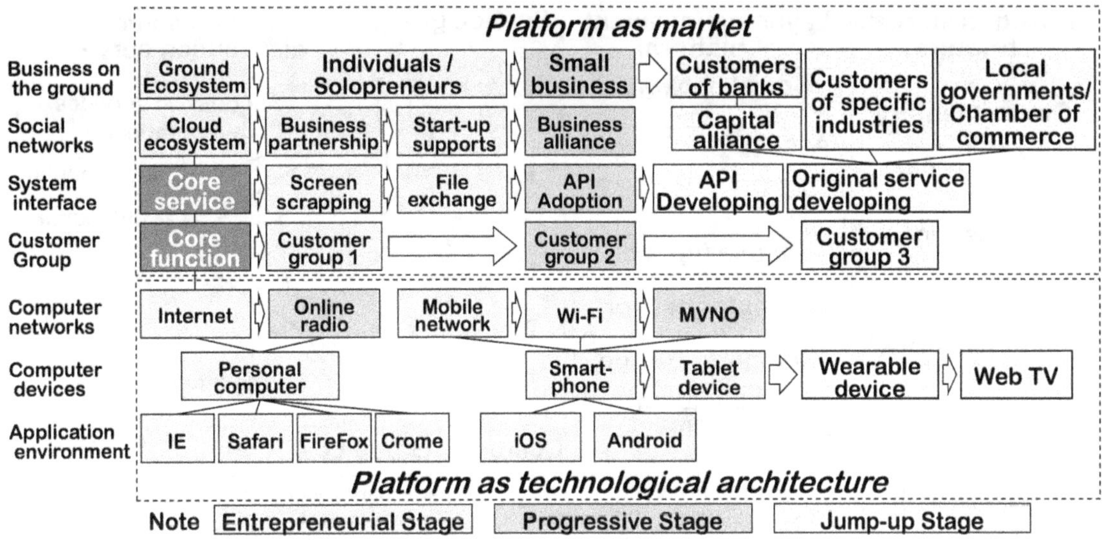

Figure 16. Layer management of platform (Source: Authors, based on Mizuno and Odake (2016a))

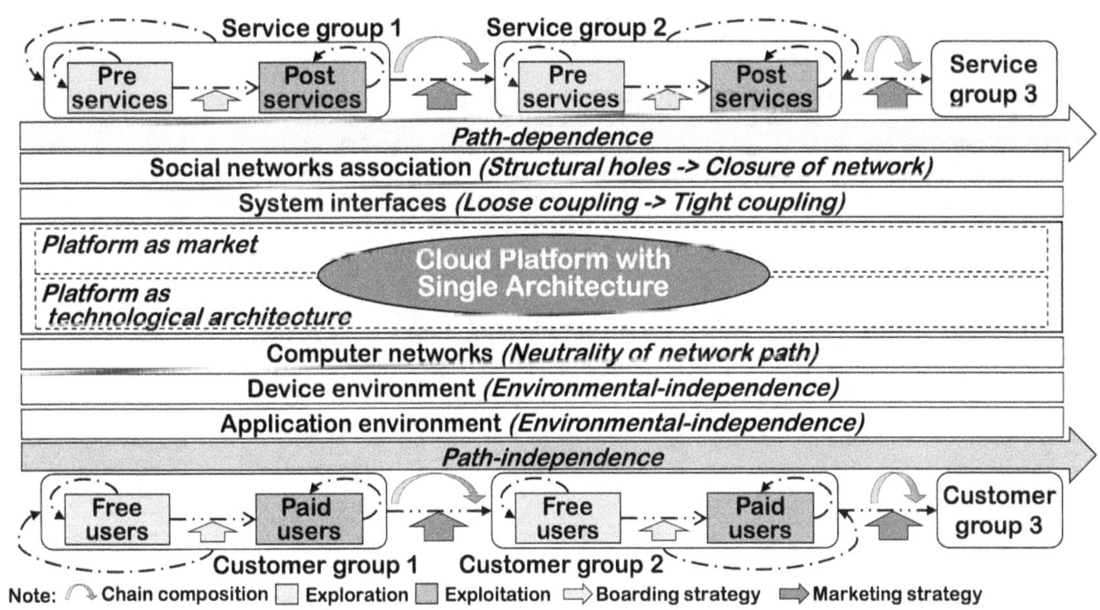

users on board widely. Thus, the service provider realizes economies of scale in the service platform and reduces its service costs. The reduction of its service costs realizes the successive developments of the service platform.

Third, the service provider gathers paid users. Paid users defray all service costs of the accounting cloud service. It develops its system interfaces as tight coupling and realizing closure of network of its cloud ecosystem. In addition, the boarding strategy helps to migrate from free users to paid users with

Figure 17. Added value distribution (Source: Authors, based on Tatemoto and Takanashi (2010))

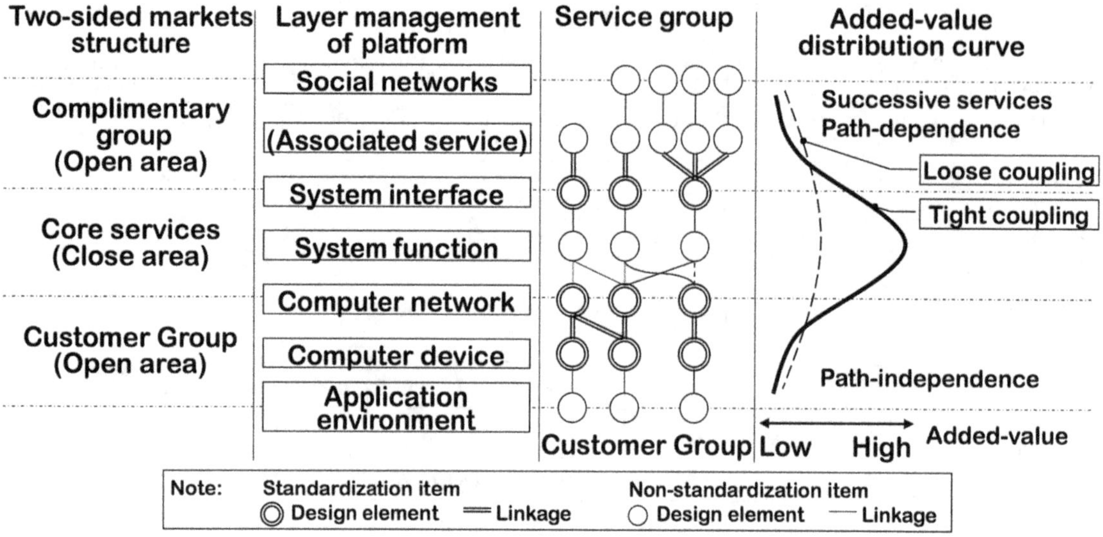

Figure 18. Boarding strategy (Source: Authors, based on Mizuno and Odake (2015))

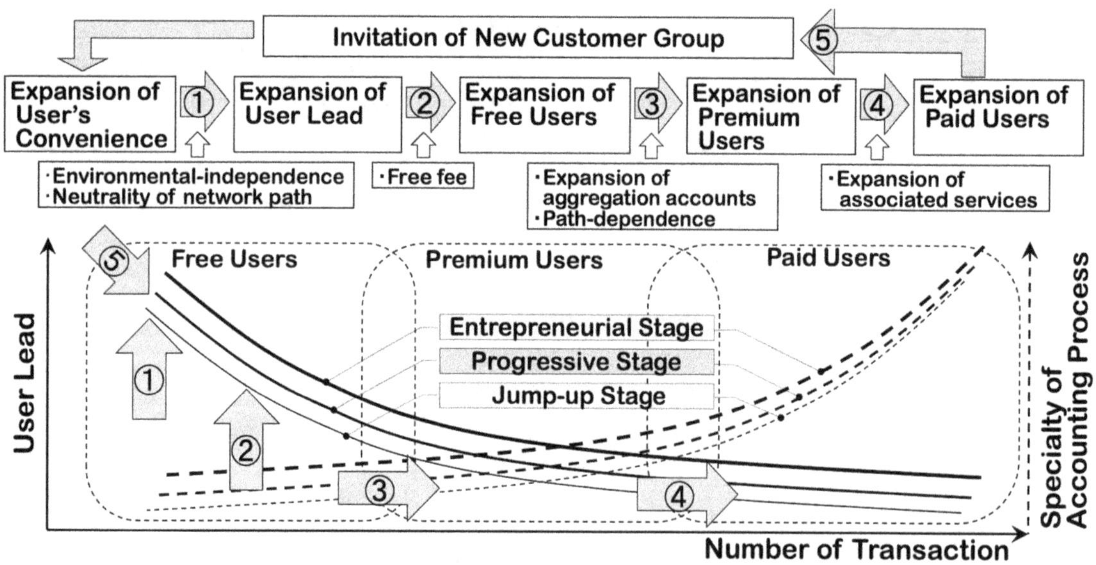

path-dependence. The path-dependence in market realizes economies of scope in the service platform, which realizes the successive service formation.

Fourth, the service provider gathers a new customer group on board. It operates the spiral-structured marketing strategy and invites a new customer group. The new customer group on board brings a huge variety of service transactions and a large amount of service transactions at the same time. These service transactions drive sophistication and standardization of services.

Figure 19. Spiral-structured marketing strategy (Source: Authors, based on Mizuno and Odake (2016b))

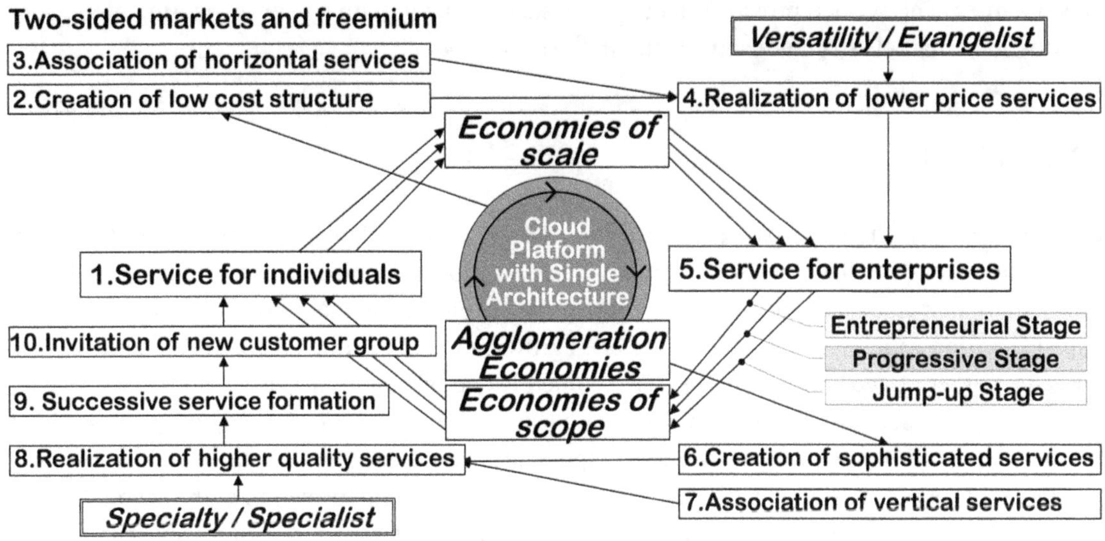

Figure 20. Three-part entrepreneurial strategy (Source: Authors)

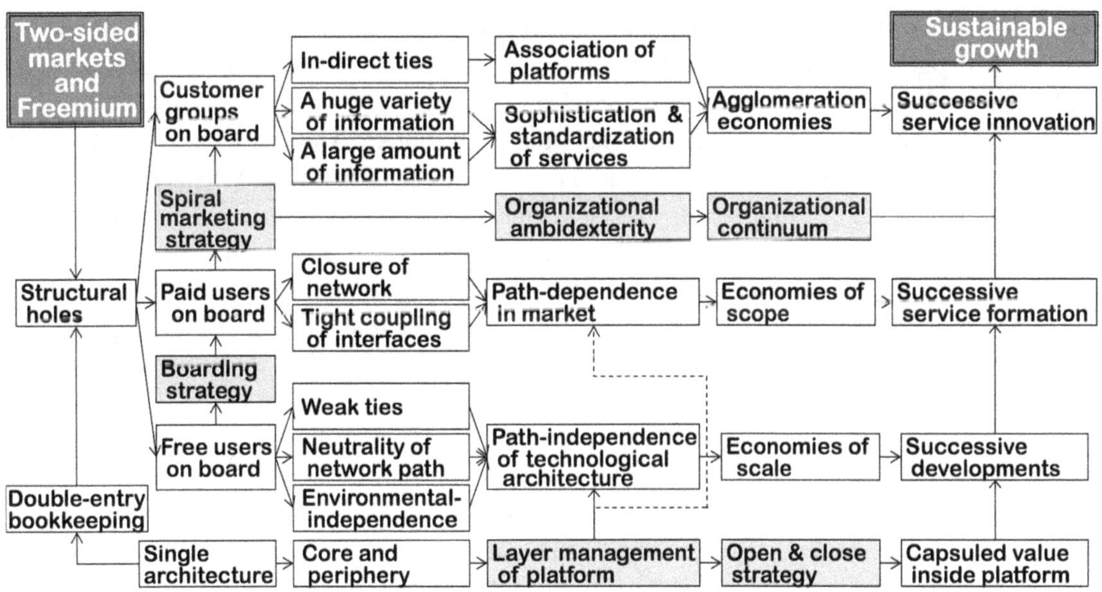

Fifth, the service provider associates its service platform with its business partner's platforms utilizing indirect-ties. The spiral-structured marketing strategy tightens up on the organizational ambidexterity and invites more customer groups on board. More customer groups on board accelerate the sophistication and standardization of services. The service provider realizes agglomeration economies in the service platform. The organizational ambidexterity drives the organizational continuum and accelerates successive service innovation of the service provider.

Finally, this three-part entrepreneurial strategy realizes economies of scale and scope, and agglomeration economies. The service provider can drive its sustainable growth of the accounting cloud service with two-sided markets structure and freemium business model.

CONCLUSION

This study clarified the three-part entrepreneurial strategy in the Japanese accounting cloud service with two-sided markets structure and freemium business model in "Fintech" service industry in Japan.

First, this paper elucidated the current status of domestic Fintech market and accounting cloud services in Japan. The domestic Fintech markets have been growing rapidly and will become 80.8 billion JPY in 2021. The market of accounting cloud service, which is one of Fintech service, also has been growing rapidly. This paper selected one of accounting cloud service as a case study. Then, this paper divided its activities until its IPO into three stages: the entrepreneurial stage, progressive stage, and jump-up stage.

Second, this paper analyzed platform development and ecosystem formation of the accounting cloud service until its initial public offering. In the entrepreneurial stage, the service provider has realized B to B connections with loose coupling system interfaces within net-firms utilizing weak ties. Its service platform has become the internal platform. In the progressive stage, it realized the cloud ecosystem with its system interfaces from loose to tight coupling within its business alliances utilizing closure of network. Its service platform became the supply-chain platform. In the jump-up stage, it realizes ground ecosystem with tight coupling of its system interfaces utilizing indirect ties. Its service platform becomes one of industry platform in Japanese Fintech service.

Third, this paper discussed the three-part entrepreneurial strategy. The service provider designs two business strategies: the organizational continuum and the organizational ambidexterity. It operates two marketing strategies: the boarding strategy and spiral marketing strategy. It manages its service platform with the layer management of platform and the open & close strategy. The chain of these strategies becomes a three-part entrepreneurial strategy. The three-part entrepreneurial strategy realizes economies of scale and scope cyclically. Finally, the three-part entrepreneurial strategy realizes agglomeration economies in the service platform.

In conclusion, the three-part entrepreneurial strategy drives the successive service innovation of the accounting cloud service with two-sided markets structure and freemium business model, and contributes to grow the service provider sustainably in the Japanese Fintech market. Therefore, start-up of cloud service with two-sided markets and freemium, not only dares to design its single service platform as technological architecture, but also should build up the three-part entrepreneurial strategy to formulate service platform as market before building up its business ecosystem.

REFERENCES

Arthur, W. B. (1994). *Increasing returns and path dependence in the economy.* University of Michigan Press. doi:10.3998/mpub.10029

Bass, F. (1969). A New Product Growth Model for Consumer Durables. Management Sciences, 15(5).

Burt, R. S. (2000). The network structure of social capital. *Research in Organizational Behavior, 22,* 345–423. doi:10.1016/S0191-3085(00)22009-1

Burt, R. S. (2001). Structural Holes versus Network Closure as Social Capital. In N. Lin, K. S. Cook, & R. S. Burt (Eds.), *Social Capital: Theory and Research* (pp. 31–56). Transaction Publishers.

Chesbrough, H. (2003). *Open Innovation: The New Imperative for Creating and Profiting from Technology.* Harvard Business Press.

Chesbrough, H. (2010). *Open services innovation: Rethinking your business to grow and compete in a new era.* John Wiley & Sons.

Coleman, J. S. (1988). Social capital in the creation of human capital. *American Journal of Sociology, 94,* S95–S120. doi:10.1086/228943

Dapp, T. F. (2014). *Fintech - The digital (r)evolution in the financial sector: Algorithm-based banking with the human touch.* Retrieved 2015/1/6 from https://www.dbresearch.com/PROD/DBR_INTERNET_EN-PROD/PROD0000000000345837.pdf

de la Iglesia, J. L. M., & Gayo, J. E. L. (2009). *Doing business by selling free services. In Web 2.0* (pp. 1–14). Springer.

Duncan, R. B. (1976). The ambidextrous organization: Designing dual structures for innovation. *The Management of Organization, 1,* 167-188.

Eisenhardt, K. M. (1989). Building theories from case study research. *Academy of Management Review, 14*(4), 532 550.

Eisenhardt, K. M., & Graebner, M. E. (2007). Theory building from cases: Opportunities and challenges. *Academy of Management Journal, 50*(1), 25–32. doi:10.5465/AMJ.2007.24160888

Eisenmann, T., Parker, G., & Van Alstyne, M. W. (2006). Strategies for two-sided markets. *Harvard Business Review, 84*(10), 92. PMID:16649701

Gawer, A. (2014). Bridging differing perspectives on technological platforms: Toward an integrative framework. *Research Policy, 43*(7), 1239–1249. doi:10.1016/j.respol.2014.03.006

Gawer, A., & Cusumano, M. A. (2002). *Platform leadership: How Intel, Microsoft, and Cisco drive industry innovation.* Harvard Business School Press Boston.

Gawer, A., & Cusumano, M. A. (2008). How companies become platform leaders. *MIT Sloan Management Review, 49*(2), 28.

Gibson, C. B., & Birkinshaw, J. (2004). The antecedents, consequences, and mediating role of organizational ambidexterity. *Academy of Management Journal*, *47*(2), 209–226. doi:10.2307/20159573

Granovetter, M. (1983). The strength of weak ties: A network theory revisited. *Sociological Theory*, *1*(1), 201–233. doi:10.2307/202051

Grönroos, C. (1994). From scientific management to service management: A management perspective for the age of service competition. *International Journal of Service Industry Management*, *5*(1), 5–20. doi:10.1108/09564239410051885

Hagiu, A. (2007). Merchant or two-sided platform? *Review of Network Economics*, *6*(2). doi:10.2202/1446-9022.1113

Hagiu, A. (2009). *Multi-sided platforms: From microfoundations to design and expansion strategies.* Harvard Business School Strategy Unit Working Paper (09-115).

Iansiti, M., & Levien, R. (2004). *The keystone advantage: what the new dynamics of business ecosystems mean for strategy, innovation, and sustainability*. Harvard Business School Press.

Kim, P., & Aldrich, H. (2005). *Social capital and entrepreneurship*. Now Publishers Inc.

Kumar, V. (2014). Making" freemium" work. *Harvard Business Review*, *92*(5), 27–29.

March, J. G. (1991). Exploration and exploitation in organizational learning. *Organization Science*, *2*(1), 71–87. doi:10.1287/orsc.2.1.71

Mizuno, Y., & Odake, N. (2015). A Case Study of Progressive Formation of Accounting Cloud Services in JAPAN. *International Journal of Service Science, Management, Engineering, and Technology*, *6*(1), 1–21. doi:10.4018/ijssmet.2015010101

Mizuno, Y., & Odake, N. (2016a). A Case Study of Progressive Formation of Accounting Cloud Service in Japan: The Second Stage. *International Journal of Service Science, Management, Engineering, and Technology*, *7*(3), 19–37. doi:10.4018/IJSSMET.2016070102

Mizuno, Y., & Odake, N. (2016b). A Study of Diffusion of Cloud Service with Two-sided Markets Structure and Freemium Business Model. *Production management: journal of Japan Society for Production Management, 23*(2), 7-15. (in Japanese)

Mizuno, Y., & Odake, N. (2017). A Study of Development and Formation of Personal Information Trust Service in Japan. *International Journal of Service Science, Management, Engineering, and Technology*, *8*(3), 108–129. doi:10.4018/IJSSMET.2017070107

Moore, J. F. (1993). Predators and Prey. *Harvard Business Review*, *71*(3), 75–83. PMID:10126156

Ogawa, K. (2014). *An Open & Close strategy: Condition of the Japanese company revival.* Tokyo: Shoeisha Publisher. (in Japanese)

Rochet, J. C., & Tirole, J. (2003). Platform competition in two-sided markets. *Journal of the European Economic Association*, *1*(4), 990–1029. doi:10.1162/154247603322493212

Rogers, E. M. (2003). *Diffusion of innovations*. Simon and Schuster.

Rosenthal, S. S., & Strange, W. C. (2004). Evidence on the nature and sources of agglomeration economies. Handbook of regional and urban economics, 4, 2119-2171.

Spohrer, J. (2007). Service Science, Management, and Engineering and Its Relation to Academic Disciplines. In B. Stauss, K. Engelmann, A. Kremer, & A. Luhn (Eds.), *Services science: Fundamentals, challenges and future developments* (pp. 11–40). Springer.

Stare, M., & Rubalcaba, L. B. (2007). Research on Services: From Exploring the "Residual" to Services Science. In B. Stauss, K. Engelmann, A. Kremer, & A. Luhn (Eds.), *Services science: Fundamentals, challenges and future developments* (pp. 71–78). Springer.

Stauss, B., Engelmann, K., Kremer, A., & Luhn, A. (2007). *Services science: Fundamentals, challenges and future developments*. Springer.

Tatemoto, H., & Takanashi, T. (2010). Competitive strategies on standardization: Establishing and exploiting consensus standardization [in Japanese]. *Journal of Japan Association for Management Systems*, *26*(2), 67–81.

Vargo, S. L., & Lusch, R. F. (2004). Evolving to a new dominant logic for marketing. *Journal of Marketing*, *68*(1), 1–17. doi:10.1509/jmkg.68.1.1.24036

Yano Research Institute. (2017). *Forecasts the domestic market scale of Fintech Venture Sales volume in Japan*. Retrieved 2017/11/2 from http://www.yano.co.jp/market_reports/C58122500 (in Japanese)

Chapter 5
A Cloud Broker for Service Adaptation Based on Interface Localization

Claus Pahl
Free University of Bozen-Bolzano, Italy

Luke Collins
Dublin City University, Ireland

ABSTRACT

The aim of the Cloud "as-a-service" concept is provisioning of software services to facilitate access to resources to a range of different users in different locations. Service adaptation through localization in this context can facilitate the internationalization and localization of services by allowing their adaption to different locales. Three localization dimensions are investigated: (1) lingual localization by providing service-level language translation techniques to adopt services to different languages; (2) regulatory localization by providing standards-based mappings to achieve regulatory compliance with regionally varying laws, standards, and regulations; and (3) social localization by taking into account preferences and customs for individuals and the groups or communities in which they participate. The objective is to support and implement an explicit modelling of aspects that are relevant to localization and runtime support consisting of tools and middleware services to automating the deployment based on models of locales, driven by the two localization dimensions.

INTRODUCTION

Software services, particularly in the cloud computing context, can support users on a global scale (Armbrust, M., Fox, A., Griffith, R., Joseph, A.D., Katz, R., Konwinski, A., Lee, G., Patterson, D., Rabkin, A., Stoica, I., & Zaharia, M., 2009; Buyya, Broberg, & Goscinski, 2011). In economically tightly integrated regions like Europe, where a multitude of languages are spoken, services are often only deployed to support a single language or region (EU Commission, 2010). Often, smaller organisations do not have the capacity or capability to do multi-lingual and multi-regional development. Localisation

DOI: 10.4018/978-1-5225-5951-1.ch005

is the process of adapting digital resources like services and associated data and content to a locale, i.e. the lingual, regulatory and social environment (restrictions, rules, settings) of a location or region. With the emergence of Web and cloud services and a trend towards end-to-end personalisation of service offerings (451 Group, 2010), the need to address this wider understanding of locale and localisation in a dynamic service context is evident.

Cloud Service Brokerage (CSB) is a mediation approach between users and providers of services. Brokers can integrate services offerings or adapt these to the specific needs of users. This adaptation can be in the form of functionality adaptation, e.g., selecting relevant features only. However, interface-level adaptation is another option that will be address here. The aim is to localise interfaces, i.e., adapt these to different locales.

The objectives of service localisation are, firstly, to introduce service-based localisation techniques that localise software and the interaction at the service interface level, and, secondly, provide localisation techniques at runtime for dynamic service localisation and end-to-end personalisation as an adaptation technique. As there are only a very few related works, the focus here is on defining a conceptual information model as the backbone of a wider localisation solution (Pahl & Collins, 2015). Based on this, further challenges will be outlined and possible architectural solutions for this cloud service brokerage (CSB) presented context (Pahl & Jamshidi, 2015; Pahl, C., Jamshidi, P., & Zimmermann, O., 2018). A significant part here is dedicated to illustrating service localisation as a new concept.

In the literature, localisation often refers to either languages or physical locations only. Three different locale dimensions are the focus of our investigation that embrace these and widen the concepts of localisation and locale for services:

- **Lingual Localisation:** By enabling service-level language translation techniques to adopt services (including API, description, models) to different languages,
- **Regulatory Localisation:** By realising standards-based mappings to achieve regulatory compliance with laws and regulations that might vary regionally (business rules, standardised name/value mappings, currencies and units, and legal governance/compliance rules in relation to different locations or regions),
- **Social Localisation:** By considering preferences and customs for individuals and groups or communities in which they participate (preferred media, forms of interaction and communication).

These localisation features are embedded into an adaptation broker architecture (Elango, D.M., Fowley, F., & Pahl, C., 2017; Elango, D.M., Fowley, F., & Pahl, C., 2018) that suits the needs of a cloud service broker.

Progress beyond the state-of-the-art in *service adaptation* with respect to the following aspects is aimed at:

- **Localisation at the Service Interface (API) Level:** Classical concepts of software localisation will be repurposed to address internationalisation at the interface level. Model-driven development including model-based mapping and translation are the techniques to develop a coherent and integrated solution across the locale dimensions here. The challenge is to define a semantic model integrating heterogeneous translation, mapping and adaptation needs within one dynamically processable format.

- **Adaptation and Integration:** Software adaptation at service-level can be considerably improved by integrating linguistic and regulatory dimensions. This is a significant new direction as the current adaptation concern is on functional and software quality aspects, which are software-technical in nature.
- **Semantics-Enhanced Brokering and Mediation:** The matching of services is equally improved to encompass linguistic, regulatory and social aspects included in a locale-centric negotiation process and infrastructure, which requires respective coordination processes (Di Martino, B., Cretella, G., & Esposito, A., 2017).

We first introduce cloud service brokerage. We then motivate service localisation afterwards. The following section introduces a conceptual information model and a supporting architecture. Then, we define the localisation techniques, before looking at their implementation in the form of locale-specific process adaptation. We evaluate our solution and discuss future challenges. Finally, we end with a discussion of related work and some conclusions.

Cloud Service Brokerage

Service brokerage becomes of significance over the past years in the context of cloud computing, where many software systems are migrated to and re-architected as service-based ones (Jamshidi, P., Ahmad, A., & Pahl, C., 2013; Pahl, C., & Xiong, H., 2013; Jamshidi, P., Pahl, C., & Mendonca, N.C., 2017). Cloud service brokers (CSBs) are needed to provide adaptation or integration solutions.

A service broker (Ried, 2012) is an intermediary software application that sits between a client and provider service. A benefit of service brokerage is a reduction of the time needed by a client in finding and comparing a range of services provided by different service providers.

The focus is on adaptation brokers (Fowley, F., Pahl, C., & Zhang, L., 2013; Fowley, F., Pahl, C., Jamshidi, P., Fang, D., & Liu, X., 2017) here that adjust existing provided services to match their consumers' need, thus mediating between provider and consumer. Service brokerage enables a broker platform to offer the client a customised service. This results in cost optimization and reduced level of back-end data management requirements, but also enables easy migration through an abstracted interface (Fehling, 2011). Changeability and extensibility are central requirements for our broker as interfaces might change (Pahl, C., & Xiong, H., 2013b).

The guiding architecture principles will be discussed, before introducing our abstraction ontology that organises the architecture and showing how provider functionality is mapped onto this. The architecture of a broker is typically built on multiple layers of abstraction. Abstraction serves to reduce complexity. It provides for service-neutral functional logic which also realises extensibility, i.e., allows additional vendors to be supported without changing the underlying core functional logic of the design. Thus, interoperability and portability can be achieved.

The advantage of bringing these functionalities to an interoperable broker application provides 1) an easy way of processing data; 2) choice over price; 3) enhanced SLA; and 4) the elimination of vendor lock-in.

In many CSBs, an ontology-based layered architecture can serves to provide interoperability and extensibility (Elango, D.M., Fowley, F., & Pahl, C., 2017; Elango, D.M., Fowley, F., & Pahl, C., 2018).

At the core of a broker architecture can be an abstraction ontology that describes the common service concepts across the abstraction layers (Pahl C., 2005).

A model-driven software engineering technique can simplify the process of design from concept modelling to implementation. This can be utilised at each level in the abstraction ontology by breaking entities into single components. Adopting this kind of best-practice in software design is important in order to reduce the development overhead and produce a quality library that can be extended easily. Two types of modelling approaches could have been used here.

- Firstly, a provider-specific model, in which the provisioning and deployment of the abstraction is defined for each provider.
- Secondly, a provider-independent model, which defines the provisioning and deployment of the abstraction library in a platform-agnostic way.

LOCALIZATION-BASED SOFTWARE SERVICE ADAPTATION

Cloud Computing is a technology that greatly impacts on the use of software online (Fingar, 2009). The focus here is a platform for cloud service localisation that makes a step from one-size-fits-all services towards end-to-end personalised service offerings based on different locales. Current service computing for international settings has limitations for localisation and adaptability for multiple but different users (Mietzner, R., Unger, T., Titze, R., & Leymann, F., 2009; Wang, Bandara & Pahl, 2010; Metzger, A., Pohl, K., Papazoglou, M., Di Nitto, E., Marconi, A., & Karastoyanova, D., 2012), which can, however, be overcome by multi-lingual and multi-regional localisation.

Some application scenarios can highlight the benefits of a localisation solution:

- **Media Store and Play Services:** Some cloud media players are currently available in only one language. This form of end-user oriented service interaction would benefit from multilingual access. It would involve the localisation (i.e., translation) of service description values, interaction text, and other auxiliary text, which can take place statically (prepared localisation) or dynamically (ad-hoc localisation).
- **Business-Targeted Application Services:** An analysis and storage service for data as a business-oriented offering could be adapted (localised) to be compliant with different standards and regulations for different regions or legal environments.
- **End-User Composed Service Processes:** Take public sector applications where governance and regulations are as important as lingual aspects as an example that can be composed/configured based on individual (possibly cloud-based) services, which are adapted (possibly even dynamically) by individual users to their specific needs.
- **Mediated E-Commerce Offerings:** Telecoms-supported consumer access services are examples, where online shops use service-based support infrastructure to manage their business online.

The second scenario shall now be detailed. A data archiving service in the cloud might require business-oriented service localisation. Sales data from local subsidiaries shall be stored centrally (e.g. in a private cloud) or alternatively the use of a low-cost storage provider abroad (in a public cloud) can be considered. Focusing on regulatory (REG) and lingual (LING) localisation, but also considering social (SOC) localisation, the following is needed:

- **LING:** To translate service data between different languages, e.g. from English into German - "Quote" to "Angebot" - based on business standards such as EANCOM or document attributes based on e.g. the GS1 standard for documents. Or see the EDIFACT (United Nations rules for Electronic Data Interchange for Administration, Commerce and Transport, see http://www.unece.org/trade/untdid/welcome.html) or GS1 standards (supply and demand chains globally and across multiple sectors, see http://www.gs1.org/) for illustration.
- **REG:** To transform data between standards or respective variants, e.g. "Quote" translates to "FullQuotation" based on a transformation between EDIFACT variants (subsets like EANCOM, EDIKEY, or EDIFICE). Other examples are currency conversions or the transformation of rules and procedures, such as access rights to enable regulatory compliance by enabling legally required recording of activities through service adaptation.
- **SOC:** To select and switch between content presentations, like changing between a text-based and graphics-based presentation of data.

These data validation and archiving services might be composed by a service broker that mediates this localisation based on integrated semantically enhanced locale policies. The requestor provides an abstract process that the broker implements - for example to offer locale-specific services abroad for roaming network customers (if compatible) using predefined mappings that are then deployed dynamically. End-user composition can be used where the end-user configures a process of different services in different languages. The potential user searches for best provider internationally for both services, but not necessarily as a package. Sales records where data is validated for sector categories can serve as an example, e.g. GS1 for sector categorisations across different languages.

The following examples illustrate localisable service artefacts with some specifications for locales US (USA) and DE (Germany):

- **API:** *Validate(dataset):result* in US maps to *Validiere(Datensatz):Resultat* in DE
- SemanticDescription: *compliant(GS1,dataset)->validated* in US maps to *einhaltend(GS1,Datensatz) -> validiert* in DE
- **Contract:** *Per_activation = 1 USD (incl. 18% VAT)* and *availability = 98% in US* maps to *pro_Aufruf = 1.35 EUR (inkl. 21% MwSt)* and *Verfügbarkeit = 98%* in DE
- **Documentation:** *Service validates dataset for standards compliance* in US maps to *Service validiert Datensätze und Einhaltung von Standards* in DE

Here inter-artefact relationships exist. For instance validate/validated/validates and complaint/compliance are two sets of terms that are internally cross-referenced. Some progress has been made for Web service internationalisation (Phillips, 2005; W3C, 2005). This has resulted in a broader framework, particularly addressing data formatting and conversion aspects. More complex, rule-based translations are, however, not included but will be addressed by us.

A SERVICE LOCALISATION ONTOLOGY

Semantic technologies can be applied for our localisation solution. Multi-locale modelling and service adaptation needs to supported by a new approach to semantic service description through locale-specific domain ontologies. Locale-specific domain models are needed. Ontologies provide our formal framework. In order to define these, there is a need to consider a data layer and a services layer, each with respective semantic annotations. A *multi-lingual, multi-ontology framework* with corresponding facts, rules and mappings is needed (Pahl, 2012). The authors propose adopting the *linked data* approach. The XLIFF translation mapping standard, a widely used standard in the translation and localisation industry, is XML-based and can be converted into RDF. The authors aim to support a multi-local ontology model for data and services and annotation and discovery support.

The localization of brokered/mediated services is the context in which service localization takes place – which necessitates a coherent integrated information model for multi-dimensional localisation and a supporting cloud infrastructure to deal with localization statically and dynamically (Grozev & Buyya, 2012). The conceptual framework presented here consists of an information architecture, i.e. an ontology, a systems architecture and an abstract process that would be enacted on the architecture (Pahl & Zhu, 2005). An ontology is a representation of concepts (or classes) defined in terms of their properties, i.e. relationships to other concepts. These concepts can be instantiated into concrete values. Properties can be object or data valued, depending on whether they refer to other complex objects (concept – e.g. the relationship of a service to another service through invocation or inheritance) or data (e.g. the performance of a service as a numeric value (Elango, D.M., Fowley, F., & Pahl, C., 2018,b). Concepts can be arranged into a hierarchy using the sub/superconcept relationship, the subsumption $\sqsubseteq$. This relationship forms the taxonomy of the ontology. Concepts can also be organised in a part-of relation, the composition $\rhd$. Concept descriptions are logically combined using conjunction $\sqcap$ and disjunction $\sqcup$.

Localisable artefacts are service specifications and models, but also service contracts and other documentation addressing e.g. interaction in the form of messages and also licensing, usage and access rights, see Figure 1 for an overview.

The Ontology Expression

$ServiceDesc \sqsubseteq ServiceAPI \sqcap SemanticServiceDesc \sqcap ServiceSLA \sqcap ServiceDoc \sqcap ServiceHCI$

defines a hierarchy of concepts that can be further detailed as a composition:

$ServiceAPI \rhd Operations \sqcap Data$ and $Data \rhd Names \sqcup Values$

A rich ontology notation based on description logics with subsumption (subclasses) and composition (part-of relationship) is used here – introduced in (Pahl, Giesecke & Hasselbring, 2009). This rich language allows a layered organisation of different functional aspects. The upper subsumption part defines a taxonomy, which acts as a framework to define structural composition.

As part of localisation, translation between controlled vocabularies is necessary. Underlying data schemas are based on standards and mappings that need to be formalised. If services are not locale-independent, localisation is needed. A *locale model* is based on locale-affected attributes like time/data format (the localisable artefact is data), language (the localisable artefact is interaction like text and/or

Figure 1. Localisation Artefact Ontology with Taxonomy and Composition - Excerpt with Focus on Service API

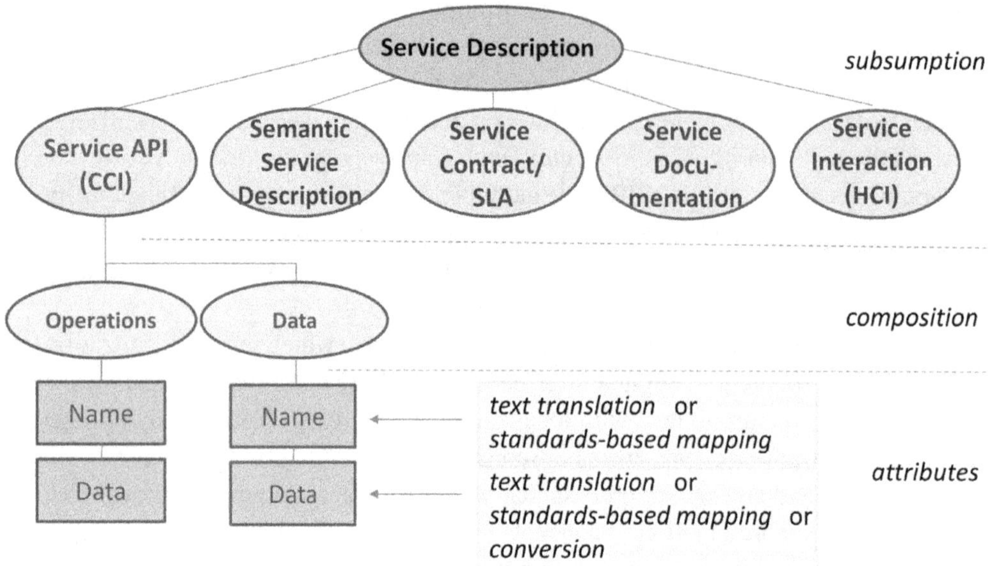

dialogues), and collation/ordering (the localisable artefact is generally data). User-specific locale models (realised as user locale policies) are applied at the service endpoints after a locale negotiation process with the provider. Services can be functionally described in terms of their interfaces (in WSDL): data types, data, operations. Other, more comprehensive models might support functionality, quality, locale (language, currency, units, legal/tax), context (location, device, platform), and documentation (manual, help). Thus, the localisable artefacts are (as in Figure 1):

- **Service Interface (API):** Data-level internationalisation and transformation for the different locale dimensions that apply in particular to data (translation based on e.g. EAN Identification or EANCOM/EDIFACT at schema or data-level).
- **Semantic Service Description and Contract (SLA):** Aspects like licensing, usage and access rights, and other metadata aspects are subject to translation between controlled vocabularies (which is data-level translation).

The ontology layers shall be further explained:

- **Taxonomy Level (Subsumption):** Service descriptions are linked to the generic service description concept through subsumption.
- **Composition Level:** Each of the description types is then decomposed (using the is_part_of composition) into individual elements, which can then be localised. Composition is a necessary as subsumption (subtype) does not apply between an artefact and its parts. For the service API, operations and data (in the form of input and output parameters or messages) are the individual, localisable elements.

- **Functionality Level (Attributes):** Each localisable element has attributes that characterise the localisation technique to be applied. The localisation type can be text (translated using usual MT techniques), standards-based data (the translation is defined in a predefined glossary) or conversions (e.g. for statically defined measurements or dynamically defined currency conversions). For the given ontology excerpt above, the translation type attribute is given.

Composition and attributes are intra-artefact properties. In addition, there are inter-artefact properties in the ontology. Two examples are is_defined_in and is_explained_in for cross-artefact properties, e.g. a contract refers to a definition of an operation in an API or data explained in documentation.

A *locale policy language* for locale description (for both provider and user) needs to be supported. The suggestion is a (locale) policy language that captures required localisations. This language needs to consider interoperability (e.g. with respect to WS-Policy or XACML). A further requirement is to dynamically instrument service processes with policy-based localisation functions. The PCPL policy language supports this (Wang, Bandara, & Pahl, 2010). The localisation ontology is the foundation, on which then rules (defining localisation policies and actions to be executed for localisation) are introduced.

SERVICE LOCALIZATION BROKERAGE: ARCHITECTURE AND PROCESS

Two infrastructure and mediation techniques are needed to realise a service localisation architecture and processes enabled by this architecture:

- **Coordination:** A localisation coordination strategy and respective protocol to implement the locale negotiation and coordination processes for different locales - essentially a rule-based solution with conflict resolution.
- **Service Mediation:** A mediation architecture where an intermediary/broker mediates between clients and providers at different locations. Solution components are based on hybrid techniques for translation and model mappings (across languages, regulations and social contexts) and process adaptation techniques.

The principles of a service brokering solution for the adaptation and mediation of services are now discussed, see Figure 2. Multi-dimensional user and service models and mappings require techniques for adaption and personalisation. A *user model* developed around a notion of locale achieves cross-lingual, cross-regional and cross-social coordination. To match user models and *service profiles*, there is a need to formalise and automate mappings and mediation between the models. On the regulatory localisation side, conversions between formats and regulations need to be defined. On the lingual localisation side, meta data translations and translation of values and text need to be facilitated, in both guided and unguided forms. Equally, social mappings based on user profiles is required.

A mediation architecture is a process framework for locale mediation based on negotiation and localisation implementation activities (Pahl, Giesecke & Hasselbring, 2009; Pahl, C. & Jamshidi, P., 2015). The authors propose a *dynamic localisation engine* made up of services providing discovery, locale negotiation (static and dynamic, to generate SLAs and BPEL code, respectively), localization (lingual, regulatory social, prepared and dynamically). This engine enacts a *localization process* to govern individual activities (Karastoyanova & Leymann, 2009). A repository is needed to ensure the efficient and

Figure 2. Localisation Systems Architecture

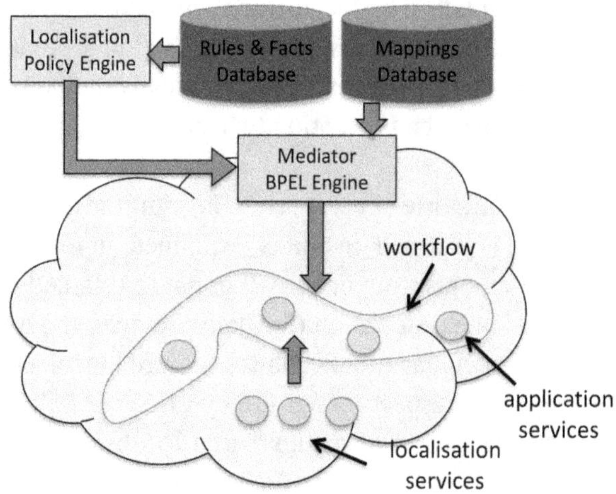

effective operation of the engine, which maintains pre-translated content and data/format mappings as translation memories. A locale can be seen as a specific type of context. *Context- or locale-driven service provision* is our aim. From a cloud perspective, this can also be seen as a *multi-tenancy* problem, where a cloud tenant is defined by its locale. These aspects are further discussed later on.

SERVICE LOCALIZATION BROKERAGE: TECHNIQUES

Our broker architecture is based on three core concerns. A core model captures the policy language and the localisation model. Lingual localisation and regulatory localisation are specific aspects with dedicated mechanisms.

Policy Language and Localization Model

The core of the proposed localisation techniques is a rule language (Kharbili & Keil, 2010; Rosenberg & Dustdar, 2005; Weigand, van den Heuvel & Hiel, 2008), which is an interoperable locale policy language that is needed here (in addition to languages like WSDL that only capture the functionality side of the service API) to define locale descriptions for provider and user. This policy language is based on rules with underlying ontology support (OWL) for the conceptual aspects. A locale policy defines the individual rules and instrumentations that characterise a locale. In Figure 3, a layered localisation model is defined that connects ontology-based locales (from Figure 1) and rule-based locale mappings:

1. An ontology-based base layer captures the different translation types. The ontology defines the localisable artefacts and their relationships (see Figure 1). Each unit is characterised in terms of a number of attributes:
    a. **Translation Unit Type:** API (data, operation), semantic description, SLAs, or documentation are examples.

b. **Translation Type:** Standards-based, ad-hoc text translation, conversion based on distinct repositories, called translation memories (TMs) here.

An example for a unit definition in terms of the artefact ontology is:

Currency hasType API.Data or *SDescr hasType Text*

This adds functionality (processing) to the taxonomy/composition ontology (Figure 1).

2. Basic rules are provided, including *Locale* or *hasCurr* based on which specific locales can be defined. An example is for a *Currency c*

UKLocale(?l) <- Locale(?l) ∧ hasCurr(?l,?c) ∧ ?c=GPB

Higher level rules allow a locale to be inferred from partial knowledge, e.g.

?c=GBP -> ?l=UKLocale

to detect inconsistencies.

3. Depending on provider and consumer locales, translations might need to be performed. Translations are guided by the locale definitions and utilise mappings defined in the translations layer (either as pairwise translation units (e.g. text) or as executable conversions (e.g. for measurements or currencies). The overall translation is dynamically assembled from the translation base (translation memory). The conversion rule for currencies can be dynamically created:

UKLocale2DELocale(?l1,?l2) <-

hasCurr(?l1,?c1) ∧ hasCurr(?l2,?c2) ∧ ?c2 = convertCurr(GBP,EUR,?c1)

Mappings appear primarily as translations between locales. In some cases, consequential actions need to be added. For example, dynamic currency conversions can be added or logging of activities for compliance reasons is added. Actions consequently comprise another type of activity in addition to translation. Localisation might entail additional activities like adding logs for activities.

Examples of locale configuration are the metric system of units or a tax system. A locale mapping is compliant if it does not violate any locale configuration. While a coherent conceptual model is presented in Figure 3, some support is needed for the localisation dimensions:

- **Linguistic Localisation and Machine Translation:** The challenge is localisation of service artefacts with little textual context - at design and run-time.
- **Regulatory Localization and Governance:** Integrated and coherent adaptation to different regulatory standards and procedures based on semantic, rule-based techniques using multi-locale modelling and discovery using ontologies; and in addition multi-dimensional user and service models and mappings for adaptation.

Figure 3. Conceptual Localisation Model – Rules and Mappings over Ontology

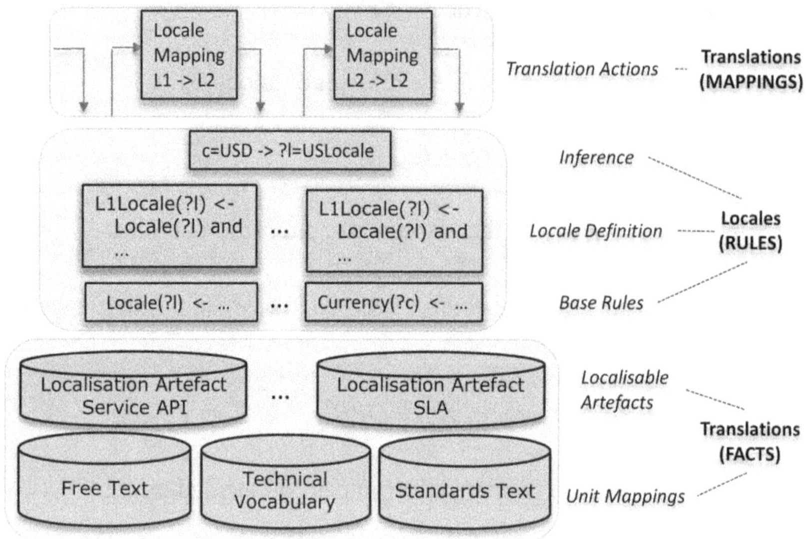

- **Social Localization:** Preference identification using rule-based techniques; conflict identification.

Consistency is a key requirement behind the rule-based approach. Locale mappings need to be consistent. A translation (top layer Mappings) is *consistent*, if two (or more) rules apply to different translations or mappings to the same unit. An example is currency conversion. Inference rules in the rule layer provide the mechanism to detect inconsistencies and enforce correct mappings. An example are countries that use a coherent set of measurements, e.g. the metric system. Also, currencies are linked to fiscal systems defined in a specific language. Since locale mappings are often dynamically assembled from translations, their composition needs to be checked for consistency.

Lingual Localization and Translation

Localisation of artefacts with limited textual context – in both prepared and dynamic settings - is the focus. Translation of technical content based on reduced-context machine translation techniques is an avenue (van Genabith & Crouch, 1999). Furthermore, tailored techniques are needed to facilitate reliable translations between technical content representations, e.g. formalised, technical content such as ontologies, service API and service models (Munday, 2016). Multi-lingual ontologies form the centrepiece of the solution. They provide a mapping and glossary (as core translation memory). For enable effective processing, it needs to considered a merging of prepared translation material with other translations generated dynamically. The key challenge is the translation accuracy and quality, which needs to be trustworthy in automated environments. Ontology translations of technical and business application domains and translation of controlled technical context with little contextual information are the solution (Bella, G., Giunchiglia, F., & McNeill, F., 2017; Helou, M. A., Palmonari, M., & Jarrar, M., 2016), consisting of translation techniques and supporting translation memories using predefined units (words/phrases) for standardised multilingual glossaries and variable units (words/phrases) involving statistical machine translation techniques based on ontological relatedness to guide the translation. The XLIFF format (XML Localization Interchange File Format) is an OASIS standard for localisation exchange - https://www.oasis-open.org/committees/tc_home.php?wg_abbrev=xliff – used to capture translations.

Translations units (pairs here) are kept in translation memories, formalised as:

hasTransl(?t1.?t2)

where *t1* and *t2* are the XLIFF source and target translation units. This is an ontology-level representation of a localisation activity. Rules that guide the application of localisation activities are defined in terms of conversion and translation rules, similar to those for currency conversion. A deeper investigation of computational linguistics techniques is, however, beyond the scope here.

Regulatory Localization and Governance

Regulatory localisation and governance through the adaptation of artefacts to different regulatory standards and procedures is based on localising regulatory concerns, which are often captured in (legal) terms and conditions. Regulations apply for the identification and description of business entities or the way service processes are handled (Leusse, Dimitrakos & Brossard, 2009; Wang, Bandara & Pahl, 2009). Furthermore, regulatory locale aspects like tax, currencies or units need to be considered (Bandara, Wang & Pahl, 2009; Phillips, 2005).

The objective here are specifically multi-facetted rules, which are modular and composable and that enable interference checking. For instance for data object localisation, a multilingual, multiregional schema mapping and integration technique is needed to adapt information to regulations in different regions, in possibly different languages (Fu, Brennan, O'Sullivan, 2009). For SLA translation as another concern, enabling mappings of a SLA into a different locale that needs to consider the respective standards and procedures is a solution. A translation technique needs to implement quality assurance and the accountability of the translation. One can distinguish physical units like length, size, weight, and also colour and financial aspects like pricing, payment, currency or tax. Consequently, regulatory rule-based information translation techniques are required based on predefined mappings with fixed conversions between lengths and other units and consequential actions (instrumentations) that are needed when dynamic conversions (an example is currency) or auxiliary actions (logging for accountability) are entailed. Individual definitions are not the core problem – which is rather keeping consistency.

Social Localization

Groups of individual differ, depending on common characteristics of their members. These groups might form social groups of friends, e.g. young adults at school or university, or groups of team members in organisations, e.g. which could be sports clubs as well as companies. An example of a social group that consumes content differently from others are young, media-competent adults, which might prefer video-based content over text.

YoungAdultsLocale(?l) <- Locale(?l) ∧ hasMediaPref(?l,?m) ∧ ?m=video

Specifically, (social) media services would exploit this type of adaptation setting.

The technical problems are preference identification using rule-based techniques and conflict identification. A range of preference could result in conflicts if they were to be met for a specific application service.

LOCALIZATION BROKERAGE: PROCESS AND ADAPTATION ARCHITECTURE

Interface localization is a process that applies guided translations. This is described this first before addressing architecture implications.

Localization Process

A service localisation framework is based on guided translations (following ontology structures), which may be pre-translated. Static value mappings (cross-language or cross-regulation) and also dynamic value mappings need to be implemented. A static setting means that localisations are prepared. The solution architecture needs to enable providers to prepare material for a range of locales in advance and to check their quality. An intermediary (like a mediator) deploys and executes these techniques depending on user profile and a negotiation process, from which then SLAs are formed. These are used to govern the invocation of services. In a dynamic setting, a mediator selects the required services (involving query translation), implements the negotiation using mappings (based on closest profiles) by using localization and quality assurance services. The overall quality will, of course, vary between in-advance and fully automated scenarios.

An orchestrated service process in WS-BPEL integrates application services and localisation support services. These techniques can be facilitated for service localisation through an adaptation and instrumentation of an application service process (Wang, Bandara & Pahl, 2009). The mediator links clients (at different locations) and providers (at different locations) by enacting this core *orchestrated process with localization* adaptations (Pahl, 2010).

Different localisation orchestration *patterns* emerge. For a *single-provider setting*, one provider supports $n > 0$ locales; for a *market-place setting*, a client uses a composition of different provided services $n > 1$. Both actually enable a localisation process with locale negotiation (brokering) and localisation (mediation) implementing a localisation coordination protocol (e.g. in the form of a BPEL process).

1. **Dynamic Generation:** For regulatory localisation, an aspect-based instrumentation can be generated on-the-fly for each defined locale, allowing the user the control over locale definition
2. **Configuration Management and Generation:** Different endpoints and their respective bindings for different locales are generated for a localised API. Then localisations (translations and mappings) are applied. Which of these to use can be dynamically decided.
3. **Negotiation and Coordination:** An exchange and agreement on locale policies through SOAP headers for coordination based on (Wang, Bandara, & Pahl, 2010).
4. **Architecture:** The data integration layer based on ontologies; the service localisation layer is based on adaptation (user model) and regulatory, lingual and social localisation; and management through a locale negotiation process.

The orchestration process that governs the activities is a basic sequence with *Negotiation*, followed by *Mediation and Localization* and *Service Execution*. Abstract services and locale policies are the basis of negotiation. From these service endpoints give access to the application services or instrumented processes. These in turn need a selection of core services for translation LING, adaptation REG, and personalisation SOC and also the necessary instrumentation, i.e. localisation as a service implemented as a BPEL process. Take a sample application service:

Provider-Service

That is adapted into a service orchestration:

if negotiate(locale(user-US) locale(provider-DE))

then localised-Interface:= localise(US-locale, provider-Service);

result:= invoke(localised-Interface);

de-localise(result)

of application service *provider-Service* and the respective localization support services.

Architecture Implementation

The architecture for locale adaptation is based on the localisation process, see Figure 4. The main components of the architecture are:

- A rule definition editor to define the translation and conversion rules,
- A process engine (jBPM and Tomcat) to execute application and localisation services,
- A monitoring system to determine current settings (based on PostgreSQL) and,
- A rule engine to execute the rules.

Figure 4 illustrates the interaction between rule engine and process. The rule engine is decomposed into localisation policy definition, monitoring support and policy validation. The implementation is assumed to be a BPEL engine, which needs to be combined with policy weaving to allow the localisation policies to be automatically added (i.e. adapted) to an application process. The authors have used Java as the implementation language.

Enhanced adaptivity is addressed here. The architecture allows localisation policies to be defined at the client side. Then the process is adapted to client domain needs and enacted by a central process engine. This allows easy adaptation to specific locale settings. Details about the architecture without the localisation domain extension are provided elsewhere (Wang, Bandara, & Pahl, 2010). There are two major components. Firstly, a policy language to define localisation policies and, secondly, a coordination framework based on WS-COORDINATION allowing client-side specified policies to be communicated and woven into the server-side process. In our implementation of the service process customisation and adaptation prototype illustrated in Figure 4, the authors used Jersey, Tomcat, PostgreSQL, jBPM and

Figure 4. Localization Policy Orchestration Architecture

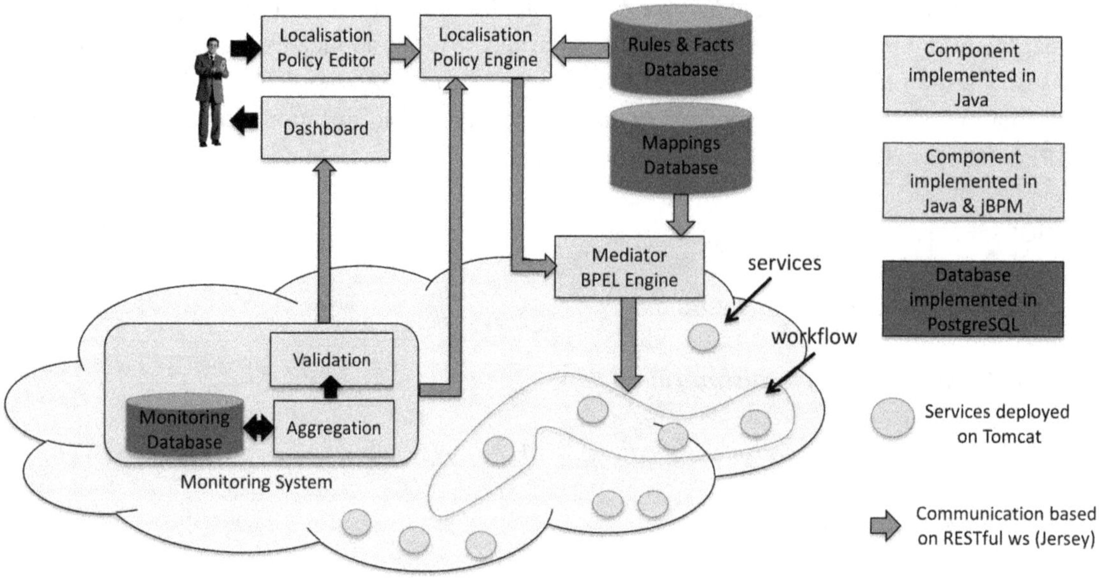

Eclipse. This prototype is a generic process adaptation infrastructure, described in (Wang, Bandara, & Pahl, 2010). This architecture is implemented around the generic PCPL policy language that was developed for constraints definition. This generic platform for service processes was adopted to configure user-specific localisation constraints following the localisation model. Our solution is an extension of the general-purpose policy management solution for domain-specific customisation. Consequently, here in this investigation, our focus has been conceptual rather than on infrastructure aspects. Future work in the implementation context will concentrate on domain-specific implementations. A focus is on translation, where content is marked up in the XLIFF format and respective processing and quality constraints are implemented.

VALIDATION AND EVALUATION

A use case-based evaluation shall help us to validate the utility and efficacy of the localisation mechanisms in our adaptation broker. This shows that the proposed techniques are indeed necessary and beneficial. A use case scenario is introduced, which extends the earlier justification of the need for a service localisation solution as a domain-specific adaptation technique. Technical concerns such as performance overhead at runtime are also addressed as dynamicity was a concern.

Use Case Validation

In addition to the demonstration of the need for localisation, the authors chose another use case to discuss more detailed requirements. An environment was used that provides service-level access to stock market information and respective analyses. This is based on a case study using real financial stock market information and analysis services from http://xignite.com/ and http://deutsche-boerse.com . A German user might want to access data from the New York stock exchange (NYSE), which is provided in an English format. The authors present a scenario in which the user can implement a locale-compliant service interface that would allow technical interaction of service interface and description aspects in German. At the application-level, two sample calls of a data analysis service for the two chosen locales (here US-locale with English as language and USD as currency and DE-locale with German as language and EUR as currency) could be:

Analyse(10/30/2011, logistics) -> 3.82 USD and *Analysis(30.10.2011, Logistik) -> 4.23 EUR*

The localisable artefacts in this scenario are the following:

- **Date:** A format change is needed (Javed et al., 2013) - which also applies to time and collation aspects,
- **Sector:** Data values describing an industry sector are localised based on a translation between standardised terminologies - which also applies to product categories,
- **Language:** Operation names (and possibly other interface and software model elements) are translated between languages,
- **Currency:** Values are converted – in the same way as other measurements and units.

This list can be extended. For instance, different regulatory environments can be based on multilingual and standardised glossaries and dictionaries or calculations and conversions can be based on rules (fixed) or repositories (dynamic); tax rates and customs duties can be added if products are sold; any messages including help and error messages to which text translation would be applied. Typical examples for technical concepts (terms) that require translation in the stock markets (or other financial) context are (*average price - Durchschnittspreis*), (*main trading phase - Haupthandelsphase*), or (*volume weighted average - volumengewichteter Durchschnitt*). These are based on accepted, in some cases standardised vocabularies. Some examples might be defined in terms of classification and categorisation standards: (*logistics Logistik*) for a sector or (*dairy - Milchprodukte*) for product categories.

In the example illustrated in Figure 5, the user-DE can discover services based on a German specification and can invoke them based on a German interface. An analysis service provider can add a DE-locale to its default US-locale policy. This would allow a correct match in a full negotiation process in which a user searches for services that are provided in a locale-specific way. Here, the provider is able to support US-to-DE locale mappings if necessary. In an architecture implementing this, the service instrumentation would yield a process to be enacted, rather than just a single SOAP request (see Figure 5). This process could for instance comprise service invocation and logging (of locations) for accountability purposes where the location is a parameter. This then indicates where and how records are kept (an important option if required by privacy laws). A coordination protocol governs the exchange of locales and the subsequent SOAP-based application interaction.

The scenario could be further utilised to allow an American user (with the locale US) to access a German-language stock market information provider, e.g. for Deutsche Börse, Frankfurt (DB-FRA). Localisation can be provided on a SOAP level, see Figure 6, for automated service activations and interactions.

Figure 5. Localization of Stock Market Analysis Feature - Focus on Service API

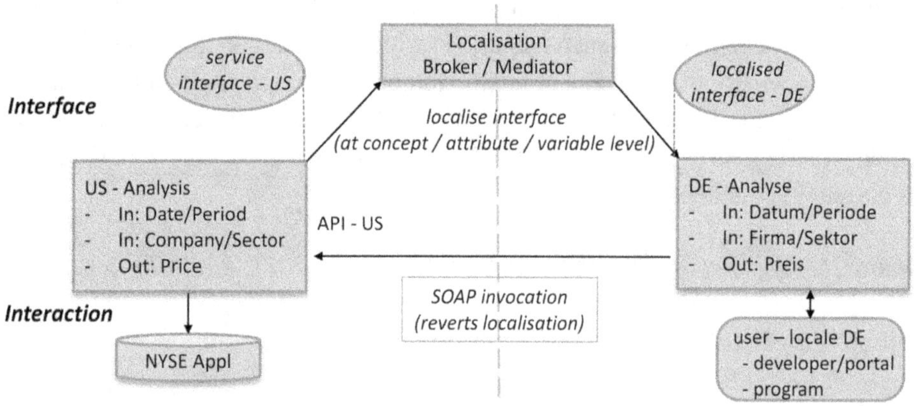

Figure 6. Localisation of Stock Market Analysis Feature - extends NYSE-specific Analysis Service to include DB-FRA

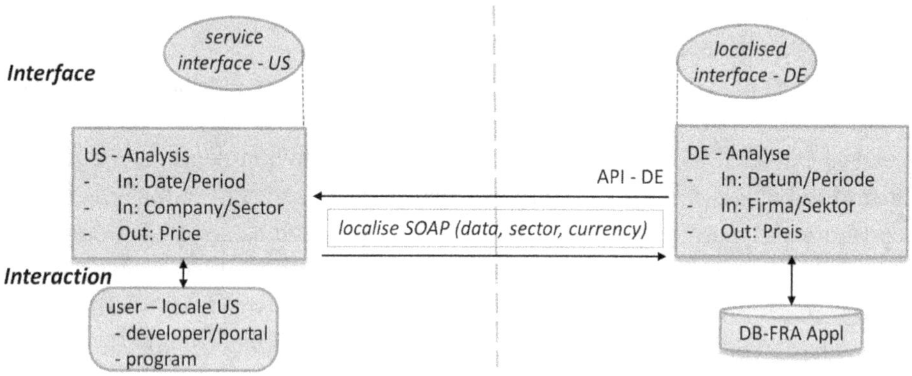

The discussion of this real-world scenario demonstrates the utility and efficacy of our localisation solution as a domain-specific adaptation framework, which benefits from a targeted framework of specific models, rules and support services.

Technical Evaluation

Besides utility and efficacy, the central technical concern of particularly dynamic localisation is runtime performance, as only some aspects can be pre-computed and these need to be integrated at runtime with other dynamic concerns. Scalability is another concern to be considered (Arabnejad, H., Jamshidi, P., El Ioini, N., Estrada, G., & Pahl, C., 2016; Jamshidi, P., Sharifloo, A., Pahl, C., Arabnejad, H., Metzger, A., & Estrada, G., 2016).

The authors implemented a prototype of the service adaptation broker. The localisation architecture presented in Figure 4 was fully implemented as a Java solution, which has the objective of studying the feasibility of the conceptual solution. This was assessed based on the criteria performance and extensibility. These have different effects on the end-user experience of the product. These criteria are key performance indicators and essentially critical success factors of the localisation platform described.

Insufficient performance often affects software significantly as different users consume a service at the same time. The core problem is the overhead created by adding localisation support dynamically to service provisioning. This is impacted on by the policy engine, which was separated from the process engine to allow client-side constraints to be facilitated. The remote instrumentation of processes with these constraints slightly impacts on performance. Our evaluation shows an acceptable overhead of about 10-15% additional execution time for fully localised services (i.e., the localisation involving different localisation aspects). This overhead is still relatively low compared to other performance factors like network latency and the average service execution time anyway (Wang, Bandara, & Pahl, 2010). As the application deals with different users, the latency would also increase as a consequence of extra load put on the platforms services. This clearly defines latency as a KPI. Latency is also an area to be looked at as integrating the localisation solution into the workflow of an existing process has the potential to add lag-time. This lag-time arises due to time required to compute and also initialise the various variables. The propagation latency is presented in Table 1. Please note that the numbers can be affected by environmental changes or the locale being transformed from and the locale being transformed into (Pahl, C., Jamshidi, P., & Weyns, D., 2017b).

The chosen software architecture has a major influence on the performance. The authors aimed to improve performance by using pre-translated aspects (using stored mappings). A closely related aspect is scalability. This becomes more important when a service may have large numbers of users. Scalability has not been empirically addressed for this phase of research and will be evaluated in later prototypes.

- Some platform components would need modification to efficiently enable the infrastructure to vertically scale-up or scale-out. Solutions here are stateless programs and data externalisation. Using the rule base and the pre-translation repositories, some beneficial architectural decisions have already been implemented.
- Horizontal scalability - i.e., the addition of more localisation concerns - is easily be supported by the modular mediator architecture (to be addressed further below as part of the extensibility context from an implementation view).

Another approach to address scalability is a tuple space-based coordination (Pahl, 2010), which allows a flexible and elastic assignment of localisation services to multiple requests.

Extensibility becomes an important concern when dealing with complete solutions like a localisation platform. During an initial development of a system, often features need to be included due to various constraints. For the localisation platform described here, some localisation services where not developed, some of which include a service to handle taxation. However, the platform was designed to be extendable. At a platform level, this is extensible, allowing the addition of further services and the support for more locales.

Table 1. Latency Table - Localization of Service

Service	Prior (ms)	Post (ms)	Delta t (ms)
NASDAQ	132	182	50
FTSE	110	152	42

Discussion: Merits, Limitations and Applicability

A discussion and summary of the merits, limitations and general applicability of the solution follows now. The benefits have been highlighted throughout, but a limitation is the possible diversity of locale settings. Locale definitions might vary from domain to domain. For business applications, in finances or e-commerce, lingual and various regulatory constraints apply. For social media application, some of the latter (e.g. standards) are likely to be replaced by social constraints. A wider range of different applications from social networking to entertainment and gaming to education might require the definition of more strongly group-oriented than region-oriented locale definitions.

Based on the evaluation carried out in this section, the authors have demonstrated the overall applicability and efficacy of the solution, using a dynamic, real-world scenario. However, further validations would need to be carried out, e.g. regarding the data model and an additional empirical validation of a social media application (focusing in this presentation on business applications). However, a more complete model would need to be discussed - while here the focus has been on the higher-level structure of the ontology model.

The authors implemented a proof-of-concept solution that demonstrates the efficacy of the broker to provide mappings for the different localisation concerns. More experience is needed from a practical perspective. With the emergence of microservices as a specific service style, that inherently is based on the need to interface these microservices to communicate with each other, a broker top on top a service repository can be a valuable improvement. This is discussed in the future research direction section below.

RELATED WORK

Several aspects are related to the adaptivity context here, although the authors here approach the problem in a different way compared to other research publications (Fuji & Suda, 2009; Kharbili & Keil, 2010; Mietzner, Leyman & Papazoglou, 2008; Weigand, van den Heuvel & Hiel, 2008). Work covered by for example EU-supported research projects, like SOA4All (www.soa4all.eu), ACSI (www.acsi-project.eu), 4Caast (4caast.morfeo-project.org) or mOSAIC (www.mosaic-cloud.eu) address end-user adaptivity using generic semantic models (SOA4All), software-centric coordination and marketplaces (ACSI, 4Caast), or multi-cloud provisioning (mOSAIC). Our framework is orthogonal to these works on end-to-end offerings. It is different in its multi-facetted character focusing on linguistic and regulatory aspects (Di Martino, B., Cretella, G., & Esposito, A., 2016). More general, our proposal relates to the following aspects:

- Adaptation and Data Integration: Software adaptation at service-level and data integration are common techniques. Examples are schema mappings for data integration, where consistency and semantic preservation are key concerns. The authors followed here ontology-based approaches, using these for semantics preservation, but added a multi-lingual layer using ontology mappings. In a similar approach, service/process instrumentation is used to add enhanced processing abilities, but our work focuses on lingual, regulatory and social concerns within a rule-based framework is novel.

- Service Context-awareness has been covered by a range of contributions such as (Bandara, Wang & Pahl, 2009). The notion of context is similar to that of locales, reflecting properties of the execution or client environment. Context, however, generally does not include lingual, regulatory and social aspects. The authors suggest to adopt and extend common adaptation and instrumentation techniques (Baresi & Guinea, 2005; Kapitsaki, G., Kateros, D., Prezerakos, G., & Venierris, I., 2009).
- Multi-tenancy is a cloud computing concern (Mietzner, R., Unger, T., Titze, R., & Leymann, F., 2009; Wang, Bandara & Pahl, 2010) that requires in a fully customised framework different solutions for users with different needs, to keep them separate.
- Semantic Technologies: matching of services and supporting the negotiation process and infrastructure (Doulkeridis, Loutas & Vazirgiannis, 2006; Fuji & Suda, 2009; Pahl & Zhu, 2005; Gacitua-Decar, V., & Pahl, C., 2017). Through ontology mappings, multi-lingual terminologies and multi-regional regulations can be captured and dynamically processed.
- Software Localisation: localisation of software normally addresses the human interaction with software, i.e. messages and dialogues produced by the software. The authors focus on localisation at the service interface level (i.e. machine-based interaction) through internationalisation and ontology mapping and translation. Current technology on service internationalisation, which is the closest, is supported by the W3C Service Internationalisation activity (Phillips, 2005; W3C, 2005). The focus in that initiative is data localisation, specifically for dates, currencies and units and common approaches to collation. The respective solutions are conversions - statically defined between units or dynamically defined between currencies as examples. Lingual aspects or more complex regulatory or business aspects are not addressed, although in the documentation for instance taxation as a sample concern is mentioned.

A basic localisation conceptualisation was presented in (Pahl, 2012). Here, this was extended for this model by adding a third dimension (SOC) to provide a solution beyond the business application domain. This investigation here also formalises the conceptual model as an ontology. Furthermore, the process aspect was deepened and also implemented and the solution systematically evaluated.

FUTURE RESEARCH DIRECTIONS

With the merging topic of microservices (Aderaldo, C.M., Mendonca, N.C., Pahl, C., & Jamshidi, P., 2017; Pahl & Jamshidi, 2016; Taibi, D., Lenarduzzi, V., & Pahl, C., 2017) the granularity of services moves towards smaller entities, thus increasing the need for more automated processing. An ontology-based model-driven approach can help to automate the mapping between interfaces.

Microservices are independently deployable artifacts, typically deployed as containers (Pahl, C., Jamshidi, P., & Weyns, 2017). Docker as the most successful container platform today is based on a repository architecture (called the Hub). This create an opportunity to provide brokerage services on top of this repository architecture.

This creates also the need to look at performance (Heinrich, R., van Hoorn, A., Knoche, H., Li, F., Lwakatare, L.E., Pahl, C., Schulte, S., & Wettinger, J., 2017), specifically given that interactions between services will increase.

CONCLUSION

Service interface localisation is a form of service personalisation and adaptation to specific use needs. Our investigation addresses service engineering methods and tools allowing a service to be adapted to different locales. Services are readied for their integration by providing localisation of services as software in combination with related content and data processed and communicated by these services. The authors have discussed a range of techniques to enable the extension of service offerings for different markets – something that is of significant economic benefit. Localisation can bring products and here particularly services to markets that are otherwise not accessible.

Multi-locale services can support interoperable services and thus contribute to a market of services by allowing services to be internationalised and localised. This is of importance for regions such as the EU which has 27 members and an even higher numbers of local languages and regulatory systems or even different social context. Service localisation adds to the availability of platforms for simplified development and deployment of value-added services through service front-ends to localise and manage multi-locale services. Localisable services enable users to develop, select, combine and use value-added services and to allow providers to provide new solutions without in-house localisation capabilities.

The objectives of this investigation were two-fold. Firstly, the authors have introduced a conceptual framework to capture the key concerns of multi-lingual, multi-regulatory and multi-context service localisation as a rule- and ontology-based information framework. Secondly, our investigation aimed at justifying the need for service localisation, e.g. for Web and cloud based services. In this vein, current concerns and directions to be addressed were identified. The focus of our paper was, however, limited. Aspects not included like program-level localisation, end-to-end cloud personalisation and multi-lingual clouds are examples of aspects left out at this stage. It is also evident that more implementation work needs to be done to aid further empirical evaluations. So far, the conceptual solution is only motivated and justified through the discussed case studies and the implementation and only been used for performance evaluation.

ACKNOWLEDGMENT

This work was in part supported by the Free University of Bozen-Bolzano through the ECO and ECORE projects.

REFERENCES

451 Group. (2010). *Report on Cloud Computing 'As-a-service' market sizing - Report II.* 451 Research.

Aderaldo, C. M., Mendonca, N. C., Pahl, C., & Jamshidi, P. (2017). Benchmark requirements for microservices architecture research. In *1st International Workshop on Establishing the Community-Wide Infrastructure for Architecture-Based Software Engineering*. IEEE.

Arabnejad, H., Jamshidi, P., El Ioini, N., Estrada, G., & Pahl, C. (2016). An auto-scaling cloud controller using fuzzy q-learning-implementation in openstack. *European Conference on Service-Oriented and Cloud Computing ESOCC.* 10.1007/978-3-319-44482-6_10

Armbrust, M., Fox, A., Griffith, R., Joseph, A. D., Katz, R., Konwinski, A., ... Zaharia, M. (2009). A view of cloud computing. *Communications of the ACM*, *53*(4), 50–58. doi:10.1145/1721654.1721672

Bella, G., Giunchiglia, F., & McNeill, F. (2017). Language and domain aware lightweight ontology matching. *Journal of Web Semantics*, *43*, 1–17. doi:10.1016/j.websem.2017.03.003

Bandara, K. Y., Wang, M. X., & Pahl, C. (2009). Context modeling and constraints binding in web service business processes. In *Proceedings of the Workshop on Context-Aware Software Technology and Applications CASTA* (pp. 29-32). ACM Press. 10.1145/1595768.1595780

Baresi, L., & Guinea, S. (2005). Towards Dynamic Monitoring of WS-BPEL Processes. In *Proceedings International Conference on Service-Oriented Computing ICSOC'05*. Springer-Verlag.

Boukadi, K., Ghedira, C., Chaari, S., Vincent, L., & Bataineh, E. (2008). CWSC4EC: how to employ context, web service, and community in enterprise collaboration. In *Proceedings of the 8th International Conference on New Technologies in Distributed Systems (NOTERE '08)*. ACM. 10.1145/1416729.1416737

Buyya, R., Broberg, J., & Goscinski, A. (2011). *Cloud Computing - Principles and Paradigms*. New York: Wiley. doi:10.1002/9780470940105

Di Martino, B., Cretella, G., & Esposito, A. (2017). Legislation-Aware Cloud Computing: An Overview. In Reshaping Accounting and Management Control Systems (pp. 77-87). Springer.

Di Martino, B., Cretella, G., & Esposito, A. (2017). Cloud services composition through cloud patterns: A semantic-based approach. *Soft Computing*, *21*(16), 4557–4570. doi:10.100700500-016-2264-1

Doulkeridis, C., Loutas, N., & Vazirgiannis, M. (2006, January). A system architecture for context aware service discovery. *Electronic Notes in Theoretical Computer Science*, *146*(1), 101–116. doi:10.1016/j.entcs.2005.11.010

Elango, D. M., Fowley, F., & Pahl, C. (2017). Pattern-driven Architecting of an Adaptable Ontology-driven Cloud Storage Broker. University of Oslo.

Elango, D. M., Fowley, F., & Pahl, C. (2018 To appear). An Ontology-based Architecture for an Adaptable Cloud Storage Broker. In *Advances in Service-Oriented and Cloud Computing*. Springer CCIS.

Elango, D. M., Fowley, F., & Pahl, C. (2018 To appear). Testing and Comparing the Performance of Cloud Service Providers using a Service Broker Architecture. In *Advances in Service-Oriented and Cloud Computing*. Springer CCIS.

EU Commission (2010). Report on The Future of Cloud Computing - Opportunities for European Cloud Computing Beyond 2010. European Commission.

Fingar, P. (2009). Cloud computing and the promise of on-demand business innovation. *Intelligent Enterprise*.

Fowley, F., Pahl, C., & Zhang, L. (2013). A Comparison Framework and Review of Service Brokerage Solutions for Cloud Architectures. *1st International Workshop on Cloud Service Brokerage*.

Fowley, F., Pahl, C., Jamshidi, P., Fang, D., & Liu, X. (2017). A Classification and Comparison Framework for Cloud Service Brokerage Architectures. IEEE Transactions on Cloud Computing.

Fu, B., Brennan, R., & O'Sullivan, D. (2009). Multilingual Ontology Mapping: Challenges and a Proposed Framework. *Proceedings of the 23rd Convention of the Society for the Study of Artifitial Intelligence and Simulation of Behaviour - Adaptive & Emergent Behaviour & Complex Systems*, 32 - 35.

Fujii, K., & Suda, T. (2009). Semantics-based context-aware dynamic service composition. *ACM Transactions on Autonomous and Adaptive Systems*, *4*(2), 1–31. doi:10.1145/1516533.1516536

Gacitua-Decar, V., & Pahl, C. (2017). Structural Process Pattern Matching Based on Graph Morphism Detection. *International Journal of Software Engineering and Knowledge Engineering*, *27*(2), 153–189. doi:10.1142/S0218194017500073

Grozev, N., & Buyya, R. (2012). InterCloud architectures and application brokering: taxonomy and survey. Software: Practice and Experience.

Heinrich, R., van Hoorn, A., Knoche, H., Li, F., Lwakatare, L. E., Pahl, C., . . . Wettinger, J. (2017). Performance Engineering for Microservices: Research Challenges and Directions. *Proceedings of the 8th ACM/SPEC on International Conference on Performance Engineering Companion*.

Helou, M. A., Palmonari, M., & Jarrar, M. (2016). Effectiveness of Automatic Translations for Cross-Lingual Ontology Mapping. [JAIR]. *Journal of Artificial Intelligence Research*, *55*, 165–208.

Jamshidi, P., Sharifloo, A., Pahl, C., Arabnejad, H., Metzger, A., & Estrada, G. (2016). Fuzzy self-learning controllers for elasticity management in dynamic cloud architectures. *12th International ACM SIGSOFT Conference on Quality of Software Architectures QoSA*. 10.1109/QoSA.2016.13

Jamshidi, P., Ahmad, A., & Pahl, C. (2013). Cloud Migration Research: A Systematic Review. *IEEE Transactions Cloud Computing*.

Jamshidi, P., Pahl, C., & Mendonca, N. C. (2017). Pattern-based multi-cloud architecture migration. *Software, Practice & Experience*, *47*(9), 1159–1184. doi:10.1002pe.2442

Javed, M., Abgaz, Y. M., & Pahl, C. (2013). Ontology change management and identification of change patterns. *Journal on Data Semantics*, *2*(2-3), 119–143. doi:10.100713740-013-0024-2

Kapitsaki, G., Kateros, D., Prezerakos, G., & Venierris, I. (2009). Model-driven development of composite context-aware web applications. *Information and Software Technology*, *51*(8), 1244–1260. doi:10.1016/j.infsof.2009.03.002

Karastoyanova, D., & Leymann, F. (2009). BPEL'n'Aspects: Adapting service orchestration logic. In *International Conference on Web Services* (pp. 229-229). IEEE Press.

Kharbili, M. E., & Keil, T. (2010). Bringing agility to business process management: Rules deployment in an SOA. Emerging Web Services Technology: Vol. 3. *Whitestein Series in Software Agent Technologies and Autonomic Computing*. Birkhaeuser Verlag.

Leusse, P. D., Dimitrakos, T., & Brossard, D. (2009). A governance model for SOA. In *IEEE International Conference on Web Services* (pp. 1020-1027). IEEE Press.

Lewis, D., Curran, S., Feeney, K., Etzioni, Z., Keeney, J., Way, A., & Schäler, R. (2009). Web service integration for next generation localisation. *Workshop on Software Engineering, Testing, and Quality Assurance for Natural Language Processing*, 47-55. 10.3115/1621947.1621956

Mahmudul Hasan, M. (2016). Regulatory Requirements Compliance in Requirements Engineering: A Systematic Classification and Analysis. *International Journal of Systems and Service-Oriented Engineering*, *6*(4). doi:10.4018/IJSSOE.2016100102

Mietzner, R., Unger, T., Titze, R., & Leymann, F. (2009). Combining different multi-tenancy patterns in service-oriented applications. In *IEEE International Enterprise Distributed Object Computing Conference* (pp. 131-140). IEEE Press. 10.1109/EDOC.2009.13

Mietzner, R., Leymann, F., & Papazoglou, M. P. (2008). Defining composite configurable SaaS application packages using SCA, variability descriptors and multi-tenancy patterns. In *International Conference on Internet and Web Applications and Services* (pp. 156-161). IEEE Press. 10.1109/ICIW.2008.68

Metzger, A., Pohl, K., Papazoglou, M., Di Nitto, E., Marconi, A., & Karastoyanova, D. (2012). Research challenges on adaptive software and services in the future internet: towards an S-Cube research roadmap. *Workshop on European Software Services and Systems Research-Results and Challenges*, 1-7. 10.1109/S-Cube.2012.6225501

Munday, J. (2016). *Introducing translation studies: Theories and applications*. Routledge.

Pahl, C., Xiong, H., & Walshe, R. (2013). A Comparison of On-premise to Cloud Migration Approaches. *European Conference on Service-Oriented and Cloud Computing ESOCC*. 10.1007/978-3-642-40651-5_18

Pahl, C. (2005). Layered Ontological Modelling for Web Service-oriented model-Driven Architecture. *European Conference on Model-Driven Architecture ECMDA'05*.

Pahl, C., & Xiong, H. (2013). Migration to PaaS Clouds – Migration Process and Architectural Concerns. *IEEE 7th International Symposium on the Maintenance and Evolution of Service-Oriented and Cloud-Based Systems MESOCA*.

Pahl, C., Brogi, A., Soldani, J., & Jamshidi, P. (2017). Cloud Container Technologies: a State-of-the-Art Review. IEEE Transactions on Cloud Computing. doi:10.1109/TCC.2017.2702586

Pahl, C., & Jamshidi, P. (2015). Software architecture for the cloud – a roadmap towards control-theoretic, model-based cloud architecture. *European Conference on Software Architecture*. 10.1007/978-3-319-23727-5_17

Pahl, C., & Collins, L. (2015). Software Service Adaptation Based on Interface Localisation. *International Journal of Systems and Service-Oriented Engineering, 5*(1). DOI: 10.4018/IJSSOE.2015010102

Pahl, C., & Jamshidi, P. (2016). Microservices: A Systematic Mapping Study. *Proceedings CLOSER Conference*, 137-146.

Pahl, C., Jamshidi, P., & Weyns, D. (2017). Cloud architecture continuity: Change models and change rules for sustainable cloud software architectures. *Journal of Software: Evolution and Process*, *29*(2).

Pahl, C., Jamshidi, P., & Zimmermann, O. (2018). . Architectural Principles for Cloud Software. *ACM Transactions on Internet Technology*.

Pahl, C., & Zhu, Y. (2005). A Semantical Framework for the Orchestration and Choreography of Web Services. *International Workshop on Web Languages and Formal Methods WLFM'05. ENTCS 151*(2), 3-18.

Pahl, C. (2010). Dynamic adaptive service architecture - towards coordinated service composition. In *European Conference on Software Architecture ECSA'2010* (pp. 472-475). Springer-Verlag. 10.1007/978-3-642-15114-9_43

Pahl, C., Giesecke, S., & Hasselbring, W. (2009). Ontology-based Modelling of Architectural Styles. *Information and Software Technology*, *1*(12), 1739–1749. doi:10.1016/j.infsof.2009.06.001

Pahl, C. (2012). Cloud Service Localisation. In *European Conference on Service-Oriented and Cloud Computing ESOCC 2012* (pp. 138-153). Springer-Verlag.

Phillips, A. P. (2005). *Web Services and Internationalization.* Whitepaper. Retrieved October, 27, 2017 from http://www.inter-locale.com/whitepaper/multilingual/ml73-ws-20050524.xml

Ried, S. (2011). *Cloud Broker - A New Business Model Paradigm.* Forrester.

Rosenberg, F., & Dustdar, S. (2005). Business rules integration in BPEL - a service-oriented approach. In *7th IEEE Intl Conference on E-Commerce Technology* (pp. 476 – 479). IEEE Press.

Taibi, D., Lenarduzzi, V., & Pahl, C. (2018). Processes, Motivations and Issues for Migrating to Microservices Architectures: An Empirical Investigation. IEEE Cloud Computing.

W3C. (2005). *Web Services Internationalization Usage Scenarios.* Retrieved October, 27, 2017 from http://www.w3.org/TR/ws-i18n-scenarios

Wang, M. X., Bandara, K. Y., & Pahl, C. (2010). Process as a Service - Distributed Multi-tenant Policy-based Process Runtime Governance. In *IEEE International Conference on Services Computing SCC 2010* (pp. 578-585). IEEE Press. 10.1109/SCC.2010.33

Weigand, H., Heuvel, W.-J. d., & Hiel, M. (2008). Rule-based service composition and service-oriented business rule management. *Interdisciplinary Workshop Regulations Modelling and Deployment*, 1-12. Retrieved October, 27, 2017 from http://ceur-ws.org/Vol-342/

Chapter 6
The Role of Value Facilitation Regarding Cloud Service Provider Profitability in the Cloud Ecosystem

Alexander Herzfeldt
Technische Universität München, Germany

Sebastian Floerecke
Universität Passau, Germany

Christoph Ertl
Technische Universität München, Germany

Helmut Krcmar
Technische Universität München, Germany

ABSTRACT

With the increasing maturity of cloud technologies and the growing demand from customers, the cloud computing ecosystem has been expanding continuously with both incumbents and new entrants, whereby it has become more distributed and less transparent. For cloud service providers previously focusing on growth strategies, it is now necessary to shift the attention to providing service efficiently, as well as profitably. Based on 14 explorative interviews with cloud service experts, the relationship between cloud service provider profitability and value facilitation, which stands for the capability to build up resources in advance of future customer engagements, is investigated. The results indicate a positive relationship between cloud service profitability and value facilitation and deliver valuable insights for both researchers and practitioners. In particular, guidelines on how to design profitable cloud service offerings are discussed.

INTRODUCTION

A cloud service represents an example for the shift from selling products to providing integrated combinations of products and services that deliver value in use (Floerecke, Wolfenstetter, & Krcmar, 2015; Sultan, 2014). The vision of cloud computing is IT becoming commoditized and thus delivered in a

DOI: 10.4018/978-1-5225-5951-1.ch006

manner similar to traditional utilities like water, gas, electricity and telephony (Buyya, Yeo, Venugopal, Broberg, & Brandic, 2009). Although the underlying technology of cloud-based IT solutions is not new, it is a new IT operations model that combines a set of existing technologies and concepts such as virtualization, autonomic computing, grid computing and utility-based pricing (Zhang, Cheng, & Boutaba, 2010). This new technology mix-based operations model has radically changed the way IT resources are offered and consumed (Marston, Li, Bandyopadhyay, Zhang, & Ghalsasi, 2011).

Due to this fact, cloud service providers face major challenges (Böhm, Leimeister, Riedl, & Krcmar, 2009). Firstly, they generally require profound expertise with regard to both technical infrastructure concepts and to the development and management of service-oriented business models (Iyer & Henderson, 2012). In practice, it can be observed that both established IT vendors and start-ups in the field of cloud computing have particular difficulties to design suitable business models. That is why they are still experimenting with a variety of business models in order to achieve a sustainable and profitable position in the cloud ecosystem (Clohessy, Acton, Morgan, & Conboy, 2016). Additionally, a clear understanding of the customer business needs is a mandatory prerequisite to design a service allowing customers to easily integrate it in their existing business processes (Böhm, Koleva, Leimeister, Riedl, & Krcmar, 2010).

As cloud providers frequently act as system integrators and resellers bundling offerings from multiple providers, the classical distinction between customers and providers is blurring (Floerecke & Lehner, 2016). This is caused by the multi-layered architecture of cloud services (Infrastructure-as-a-Service, Platform-as-a-Service, Software-as-a-Service) and the special cloud computing characteristic on-demand self-service (Floerecke & Lehner, 2016). From the end customer's point of view, the traditional model of a single provider one-stop provision of IT outsourcing is, thus, replaced by a web of different vendors (Böhm, M., Leimeister, S., Riedl, C., & Krcmar, H., 2009). Hence, end customers commonly do not even know which component of a service package is provided by which specific actor (Floerecke & Lehner, 2016). However, for providers, it is an enormous challenge to ensure interoperability between several cloud services offered by different providers due to missing global interface standards in the ecosystem (Fischer, R., Janiesch, C., Strach, J., Bieber, N., Zink, W., & Tai, S., 2013).

Furthermore, a cloud service has to feature special characteristics such as self-service, pay-as-you-go pricing and convenience to scale the service according to consumers' needs that distinguish it from traditional concepts like on-premise IT or hosting (Clohessy, T., Acton, T., Morgan, L., & Conboy, K., 2016). One main driver for customers to adopt cloud services is certainly the expectation of cost savings in comparison to traditional IT services designed for one individual customer (Oliveira, Thomas, & Espadanal, 2014). However, in practice, this is not always achievable within a permanent cloud operation in all conceivable service scenarios. In general, the benefits associated with cloud computing such as flexibility, scalability and cost reduction are mainly realized on the customer's side, while the risks involved are shifted to the provider (Floerecke, Herzfeldt, & Krcmar, 2012). This high-risk situation also comes as a result of the current high performance and availability guarantees within the service level agreements that cloud providers have to offer in order to attract customers.

As a result of all these above-mentioned factors, cloud service providers face a strong price competition and are continuously threatened by new market entrants. These dynamic market conditions have led to the situation that providers which previously focused on growth strategies, need to shift their attention to providing cloud services efficiently and profitably. Indeed, studies show that there are enormous differences in the level of profitability in the cloud ecosystem. Whereas some cloud service providers are highly successful in terms of profitability, others lag behind (Velten & Janata, 2012). The literature's understanding of success factors of cloud service providers and therefore the reasons for this discrepancy is, however, still very limited (Labes, Hanner, & Zarnekow, 2016).

Reviewing the existing scientific literature on efficiently providing a cloud service, and in particular, concerning the concepts of value facilitation (Anker, Sparks, Moutinho, & Grönroos, 2015; Grönroos & Ravald, 2011; Grönroos & Voima, 2013) and the resource-based view (RBV) of a firm (Wade & Hulland, 2004) reveals that successful organizations need to build up resources in advance of future customer engagements. According to Anker et al. (2015) and Grönroos and Voima (2013), the value creation process is taking place through a chain of three successive spheres: provider sphere (value facilitation), joint sphere (value co-creation) and consumer sphere (value-in-use). In this contribution, the main focus is placed on the provider sphere, where the provider's role is to facilitate the customer's value-creating process. This is done by building up resources such as software, hardware, services, knowledge and capabilities in advance that have the potential to support the customer's creation of value-in-use. Whereas the value facilitation usually occurs without an interaction with the customer, the actual value creation process happens subsequently in the joint sphere where both the customer and the provider integrate resources and generate value. Finally, in the customer sphere, the provider is yet again an active, although indirect, value facilitator as the independent value creation of the consumer is significantly influenced and shaped by corporate agency and provider co-creation (Anker, Sparks, Moutinho, & Grönroos, 2015; Grönroos & Voima, 2013).

The goal of this research is to investigate the effect of building up resources in advance of future customer engagements as antecedent for cloud service provider profitability, building on the value facilitation concept, as well as on the resource concept from the RBV. This study is carried out as exploratory qualitative research based on 14 expert interviews with cloud service project managers, cost experts and board members from ten different organizations.

The rest of this contribution is organized as follows: First, the theoretical background with regard to the current state of research in cloud service profitability and ecosystems, the necessity to discuss cloud services from a service-dominant logic (SDL) perspective and the link between value facilitation and profitability are explained. The next section deals with the research design, namely the preparation, the conduct and the analysis of the expert interviews. Afterwards, the outcomes of these expert interviews are presented. Subsequently, the results, as well as theoretical and practical limitations are discussed. This contribution is concluded by summarizing the main findings.

BACKGROUND

What Is the Current State of Profitability Research in the Cloud Service Domain?

The field of cloud computing has become a hyped domain among researchers, for instance, with regard to topics such as cloud service adoption (Haag & Eckhardt, 2014) or innovation (Barrett, Davidson, Prabhu, & Vargo, 2015). Nevertheless, a review of the scientific literature concerning the influencing factors of the profitability distribution among cloud computing providers reveals that cloud service profitability has not yet been addressed systematically from the provider's perspective.

The authors' literature review came to the conclusion that three classes of research contributions from the cloud provider's business perspective have been advanced so far: (1) papers suggesting models or algorithms focusing on the optimization of costs or other resources; (2) papers dealing with business models and business ecosystem models and; (3) papers discussing fundamental business benefits and challenges of cloud computing.

To give some examples for the first class of publications, Brumec and Vrček (2013) developed a model to compare costs of leasing IT with costs of owning IT based on computing and storage resources required for the cloud service to be delivered to the customer. Chaisiri, Bu-Sung and Niyato (2012) similarly developed a model for the resource-optimal provision of cloud services. In contrast to Brumec and Vrček (2013), Chaisiri et al. (2012) proposed a stochastic programming model and added network bandwidth as a third parameter. While Brumec and Vrček (2013) and Chaisiri et al. (2012) focused on the relationship between providers and customers, Li (2012) extended these views and developed a resource provisioning algorithm optimizing costs in a value network. Furthermore, Alkhanak, Lee and Khan (2015) suggested cost-aware workflow scheduling approaches in cloud computing. Azad and Navimipour (2017) proposed one of the numerous further efficient task scheduling algorithms which include, in this case, two important criteria, the makespan and energy consumption.

The second class of publications deals with business models and business ecosystem models. The discussion on business ecosystems was initiated by Moore (1993) in analogy to biological ecosystems in which:

[…] companies coevolve capabilities around a new innovation: they work cooperatively and competitively to support new products, satisfy customer needs, and eventually incorporate the next round of innovations. (Moore, 1993)

Accordingly, a business ecosystem consists of a large number of companies that are loosely coupled and, thus, have an effect on each other so that failures of one firm can lead to failures of another (Peltoniemi, 2006). In order to understand the relationships and interdependencies between roles of market players in a business ecosystem, the cross-domain concept of ecosystem mapping has been widely used as a visual tool in the literature (Bahari, Maniak, & Fernandez, 2015). Within such role-based ecosystem models, each company acts as an independent agent that can fulfill one or multiple roles (Böhm, M., Koleva, G., Leimeister, S., Riedl, C., & Krcmar, H., 2010). To give an example of a role-based ecosystem model in the field of cloud computing, the recently published *Pa*ssau *C*loud Computing *E*cosystem Model (PaCE Model) is briefly introduced (Floerecke & Lehner, 2016, see Appendix, Figure 2). This model has been developed at the University of Passau within a comparative analysis of the previously proposed models according to the design science paradigm. The PaCE Model comprises 26 roles of market actors that are grouped into five categories: (1) client; (2) vendor; (3) hybrid role; (4) support and; (5) environment – and entails the basic service flows between the roles (see Appendix) (Floerecke & Lehner, 2016). As each role provides specific business opportunities, they demand different business models (Iivari, Ahokangas, Komi, Tihinen, & Valtanen, 2016). Business models can, therefore, be considered as a specification of how individual roles are concretely played by market actors (Zott, Amit, & Massa, 2011). Generally, a business model is made up of several constitutive components acting as a tool for depicting, innovating and evaluating the business logic of an organization (Veit et al., 2014). One fundamental component is key resources that are required to provide an adequate value proposition for customers in a cost-efficient way (Osterwalder, Pigneur, & Tucci, 2005), which is in line with the value facilitation concept. However, within this research strand, there only exists one first extensive analysis concerning the characteristics of cloud business model components and their link to business success up to now. In their work, Labes et al. (2016) derived abstract success factors of cloud computing business models by relating publicly available characteristics of the business model components to a firm's web visibility and profit.

Within the third class of publications, the benefits and challenges of cloud service models are discussed from different perspectives, for example, controlling (Hoberg, Wollersheim, Böhm, & Krcmar, 2012) and data security (Tiwari & Joshi, 2015). Further publications deal with the utilization potentials of cloud services for specific industries such as the health-care (Neagu, Preda, Stanciu, & Florian, 2017) or the public sector (Jones, Irani, Sivarajah, & Love, 2017).

The overview above shows that none of the contributions systematically researched the profitability of cloud service offerings with regard to value facilitation among different cloud service providers. Generally, it is remarkable that research on this topic is seldom grounded in theory. A reason for this is the rapid pace of growth in the cloud service domain. By contrast, this study is consciously built on service research, especially on the SDL and the RBV.

Why Research Cloud Service From a Service (-Dominant) Logic Perspective?

The SDL, introduced by Vargo and Lusch in 2004, is a paradigm applied to discuss the nature of a service and how organizations provide a service to customers (Vargo & Lusch, 2004). This paradigm is based on ten fundamental premises (Vargo & Lusch, 2008), whereby service can be considered as the application of resources for others. In this understanding, a service is not an alternative form of a product. Instead, a service is what is exchanged and a product, when employed, is an aid to the service process (Lusch & Nambisan, 2016). Moreover, service should be conceptualized as relational processes focused on fulfilling the business needs of customers (Vargo & Lusch, 2016). With cloud services comprising virtualized IT resources in terms of infrastructure, applications and data, made available by a provider to customers, the cloud computing concept aligns with the service perspective expressed in the SDL (Zhao, Tanniru, & Zhand, 2007).

SDL makes a distinction between operand resources and operant resources: Operand resources are static, tangible and inactive resources, such as hardware and software, which are used as input and require other resources to be of value. Operant resources, in turn, are intangible and dynamic resources, such as knowledge and skills, which are utilized to act on operand resources and which create value themselves (Vargo & Lusch, 2004). The more decisive resources are operant resources as they are often difficult to transfer. They are therefore a source of sustained competitive advantage that ensures the firm's position in the ecosystem (Lusch & Nambisan, 2016).

Furthermore, the SDL views all social and economic actors as resource integrators since resources are integrated from the provider and the customer when rendering the service (Vargo & Lusch 2016). This process of integrating resources and rendering the service to the customer in order to enable the customer to make use of it is referred to as the value co-creation process (Cesarotti, Giuiusa, Kwan, Introna, & Spohrer, 2014). Based on this definition, actors are not viewed separately in their dyadic roles as producers that create value and consumers that destroy or use up value. Instead, they are considered in a more generic sense as actors in a system of other actors co-creating value through resource integration and service provision (Vargo & Lusch, 2016). The goal of the value co-creation process is to increase the customer's well-being. The customer perceives this increase in well-being through the process of acquisition, usage and disposal of the service, which is referred to as value-in-use (Vargo, Maglio, &

Akaka, 2008). Therefore, the provider's role in the value co-creation process is the proposition of value and the provision of the service in an interactive and mutual exchange (Vargo, Maglio, & Akaka, 2008).

Although the value co-creation process is intensively discussed in the literature (Cesarotti, Giuiusa, Kwan, Introna, & Spohrer 2014), researchers stress the fact that the processes involved in the value co-creation are not well known. In order to overcome this shortage, some researchers (Anker, Sparks, Moutinho, & Grönroos, 2015; Grönroos, 2011; Grönroos & Ravald, 2011; Grönroos & Voima, 2013) emphasize the role of building up resources in advance for future customer engagements. Grönroos and Voima (2013) state that value creation should not be seen as an all-encompassing process. Instead, it should be considered as taking place through a chain of three successive spheres: provider sphere (value facilitation), joint sphere (value co-creation) and consumer sphere (value-in-use) (Anker, Sparks, Moutinho, & Grönroos, 2015; Grönroos & Voima, 2013). In the provider's sphere, the firm's role is to facilitate the customer's value-creating process. This is done by building up resources such as software, hardware, services, knowledge and capabilities in advance that offer the potential to support the customer's creation of value-in-use (Anker, Sparks, Moutinho, & Grönroos, 2015; Grönroos & Voima, 2013). These resources can also be reused in other future customer engagements (Grönroos, 2011). Grönroos and Ravald (2011, 10) further explains that:

[i]n a supplier-customer relationship, value facilitation can be regarded as a prerequisite or foundation for value creation, and is a reason why customers seek a relationship with a supplier.

Whereas value facilitation usually occurs without an interaction with the customer, the actual value creation process takes place in the joint sphere where both the customer and the provider integrate resources and generate value. Finally, in the customer sphere, the provider is yet again an active, although indirect, value facilitator as the independent value creation of the consumer is significantly influenced and shaped by corporate agency and provider co-creation (Anker, Sparks, Moutinho, & Grönroos, 2015; Grönroos & Voima, 2013).

What Is the Link Between Value Facilitation and Profitability?

According to the RBV, a resource represents a strength or weakness of an organization (Wernerfelt, 1984). Wade and Hulland (2004) summarize that resources which are rare, appropriate, non-imitable, non-substitutable and not perfectly mobile have the potential to provide the organization a competitive advantage. SDL states in a similar manner in its fourth fundamental premise, that *[o]perant resources are the fundamental source of strategic benefit* (Vargo & Lusch, 2016). Therefore, the idea of operant resources fundamentally aligns with the resource concept within the RBV. Moreover, as the goal of value facilitation is to build up resources in advance of future customer engagements, it can be argued that the outcome of value facilitation is operant resources. Value facilitation is accordingly the process of building up operant resources.

Bharadwaj (2000) argues that although many authors define resources broadly, there is a need to differentiate between resources and (dynamic) capabilities: While resources can be seen as a basic unit of analysis, capabilities combine resources to make them an organizational capability. Capabilities thus *[...] refer to an organization's ability to assemble, integrate, and deploy valued resources* (Bharadwaj, 2000).

Capabilities can be extended to be dynamic capabilities when they are considered from a longitudinal perspective and include the ability of the provider to adapt, integrate, reconfigure, renew and recreate his resources and capabilities dynamically to the ever-changing market environment and customer needs (Wang & Ahmed, 2007).

Several authors in the information systems research domain adopted this point of view and developed and defined so-called IT capabilities. By building on previous studies and empirical evidence for these dimensions, Kim, Shin, Kim and Lee (2011), for example, found IT infrastructure flexibility, IT personnel expertise and IT management capabilities to be the main organizational capabilities among IT organizations: IT infrastructure flexibility is defined as the ability of an organization's IT infrastructure to enable quick development and support various system components. IT personnel expertise refers to the level of professional skills or knowledge of the IT staff and IT management capabilities relate to the ability of an organization to manage IT resources to deliver business value (Kim, Shin, Kim & Lee, 2011).

With regard to the definition of value facilitation and by building on the similarities between the value facilitation concept from the SDL and the resource concept from the RBV, it can be argued that the idea of value facilitation should be meaningfully extended to also account for dynamic capabilities. On that basis, the following working definition for value facilitation is suggested:

Value facilitation is the buildup of operand and operant resources (hardware, software, infrastructure, knowledge and capabilities) to prepare for future customer engagements and to keep both operand and operant resources valuable on the market.

Furthermore, the RBV is a useful concept to research profitability among cloud service providers as the dynamic capabilities perspective is an applied paradigm to explain variance in performance across competing firms (Liu, Ke, Wei, & Hua, 2013). In general, building up dynamic capabilities helps an organization to make operational process change better, faster and more cost-efficient than other process changes initiated by the competition and helps an organization to be more responsive to environmental changes (Kim, Shin, Kim & Lee, 2011). This, in turn, helps to improve, adapt, adjust, reconfigure, refresh and renew business processes continuously when changes occur (Kim, Shin, Kim, & Lee, 2011). Moreover, Kim et al. (2011) state that dynamic capabilities improve communication, coordination and information sharing in an organization and Cohen and Levinthal (1990) argue that dynamic capabilities improve the acquisition of internal and external knowledge, the configuration and reconfiguration of the resources base and the deployment and redeployment of resources for the customer. However, according to Cheng, Chen and Huang (2014), few empirical studies have examined the possible relationships between IT capability, business process capability and firm performance. Nevertheless, with regard to IT resources and capabilities, Bharadwaj (2000), Kim et al. (2011) and Liu et al. (2013) point out that IT competencies in mobilizing and deploying IT-based resources can be a source of competitive advantage and lead to superior performance.

Hence, the RBV concept supports the statement that successful cloud service providers build up resources in advance of future customer engagements. In turn, it can be concluded that a low degree of value facilitation leads to comparable low cloud service profitability. Therefore, the following hypothesis is proposed:

H1: The degree of value facilitation is positively associated with cloud service profitability.

INSIGHTS FROM 14 IN-DEPTH INTERVIEWS

Research Design

In order to find practical evidence for the conclusions reached above that profitability among cloud service providers is positively associated with the degree of value facilitation, 14 semi-structured and semi-open interviews on value facilitation and service profitability were conducted with cloud service experts. The interviews were analyzed using qualitative analysis techniques.

An exploratory qualitative research approach was chosen to conduct this study as research on profitability among cloud service providers and service research is nascent. An in-depth understanding of factors contributing to the profitability of cloud service providers is, however, required to better understand the dimensions and the significance of this topic. In line with the goals of this study, semi-structured interviews allow the interviewees to answer freely instead of choosing predefined answers from a list (Myers & Newman, 2007). Semi-open questions, furthermore, support collecting additional and new information and leave room for flexibility during the interviews (Myers & Newman, 2007). Thus, the interviews were conducted as guided conversations rather than structured interviews by which the authors were able to make use of naturally occurring data.

Primarily, the experts were asked about the impact of costs and value facilitation on the lifecycle profitability of service offerings. In total, seven interview questions were prepared to guide the sessions: First, the interviewees were asked to explain their companies' IT service. Then, multiple questions related to costs and cost drivers occurring over the entire lifecycle of the cloud service were posed. The lifecycle was in accordance with the literature separated into a value facilitation phase and a value co-creation phase to find out whether the respective organization is actively facilitating its cloud service and is able

Table 1. Overview of experts interviewed in our study

Expert	Professional Title	Company
1	Solution manager	Multinational IT consulting company
2	Head of Outsourcing	Multinational IT consulting company
3	Head of IT Controlling	IT service provider for a major German airport
4	Partner	Multinational IT consulting company
5	Transformation Manager	Multinational IT consulting company
6	Management Advisory	IT department of a "big four" auditor
7	CEO	Medium-sized German IT service provider
8	Self-employed	Self-employed IT consultant
9	Co-founder	Multinational IT consulting company
10	Founder	Small German IT service provider
11	Senior Project Manager	Multinational IT consulting company
12	Manager	IT department of a "big four" auditor
13	IT project manager	Small German IT service provider
14	IT project manager	Medium-sized German IT service provider

to determine costs with regard to value facilitation and value co-creation. At the end, the interviewees were asked to rank the most important factors in cloud service lifecycle profitability.

The interview sessions took place in 2012 and 2013. All interviews were done face-to-face. The professional experience of the interviewees was ranging from five to 30 years leading to an average of 16.6 years. The experts held different jobs in their organization, including controller, solution manager, management advisory, founder, chairman and self-employed. All of the experts were actively involved in cloud service projects. The interviews lasted between 45 and 90 minutes and were held in German. All expert interviews were tape-recorded and transcribed afterwards. The quotations cited in this contribution were translated into English by the authors and cross-checked by an English language editor.

The experts worked in ten different German organizations with one expert being a self-employed IT consultant. All of the experts' organizations were involved in the cloud computing business. Organizations of all sizes (small, medium and large organizations) as well as organizations only serving the German market and multinational organizations were deliberately included in the sample. General IT consulting organizations and organizations providing IT service to one major customer were included as well. This sample was chosen as it can be considered as a sufficient representation of the cloud service industry. Table 1 presents an overview of the experts and their organizations' service model coverage.

With regard to the data analysis, the software Atlas.ti (version 6.2.15) was used for encoding and structuring. Coding was an iterative process of (re-) coding data, splitting and combining categories, and generating new or dropping existing categories in line with content analysis techniques. Each transcript was coded by two researchers. The inter-coder reliability was 74 percent, whereby the data analysis quality can be seen as reliable. The outcome of the data analysis was a table of 190 text passages and 510 codes. Some of the statements could be attributed to multiple codes.

Results

Before researching the relationship between value facilitation and cloud service profitability in detail, firstly, the data was analyzed with regard to the occurrence of operand and operant resources to crosscheck if the resource-based view holds the field. Thereby, it was found that organizations build up operand resources in terms of a technical infrastructure and operant resources in terms of personal experience, know-how and an organizational structure (Table 2). One of the surveyed experts even used the term capabilities. Interestingly, while a well-functioning infrastructure and a properly working hardware and software environment might be of great importance for cloud service providers, the experts instead seemed to focus on operant resources.

The data was inductively coded for different parameter values of value facilitation to explore the relationship between value facilitation and cloud service profitability. The result included three parameter values: (1) low value facilitation; (2) adequate value facilitation and; (3) over-facilitation. Low value facilitation occurs when a service provider agrees to carry out a project in a setting where he has little experience and knowledge in the service domain and the customer environment. From the set of data, adequate value facilitation can be defined as a provider-customer setting where designing and delivering the cloud service fits the capabilities of the service provider. Over-facilitation is the buildup of resources for a specific use case. These resources are difficult to be reused within other customer engagements.

With regard to a low degree of value facilitation (cf. Table 3), a misalignment of the provider's offering portfolio, with regard to customer needs (e.g., inadequate customization options) or inexperienced personnel, was found. This incurs high costs and consequently leads to relatively low cloud service

Table 2. Operand and operant resources

Parameter Value	Representative Quotes
Operand and operant resources	• *"Indirect investment costs arise, for instance, if we have to fulfill certain infrastructural preconditions. I have to ensure the presence of certain air-conditioning or cabling systems. I mean the establishment of a certain basic infrastructure for the service." (Head of Outsourcing)* • *"I need to maintain my basic organization [to provide a cloud service]." (Manager)* • *"At the least, the capabilities, personal experience and know-how of those implementing the solution come into play. That's important for both those in charge of the functional combination, as well as for the people involved in the project technical implementation." (Management Advisory)*

Table 3. Low facilitation and cloud service profitability

Parameter Value	Representative Quotes
Low value facilitation	**Effects of low value facilitation** • *"The more possibilities to parameterize, the lower the costs. Parameterization is quite a simple adjustment. But if certain things are not feasible to configure, I have to adjust. That's very expensive. If I can change the parameter, that's very easy [...] it may be just a matter of changing an interface. But if it isn't possible, it becomes expensive." (Freelancer)* • *"So it's simply the time the people take to get the information in the beginning or to read the documentation. If my people need a week to read all the documentation and the specifications, I have to take this into account. If they already know everything, it's great." (Solution Manager)* • *"The most important cost driver is to find and accept only those projects that fit our offering. The worst case is to win projects just to generate revenue, but they imply we have to start from zero. A lack of expertise means that everything has to be rebuilt completely. That's the worst cost driver." (Freelancer)* **Reasons for low value facilitation** • *"I need to meet certain margin goals to earn money. The margin I can achieve is higher than what we can account for in the end. The reason lies in, for example, business development... Of course we are very good, but unfortunately we do not win every project, i.e., we have to cope with business development costs where no corresponding contract can be found." (Partner)* • *"There is a major cost group, which we refer to as other operating expenses. Well, that can be misleading; this group involves leasing costs as well as project costs. Project costs are not eligible for capitalization. And other costs, which can be directly related to turnover, are also part of this group." (Head of IT Controlling)*

profitability. The practitioners named three reasons for a low degree of value facilitation: First, it is risky to invest in value facilitation because the cloud service cannot always be marketed successfully. Second, customer requirements are difficult to forecast and projects might fail. Third, providers are sometimes discouraged to invest in value facilitation as, according to (German) accounting principles, value facilitation cannot be capitalized on the balance sheet. From a financial and non-cost management perspective, value facilitation therefore is hard to exploit.

Nonetheless, it was found that practitioners are well aware of the importance of an adequate degree of value facilitation (cf. Table 4): Experts highlighted that providers need to build up operant resources in advance of customer engagement to support the actual service delivery. In line with an adequate degree of value facilitation, experts discussed the idea of standardization. Standardization can be achieved by the modularization of infrastructure, hardware and software, as well as processes. This allows cloud service organizations to share costs among multiple customers and therefore to profit from economies of scale. However, practitioners also stated that customers have individual requirements which cannot be met by fully standardized software. Hence, providers additionally need to possess know-how for the required cloud service setting. Self-evidently, offering individualized service is more expensive for providers.

Table 4. Adequate value facilitation and cloud service profitability

Parameter Value	Representative Quotes
Adequate value facilitation	**Building up resources in advance of future customer engagements** • *"We definitely have the ambition. It is referred to as industrialization. So we have standardized components to make use of scale economics and then combined the components in the end. Actually, the endpoint is to have a certain application to use modular infrastructure components for the implementation." (Solution Manager)* • *"So, for example, within application management, a field I know very well, we use industrialized methods. We have incident management, problem management, change management etc. […]. It can be called industrialization." (Solution Manager)* • *"The question is to what extent individual customer requirements have to be realized. Put the other way round, when the customer says, "I use standard software and adapt my processes to the standard as far as possible", that's cheaper for us. Or, the customer says:" I want to keep what I had until now and want to change as little as possible." Then, the customer wants to expand the standard software so that he can still keep his own individual world." (Co-founder)* • *"During the preparation or contract phase, it's very important that the project fits our potential services. And that all the agreements that are set in the contract are realizable." (Freelancer)* • *"We have made a draft for our portfolio management; in addition to the topic of strategic fit, risk plays an important role. It is also important that we have the required knowhow for a particular development environment." (Head of IT Controlling)* **Cost sharing among customers** • *"A very important point is cost sharing – to sell same things to different customers. Otherwise it isn't affordable." (Founder)* • *"How well do I control my organization to recognize recurrent projects and learn from them, which results in standardization […]? That's a direct cost factor which is turned in an indirect cost saving factor." (Manager)*

The experts named several causes for over-facilitation which reduce potential cloud service profitability due to additional costs for not required resources (cf. Table 5). One interviewee stated that the offering portfolio should be parsimonious and not include excess functionality. Excess functionality causes costs for implementation and maintenance. Second, providers should find a trade-off between cutting edge technologies and a stable offering portfolio. That means, the offering portfolio should be attractive for customers, while problems with new and immature technologies should be avoided. Third, the cloud service market is from the viewpoint of the experts volatile. If a provider had invested in a certain technology or service and suddenly customer business needs changed, the provider would incur sunk costs. Finally, one expert stressed that over-facilitation can be avoided by the provider focusing on its core competencies and reverting to third-party personnel when required.

Table 5. Overfacilitation and cloud service profitability

Too high degree of value facilitation	**Reasons of over-facilitation** • *"If I support several types of operating system platforms, then they too can become a cost driver." (Freelancer)* • *"If we want the newest technology everywhere, pioneer problems can arise because the risk of suffering from setbacks is quite large. We have to find an optimum between marketable state-of-the-art and stabile solutions." (Freelancer)* • *"If I'm good at my business, I have a high hit rate between the requirements and the solution I offer. But the critical point is whether I have been mistaken or the other party misunderstood what I have presented – and that becomes expensive." (Freelancer)* • *"Usually one should know that not all projects can be completed. That means there are a lot of expenses where no equivalent value will ever be paid back, ever." (Freelancer)* **How to avoid over-facilitation** • *"We have subcontractors of course. When we have certain special topics and/or certain specific skill profiles in PLM, for example. And we don't have the required people. The customer doesn't expect us to know everything. In this case, we act as general contractor." (Solution Manager)*

DISCUSSION

Value Facilitation and Cloud Service Profitability

This study offers first insights into the role of value facilitation as antecedent of cloud service profitability. The interview data supports the hypothesis that value facilitation in terms of building up IT infrastructure, IT personnel and IT management capabilities in advance of future customer engagements is positively associated with cloud service profitability. Based on the interview results, this is due to the fact that providers which are well-facilitated incur lower costs in the value co-creation process as they can revert to already existing resources, which, in turn, leads to higher service profitability.

The findings from the interviews further show that providers which do not build up reusable resources in advance to future customer engagements experience high value co-creation costs and low profitability as they need to design and implement resources on a project-specific basis for each customer. Accordingly, this setting does not allow providers to profit from economies of scale and scope. However, a too high degree of value facilitation (e.g., excess software features) also leads to low profitability. When over-facilitated, providers have invested time and money in building up resources they cannot make use of during value co-creation processes. Thus, they have misaligned their value facilitation efforts with customer needs.

To achieve an adequate degree of value facilitation, providers should project future customer needs, as well as learn from previous customer engagements. For this purpose, customer needs should be translated into IT resources that are built up in preparation for new customer engagements. In order to reduce costs, experts recommend standardizing and industrializing resources which allow providers to profit from economies of scale and scope. Furthermore, it can be derived from the findings that providers should stay in their market or market niche after having built up IT resources in advance of future customer engagements. Entering a new customer engagement without the possibility of making use of facilitated resources would make existing value facilitation useless and lead to sunk costs. But when providers start taking over an additional ecosystem role, they have to examine, which existing resources can be reused and which necessarily have to be built up. The additional role has to be implemented in the form of a business model. However, the coexistence of different roles and thus of different business models on a corporate level increases the complexity for a provider substantially. In this context, it is important to understand that not only the resource component of a business model influences the success of a cloud service provider. Also, the other components of a business model, such as the value proposition, the partner network, the pricing model and the cost structure must adequately be designed since a business model as a whole system decides between success and failure. Thereby, in accordance with Osterwalder et al. (2005), it is important to keep in mind that a business model cannot be successful per se as it still has to be implemented. This means that a strong business model can be managed badly and therefore fail, whereas a weak business model may succeed because of strong management and implementation skills.

Study Limitations

First, to test the research model, a qualitative content analysis approach was used. All of the shortcomings of this data analysis approach apply to this study (cf. Mayring, 2010). Second, only a relative small number of experts were interviewed. However, due to the variety of the interviewed experts, it can be argued that the experts consulted for this study are likely to represent the German cloud service market.

After ten expert interviews a theoretical saturation was reached. The focus was on value facilitation as an antecedent of cloud service profitability – a possible third limitation of this study: In a continuation of this study, revenues and other cost drivers, e.g., bargaining power and uncertainty should be taken into account. Furthermore, as the state of research concerning success factors of entire cloud business models is very limited, it is suggested to study this aspect within an exploratory qualitative research, for instance, based on expert interviews.

Moreover, no distinction was made between the various cloud computing ecosystem roles such as application provider, platform provider, infrastructure provider, aggregator, consultant or market place operator as given in the PaCE Model (Floerecke & Lehner, 2016). This aspect should be integrated into further research as the impact of different parameter values of value facilitation on the provider's profitability might vary between these roles. It is also needed to consider to what extent various roles differ in the type of their relevant resources. For example, an infrastructure provider may tend to focus more on operand resources such as server hardware and virtualization software. Aggregators or consultants, in turn, should be comparatively more reliant on operant resources like integration and industry expertise.

In addition, this contribution is among the firsts to combine the theoretical approaches of the SDL and the RBV. Therefore, there are only a few studies in the literature to which these results can be compared. To the best of the authors' knowledge, the profitability of cloud service providers has not yet been discussed systematically from a provider and a business viewpoint. The authors, however, believe that profitability among cloud service providers is much lower than generally assumed. This is reflected in the fact that many providers do not release accurate revenue-expense estimates, which might indicate their reluctance to publish poor or unfavorable profitability figures.

Theoretical Implications

This study contributes to the refinement of the value co-creation concept as described in the SDL. The value co-creation process is still rather unspecified and Vargo et al. (2008) state that *[…] the exploration of value co-creation raises as many questions as answers. What exactly are the processes involved in value creation?* Empirical support was found for the argumentation of Anker et al. (2015) and Grönroos and Voima (2013) that the value creation process is taking place through a chain of three successive spheres: provider sphere (value facilitation), joint sphere (value co-creation) and consumer sphere (value-in-use). Differentiating value facilitation and value co-creation, this study empirically confirms two spheres of the three-sphere concept.

With regard to profitability of cloud service provision, it could be found that providers do not always have the needed resources in place and only those cloud service providers that actually manage to have the adequate resources in place are highly profitable. In practice, for example, only a minority of the platform providers in the cloud ecosystem does not have the underlying infrastructure in-house and instead, build their offerings on the infrastructure of another provider.

Furthermore, and contrary to Grönroos (2011) and Grönroos and Ravald (2011), the results show that value facilitation should not be equated with production. Rather, it is suggested that value facilitation includes the buildup and standardization of operand resources (e.g., infrastructure, hardware and software), as well as the buildup of operant resources (e.g., knowledge and dynamic capabilities). This is of special importance for a cloud service as customers require cloud service providers to both build up infrastructure, hardware and software resources and at the same time to master all the business processes to offer a cloud service cost-efficiently.

Based on the findings from the literature and the field, the authors have come to the conclusion that the concept of the buildup of operand and operant resources is similar to the concept of dynamic capabilities in the RBV. Indeed, value facilitation can be seen as the process and dynamic resources as the outcome of the value facilitation process. In addition, reverting to the RBV, it is proposed not to view value co-creation exclusively from a dyadic perspective or solely on a provider-customer relational base. Rather, it is suggested that cloud service providers should facilitate future customer engagements using a general or global approach taking multiple customer engagements into consideration. In line with that, the authors regard value facilitation as useful for multiple customer engagements in terms of building up resources such as IT infrastructure flexibility, IT personnel expertise and IT management.

To ensure and to improve the profitability of cloud business models, cloud service providers should identify technical modules which can be reused in various settings and serve as building blocks for future customer engagements (Herzfeldt, Schermann, & Krcmar, 2010). The interview data shows that value facilitation from scratch is hardly possible as cloud service providers do usually not have the knowledge and expertise to predict their customers' demands and requirements in the first place. Cloud service providers can only obtain an understanding of these over time through customer engagements and by performing strategic projects. Value facilitation should therefore always be preceded by learning (Herzfeldt, Schermann, & Krcmar, 2010). Nevertheless, cloud service providers have to be aware of the fact that a change in one component of a business model, in this case within the infrastructure component, implies in turn changes in other business model components (Osterwalder, Pigneur, & Tucci, 2005). Business model innovations are thus considered as a major challenge and should always be carefully planned and managed (Zott, Amit, & Massa, 2011).

Practical Implications

The authors believe that there are numerous practical implications of this research. While it is common sense in the field that lifecycle costs usually split up into project costs and operational costs according to the 30/70 rule, the authors found that value facilitation has a huge impact on the overall lifecycle costs, i.e. both project and operational costs. Thus, a high value task in ensuring or improving cloud service profitability would determine the right degree of value facilitation and produce measures to reduce operational costs. Furthermore, the authors clearly identified which IT resources, namely IT infrastructure flexibility, IT personnel expertise and IT management capabilities, should be built up in advance to future customer engagements. The identification of factors either contributing to costs or decreasing profitability should provide a basis for future research in the field.

Future Research

As it was already pointed out, many experts relate value facilitation to learning. Many experts stated concretely that organizations should learn from previous customer engagements to increase the profitability of future customer engagements. Learning is mainly considered as the building up of IT capabilities and experience among employees. Experts reported that they have introduced informal and formal processes at the end of each project to capitalize past experiences. Practitioners argued that learning from previous projects helps to facilitate and to support decisions on building up resources for future customer engagements.

The RBV suggests that acquiring dynamic capabilities requires time and experience (Wade & Hulland, 2004). Thus, to build up value facilitation resources, cloud service providers need to implement processes which allow them to learn as an organization. According to literature, learning as an organization can be defined as the process of encoding, storing and retrieving lessons from history (Huber, 1991). This process of engaging in dedicated projects with individual customers and gradually translating the knowledge gained from these projects into value facilitation capabilities relates this contribution to the literature on service engineering and service modularization. The findings of this study suggest an important underlying mechanism that may guide the advancement of methods in the fields of service engineering and service modularization (cf. Herzfeldt, Schermann, & Krcmar, 2010).

The authors propose the following research model for future research: It is suggested to view the value co-creation process in a broad sense as embedded in a circular process consisting of value co-creation in the narrow sense, learning and value facilitation. This process should be iterated multiple times. After each iteration, the provider should be more attuned to customer needs. It is of utmost importance that successful cloud service providers should first learn from previous customer engagements and only then start facilitating new customer engagements. Only by knowing customer demands and by being able to assess future market demands and trends correctly, value facilitation is a promising approach to create value for the customer and, consequently, financial value for the provider. This process should be instantiated for each occupied role of the PaCE Model (Floerecke & Lehner, 2016). Thereby, it is needed to specify which operand and operant resources are required for each of the roles.

The authors find that this circular process has many similarities with the spiral model in software development suggested by Berry W. Boehm (1988). For the illustration of this idea, Figure 1 is proposed as a graphical representation of the spiral model (Boehm, 1988).

Figure 1. The circular process of value co-creation, learning and value facilitation

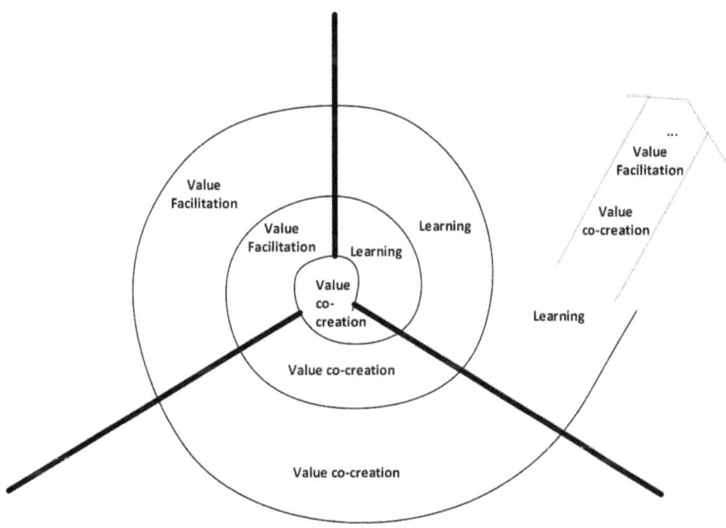

Figure 2. The Passau Cloud Computing Ecosystem model (PaCE model)
(Data Source: Floerecke & Lehner, 2016)

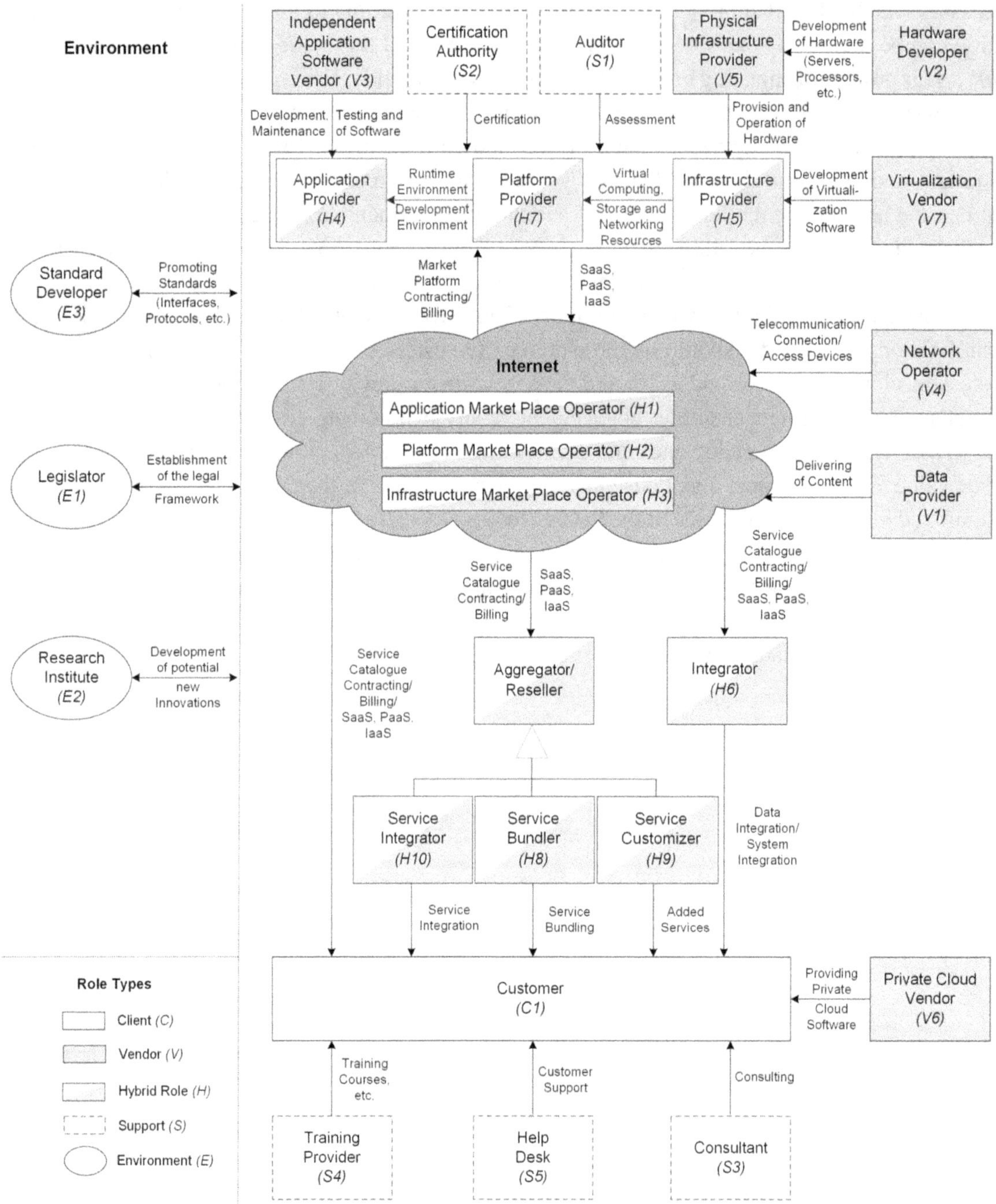

CONCLUSION

This research was motivated by the need to improve the understanding of profitability variations between cloud service offerings and how cloud service providers can ensure and improve profitability in a maturing market. While the previous cloud service literature has focused mainly on technical aspects and on the customer's perspective, this study explicitly examined the provider's perspective including the cloud service ecosystem. For this purpose, the current literature on cloud computing, the SDL and the RBV was analyzed to better understand the provider's role in facilitating and providing a cloud service. In addition, 14 qualitative interviews were conducted with experts from ten different cloud computing providers to examine value facilitation and service profitability from a practical perspective.

The data analysis indicated that cloud service profitability is positively associated with the degree of value facilitation. Moreover, based on the RBV, it could be derived that value facilitation for cloud service providers also consists of building up operant resources such as infrastructure flexibility, IT personnel expertise and IT management capabilities in advance of future customer engagements.

This contribution is intended to fill a gap in research where increasing competition and price pressures are described but where insights into value facilitation and cloud service profitability are missing. The authors hope that practitioners find these insights useful when occupying further ecosystem roles and developing new or innovating existing business models. Researchers can utilize the insights as a starting point to further study the optimal balance between value facilitation and value co-creation within different cloud ecosystem roles and to understand the interplay of value facilitation and learning. Finally, this research adds to a better understanding of the value co-creation process and the resource integration concept, which both require further elaboration in the future.

REFERENCES

Alkhanak, E. N., Lee, S. P., & Khan, S. U. R. (2015). Cost-Aware Challenges for Workflow Scheduling Approaches in Cloud Computing Environments: Taxonomy and Opportunities. *Future Generation Computer Systems*, *50*, 3–21. doi:10.1016/j.future.2015.01.007

Anker, T. B., Sparks, L., Moutinho, L., & Grönroos, C. (2015). Consumer Dominant Value Creation: A Theoretical Response to the Recent Call for a Consumer Dominant Logic for Marketing. *European Journal of Marketing*, *49*(3/4), 532–560. doi:10.1108/EJM-09-2013-0518

Azad, P., & Navimipour, N. J. (2017). An Energy-Aware Task Scheduling in the Cloud Computing Using a Hybrid Cultural and Ant Colony Optimization Algorithm. *International Journal of Cloud Applications and Computing*, *7*(4), 20–40. doi:10.4018/IJCAC.2017100102

Bahari, N., Maniak, R., & Fernandez, V. (2015). *Ecosystem Business Model Design*. Paper presented at the 16th Conférence Internationale de Management Stratégique, Paris, France.

Barrett, M., Davidson, E., Prabhu, J., & Vargo, S. L. (2015). Service Innovation in the Digital Age: Key Contributions and Future Directions. *Management Information Systems Quarterly*, *39*(1), 135–154. doi:10.25300/MISQ/2015/39:1.03

Bharadwaj, A. (2000). A Resource-Based Perspective on Information Technology Capability and Firm Performance: An Empirical Investigation. *Management Information Systems Quarterly*, *24*(1), 169–196. doi:10.2307/3250983

Boehm, B. W. (1988). A Spiral Model of Software Development and Enhancement. *Computer*, *21*(5), 61–72. doi:10.1109/2.59

Böhm, M., Koleva, G., Leimeister, S., Riedl, C., & Krcmar, H. (2010). *Towards a Generic Value Network for Cloud Computing*. Paper presented at the 17th Workshop on Economics of Grids, Clouds, Systems, and Services, Ischia, Italy. 10.1007/978-3-642-15681-6_10

Böhm, M., Leimeister, S., Riedl, C., & Krcmar, H. (2009). Cloud Computing: Outsourcing 2.0 oder ein neues Geschäftsmodell zur Bereitstellung von IT-Ressourcen?. *IM – Die Fachzeitschrift für Information Management und Consulting*, *24*(2), 6-14.

Brumec, S., & Vrček, N. (2013). Cost Effectiveness of Commercial Computing Clouds. *Information Systems*, *38*(4), 495–508. doi:10.1016/j.is.2012.11.002

Buyya, R., Yeo, C. S., Venugopal, S., Broberg, J., & Brandic, I. (2009). Cloud Computing and Emerging IT Platforms: Vision, Hype, and Reality for Delivering Computing as the 5th Utility. *Future Generation Computer Systems*, *25*(6), 599–616. doi:10.1016/j.future.2008.12.001

Cesarotti, V., Giuiusa, A., Kwan, S. K., Introna, V., & Spohrer, J. (2014). Designing Multichannel Value Propositions to Enhance Value-Cocreation Phenomenon. *International Journal of Service Science, Management, Engineering, and Technology*, *5*(1), 14–44. doi:10.4018/ijssmet.2014010102

Chaisiri, S., Bu-Sung, L., & Niyato, D. (2012). Optimization of Resource Provisioning Cost in Cloud Computing. *IEEE Transactions on Services Computing*, *5*(2), 164–177. doi:10.1109/TSC.2011.7

Cheng, J. H., Chen, M. C., & Huang, C. M. (2014). Assessing Inter-Organizational Innovation Performance through Relational Governance and Dynamic Capabilities in Supply Chains. *Supply Chain Management*, *19*(2), 173–186. doi:10.1108/SCM-05-2013-0162

Clohessy, T., Acton, T., Morgan, L., & Conboy, K. (2016). *The Times They are A-Chaning for ICT Service Provision: A Cloud Computing Business Model Perspective*. Paper presented at the 23rd European Conference on Information Systems, Istanbul, Turkey.

Cohen, W. M., & Levinthal, D. A. (1990). Absorptive Capacity: A New Perspective on Learning and Innovation. *Administrative Science Quarterly*, *35*(1), 128–152. doi:10.2307/2393553

Fischer, R., Janiesch, C., Strach, J., Bieber, N., Zink, W., & Tai, S. (2013). *Eine Bestandsaufnahme von Standardisierungspotentialen und -lücken im Cloud Computing*. Paper presented at the 11. Internationale Tagung Wirtschaftsinformatik, Leipzig, Germany.

Floerecke, S., Herzfeldt, A., & Krcmar, H. (2012). Risiken bei IT-Lösungen – Ein Risikokatalog für Praktiker aus Anbietersicht. *IM – Die Fachzeitschrift für Information Management und Consulting, 27*(4), 22-30.

Floerecke, S., & Lehner, F. (2016). *Cloud Computing Ecosystem Model: Refinement and Evaluation.* Paper presented at the 23rd European Conference on Information Systems, Istanbul, Turkey.

Floerecke, S., Wolfenstetter, T., & Krcmar, H. (2015). Hybride Produkte – Stand der Literatur und Umsetzung in der Praxis. *IM+io – Magazin für Innovation. Organisation und Management, 30*(2), 61–66.

Grönroos, C. (2011). Value Co-Creation in Service Logic: A Critical Analysis. *Marketing Theory, 11*(3), 279–301. doi:10.1177/1470593111408177

Grönroos, C., & Ravald, A. (2011). Service as Business Logic: Implications for Value Creation and Marketing. *Journal of Service Management, 22*(1), 5–22. doi:10.1108/09564231111106893

Grönroos, C., & Voima, P. (2013). Critical Service Logic: Making Sense of Value Creation and Co-Creation. *Journal of the Academy of Marketing Science, 41*(2), 133–150. doi:10.100711747-012-0308-3

Haag, S., & Eckhardt, A. (2014). Organizational Cloud Service Adoption: A Scientometric and Content-Based Literature Analysis. *Journal of Business Economics, 84*(3), 407–444. doi:10.100711573-014-0716-6

Herzfeldt, A., Schermann, M., & Krcmar, H. (2010). *Towards A Set of Requirements for A Holistic IT Solution Engineering Approach.* Paper presented at the 21st Australasian Conference on Information Systems, Brisbane, Australia.

Hoberg, P., Wollersheim, J., Böhm, M., & Krcmar, H. (2012). Cloud Computing – Überblick und Herausforderungen für das Controlling. *Controlling – Zeitschrift für erfolgsorientierte Unternehmensführung, 24*(6), 294-300.

Huber, G. P. (1991). Organizational Learning: The Contributing Processes and the Literatures. *Organization Science, 2*(1), 88–115. doi:10.1287/orsc.2.1.88

Iivari, M. M., Ahokangas, P., Komi, M., Tihinen, M., & Valtanen, K. (2016). Toward Ecosystemic Business Models in the Context of Industrial Internet. *Journal of Business Models, 4*(2), 42–59.

Iyer, B., & Henderson, J. C. (2012). Business Value from Clouds: Learning from Users. *MIS Quarterly Executive, 11*(1), 51–61.

Jones, S., Irani, Z., Sivarajah, U., & Love, P. E. (2017). Risks and Rewards of Cloud Computing in the UK Public Sector: A Reflection on Three Organisational Case Studies. *Information Systems Frontiers*, 1–24.

Karunagaran, S., Mathew, S., & Lehner, F. (2016). *Differential Adoption of Cloud Technology: A Multiple Case Study of Large Firms and SMEs.* Paper presented at 36th International Conference on Information Systems, Dublin, Ireland.

Kim, G., Shin, B., Kim, K. K., & Lee, H. G. (2011). IT Capabilities, Process-Oriented Dynamic Capabilities, and Firm Financial Performance. *Journal of the Association for Information Systems, 12*(7), 487–517.

Labes, S., Hanner, N., & Zarnekow, R. (2017). Successfull Business Model Types of Cloud Providers. *Business & Information Systems Engineering*, *59*(4), 223–233. doi:10.100712599-016-0455-z

Li, C., & Li, L. Y. (2012). Optimal Resource Provisioning for Cloud Computing Environment. *The Journal of Supercomputing*, *62*(2), 989–1022. doi:10.100711227-012-0775-9

Liu, H., Ke, W., Wei, K. K., & Hua, Z. (2013). The Impact of IT Capabilities on Firm Performance: The Mediating Roles of Absorptive Capacity and Supply Chain Agility. *Decision Support Systems*, *54*(3), 1452–1462. doi:10.1016/j.dss.2012.12.016

Lusch, R. F., & Nambisan, S. (2015). Service innovation: A Service-Dominant Logic Perspective. *Management Information Systems Quarterly*, *39*(1), 155–175. doi:10.25300/MISQ/2015/39.1.07

Marston, S., Li, Z., Bandyopadhyay, S., Zhang, J., & Ghalsasi, A. (2011). Cloud Computing – The Business Perspective. *Decision Support Systems*, *5*(1), 176–189. doi:10.1016/j.dss.2010.12.006

Mayring, P. (2010). *Qualitative Inhaltsanalyse – Grundlagen und Techniken*. Weinheim, Germany: Beltz Verlag.

Moore, J. F. (1993). Predators and Prey: A new Ecology of Competition. *Harvard Business Review*, *71*(3), 75–83. PMID:10126156

Myers, M. D., & Newman, M. (2007). The Qualitative Interview in IS Research: Examining the Craft. *Information and Organization*, *17*(1), 2–16. doi:10.1016/j.infoandorg.2006.11.001

Neagu, G., Preda, Ş., Stanciu, A., & Florian, V. (2017). *A Cloud-IoT Based Sensing Service for Health Monitoring*. Paper presented at the 6th IEEE International Conference on E-Health and Bioengineering Conference (EHB), Sinaia, Romania. 10.1109/EHB.2017.7995359

Oliveira, T., Thomas, M., & Espadanal, M. (2014). Assessing the Determinants of Cloud Computing Adoption: An Analysis of the Manufacturing and Services Sectors. *Information & Management*, *51*(5), 497–510. doi:10.1016/j.im.2014.03.006

Osterwalder, A., Pigneur, Y., & Tucci, C. L. (2005). Clarifying Business Models: Origins, Present, and Future of the Concept. *Communications of the Association for Information Systems*, *16*(1), 1–43.

Peltoniemi, M. (2006). Preliminary Theoretical Framework for the Study of Business Ecosystems. *Emergence*, *8*(1), 10–19.

Simons, A. J., Bratanis, K., Kourtesis, D., Paraskakis, I., Veloudis, S., Verginadis, Y., . . . Rossini, A. (2014). *Advanced Service Brokerage Capabilities as the Catalyst for Future Cloud Service Ecosystems*. Paper presented at the 2nd International Workshop on CrossCloud Systems, Bordeaux, France. 10.1145/2676662.2676675

Sultan, N. (2014). Servitization of the IT Industry: The Cloud Phenomenon. *Strategic Change*, *23*(5-6), 375–388. doi:10.1002/jsc.1983

Tiwari, P. K., & Joshi, S. (2015). Data Security for Software as a Service. *International Journal of Service Science, Management, Engineering, and Technology*, *6*(3), 47–63. doi:10.4018/IJSSMET.2015070104

Vargo, S. L., & Lusch, R. F. (2004). Evolving to a New Dominant Logic for Marketing. *Journal of Marketing*, *68*(1), 1–17. doi:10.1509/jmkg.68.1.1.24036

Vargo, S. L., & Lusch, R. F. (2008). Service-Dominant Logic: Continuing the Evolution. *Journal of the Academy of Marketing Science*, *36*(1), 1–10. doi:10.100711747-007-0069-6

Vargo, S. L., & Lusch, R. F. (2016). Institutions and Axioms: An Extension and Update of Service-Dominant Logic. *Journal of the Academy of Marketing Science*, *44*(1), 5–23. doi:10.100711747-015-0456-3

Vargo, S. L., Maglio, P. P., & Akaka, M. A. (2008). On Value and Value Co-Creation: A Service Systems and Service Logic Perspective. *European Management Journal*, *26*(3), 145–152. doi:10.1016/j.emj.2008.04.003

Velten, C., & Janata, S. (2012). *Cloud Vendor Benchmark 2012*. München, Germany: Experton Group AG.

Wade, M., & Hulland, J. (2004). The Resource-Based View and Information Systems Research: Review, Extension, and Suggestions for Future Research. *Management Information Systems Quarterly*, *28*(1), 107–142. doi:10.2307/25148626

Wang, C. L., & Ahmed, P. K. (2007). Dynamic Capabilities: A Review and Research Agenda. *International Journal of Management Reviews*, *9*(1), 31–51. doi:10.1111/j.1468-2370.2007.00201.x

Wernerfelt, B. (1984). A Resource-Based View of the Firm. *Strategic Management Journal*, *5*(2), 171–180. doi:10.1002mj.4250050207

Zhang, Q., Cheng, L., & Boutaba, R. (2010). Cloud Computing: State-of-the-Art and Research Challenges. *Journal of Internet Services and Applications*, *1*(1), 7–18. doi:10.100713174-010-0007-6

Zhao, J. L., Tanniru, M., & Zhand, L. (2007). Services Computing as the Foundation of Enterprise Agility: Overview of Recent Advances and Introduction to the Special Issue. *Frontiers of Information Systems*, *9*(1), 1–8. doi:10.100710796-007-9023-x

Zott, C., Amit, R., & Massa, L. (2011). The Business Model: Recent Developments and Future Research. *Journal of Management*, *37*(4), 1019–1042. doi:10.1177/0149206311406265

ADDITIONAL READING

Baines, T., Ziaee Bigdeli, A., Bustinza, O. F., Guang Shi, V., Baldwin, J., & Ridgway, K. (2017). Servitization: Revisiting the State-of-the-Art and Research Priorities. *International Journal of Operations & Production Management*, *37*(2), 256–278. doi:10.1108/IJOPM-06-2015-0312

DaSilva, C. M., Trkman, P., Desouza, K., & Lindič, J. (2013). Disruptive Technologies: A Business Model Perspective on Cloud Computing. *Technology Analysis and Strategic Management*, *25*(10), 1161–1173. doi:10.1080/09537325.2013.843661

Grönroos, C., & Gummerus, J. (2014). The Service Revolution and its Marketing Implications: Service Logic vs Service-Dominant Logic. *Managing Service Quality*, *24*(3), 206–229. doi:10.1108/MSQ-03-2014-0042

Hossain, M., & Hossain, M. (2017). Business Model Innovation: Past Research, Current Debates, and Future Directions. *Journal of Strategy and Management*, *10*(3), 342–359. doi:10.1108/JSMA-01-2016-0002

Kellermanns, F., Walter, J., Crook, T. R., Kemmerer, B., & Narayanan, V. (2016). The Resource-Based View in Entrepreneurship: A Content-Analytical Comparison of Researchers' and Entrepreneurs' Views. *Journal of Small Business Management*, *54*(1), 26–48. doi:10.1111/jsbm.12126

Lappi, T., Lee, T. R., & Aaltonen, K. (2017). Assessing the Health of a Business Ecosystem: The Contribution of the Anchoring Actor in the Formation Phase. *International Journal of Management. Knowledge and Learning*, *6*(1), 27–51.

Sadiku, M. N., Musa, S. M., & Momoh, O. D. (2014). Cloud Computing: Opportunities and Challenges. *IEEE Potentials*, *33*(1), 34–36. doi:10.1109/MPOT.2013.2279684

Vargo, S. L., & Lusch, R. F. (2017). Service-Dominant Logic 2025. *International Journal of Research in Marketing*, *34*(1), 46–67. doi:10.1016/j.ijresmar.2016.11.001

Chapter 7
A Comprehensive Report on Security and Privacy Challenges in Software as a Service

Pradeep Kumar Tiwari
Manipal University Jaipur, India

Sandeep Joshi
Manipal University Jaipur, India

ABSTRACT

Researchers have done tremendous works for data security, but a robust security mechanism is not available yet. Researchers are doing continuous work to build robust SaaS mechanism. SaaS has several security vulnerabilities. Data security is still the most important challenge to researcher and they can constantly do research to protect the data over the network but they are facing numerous technical challenges to completely secure the cloud network and cloud storage. This work would be helpful to understand data security and privacy problems. Researchers can find the new way to understand SaaS security vulnerabilities and currently available solutions.

INTRODUCTION

In new computing paradigm cloud computing is most popular cost effective, flexible, highly available, pay per use computing web based service, which provides three service models (SaaS (Software as a Service), PaaS (Platform as a Service), IaaS (Infrastructure as a Service) four Deployment model (Private, Public, Hybrid, and Community) and five essential characteristics (On demand self service, Broad network access, Resource pooling, Rapid elasticity, Measured service) (Jansen, 2011; Paul, 2014; Tiwari, 2012).

Traditional storage, data management scheme is not superior enough to store and analyzing the big data. Cloud virtual information framework system has the capacity to handle the enormous information issue, yet it is insufficient great in security of information (Hassanien, A.E., Azar, A.T., Snasel, V., Kacprzyk, J., & Abawajy, J.H., 2015). Providers to ensure robust security system to users. Service providers used third party security and security audit systems. Service providers provide security, SecaaS

DOI: 10.4018/978-1-5225-5951-1.ch007

(Security as a Service). SecaaS includes authentication, antivirus, anti malware, intrusion detection and security management at different level. SecaaS control the data loss prevention, web security, encryption, network security and disaster recovery (Alliance, 2011b; Pearson, 2013).

Cloud Users can access computing resources via internet. Security is the main concern for cloud users. Security is dived mainly seven categories: (1) Legal issues; (2) Network; (3) Interface; (4) Information (data security); (5) Compliance; (6) Virtualization and; (7) Governance (Gonzalez, N., Miers, C., Redígolo, F., Simplicio, M., Carvalho, T., Naslund, M., & Pourzandi, M., 2012).

The Result shows the legal issues and compliance are major security issues is shown in figure 1. Pi chart shows virtualization has greater security vulnerabilities then network security. Virtualization gives the elasticity, resource pooling and multi tenancy facility in cloud computing (David, 2009; Luo, 2011).

Amazon AWS provides EC2 (Elastic Cloud Computing) and S3 (Simple Storage Service) with secure system. Amazon AWS uses multiple third party authorized audit security system (ISO/IES-27002 control frame work). Salesforce.com provides secure SaaS services with CRM (Customer Resource Management) features. HP, IBM, Google, Window Azure also gives security assurance to users (Amazon, 2014; Azure, 2014; Somorovsky, J., Heiderich, M., Jensen, M., Schwenk, J., Gruschka, N., & Lo Iacono, L., 2011).

Cloud providers take care on network security, IT system (system security), information (data security) and application security. Security responsibility ensures from client side and server side (Ma, 2012). Service provider provides several new security trends, but they are still not providing full robust security mechanism till now. Amazon AWS EC2 offers physical security, environmental security and virtualization security. Salesforce.com CRM (Customer Resource Management) offers security responsibilities in physical and environmental security control. Microsoft AZURE uses token based users' authentication policy (Amazon, 2014; Salesforce, 2014; Somorovsky, J., Heiderich, M., Jensen, M., Schwenk, J., Gruschka, N., & Lo Iacono, L., 2011).

Figure 1. Security problems in grouped categories (Chen, P. M., Lee, E. K., Gibson, G. A., Katz, R. H., & Patterson, D. A., 1994)

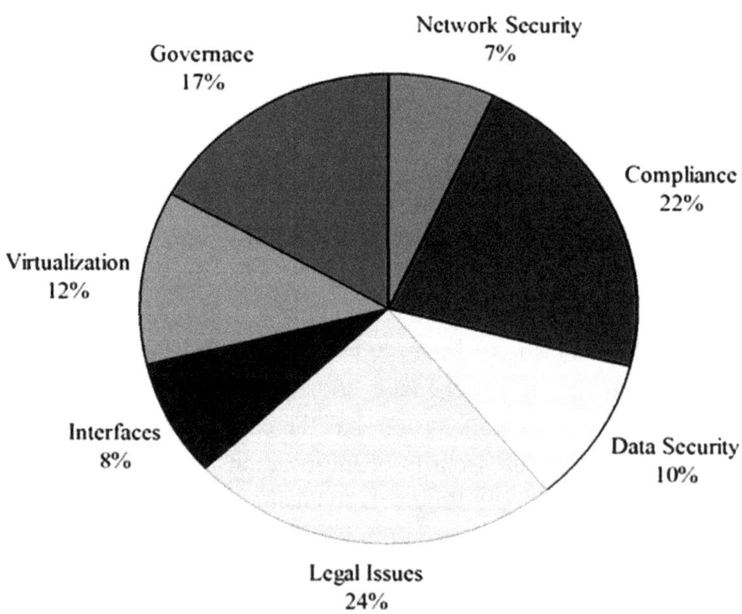

Worldwide countries (Japan, Australia, New Zealand, and Asia specific countries) have made security policies. They have mentioned the privacy and security guidelines, OECD (Organizations for Economic Cooperation and Development) and the APEC (Asia Pacific Economic Co-operation) and EEA (European Economic Area) to cloud service providers (Gonzalez, N., Miers, C., Redígolo, F., Simplicio, M., Carvalho, T., Naslund, M., & Pourzandi, M., 2012).

Several security systems (SSL/TCL, xACML, and TCCP) are available, but security attacks on cloud are increasing day by day. Most important questions are how to secure data from intruder (Gonzalez, N., Miers, C., Redígolo, F., Simplicio, M., Carvalho, T., Naslund, M., & Pourzandi, M., 2012; Ma, 2012; Ko, S. R., & Lee, S. S., 2013). Ko et al (2013) report showed incidents of attacks increased from previous years. Figure 2 shows a top three cloud service provider mostly attacks form unknown vulnerability.

Ko et al., (2013) shows unknown incidents, attacks mostly happen in Google, Amazon, Microsoft and Salesforce.com. They are mostly cloud service providers. Figure 3 shows unknown causes breakdown attacks on user's service providers.

Ko et al. (2013) shows 25% unknown and 75% known attacks happen in last previous years. Cloud providers have to need a more secure mechanism to secure. Sony picture attacks were viral news in the security world. During the transmission of information or accessing the data from the cloud, cryptography is widely used technology to protect sensitive data and storage data (Tang, Y., Lee, P. P., Lui, J. C., & Perlman, R., 2010; Yong, P. E. N. G., Wei, Z. H. A. O., Feng, X. I. E., DAI, Z. H., Yang, G. A. O., & CHEN, D. Q., 2012).

Data security is still major challenge to researchers. Researchers continuously doing research to secure data over the network, but they are facing many major technical challenges to completely secure the cloud network and cloud storage.

This paper broadly discussed protection of data over SaaS. SaaS users can understand different security aspect and vulnerabilities of SaaS.

Figure 2. Cloud provider breakdown for unknown vulnerability

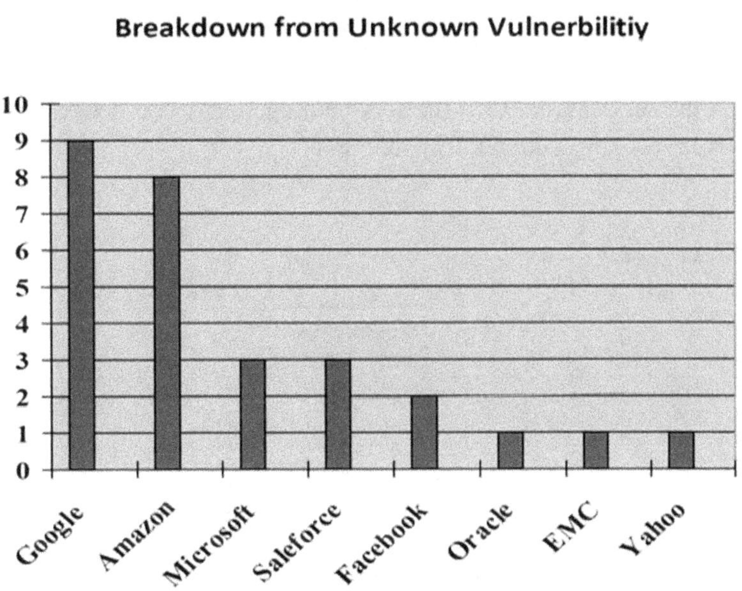

Figure 3. Transparency of causes of cloud vulnerabilities (Ko & Lee 2013)

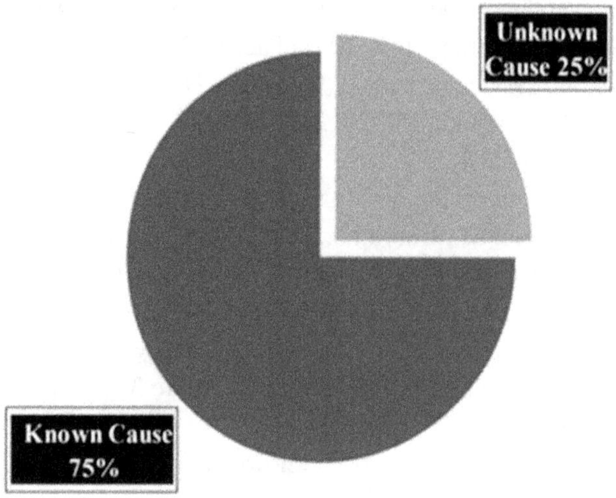

KEY FEATURES OF SOFTWARE AS A SERVICE

In order to define an excellent model for comparing SaaS, it's far a required to perceive key features of SaaS. From our assessment on current references in cloud computing (Subashini & Kavitha, 2011; Jansen & Grance, 2011), the key capabilities are given in the Figure 4.

1. **Reusability:** In software program engineering, reusability is the potential of software program elements to serve for the development of many wonderful packages. The essential motivation of CC is to reuse various kinds of net based services.
2. **Data Managed by Provider:** This service providers are responsible for service installation and data management on their own server. So, most of the data which consumers produce is stored on service provider's data center and managed by them. Thus customers do not perceive two things in detail; where is their data stored and how the data can be managed. So that, the clients but can also don't trust the services and then the service usage turns into decrease, if service providers don't provide data security and reliability function.

Figure 4. Key features of SaaS

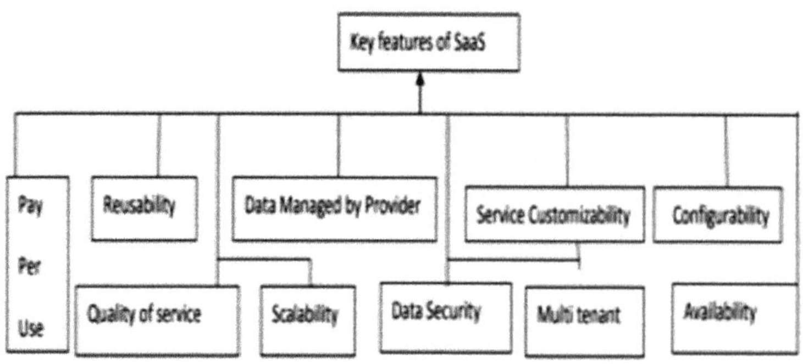

3. **Service Customizability:** Service customizability defined by using the ability for services to be modified by service customers based on their necessities. This function allows service providers to meet the different desires of every consumer. According to the characteristics of cloud services, numerous clients can become potential users of the cloud services. Because of that, the important drawback is, it is not viable that service providers customise their cloud offerings for all service clients. So, service consumers need to customize their services for own purposes.
4. **Availability:** In cloud computing, the service consumers are able to access SaaS service from a Web browser thru the Internet and that they do no longer have any ownership for the SaaS that is deployed and runs at the provider's server.
5. **Quality of Service:** QOS is related to the provider's service level and service skills.
6. **Scalability:** In software program engineering, scalability is a belongings that's vital for a system, a network, or a procedure, this means that its ability to both cope with growing quantities of work or can be larged. Due to the black- box nature of cloud computing services, service consumers can't control resources which might be used by the services, inclusive of reminiscence, network, or CPU utilization. That is, a service company is a chargeable for rescaling the resources according to consumer's requests without notifying the customers in element.
7. **Data Security:** The data security headache are the most essential motives why corporations are un-interested by software program as a service. Moreover, fixing the organizations security issues has came up as the biggest project for adopting Subservice in the cloud.
8. **Multi-Tenant:** Multi-tenant is noted the rule in software architecture, wherein an example of the software program executes on a server, giving services to more than one clients (tenants). In a multitenant environment, near to all agencies or tenants and their customers receives the service from the same technology, sharing almost all additives inside the approach stack together with the data model, servers, and database layers.
9. **Configurability:** Configurability is the primary function of any SaaS software program. This element of SaaS software consists of organizational shape, user interface, facts, data, access control workflow, and commercial logic.
10. **Pay per Use:** The expenses for SaaS are envisioned through the use of the usage of services at the side of the wide variety of service invocations or duration which services are applied. That is, service consumers can connect and use the service each time they want, and then pay for just amount of usage. So, the customer takes an activity for using the service (Banerjee, 2014).

CHALLENGES IN SAAS SECURITY

SaaS security systems depend on the SaaS service providers. SaaS users don't have technical knowledge of security issues and measurement policy. Service providers give the data security assurance in multi-tenancy environment and during live migration. Users have to know available security system to secure data (Jansen, 2011).

Figure 5 shows numbers of vulnerability attacks during 2009 to 2011. Security awareness also arises during this session, but CSA (Naruchitparames & Giine, 2011) report shows cases of attacks are increasing day by day.

Service provider not only focusing the portability of the application, but also enhance security mechanism with data integrity (Banirostam, H., Hedayati, A., Zadeh, A. K., & Shamsinezhad, E., 2013; Mah-

Figure 5. Frequency of cloud vulnerability incidents

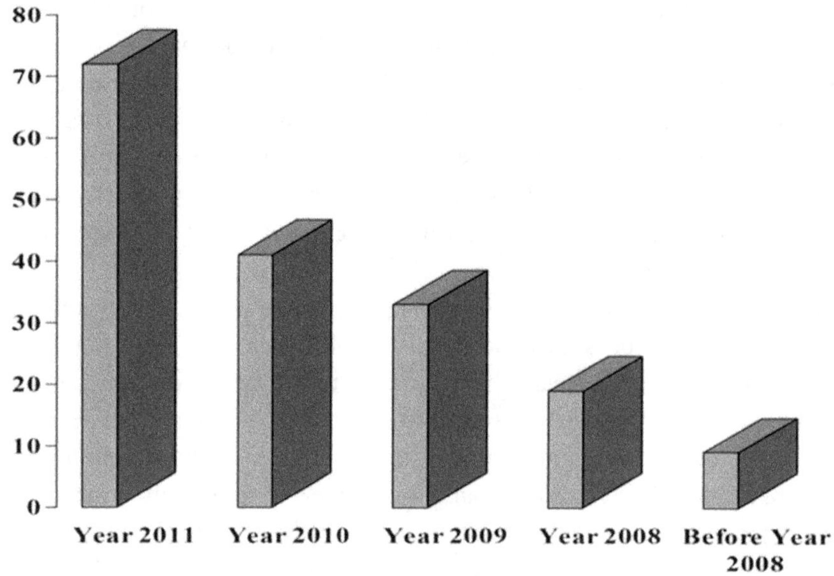

mood, 2011). The SaaS software vendor may host the application on its own private server or deploy on cloud computing infra-structure service provided by a third-party provider (e.g. Amazon, Google, etc.) (Subashini & Kavitha, 2011). Moving of data and application in an external cloud environment creates the breach and extend security risk. Service security may be/ may not be generated by service provider staff or other clients who are using SaaS services (García-Valls, M., Cucinotta, T., & Lu, C., 2014).

Major Security Issues in SaaS

Figure 6 shows major security issues in SaaS.

Data Security and Backup

SaaS vendor provides secure data configuration system with a secure storage mechanism. A cloud user uses the third party storage system, so customer always scare about data security. Data security is common issues for any technology, but it is a serious concern while using the SaaS service. Although, SaaS providers use certified and secured services, but it still has some breaches (Hashizume, K., Rosado, D. G., Fernández-Medina, E., & Fernandez, E. B., 2013; Sahoo & Mehfuz, 2013).

Users always want special and secure treatment with his processed and storage data. Vendors are responsible to secure data from MITM (Man in the Middle Attacks) and data leakage problems. SaaS providers ensure to his users, processed data and stored data have always been in the regular backup mode with quick recovery mechanism. During the any kind of disaster (server crash, data loss or fire) data will be available from backup data centre (Chen & Zhao 2012; Paul, 2014) Vendor ensure that data are in secured and manage with the RAID features. The Vendor also fallow the object storage technol-

Figure 6. SaaS security issues (Gonzalez, N., Miers, C., Redígolo, F., Simplicio, M., Carvalho, T., Naslund, M., & Pourzandi, M., 2012; Jansen, 2011; Pearson, 2013; Subashini & Kavitha, 2011)

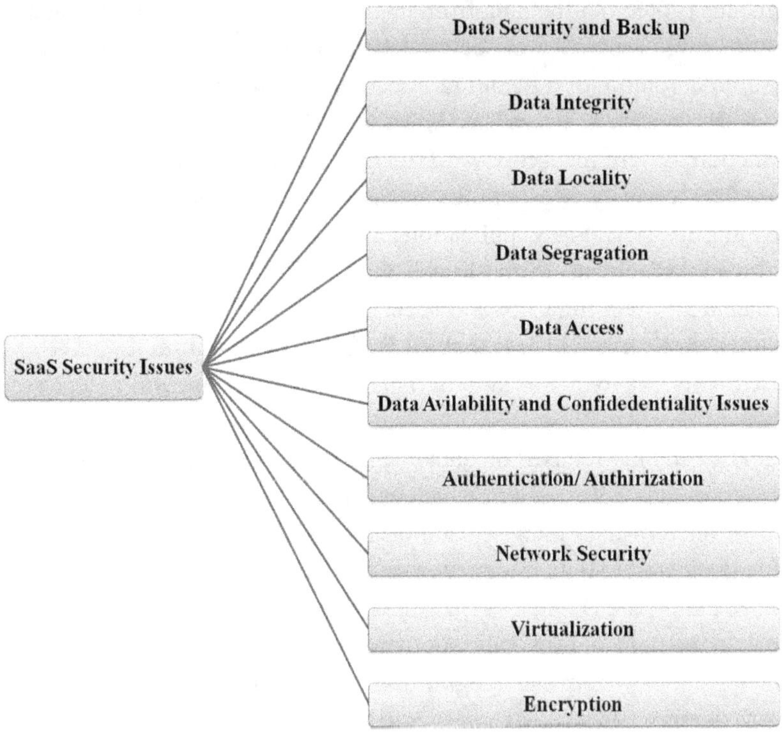

ogy like NAS (Network Attach Storage), SAN (Storage Area Network), to manage processes and stored data (Chen, P. M., Lee, E. K., Gibson, G. A., Katz, R. H., & Patterson, D. A., 1994; Jansen & Grance, 2011). TVD (Trusted Virtual Data Center) gives the assurance of secure isolation and integrity. TVD provides managed isolation among different VMs and data centers (Takabi, H., Joshi, J. B., & Ahn, G. J., 2010; Tran, 2013).

The data protection Act defines the area of sensitive data. Sensitive data can define by either client or service provider or both can decide. If data is sensitive that time apply the Act of prohibition with legal law. Laws of protecting data many times very useful, but many times they are not effective because laws change after international boundaries (Chen, P. M., Lee, E. K., Gibson, G. A., Katz, R. H., & Patterson, D. A., 1994; Gonzalez, N., Miers, C., Redígolo, F., Simplicio, M., Carvalho, T., Naslund, M., & Pourzandi, M., 2012). ENSIA (European Network and Information Security Agency) is responsible for achieving network and information security with the three main objects, policy and organization, technical and legal issues (Zissis & Lekkas, 2012).

Cross side scripting, weak access control service, insecure storage and configuration, open the back window to do fraud access and illegitimately data manipulation are main issues of data security and backup (Kaur & Pateriya 2013; Rane, 2010). Also week key management system and failures of data security mechanism increases insecurity of processed data. But still, only few methods are real and effective to secure processing data. SaaS providers generally use the SSL/TCL or PCI security control mechanism (Garfinkel, T., Pfaff, B., Chow, J., Rosenblum, M., & Boneh, D., 2003; Oleshchuk & Koien, 2011).

Data Integrity

Data integration means protecting the data from unauthorized access, deletion, and modification from the intruder. Integrity gives assurance of secure data transmission from source to destination. Service Provider must ensure that data will transmit in a secure way and data always be remain actual (David, 2009; Kandukuri, B. R., Paturi, V. R., & Rakshit, A., 2009).

The Data transaction system must follow ACID (Atomicity, Consistency, Isolation and Durability) properties to ensure data integrity. Most of the database uses ACID properties to secure data transaction and maintain data integrity. Each SaaS application has different level of data availability. It does depend on SLA (Service Level Agreement) (Subashini & Kavitha, 2011). Many standards and authorizations are available for managing data integrity and security with web service such as WS-Transaction and WS-reliability (Kumar & Kumar 2013; Rimal, B. P., Choi, E., & Lumb, I., 2009). SaaS solution uses open, standard based APIs link to on-site system. This makes integrations simple, fast, and cost-effective (Archibald, 2009).

Zetta is a system that mainly focused on data integrity for Cloud services it provides similar ideas to RAID system. Zetta provides Zetta system which considering on the data integrity in on demand storage system with assurance of data integrity and backup of data in some critical circumstances. Zetta Corporation mainly dedicated to data integrity in on demand service (Chen, P. M., Lee, E. K., Gibson, G. A., Katz, R. H., & Patterson, D. A., 1994; Zetta, 2008).

Secure storage management and firewall security can manage the data integrity. A good Encryption/ Decryption key management system can integrate data during migration of data among different servers (Naruchitparames & Giine, 2011).

Data Locality

The most common compliance issue facing by the organization is data location. Users want to know detail information of his database where data stored. Users have rights to know security mechanism, safeguard method and certification method is used to protect the data and when data relocate. Once information crosses the national borders, then difficult to protect the data because the rules and regulations of data protection change according to country judiciary laws (Alliance, 2011a; Badger, L., Grance, T., Patt-Corner, R., & Voas, J., 2012; Garfinkel, T., Pfaff, B., Chow, J., Rosenblum, M., & Boneh, D., 2003).

Amazon has its EC2 in multi-locations, and currently one in USA and the other in Europe. Google App Engine locates in many countries (i.e., it has 36 data centers across the world), such as The USA, China, and so on (Alliance, 2011a). Service providers committed to storing and processing data in specific jurisdictions, and make SLA commitment policy to fallow judiciary system, data privacy and security managements (Chen & Zhao, 2012).

Data Segregation

SaaS application required additional data security parameters. Users use data application to process his/ her data in multi-tenancy environment. Multiple users' data can reside at the same location. Multi-tenancy environments give breaches to intruder to hijack. An Intruder can find information from application or VM image (Yang & Jia, 2014).

SaaS providers ensure secure separate segregation of data at physical level. SaaS service must be intelligent enough to isolate the data among different users. Amazon S3, API provides both bucket-level and object-level access controls (Banirostam, H., Hedayati, A., Zadeh, A. K., & Shamsinezhad, E., 2013; Somorovsky, J., Heiderich, M., Jensen, M., Schwenk, J., Gruschka, N., & Lo Iacono, L., 2011).

The cloud provider committed to providing highly secured tested encryption schemes (Prince, 2009). Data can run on multiple VMs instances. These instances may be located at single physical server. SaaS providers give the proper user authorization and administrative task in a secure way and facilitate controlled secure data storage. Providers follow the available protocol standard and security standard for secure data storage. The Best suited cryptography system with practical approach fallow by the service provider to make a trust relationship with the SaaS clients (Garfinkel, T., Pfaff, B., Chow, J., Rosenblum, M., & Boneh, D., 2003; Kandukuri, B. R., Paturi, V. R., & Rakshit, A., 2009). SQL injection flaws and insecure storage system vulnerabilities do the exploitation. These flaws intruders can access user sensitive data (Banirostam, H., Hedayati, A., Zadeh, A. K., & Shamsinezhad, E., 2013; Rane, 2010).

Data Access

Access control mechanism is a confirmation of secure authorized access and prevention from illegal access. SaaS providers are responsible for secure process of data, secure network transactions, and secure application privileges. Users focused on secure data access. SaaS provider checks the key constraints of access right to secure data process and gives attention on management of applications services of SaaS.

The SaaS model is always flexible enough to incorporate the specific policies put forward by the organization (Subashini & Kavitha, 2011). SaaS provider checks the access session time and update the data access security in a proper way. OTP (One Time Password), digital signature and good algorithm of encryption and decryption system can be helpful for secure data access control Oleshchuk and Koien (2011), Sabahi (2011) and Yang et al.,(2013) give the data access idea with security mechanism this proposed solution is called DAC-MACS (Data Access Control for Multi-Authority Cloud Storage). Both are proposed effective and safe data access control policy in multi-tenancy and multi-authority cloud storage environment. Both researchers proposed multi-authority CP-ABE (Cipher text Policy- Attribute based encryption) scheme by using token based decryption method.

Data Avilability and Confidetiality Issues

Availability refers providers are responsible for 24*7 days data availability without delay and Confidentiality refers to the security of data from natural or middle man attack (Banirostam, H., Hedayati, A., Zadeh, A. K., & Shamsinezhad, E., 2013; Jansen, 2011). SaaS providers must offer on-time, secure, flexible and failover uninterrupted required services. Data availability ensures the access of data from large amount of structured, unstructured geographically distributed data, but users have rights to know server location and implemented security system (Alliance, 2011b).

Amazon S3 (Simple Storage System) provides unlimited, scalable, reliable, performance based secure storage and access facility. New York Times and Nasdaq are using S3 service of Amazon (Galen, 2008; Subashini & Kavitha, 2011). The U.S army uses salesforce.com CRM for cloud based recruiting and U.S. census bureau also uses Salesforce.com application based SaaS services (Takabi, H., Joshi, J. B., & Ahn, G. J., 2010). Amazon and salesforce.com both ensure to confidentiality and flexible access permissions in secure way (Amazon, 2014; Azure, 2014). There are some challenges to manage centralized data, scalability of data and management of network failure. SaaS provider can use the third party security services to manage the data access security and confidentiality (Zhou, M., Zhang, R., Xie, W., Qian, W., & Zhou, A., 2010). DDOS (Distributed Denial of Service) testing help to validate resilience of SaaS service and ensure service availability to end users (Amazon, 2014; Rane, 2010). Privacy act ECPA (Electronic Communication Privacy Act) of 1986 provides protection against illegal access of electronic mail or other information of users. ECPA gives the many privacy protection law of security and confidentiality of user data (Subashini & Kavitha, 2011).

Authentication/ Authorization

Authorization is the mechanism which determines levels of user access. User authorization and validation mechanism should process on controlled, secure mechanism (Yang, K., Jia, X., Ren, K., Zhang, B., & Xie, R., 2013). In SaaS service, application service is hosted outside of the corporate firewall. Many times users have multiple cloud services, customer have multiple access Id and password. Customer has to remember multiple login ID and password. It might be tough to remember belonging ID and password, if customer exchanges his/her information one login to another login. This may exploit the authorization security (Jasti, A., Shah, P., Nagaraj, R., & Pendse, R., 2010; Yang & Jia, 2014).

SaaS customer must remove or disable account of his employee who left the company and provide a new user ID password for new joined employee (Subashini & Kavitha, 2011). SAML (Security Assertion Markup Language) and WS (Web Service) Federation are widely used with SAML is being more popular. The alternative to SAML and WS Federation is an SSO (Single Sign On) solution that is deployed via a secured VPN (Virtual Private Network) tunnel (Jansen & Grance, 2011).

WS Federation provides a special type of STS (Security Token Service) that makes an authorization decision. Digital identities may be the identical authorization mechanism to identify the authorized person this identity may be based on an e-mail address or machine's unique IP address. A digital signature may be the also a good authorization technique to indentify the actual user this cryptography mechanism is a best suited method for authorization (García-Valls, M., Cucinotta, T., & Lu, C., 2014; Lockhart, H., Andersen, S., Bohren, J., Sverdlov, Y., Hondo, M., Maruyama, H., ... & Prafullchandra, H., 2006).

IDaaS (Identity as a Service) is part of SecaaS. IDaaS has a broad term (i.e. Identity, entitlement authorization and access). Authorization/Authentication is a part of IDaaS. IDaaS ensure secure access policy to the users. XACML (eXtensible Access Mark-up Language) is implemented in XML and responsible for secure authorization service (Alliance, 2011b; Ko & Lee, 2013).

Weak authentication management and unsecure network can be exploited authentication security services (Yu, Y., Ni, J., Au, M. H., Liu, H., Wang, H., & Xu, C., 2014). Login password, token authentication, key pair, digital signature are secure policy to secure authorized access (Amazon, 2014; Ko & Lee, 2013; Azure 2014; Salesforce, 2014).

Network Security

The Network is a medium to access SaaS services from cloud system. Network attracts hackers and intruders with its vulnerabilities and weak security mechanism to attack on cloud services to theft the information from cloud storage (Gonzalez, N., Miers, C., Redígolo, F., Simplicio, M., Carvalho, T., Naslund, M., & Pourzandi, M., 2012). Network security is to pay attention on data threats, access control threats, traffic monitoring and securing gateway transfer. Network security should maintain server transfer management. MFA (Multi Factor Authentication), AES-256 data encryption storage security scheme uses by Amazon AWS (Amazon, 2014). Salesforce.com uses TSL cryptographic protocol and ISO 27001 security control system (Salesforce, 2014). Microsoft Azure uses CSA control matrix, PCI DSS to secure his cloud system (Azure, 2014). The following are three main network security mechanisms:

- **Transfer Security:** Cloud computing uses the distributed architecture for sharing the data and utilize resources. These are utilized in a synchronized manner by VMs instances. Here requiring the robust security policy to prevent sniffing, spoofing in virtual network (Genoves & Akamai, 2009; Gonzalez, N., Miers, C., Redígolo, F., Simplicio, M., Carvalho, T., Naslund, M., & Pourzandi, M., 2012).
- **Firewalling:** Protect the cloud infrastructure against inside and outside attack. Its do the prevention from and DOS (Denial of Service Attack) and detect external security. It provides adapted solution from internal and external attacks (Gonzalez et al., 2012; Hulme, 2011).
- **Security Configuration:** Protocol technologies provide a robust security mechanism without compromising performance and efficiency (Gonzalez, N., Miers, C., Redígolo, F., Simplicio, M., Carvalho, T., Naslund, M., & Pourzandi, M., 2012; Oleshchuk & Koien, 2011). The SaaS service provider must implement strict regulation for data protection from any type of manipulation, theft and illicit access of data.

Figure 7. Network Security taxonomy (Genoves & Akamai, 2009; Gonzalez, N., Miers, C., Redígolo, F., Simplicio, M., Carvalho, T., Naslund, M., & Pourzandi, M., 2012; Hulme, 2011; Oleshchuk & Koien, 2011)

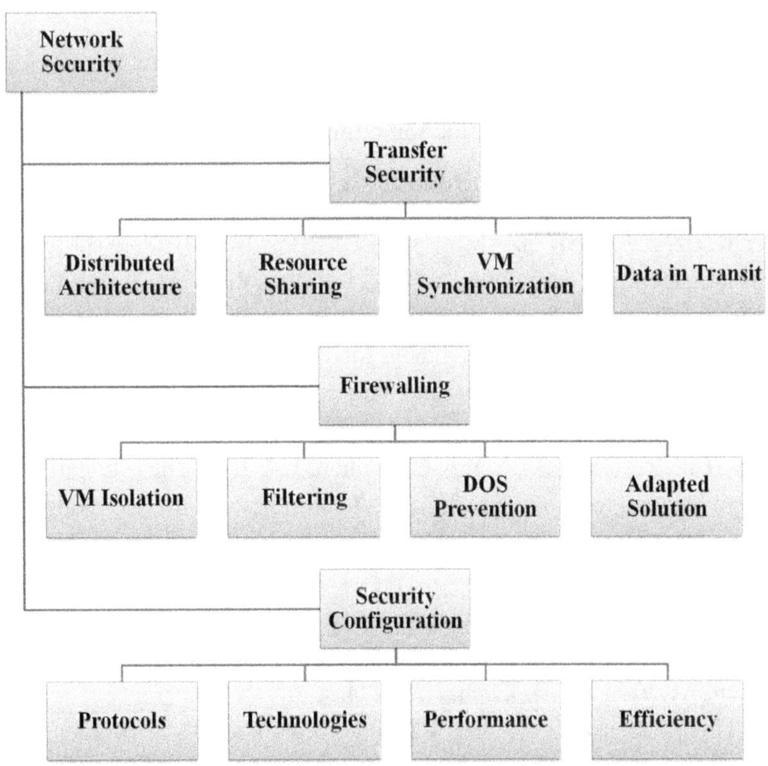

Cryptography is best suited mechanism for user authorization, but it should be certified by security service providers (Paul, 2014). A Service provider provides a robust encryption technique such as SSL (Secure Socket Layer) and TLS (Transport Layer Security) techniques for data access and data security over the network (Rane, 2010). Amazon web service provide strong security mechanism over the traditional network security issue for example MITM (Man in the Middle) attacks, IP spoofing, Port scanning, packet sniffing etc. Amazon S3 provides maximum possible security services. Process via SSL encrypted endpoint ensures secure data transfer (Subashini & Kavitha, 2011). User can connect to an AWS access point via HTTP or HTTPS using SSL mechanism to protect against data theft, tampering and message forgery (Somorovsky, J., Heiderich, M., Jensen, M., Schwenk, J., Gruschka, N., & Lo Iacono, L., 2011). *v*IDS/*v*IPS (Virtual Instruction Detection System and Virtual Protection System) protects virtual environment through collect and analyse processing information between network to host after that detect the malware and spyware code from the processed information. *v*IDS/*v*IPS monitor and analyse the virtual network behaviours, ensure security policies, detect spyware/malware, protect from virtual network attack/ network loop holes and ensure the security policies to protect the information to give assurance of unbreakable secure data transaction (Luo, S., Lin, Z., Chen, X., Yang, Z., & Chen, J., 2011; Purushothama & Amberker, 2012).

Network security mechanism must pay attention on data threats, access control, traffic monitoring and secure gateway transfer. Processed data should transfer in secure way. Virtual network has interred VM traffic and virtual segmentation of physical service security challenges. Network penetration and insecure trust configuration gives unsecure network to intruder to access user information (Alliance, 2011; Rane, 2010).

Virtualization

Virtualization has ability to emulate hardware via software. Virtualizes systems can pooling and sharing resources among multiple customers at the same time. Virtualization does the abstraction features, among different users without service interruption and explores the advantage of resources, hardware, storage and operating system (Garfinkel & Rosenblum, 2005; Hashizume, K., Rosado, D. G., Fernández-Medina, E., & Fernandez, E. B., 2013).

Virtualization is a concept and component of cloud. It does the work as middleware between server and users and gives the features of server virtualization, multi-tenancy, data isolation and resource managements. Real time challenges, multitenancy, load balancing and live migration make the most effective to virtualization, but these are also some vulnerabilities and threats. VM escape, VM hopping and sniffing/spoofing are vulnerabilities are virtualization system (García-Valls, M., Cucinotta, T., & Lu, C., 2014; Luo, S., Lin, Z., Chen, X., Yang, Z., & Chen, J., 2011). These benefits give vulnerabilities and security threats to hackers to do the attack over VM middleware for accessing data and resource or destroy services features.

- **VM Escape:** When VMM doesn't manage/control underlying infrastructure. VM Escape is one of the worst cases happen, if isolation between the host and between VMs is compromised (Luo, S., Lin, Z., Chen, X., Yang, Z., & Chen, J., 2011).
- **VM Hoping:** This problem occur when VM able to gain access to another VM (Hashizume, K., Rosado, D. G., Fernández-Medina, E., & Fernandez, E. B., 2013).

Figure 8. Virtualization vulnerbilities

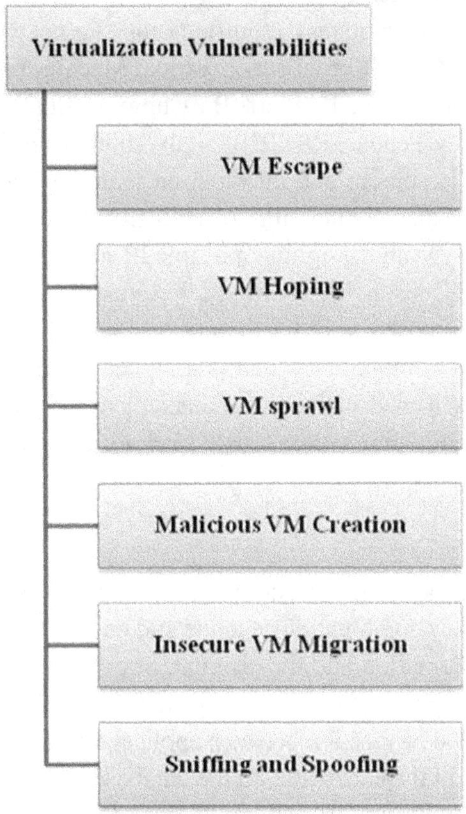

- **VM Sprawl:** Numbers of VMs are increasing continuously, but most of VMs are ideal or not in use in current scenario, which may be a cause of waste of resources and energy (Luo, S., Lin, Z., Chen, X., Yang, Z., & Chen, J., 2011).
- **Malicious VM Creation:** If any attacker has valid account, then he/she can access image. An attacker can send malicious code, virus or spyware by using this image (Hashizume, K., Rosado, D. G., Fernăndez-Medina, E., & Fernandez, E. B., 2013).
- **Insecure VM Migration:** During live migration of VMs, an attacker can access the data and transfer the VM to un-trusted host machine. Attacker also can do DOS attack (Luo, S., Lin, Z., Chen, X., Yang, Z., & Chen, J., 2011).
- **Sniffing and Spoofing:** Sniffing is passive security attack. It uses network interface to receive data. Spoofing is active security attacks (García-Valls, M., Cucinotta, T., & Lu, C., 2014; Yang & Jia, 2014).

Enforcement of security mechanism in virtual machine is difficult task. In a virtualized environment, many instances run simultaneously. Multiple guests can run different applications and different operating systems on a single host computer and provide isolation between the different users access. Virtualization allows users to create copy, share, migrate, and roll back virtual machines and allow them to run a variety of applications (Catteddu, 2010; Garfinkel & Rosenblum, 2005; Jasti, A., Shah, P., Nagaraj,

R., & Pendse, R., 2010). The other issue is the control of an administrator on host and guest operating systems. Current VMMs (Virtual Machine Monitor) do not offer perfect isolation of machine and data. Xen, an open source x86 VMM, incorporates a modified Linux kernel to implement a privilege partition for input/output operations and KVM (Kernel-based Virtual Machine) another source effort transforms a Linux kernel into a VMM (Garfinkel, T., Pfaff, B., Chow, J., Rosenblum, M., & Boneh, D., 2003; Jasti, A., Shah, P., Nagaraj, R., & Pendse, R., 2010; Shah, 2008; Tran, 2013). Multitenancy is difficult to manage in virtual machine. Resources and information share between multiusers that time attacker can access the information. It is also challenging to AWS EC2 network how can handle multiple user account domain name, service type and assigning multiple IP address to indentify to all as an individual (Somorovsky, J., Heiderich, M., Jensen, M., Schwenk, J., Gruschka, N., & Lo Iacono, L., 2011) .

Banirostam et al. (2013) proposed a TCCI (Trusted Cloud Computing Infrastructure) to ensure confidentiality and accuracy of computing, which is assigned to access software services in secured virtualized environment. Research on virtualization management and security of virtualized environment is still in progress, researchers are doing continuously work for enhancing performances and security.

Encryption

Encryption is a technique of encoding and decoding of information by using cryptographic algorithm. Cryptographic techniques use key combinations encode and decode the information. Only proper combination of key can convert information into the original. Strong encryption algorithm gives the surety of data security. AES (Advanced Encryption Standard) and SSL techniques are used by the SaaS service provider for making the encrypted connection between application and user database instances. RDS creates an SSL certificate and install the certificate on the database instance, when instance is provisioned for MySQL and SQL Server (Catteddu, 2010; Yong, P. E. N. G., Wei, Z. H. A. O., Feng, X. I. E., DAI, Z. H., Yang, G. A. O., & CHEN, D. Q., 2012).

The Cryptographic system managed by encryption and decryption of information, hashing, key exchange and digital signature. All methods are reliable to secure information. Intruders do the attacks on key management system of cryptography algorithm. Intruder always tries to find encryption key patterns to do attack on system. SaaS service provider look after data protection, data validation, data integrity and code singing forgery detection. Multitenancy, insecure VMs isolation, weak storage policies are cause of unsecure SaaS service system (Garfinkel & Rosenblum, 2005; Jasti, A., Shah, P., Nagaraj, R., & Pendse, R., 2010).

Researchers are doing work to implement SecaaS (Security as a Service), but still doing research to make robust mechanism. Security system still has some problem in encryption/ decryption delegation mechanism and computation over authenticate data. Purushothama and Amberker (2012) proposed five consecutive encryption algorithms. Algorithms belong to query processing scheme over encrypted data. Sahai and Waters (2005) proposed first ABE (Attribute Based Encryption). Kan Yang et al. Yang and Jia (2014) proposed, DAC-MACS (Data Access Control for Multi authority Cloud Storage) scheme to secure multi authority cloud storage system.

Cloud storage providers don't give the security information to his client to maintain his reputation in cloud market. Forgery attack, data isolation and data leakage are problems in data security. Cloud storage system should be inbuilt with robust encryption policy (Yu, Y., Ni, J., Au, M. H., Liu, H., Wang, H., & Xu, C., 2014). Efficient encryption/decryption mechanism can do the vital role secure data transfer. Users required different security levels. Users want sensitive data (i.e. ID, Password, Account detail) processed via secure transfer channel (i.e. TCL/SSL). Encrypt must be used AES-256 encryption mechanism. Tokenization manages integrated services of public cloud to private cloud. PII (Personal Identifiable Information) and SPI (Sensitive Personal Information) security manage by Data anonymization. Cloud data base control manages the data segregation levels. (Alliance, 2011b; Ko & Lee, 2013; Ma, 2012; Zissis & Lekkas, 2012).

CURRENT SOLUTION SCHEME PROVIDERS, ACT AND CERTIFICATE

Cloud Cube Model also highlights the challenges of understanding and mapping cloud models to control frameworks and standards such as ISO/IEC 27002, which provides: "a series of guidelines and general principles for initiating, implementing, maintaining, and improving information security management within an organization" (Gonzalez, N., Miers, C., Redígolo, F., Simplicio, M., Carvalho, T., Naslund, M., & Pourzandi, M., 2012).

CSA is an organization led by a coalition of industry practitioners, corporations, associations and other stake holders (Nahrstedt & Campbell, 2012; Pearson, 2013), NIST also issued a draft publication on cloud computing, SP 800-144, Guidelines on Security and Privacy in Public Cloud Computing, which addresses the security concerns associated with data center operations and the division of responsibilities among providers and customers (Hashizume, 2013). IBM developed a fully homomorphic encryption scheme in June 2009. This scheme allows data to be processed without being decrypted (Prince, 2009; Sahoo & Mehfuz, 2013).

CURRENT SECURITY SOLUTIONS

There are several research works taking place on in the area of cloud security. Several groups and organization are interested in developing security solutions and requirements for the cloud. The Cloud Security Alliance (CSA) is collecting solution providers, non- profits and individuals to enter into discussion approximately the current and future best practices for data approximately inside the cloud.

The Open Web Application Security Project (OWASP) continues listing of top vulnerabilities to cloud-based totally or SaaS models that is up to date as the threat landscape changes. The best security solution for SaaS applications is to expand an improvement framework that has hard security structure. Tsai W, placed forth a 4-tier framework for web-based improvement that though appears interesting, only implies a security aspect inside the process. In this work, Berre has suggested a road map in the direction of cloud-centric improvement, and the X10 language is one of the strategies to gain better use of cloud capabilities of big parallel processing and concurrency. Another approach is useful resource isolation to ensure security of data during processing, by using keeping apart the processor caches in digital machines, and

keeping apart the ones virtual caches from the hypervisor cache for Identity and access control within the SaaS, has issued an Identity and Access Management Guidance which gives a list of encouraged high-quality practices to assure identities and secure access management. Resource Locality and Data Segregation are the two key security demanding situations on which not much information is available in the existing literature, which necessitates that this could be in addition taken up for research (Rai & Mehfuz, 2013).

Table 1. Current solutions available for securing SaaS

S. No.	Security Areas	Current /Possible Solutions
5.	Information Security	Information Security Risk Management Framework.
6.	Network Security (NS)	• NS for virtual machines • Network Security Sandbox
1.	Authentication and Authorization	• Open Authorization • Two Factor Authentication • Three tier Authentication
2.	Availability	Data Dispersion
3.	Data Confidentiality	• Attribute based Proxy Re-Encryption
4.	Virtual Machine Security	• Reconfigurable distributed virtual machine • Survey on VMs
7.	Cloud Standards	• IEEE CC Standard Study Group • ITU CC Focus Group • Cloud Security Alliance (CSA)
8.	Data Access	• Multi-user access policies. • Data Access Management
9.	Web Application Security	Web Application Scanners
10.	Backup	Agent much less Method for records Backup and Recovery
11	Identity Management and Signal-On Process.	CSA's Identity and Access Management Guidance.

COMPARISONS BETWEEN SAAS AND ITS COUNTERPARTS

1. **SaaS and Legacy Software:** There are three primary variations among SaaS and legacy software: delivering distinctive products, adopting distinct pricing modes, and using one of a kind delivery strategies (Laplante, P. A., Zhang, J., & Voas, J., 2008). First, they deliver exceptional products. SaaS offers the subscribers with a collection of standard programs and services; legacy software program offers custom designed programs to its users. Second, they employ exclusive transport methods. Legacy software packages are established on users' premise; SaaS programs are positioned in services provider's valuable server and the services are delivered thru the Internet or an intranet.
2. **SaaS and SOA:** Many human beings discover it hard to distinguish among SaaS and Service-oriented structure (SOA). The misunderstanding of these phrases would possibly reasons diverse problems, from confusion to poor designs. Laplante et al. (2008) states that:

the difference among SaaS and SOA is that the former is a software delivery model while the latter is a software program production version.

Figure 9 shows the SaaS platform architecture. SaaS delivers software as utility services and costs on an according to-use basis. It separates software program possession from the user. The SaaS providers own the software, store the software system and users' data in a central server, and provide on-demand service to users via the Internet or an intranet. In a SOA model, the constituent additives of the software program device are reusable gadgets or services. Functions are separated into wonderful devices or services. Developers are allowed to get admission to the SOA offerings framework, combine and reuse available additives of the device to construct their new software applications. Today's major SOA

Figure 9. SaaS platform architecture

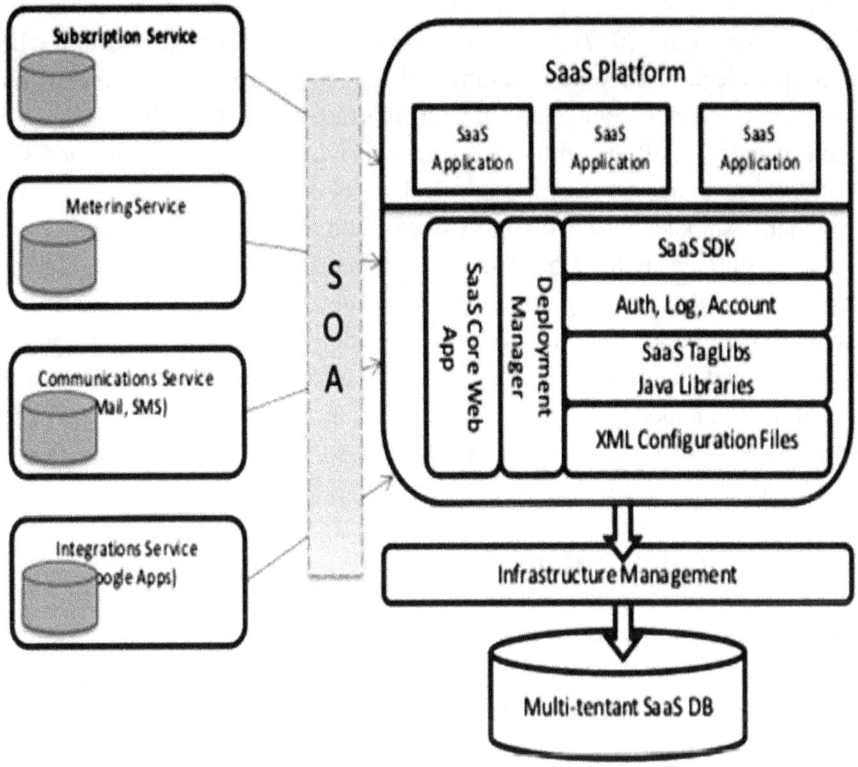

products encompass IBM Web Sphere, Microsoft .NET, Oracle Web Services Manager, and Composite Application Platform Suite. Despite the differences, SaaS and SOA complement every other inside the large-scale information systems. Laplante et al. (2008) concludes that "SaaS enables to provide components for SOA to apply, and SOA enables to quick realize SaaS."

DEVELOPMENT OF SAAS APPLICATIONS

Many unique issues should be taken into consideration whilst developing SaaS packages. This segment in particular opens the discussion from the integration, configuration and customization views.

Integration

A SaaS subscriber normally deploys on-premises legacy software as well. Furthermore, if the person subscribes to a couple of SaaS answers simultaneously, integration among unique SaaS answers is likewise favored. The integration occurs in three fundamental layers of SaaS applications, i.e. user interface, business logic, and data. According to a i.e. user interface, business logic, and data., According to a study on SaaS answer adoption trend by means of IDC in 2004, extra than 50% of the survey respondents emphasized *better integration with in-house applications*. An AMR research document in 2005 pointed out that more than 70% of the total 639 respondents in the survey expect that the SaaS answer may be incorporated with their on-premises legacy software program and other SaaS answers they subscribed or plan to subscribe. Sun et al. (2007) proposes a hybrid model of SaaS integration, wherein SaaS answers and on-premises legacy software program packages integrate on three fundamental layers: user interface, business logic (manner), and facts. In User Interface Integration, first, Single Sign On permits users to get right of entry to all the authorized user interfaces after a single unified sign on; 2nd, In Process Integration, four patterns are proposed: Push, Scheduled Pulling, Receive, and Workflow. In Data Integration, data is labeled into two kinds (master data and transactional data), proposed patterns are Batch Data Migration and Batch Data Synchronization. In Sun et al. (2007). The authors de- fine the development process of SaaS integration design and development as business process review and integration design. In their paper, the authors additionally delivered a SaaS integration framework reference architecture as an extension of net service description named SaaS-DL, and constructed a SaaSia prototype. The authors concluded that during SaaS integration, maximum practical necessities can be fulfilled through current SOA integration technology, non-functional requirements ought to be fulfilled through extending present integration technology.

Configuration and Customization

Configuration and customization is every other important issue in SaaS improvement. However, customers always request feature variations to fulfill their very own specific enterprise want. This should be executed via clean configuration and customization. Sun et al. (2008). clarify the definitions and meanings of configuration and customization of SaaS solutions. In addition to that, the authors broaden a competency model of configuration and customization, and build a technique framework for it. Configuration and customization isn't a new element. Most enterprise software applications face this trouble. They should be tailored so as to fulfill requirements' variance due to customers' significantly different background.

However, it's far obviously that SaaS providers cannot afford to broaden and maintain a version of utility code for every individual customer.

Two most important processes to resolve the requirements variance hassle are configuration and customization, which regularly make people confused. The most important differences among them are that configuration does now not contain source code changes, whereas customization does configuration uses Pre-described parameters to trade software program capabilities. As an end result, consistent with Sun et al. (2008), SaaS development need to

avoid customization with the aid of the usage of configuration to meet clients' tailoring necessities and extend configurable restrict as far as feasible toward client's unique requirements.

In Sun et al. (2008) paper, the authors introduce a SaaS Configuration Competency Model. In the model, competency of SaaS configuration is devided into 5 levels: Entry, Aware, Capable, Mature, and World Class. The variance stage supported are none, low, medium, high, and extremely high, as a consequence, which transforms from completely standardized offering to absolutely tenantized providing (Sun, 2008). The authors additionally build a methodology framework to plan and execute configuration and customization approach. The manner of this framework involves five steps:

1. Understanding environment,
2. Defining method,
3. Assessing competency,
4. Identifying hole and movements, and
5. Prioritizing and executing.

In step one, vendors should investigate the environment of SaaS service configuration and customization, specifically awareness on customer requirements and market leader competency level. In the second step, vendors make their plan to fulfill targeted customers' configuration and customization necessities. In the 0.33 step, vendors determine their personal competency level of SaaS configuration and customization. In the fourth step, the competency gaps can be identified to guide the actions' definition. In the final step, vendors prioritize and excute the actions.

CONCLUSION

Researchers have done tremendous works for data security but robust security mechanism not available yet. Researchers are doing continuous work to built robust SecaaS mechanism. Researchers investigating the unknown problems of data security, secure authentication/ authorization, and secure data segregation. Paper would be helpful to understand data security and privacy problems. Researchers can find the new way to understand SaaS security vulnerabilities and current available solutions. SaaS service model is an application based cloud service. SaaS is a most popular service among three cloud services. SaaS has several security vulnerabilities. User should understand the features and security issues before adopting to SaaS services. User must check the service and security record of SaaS service providers and also compare service and security features to others. Service provider and user make an SLA and mention data location, backup and security features. SLA terms and condition should be fallow between service provider and user. This paper represents the vulnerability and threat of SaaS security. This review analysis presents the traditional and new security problem with current available solutions. Privacy concern should be change time to time as per user requirement. Researchers can discover latest suggestion for the future research for better work and enhance the recent techniques.

REFERENCES

Act, F. C. R. (2009). *Fair Credit Reporting Act*. Flood Disaster Protection Act and Financial Institute.

Akhigbe, A., & Whyte, A. M. (2004). The Gramm-Leach-Bliley Act of 1999: Risk implications for the financial services industry. *Journal of Financial Research*, *27*(3), 435–446. doi:10.1111/j.1475-6803.2004.00093.x

Alliance, C. (2011a). *Security guidance for critical areas of focus in cloud computing v3. 0*. Cloud Security Alliance.

Alliance, C. (2011b). *Security as a Service*. Retrieved from https://cloudsecurityalliance.org/wpcontent/uploads/2011/09/SecaaS_V1_0.pdf

Amazon Web Services. (2014, Novemeber). *Overview of Security Processes*. Retrieved from http://d0.awsstatic.com/whitepapers/Security/AWS%20Security%20Whitepaper.pdf

Archibald, B. (2009). Expect More from Your SaaS Vendor. *Occupational Health & Safety (Waco, Tex.)*, *78*(10), 83–84. PMID:19856771

Badger, L., Grance, T., Patt-Corner, R., & Voas, J. (2012, May). *Cloud Computing Synopsis and Recommendations: Recommendations of the National Institute of Standards and Technology*. CreateSpace Independent Publishing Platform.

Banerjee, S., & Jain, S. (2014). A survey on Software as a Service (SaaS) using quality model in cloud computing. *International Journal of Engineering and Computer Science*, *3*(1), 3598–3602.

Banirostam, H., Hedayati, A., Zadeh, A. K., & Shamsinezhad, E. (2013, April). A Trust Based Approach for Increasing Security in Cloud Computing Infrastructure. In *Computer Modelling and Simulation (UKSim), 2013 UKSim 15th International Conference* (pp. 717-721). IEEE. 10.1109/UKSim.2013.39

Brunette, G., & Mogull, R. (2009, December). Security guidance for critical areas of focus in cloud computing. *Cloud Security Alliance*. Retrived from http://www.cloudsecurityalliance.org/guidance/csaguide.v2.1.pdf

Burnside, R. S. (1987). Electronic Communications Privacy Act of 1986: The Challenge of Applying Ambiguous Statutory Language to Intricate Telecommunication Technologies, The. *Rutgers Computer & Technology Law Journal*, *13*, 451.

Cao, F., Huang, H. K., & Zhou, X. Q. (2003). Medical image security in a HIPAA mandated PACS environment. *Computerized Medical Imaging and Graphics*, *27*(2), 185–196. doi:10.1016/S0895-6111(02)00073-3 PMID:12620309

Catteddu, D. (2010). Cloud Computing: benefits, risks and recommendations for information security. In Web Application Security (pp. 17-17). Springer Berlin Heidelberg. doi:10.1007/978-3-642-16120-9_9

Chen, D., & Zhao, H. (2012, March). Data security and privacy protection issues in cloud computing. In *Computer Science and Electronics Engineering (ICCSEE), International Conference on* (Vol. 1, pp. 647-651). IEEE. 10.1109/ICCSEE.2012.193

Chen, P. M., Lee, E. K., Gibson, G. A., Katz, R. H., & Patterson, D. A. (1994). RAID: High-performance, reliable secondary storage. *ACM Computing Surveys*, *26*(2), 145–185. doi:10.1145/176979.176981

Chowdhuri, S., Chakraborty, S., Dey, N., Azar, A. T., Salem, M. A. M. M., Chaudhury, S. S., & Banerjee, P. (2014). Recent Research on Multi Input Multi Output (MIMO) based Mobile ad hoc Network: A Review. *International Journal of Service Science, Management, Engineering, and Technology*, *5*(3), 54–65. doi:10.4018/ijssmet.2014070104

García-Valls, M., Cucinotta, T., & Lu, C. (2014). Challenges in real-time virtualization and predictable cloud computing. *Journal of Systems Architecture*, *60*(9), 726–740. doi:10.1016/j.sysarc.2014.07.004

Garfinkel, T., Pfaff, B., Chow, J., Rosenblum, M., & Boneh, D. (2003, October). Terra: A virtual machine-based platform for trusted computing. *Operating Systems Review*, *37*(5), 193–206. doi:10.1145/1165389.945464

Garfinkel, T., & Rosenblum, M. (2005, May). When Virtual Is Harder than Real: Security Challenges in Virtual Machine Based Computing Environments. In *Procedding of the 10th Conference on Hot Topics in Operating System* (Vol. 10, pp. 20-20). ACM.

Genoves, S., & Akamai. (2009). *Cloud based fire wall.* Retrived from http://cloudcomputing.sys-con.com/node/1219023

Gonzalez, N., Miers, C., Redígolo, F., Simplicio, M., Carvalho, T., Naslund, M., & Pourzandi, M. (2012). A quantitative analysis of current security concerns and solutions for cloud computing. *Journal of Cloud Computing*, *1*(1), 1–18.

Google Apps Service Level Agreement. (2010, November). Retrieved from http://www.google.com/apps/intl/en/terms/sla.html

Gruman. (2008). *Early experiments in cloud computing*. Retrieved from http://www.infoworld.com/article/08/04/07/15FE-cloud-computing-utility_1.html

Hashizume, K., Rosado, D. G., Fernández-Medina, E., & Fernandez, E. B. (2013). An analysis of security issues for cloud computing. *Journal of Internet Services and Applications*, *4*(1), 1–13. doi:10.1186/1869-0238-4-5

Hassanien, A. E., Azar, A. T., Snasel, V., Kacprzyk, J., & Abawajy, J. H. (2015). Big Data in Complex Systems: Challenges and Opportunities, Studies in Big Data. Springer-Verlag GmbH. doi:10.1007/978-3-319-11056-1

Hulme, G. V. (2011). *Cloud passsge aims to ease cloud server security management.* Available from http://www.csoonline.com/artical/658121/cloudpassage-aims-to-ease-cloud-server-security-management

Jansen, W., & Grance, T. (2011). Guidelines on security and privacy in public cloud computing. *NIST special publication, 800*, 144.

Jansen, W. A. (2011, January). Cloud hooks: Security and privacy issues in cloud computing. In *System Sciences (HICSS), 44th Hawaii International Conference* (pp. 1-10). IEEE.

Jasti, A., Shah, P., Nagaraj, R., & Pendse, R. (2010, October). Security in multi-tenancy cloud. In *Security Technology (ICCST), IEEE International Carnahan Conference* (pp. 35-41). IEEE. 10.1109/CCST.2010.5678682

Kandukuri, B. R., Paturi, V. R., & Rakshit, A. (2009, September). Cloud security issues. In *Services Computing, 2009. SCC'09. IEEE International Conference* (pp. 517-520). IEEE.

Kaur, R., & Pateriya, P. K. (2013, March). A Study on Security Requirements in Different Cloud Frameworks. International Journal of Soft Computing and Engineering, 3(1), 133-136.

Ko, S. R., & Lee, S. S. (2013). *Cloud computing vulnerability incidents: A statistical overview*. Cloud Security Alliance.

Kumar, P., & Kumar, L. (2013). Security Threats to Cloud Computing. *International Journal of IT, Engineering and Applied Sciences Research, 2*(1).

Langheinrich, M. (2001, January). Privacy by design—principles of privacy-aware ubiquitous systems. In Ubicomp: Ubiquitous Computing (pp. 273-291). Springer Berlin Heidelberg.

Laplante, P. A., Zhang, J., & Voas, J. (2008). What's in a Name? Distinguishing between SaaS and SOA. *IT Professional, 10*(3), 46–50. doi:10.1109/MITP.2008.60

Linthicum. (2009). *Approaching SaaS Integration with Data Integration Best Practices and Technology*. White Paper.

Lockhart, H., Andersen, S., Bohren, J., Sverdlov, Y., Hondo, M., Maruyama, H., . . . Prafullchandra, H. (2006). Web services federation language (WS-federation) version 1.1. *International Business Machines Corp.* Retrived from http://www-128. ibm. com/developerworks/library/specification/ws-fed/(17.08. 2007)

Luo, S., Lin, Z., Chen, X., Yang, Z., & Chen, J. (2011, December). Virtualization security for cloud computing service. In *Cloud and Service Computing (CSC), International Conference* (pp. 174-179). IEEE.

Ma, D. (2012, March). *Cryptographic Approach for Delegation and Authorization in Cloud Computing*. Presentation at *NSF Workshop on Security for Cloud Computing*.

Mahmood, Z. (2011, September). Data location and security issues in cloud computing. In *Emerging Intelligent Data and Web Technologies (EIDWT), International Conference* (pp. 49-54). IEEE. 10.1109/EIDWT.2011.16

Nahrstedt, K., & Campbell, R. (2012). *Security for Cloud Computing. In A Report: Directorate for Computer and Information Science and Engineering* (pp. 1–19). CISE.

Naruchitparames, J., & Giine, M. H. (2011, July). Enhancing data privacy and integrity in the cloud. In *High Performance Computing and Simulation (HPCS), 2011 International Conference* (pp. 427-434). IEEE. 10.1109/HPCSim.2011.5999856

Oleshchuk, V. A., & Koien, G. M. (2011, February). Security and Privacy in the Cloud-A Long-Term View. In *Security and Privacy in the Cloud-A Long-Term View. In: 2nd International Conference on Wireless Communications, Vehicular Technology, Information Theory and Aerospace and Electronic Systems Technology (Wireless VITAE), WIRELESS VITAE'11* (pp. 1-5). 10.1109/WIRELESSVITAE.2011.5940876

Paul Meinl. (2014, February). Software as a Service – *Correct Conclusion of Contracts 2nd enhanced edition*. Retrived from http://saas.clusterwien.at/fsDownload/SaaS_Guide_EN_07.pdf?forumid=342&v=1&id=5562745

Pearson, S. (2013). Privacy, security and trust in cloud computing. In *Privacy and Security for Cloud Computing* (pp. 3–42). Springer London. doi:10.1007/978-1-4471-4189-1_1

Popovic, K., & Hocenski, Z. (2010, May). Cloud computing security issues and challenges. In *MIPRO, 2010 proceedings of the 33rd international convention* (pp. 344-349). IEEE.

Prince, B. (2009). *IBM Discovers Encryption Scheme That Could Improve Cloud Security, Spam Filtering*. E-Week.com.

Privacy, and Compliance in Microsoft Azure. (2014, February). Retrieved from http://azure.microsoft.com/en-in/support/trust-center/security/

Purushothama, B. R., & Amberker, B. B. (2012, December). Efficient Query Processing on Outsourced Encrypted Data in Cloud with Privacy Preservation. In *Cloud and Services Computing (ISCOS), 2012 International Symposium* (pp. 88-95). IEEE. 10.1109/ISCOS.2012.16

Rai, R., Sahoo, G., & Mehfuz, S. (2013). Securing software as a service model of cloud computing: Issues and solutions. International Journal on Cloud Computing: Services and Architecture, 3(4).

Rane, P. (2010). Securing SaaS Applications: A Cloud Security Perspective for Application Providers. *Information Security Management Handbook, 5.*

Rimal, B. P., Choi, E., & Lumb, I. (2009, August). A taxonomy and survey of cloud computing systems. In *INC, IMS and IDC, 2009. NCM'09. Fifth International Joint Conference* (pp. 44-51). IEEE. 10.1109/NCM.2009.218

Roy, P., Goswami, S., Chakraborty, S., Azar, A. T., & Dey, N. (2014). Image Segmentation Using Rough Set Theory: A Review. *International Journal of Rough Sets and Data Analysis*, *1*(2), 62–74. doi:10.4018/ijrsda.2014070105

Sabahi, F. (2011, May). Cloud computing security threats and responses. In *Communication Software and Networks (ICCSN), 2011 IEEE 3rd International Conference* (pp. 245-249). IEEE. 10.1109/ICCSN.2011.6014715

Sahai, A., & Waters, B. (2005). Fuzzy identity-based encryption. In *Advances in Cryptology–EUROCRYPT* (pp. 457–473). Springer Berlin Heidelberg.

Sahoo, G., & Mehfuz, S. (2013). Securing Software as a Service Model of Cloud Computing: Issues and Solutions. *International Journal on Cloud Computing: Services & Architecture*, *3*(4).

Securty and Privecy Overview. (2014, January). Retrived from https://trust.salesforce.com/

Shah, A. (2008). Kernel-based Virtualization with KVM. *Linux Magazine*, *86*, 37–39.

Sharma, D. H., Dhote, C. A., & Potey, M. M. (2013). Security-as-a-Service from Clouds: A Comprehensive Analysis. *International Journal of Computers and Applications*, *67*(3), 15–18. doi:10.5120/11374-6642

Somorovsky, J., Heiderich, M., Jensen, M., Schwenk, J., Gruschka, N., & Lo Iacono, L. (2011, October). All your clouds are belong to us: security analysis of cloud management interfaces. In *Proceedings of the 3rd ACM workshop on Cloud computing security workshop* (pp. 3-14). ACM. 10.1145/2046660.2046664

Subashini, S., & Kavitha, V. (2011). A survey on security issues in service delivery models of cloud computing. *Journal of Network and Computer Applications*, *34*(1), 1–11. doi:10.1016/j.jnca.2010.07.006

Sun, W., Zhang, K., Chen, S. K., Zhang, X., & Liang, H. (2007). Software as a service: An integration perspective. *Service-oriented computing–ICSOC*, 558-569.

Sun, W., Zhang, X., Guo, C. J., Sun, P., & Su, H. (2008, September). Software as a service: Configuration and customization perspectives. In Congress on Services Part II, 2008. SERVICES-2 (pp. 18-25). IEEE.

Takabi, H., Joshi, J. B., & Ahn, G. J. (2010, July). Securecloud: Towards a comprehensive security framework for cloud computing environments. In Computer Software and Applications Conference Workshops (COMPSACW), 34th Annual (pp. 393-398). IEEE.

Tang, Y., Lee, P. P., Lui, J. C., & Perlman, R. (2010). FADE: Secure overlay cloud storage with file assured deletion. In Security and Privacy in Communication Networks (pp. 380-397). Springer Berlin Heidelberg.

Tiwari, P. K., & Mishra, B. (2012). Cloud Computing Security Issues, Challenges and Solution. *International Journal of Emerging Technology and Advanced Engineering*, *2*(8), 306–310.

Tran, H. V. (2013, June). Data Management Challenges in Cloud Computing. In *Computational Science and Its Applications (ICCSA), 2013. 13th International Conference on* (pp. 19-27). IEEE. 10.1109/ICCSA.2013.14

Xen Architcture Overview. (2008, February). Xen Wiki Whitepaper.

Yang, K., & Jia, X. (2014). Expressive, efficient, and revocable data access control for multi-authority cloud storage. *Parallel and Distributed Systems. IEEE Transactions*, *25*(7), 1735–1744.

Yang, K., Jia, X., Ren, K., Zhang, B., & Xie, R. (2013). Dac-macs: Effective data access control for multiauthority cloud storage systems. *Information Forensics and Security. IEEE Transactions*, *8*(11), 1790–1801.

Yong, P. E. N. G., Wei, Z. H. A. O., & Feng, X. I. E. (2012). Secure cloud storage based on cryptographic techniques. *Journal of China Universities of Posts and Telecommunications*, *19*, 182–189. doi:10.1016/S1005-8885(11)60424-X

Yu, Y., Ni, J., Au, M. H., Liu, H., Wang, H., & Xu, C. (2014). Improved security of a dynamic remote data possession checking protocol for cloud storage. *Expert Systems with Applications*, *41*(17), 7789–7796. doi:10.1016/j.eswa.2014.06.027

Zetta. (2008). *Zetta: Enterprise cloud storage on demand.* Retrieved from http://www.zetta.net/

Zhou, M., Zhang, R., Xie, W., Qian, W., & Zhou, A. (2010, November). Security and privacy in cloud computing: A survey. In *Semantics Knowledge and Grid (SKG), 2010 Sixth International Conference* (pp. 105-112). IEEE. 10.1109/SKG.2010.19

Zissis, D., & Lekkas, D. (2012). Addressing cloud computing security issues. *Future Generation Computer Systems*, *28*(3), 583–592. doi:10.1016/j.future.2010.12.006

Chapter 8
Semi-Automated Tool Support for Identification and Prioritization of Impacted Functions in Software Systems

Chetna Gupta
Jaypee Institute of Information Technology, India

Varun Gupta
Amity University, India

ABSTRACT

Change is an integral part of any software system. Predicting the impact of making changes through techniques of change impact analysis helps engineers identify and analyze those parts of the system that will be potentially affected by requested change(s). This chapter presents a semi-automated approach to (a) compute likelihood of impacted functions in a system through identification and analysis of functional dependencies between them and (b) assist software engineers in selective regression testing. This technique first classifies the impact set data into two categories based on their type of impact propagation. Next it performs prioritization on classified data to rank functions into higher and lower levels according to the degree of impact they will make. This prediction will help in lowering maintenance cost and effort of software engineers. Thus, a software engineer can first run those test cases which cover segments with higher impacted priority to minimize regression test selection.

INTRODUCTION

Software maintenance act as a backbone in supporting change management issues for keeping up with the quality of the software. In this era of time-to-market industry, software evolution and its maintenance emerge as one of the challenging tasks of software engineering. Studies indicate that 90% (Seacord, Plakosh & Lewis, 2003) of all software development is maintenance and more than 50% of the total maintenance cost of software lies in rework i.e. in changing the software (Lee, Offutt & Alexander, 2000; Myres, 1979). Whenever a single part of software is changed, other parts of the software are pos-

DOI: 10.4018/978-1-5225-5951-1.ch008

sibly subject to change as well. Change impact analysis (CIA) helps software developers to analyze and identify artifacts that will be potentially affected by requested or proposed change(s). CIA often called as simply impact analysis (IA), particularly addresses these problems (Bohner & Arnold, 1996; Arnold & Bohner, 1993; Loyall, Mathisen & Satterthwaite, 1997; Pfleeger, 1998; Turver & Munro, 1994). According to IEEE Standard for Software Maintenance (ISO/IEC, 2006) impact analysis is defined as:

impact analysis: Identifies all system and software products that a change request affects and develops an estimate of the resources needed to accomplish the change. This includes determining the scope of the changes to plan and implement work, accurately estimating the resources needed to perform the work and analyzing the requested change's cost and benefits.

CIA is effective mechanism in reducing the amount of corrective maintenance task by developers by paying focused attention on predicting and maintaining resource estimates, task scheduling, assisting in re-designing of some modules and helping testers in focusing on their resources and other testing activities (Moreton, 1990). Thus predicting impact of incorporating changes is effective way of maintaining software quality. Software maintenance process can only be optimized if precise and unambiguous information is available about the potential ripple effects of a change to an existing system. Whenever a change is made to any part of software system, it is important to gain complete understanding of structure, behaviour and functionality of the system being modified. Making changes to the software without understanding and knowledge of the software component can produce disastrous effects (Bohner & Arnold, 1996) and can lead to degraded software. Bohner (2002) identifies three basic software change activities: understanding software with respect to the change, implementing the change, and retesting the newly modified system.

Broadly impact analysis techniques can be partitioned into two categories: (a) traceability analysis and; (b) dependence analysis (Bohner, 2002). Traceability analysis refers to the ability to define, trace, and document relationships between software lifecycle objects such as requirements, design documents, source code, and other documentation (Cimitile, Fasolino & Visaggio, 1999). Impact analysis techniques based on dependence analysis attempt to assess the effects of change on semantic dependences between program entities, typically by identifying the syntactic dependences (chains of data and control dependences) that may signal the presence of such semantic dependences (Podgurski & Clarke, 1990). If done in a proactive manner, i.e. before the changes are made it can be helpful in predicting the effects of the proposed changes in terms of their affect on the overall system and the corresponding costs while simultaneously providing the maintainer the option to select various alternatives. On the other hand, if applied after modifications, it can help in reducing the risks associated with releasing modified software by alerting engineers to potentially affected program components.

This paper presents an approach to prioritize program segments particularly functional segments within impact set computed using functional call graph to assist regression testing. The proposed technique will be helpful in saving maintenance cost and effort by allocating higher priority to those segments which are impacted more than other segments within the impacted set. Thus a software engineer can first run those test cases which cover segments with higher impacted priority to minimize regression test selection. Most of the research in this area has focused mainly on computing impact sets whereas the proposed technique will first predict impact set and later classify and prioritize the most impacted segments from others within the impact set. The main contribution of this work is as follows:

1. A tool support for computation of impacted segments using functional call graph.
2. Categorization of impact set based on their type of impact propagation.
3. Prioritization of impact segments into higher and lower levels based on categorization results and their impact propagation.
4. Minimization of test cases during regression test selection resulting in less maintenance cost.

These contributions are helpful in estimating the impact of such changes thereby aiding in selecting useful test cases for regression testing. The remainder of this paper is organized as follows: next section discusses related work followed by proposed approach, demonstration with the example to generate results followed by conclusion.

BACKGROUND

When planning modifications, impact analysis is an effective mechanism which helps maintainers to predict effects and costs of planned changes. Impact analysis during software evolution is a labor intensive task and several authors in last few decades have proposed various techniques/approaches/methods/tool to support for CIA. Decades of research have shown that static analysis based techniques can safely estimate the impact of changes but their conservative assumptions often result in impact sets that include most of the software and in some cases it include more than 90% of the program (Apiwattanapong & Harrold, 2003). Such impact sets make the results of the impact analysis almost useless for other software-engineering tasks. For example, regression testing techniques that use impact analysis to identify which parts of the program to retest after a change would have to retest most of the program. This is because static impact analysis requires access to source code and thus the calculation of impact set is based on a static program profile synthesized with assumptions of possible system behaviours. Hence, this type of analysis technique is either imprecise, unsafe, or both, and tends to overestimate the effects of changes. Therefore, researchers have shifted their focus and have investigated and defined techniques that rely on dynamic, rather than static, information about program behaviour (Law & Rothermel, 2003; Breech, Danalis, Shindo & Pollock, 2004; Law & Rothermel, 2006). Dependency based analysis techniques (Lee, Offutt & Alexander, 2000; Arnold & Bohner, 1993; Loyall, Mathisen & Satterthwaite, 1997; Pfleeger, 1998; Ryder & Tip, 2001; Briand, Labiche & O'Sullivan, 2003; Weiser, 1981; Agrawal & Horgan, 1990; Horwitz, Reps & Binkley, 1990; Kamkar, 1995; Korel & Laski, 1990; Li & Offutt, 1996) works on the principle of analyzing program syntax for semantic dependencies among program entities such as linkages between parts, variables, logic, modules; thereafter it calculates impact sets according to dependencies. A brief summary of a comparison between these techniques is provided in (Law & Rothermel, 2003).

Dynamic analysis gathers impact sets by analyzing program behaviours for a specific set of executions (at least one of the considered program executions) during run time (i.e. data is obtained from executing a program). Dynamic IA does not require access to source code or linking process. Instrumentation and calculation of dynamic IA cause overhead in both time and space and results produced are much more precise than those produced by static impact analysis techniques. Various techniques mentioned in literature are based on collecting impact sets of dynamic program behavior. Law and Rothermel (2003) performs at the method level, based on whole path profiling (Larus, 1999). It produces traces of procedure names, function returns and program exits in the order in which they occur in multiple executions. Law & Rothermel (2006) is an extension of (Law & Rothermel, 2003) that allows an algorithm to collect data incrementally so that it becomes cost effective to re-compute (i.e. re-collect) data required for newer versions of the software. Apiwattanapong & Harrold (2003) uses light weight instrumentation and collects coverage information of methods per executions. It also works at the method level, but uses coverage, rather than trace, information to compute impact sets. The coverage information for each execution is stored in a bit vector that contains one bit per method in the program. The comparison of (Law & Rothermel, 2003; Apiwattanapong & Harrold, 2003) is given in (Orso, Apiwattanapong, Law, Rothermel, & Harrold, 2004). The generic approach of (Apiwattanapong, Orso & Harrold 2005) is based on an EA (execute after) relation for efficiently collecting and analyzing collected information dynamically. Huang& Song (2006) calculates impact sets by finding methods that are executed after methods in a changed set and includes them in the impact set. Sherriff and Williams (2008) proposes a methodology for determining the impact of new system modification(s) by analyzing software change records through singular value decomposition. It generates clusters of files that historically tend to change together to address faults and failures found in code base. Ceccarelli, Cerulo, Canfora, and Penta (2010) proposes a novel change impact analysis method based on the idea of mutual relationships between software objects that can be inferred using a statistical learning approach. Gupta, Singh and Chauhan (2010) proposes an approach to estimate impact analysis by classifying change based on type of change classification e.g. (a) nature and; (b) extent of change propagation dynamically. Gupta, Singh and Chauhan (2010) computes the impact sets by analyzing dependency tractability relations with other connected method components. Gupta, Srivastav and Gupta (2015) presents an approach that helps in minimizing software maintenance task by computing the difference between impact sets for actual logic change or just name change. Cai and Santelices (2015) presents a multiple levels cost-effective tradeoffs for dependence based dynamic analysis framework for providing flexible options to meet various cost-effectiveness needs for impact analysis. On the other hand if applied online, it can calculate impact sets concurrently with program execution (Breech, Tegtmeyer & Pollock, 2005). Online dynamic impact analysis has the same goal as dynamic impact analysis, but online impact analysis is performed concurrently with program execution rather than calculating impact sets from executing a program. Gupta, Singh and Chauhan (2009) presents a technique based on analysis of program dependency in terms of variable definition and usage to generate impact sets. Gupta, Singh and Chauhan (2010) provides framework for collecting impact traces completely online and it also provides support for impact visualization for regression testing.

Researchers have also reported that CIA can also be applied in other areas including, process automation Borg, Gotel and Wnuk, (2013), in nuclear power Chou (2011), in the automotive industry Grimm (2003), and in the aerospace domain Kilpinen, Eckert and Clarkson (2009). According to cross domain survey conducted in De la Vara, Borg, Wnuk, and Moonen (2016), CIA still lacks tool support in industry. Borg, De la Vara and Wnuk (2016) presents a discussion on study conducted on importance of CIA where authors have presented mapping of connotation versus importance to incorporate the view of researchers on CIA.

PROPOSED APPROACH

This paper presents an approach that will help in minimizing the cost related to software maintenance by allowing software engineer to select selective test cases for regression testing. This selective regression testing is performed on impact sets generated by analyzing (a) the functional dependences between them and (b) on results of classification obtained. Depending upon the type of classification different priority values are assigned to various functions. The software engineer can then analyze generated rank values and can perform selective regression testing. The whole process of impact set generation and priority computation as shown in Figure 1.

The process starts with generation of functional call graph from the source code. Functional call graph is a set of nodes and arrows where nodes represent functions extracted from the program and arrows points to control flow (relationship or dependency) among the functional nodes. Next stage is to analyze the execution traces of all functional calls generated during execution for n execution for a software program separately. This information is maintained in order to compute impact set. The change is already known and impact set is now computed for all those functions where the change has propagated in the program. The nature of this semi-automated tool support is sketched in Figure 2. As illustrated each execution is maintained as a separate data structure in maintainer after tracing all functional calls for a given execution. A counter to every functional node is also maintained which keeps the count of every node visited during execution. At the end of every execution this counter information is fetched and is stored in a database for further processing. As illustrated there are two databases, the first database consists of two parts, impact set and classifier which stores the generated impact set and corresponding results of classification into consume and produce category. This information is then processed by processor in ascending order of their functional call count fetched from maintainer. The priority is then assigned to these functional nodes for assisting better regression testing. The software engineer can at any time access this information for analysis.

Figure 1. Process flow diagram

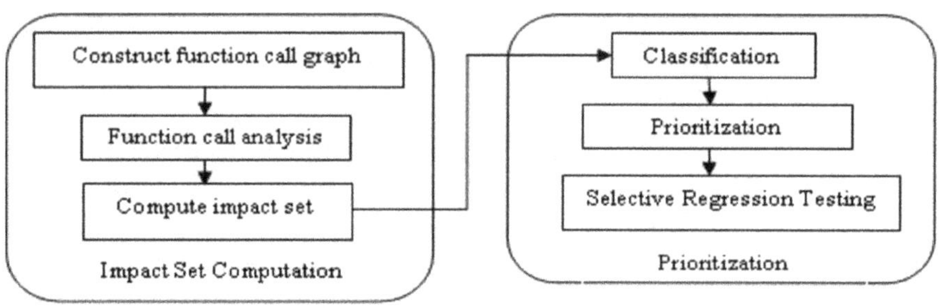

Figure 2. Tool Sketch for proposed technique

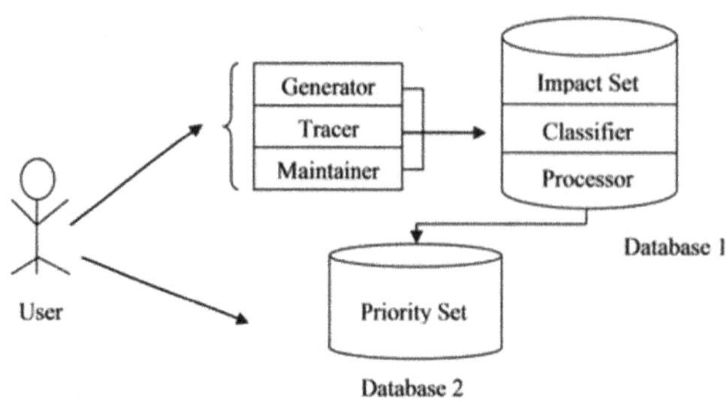

To illustrate the whole process of impact set computation, classification and priority estimation, let us consider a functional call graph in Figure 3. A function call graph lists how functions call each other. As shown in Figure 3 each node A to I represents a function for some program P and the arrows represents the flow of calls between them. M here represents the main function to which all calls return.

Figure 4 provides the list of execution calls (for this example we consider 4 such execution calls) made by methods to each other. Each call is traced as function calls (fc), returns (fr) and exit (fx). This information is stored in database as shown in Table 1 below.

Suppose that the change has occurred at node A. From Figure 4 and Table 1 it can be seen clearly that execution 1, 3 and 4 involve node A and thus impact analysis and regression testing is to be applied to execution 1, 3 and 4. Table 2 and Table 3 provide the call and return status of all executions along with functional count of each function for all executions respectively.

On careful observation of obtained results the impact set computed for further prioritization is given in Figure 5.

Figure 3. Functional Call Graph

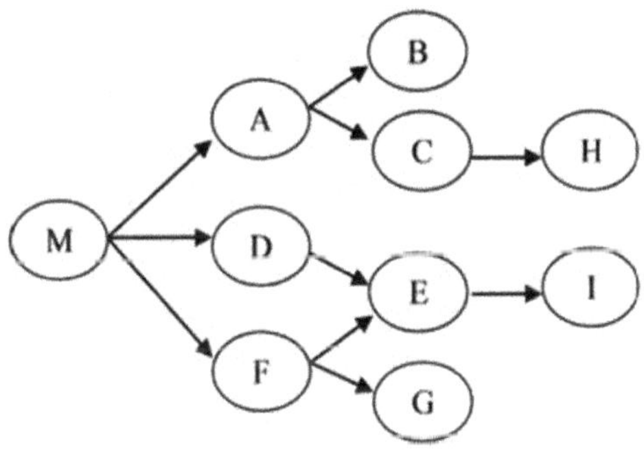

Figure 4. Execution calls

Execution 1: Mfc Dfc Dfr Afc A fr fx;
Execution 2: Mfc Ffc F fr Ffc Efc Ifc F fr Efr Gfc I fr G frf x;
Execution 3: Mfc Afc Bfc B fr A fr Afc Cfc C fr A frf x;
Execution 4: Mfc Afc Cfc Hfc H fr C fr A fr Ffc Efc E fr F fr Dfc D fr fx;

Table 1. Impact set and its corresponding count for each call, return and exit

Execution	M	A	B	C	D	E	F	G	H	I
Execution 1	fc_1fx_6	fc_4fr_6	--	--	fc_2fr_3	--	--	--	--	--
Execution 2	fc_1fx_{12}	--	--	--	--	fc_5fr_8	$fc_2fr_3fc_4fr_7$	fc_9fr_{11}	--	fc_6fr_{10}
Execution 3	fc_1fx_{10}	$fc_2fr_5fc_6fr_9$	fc_3fr_4	fc_7fr_8	--	--	--	--	--	--
Execution 4	fc_1fx_{14}	fc_3fr_7	--	fc_3fr_6	$fc_{12}fr_{13}$	fc_9fr_{10}	fc_8fr_{11}	--	fc_4fr_5	--

Table 2. Call and return status of all executions

Functions	Execution 1		Execution 2		Execution 3		Execution 4	
	Call	Return	Call	Return	Call	Return	Call	Return
A	1	1	0	0	2	2	1	1
B	0	0	0	0	1	1	0	0
C	0	0	0	0	1	1	1	1
D	1	1	0	0	0	0	1	1
E	0	0	1	1	0	0	1	1
F	0	0	2	2	0	0	0	1
G	0	0	1	1	0	0	1	0
H	0	0	0	0	0	0	0	1
I	0	0	1	1	0	0	0	0

Table 3. Functional count of each function for all executions

Function	Execution 1	Execution 2	Execution 3	Execution 4
A	2	0	4	2
B	0	0	2	0
C	0	0	2	2
D	2	0	0	2
E	0	2	0	2
F	0	4	0	2
G	0	2	0	0
H	0	0	0	2
I	0	0	0	0

Figure 5. Impact set

Impact set for execution 1: none (as no further call and return to main)
Impact set for execution 3: B, C
Impact set for execution 4: C, H, F, E, D

Classification Process

After analyzing all possible execution flows the list of all affected functions are maintained as set of impacted functions. To assist developers in selective regression testing these execution call information is further classified into (a) consume (b) produce category. If change after execution is not propagated further i.e. no function has called or returned into this changed function, after change execution then it will be categorized as consume. On the other hand if the change is propagated to other functions it will be categorized as produce.

Consume Node for Execution 1: None

Produced Node for Execution 1: None

Consume Node for Execution 3: B

Produced Node for Execution 3: C

Consume Node for Execution 4: None

Produced Node for Execution 4: C, H, F, E, D

Prioritization Process

After classification, now the next step is to assign priority to all impacted functions of impact set. Firstly estimated count of functional call, return and exit information corresponding to each function call for a given execution in ascending order is stored in a database. This will help us in identifying those functions which are called more number of times as compared to others. The functions having the maximum count are assigned higher priorities than those having lesser count. In other words, produced node will have higher priority than consumed node. A separate table is maintained to store this information. The table is then referenced by a software engineer to select test cases for regression testing. Thus nodes C, H, F, E, D are given higher priority than B. We formulate following rules for assigning priority.

Rules for priority:

Pmax = Max number of visits

Pmin = Min number of visits

From above example we get the following values:

P1 = F

P2 = C, E, D

P3 = H

Selective Regression Testing

While writing test cases for regression testing software engineer can use the above information for selective regression testing. From above discussed it is concluded that nodes that require regression testing are C, H, F, E, D and B and node F is to be tested before C, E, D followed by H and B. This information only indicates the priority to be considered while rerunning the test cases whereas all these nodes are suspicious and require re-testing.

We have implemented the whole process. Figure 6-11 depicts various stages of inputs and outputs generated by tool such as inputting the number total functional nodes and their corresponding dependency relations, inputting of execution traces, computation of impact set and its corresponding count for each call, return and exit flowed by computation of classification results into produce and consume category and finally priority estimation.

Figure 6. Snapshot to enter choice of size of dependency matrix

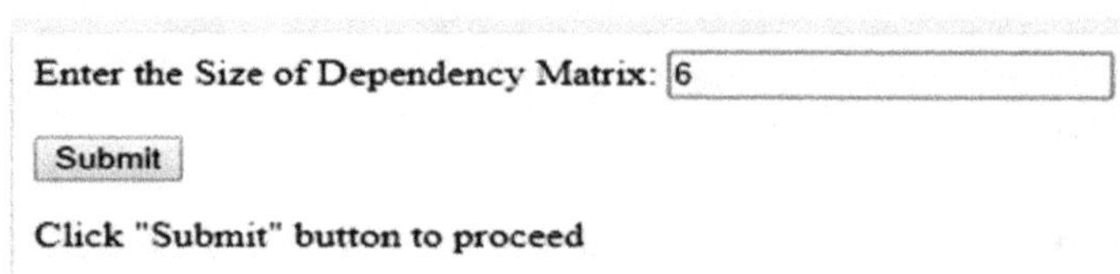

Figure 7. Functional dependency matrix

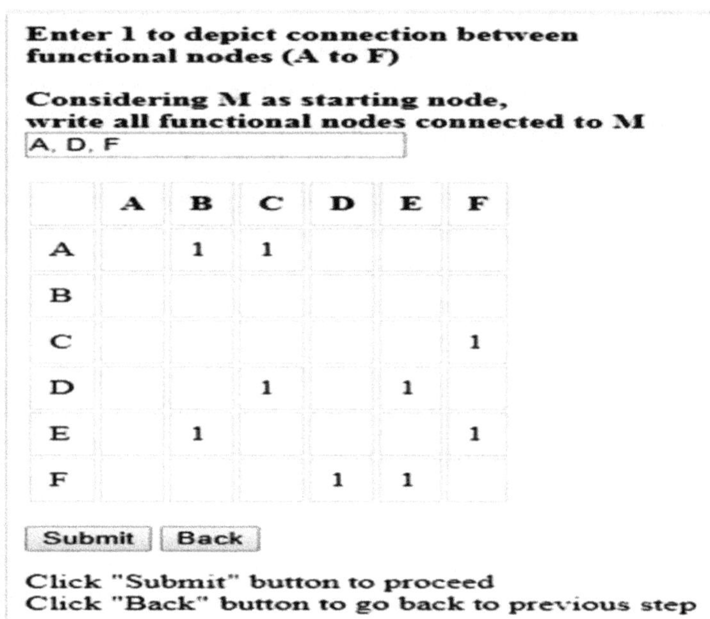

Figure 8. Snapshot to enter details of various execution calls

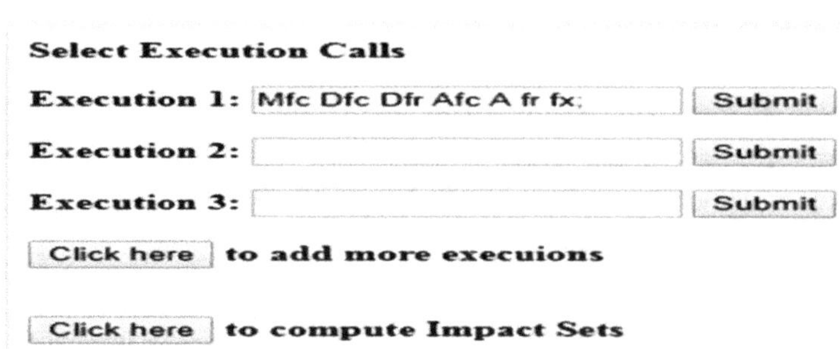

Figure 9. Snapshot to view result in desired format

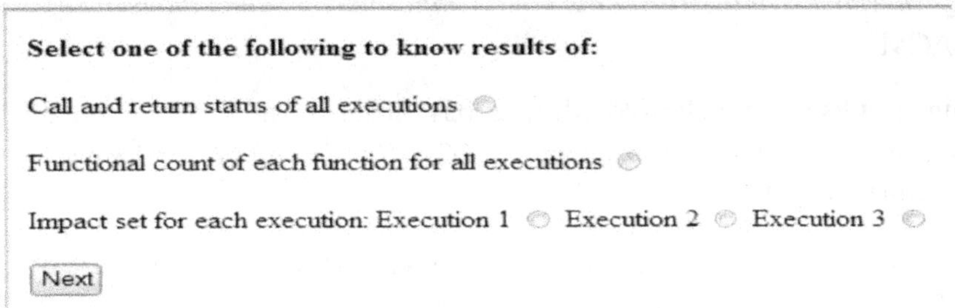

Figure 10. Snapshot to view result of classification

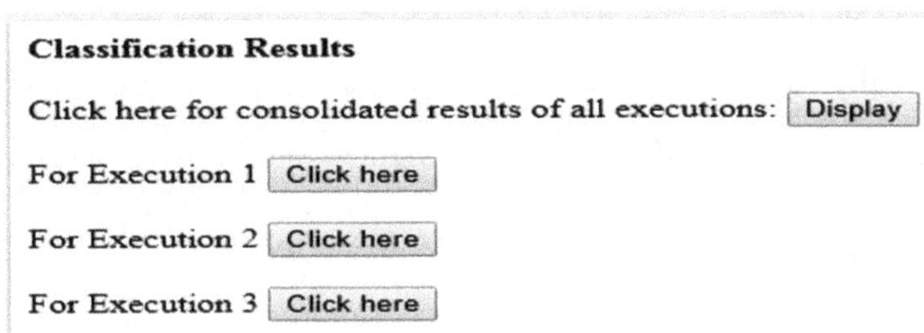

Figure 11. Snapshot to view prioritization results

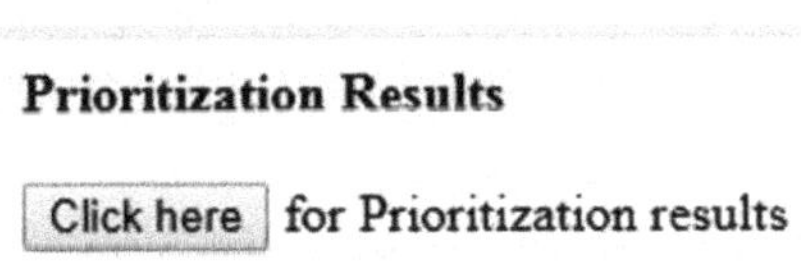

CONCLUSION

This paper presents a semi-automated tool support for identification and prioritization of impacted functions in software systems. The entire process is divided into three main parts namely; impact set computation, classification and prioritization of impacted set. It first identifies and analyses the functional dependencies and then computes impacted segments using functional call graph. This impact set is then classified into two types depending upon their type of impact propagation. Later these classification results are prioritized into impact segments of higher and lower levels. Based on this data a software engineer can perform selective regression testing covering first all those test cases which cover segments with higher impacted priority.

REFERENCES

Agrawal, H., & Horgan, J. (1990). Dynamic program slicing. *In Proceedings of SIGPLAN '90 Conference on Programming Language Design and Implementation* (pp.246-256). ACM.

Apiwattanapong, O. T., & Harrold, M. J. (2003). Leveraging field data for impact analysis and regression testing. In *Proceedings of the ACM SIGSOFT Symposium on Foundations of Software Engineering* (pp.128-137). ACM.

Apiwattanapong, T., Orso, A., & Harrold, M. J. (2005). Efficient and Precise Dynamic Impact analysis using Execute-After Sequences. In *Proceeding of ACM- International Conference on Software Engineering* (pp. 432-441). ACM.

Arnold, R. S., & Bohner, S. A. (1993). Impact analysis - towards a framework for comparison. In *Proceedings of IEEE International Conference on Software Maintenance*, (pp.292-301). IEEE. 10.1109/ICSM.1993.366933

Bohner, S. A. (2002). Software change impacts - an evolving perspective. In *Proceedings of IEEE International Conference on Software Maintenance* (pp. 263-272). IEEE. 10.1109/ICSM.2002.1167777

Bohner, S. A., & Arnold, R. S. (1996). *Software Change Impact Analysis*. IEEE Computer Society Press.

Borg, M., De la Vara, J., & Wnuk, K. (2016). Practitioners' perspectives on change impact analysis for safety-critical software—a preliminary analysis. *Proceedings of 5th International Workshop Next Generation Syst. Assurance Approaches Safety-Critical Syst.*, 346–358.

Borg, M., Gotel, O., & Wnuk, K. (2013). Enabling traceability reuse for impact analyses: A feasibility study in a safety context. *Proc. 7th International Workshop Traceability Emerging Forms Software Engineering*, 72–78. 10.1109/TEFSE.2013.6620158

Breech, B., Danalis, A., Shindo, S., & Pollock, L. (2004). Online impact analysis via dynamic compilation Technology. In *Proceedings of the International Conference of Software Maintenance* (pp. 453-457). Academic Press. 10.1109/ICSM.2004.1357834

Breech, B., Tegtmeyer, M., & Pollock, L. (2005). A Comparison of Online and Dynamic Impact Analysis Algorithms. In *Proceedings of Ninth European Conference on Software Maintenance and Reengineering* (pp. 143-152). Academic Press. 10.1109/CSMR.2005.1

Briand, L. C., Labiche, Y., & O'Sullivan, L. (2003). Impact Analysis and Change Management of UML Models. In *Proceedings of the International Conference on Software Maintenance* (pp. 256-265). Academic Press. 10.1109/ICSM.2003.1235428

Cai, H., & Santelices, R. (2015). A framework for cost-effective dependence-based dynamic impact analysis. *IEEE 22nd International Conference on Software Analysis, Evolution and Reengineering (SANER),* 231-240. 10.1109/SANER.2015.7081833

Ceccarelli, M., Cerulo, L., Canfora, G., & Penta, M. D. (2010). An Eclectic Approach for Change Impact Analysis. In *Proceedings of International Conference on Software Engineering* (pp. 163-166). Academic Press. 10.1145/1810295.1810320

Chou, I. (2011). Secure software configuration management processes for nuclear safety software development environment. *Ann. Nuclear Energy, 38*(10), 2174–2179.

Cimitile, A., Fasolino, A. R., & Visaggio, G. (1999). A software model for impact analysis: a validation experiment. In *Proceedings of the IEEE Sixth Working Conference on Reverse Engineering* (pp.212-222). IEEE. 10.1109/WCRE.1999.806962

De la Vara, J., Borg, M., Wnuk, K., & Moonen, L. (2016). An industrial survey of safety evidence change impact analysis practice. *IEEE Transactions on Software Engineering*, *42*(12), 1095–1117. doi:10.1109/TSE.2016.2553032

Grimm, K. (2003). Software technology in an automotive company: Major challenges. *Proceedings of 25th International Conference Software. Engineering,* 498–503.

Gupta, C., Singh, Y., & Chauhan, D.S. (2009). An Efficient Dynamic Impact Analysis using Definition and Usage Information. *International Journal of Digital Content Technology and its Applications*, *3*(4), 112-115.

Gupta, C., Singh, Y., & Chauhan, D. S. (2010). A Dynamic Approach to Estimate Change Impact using Type of Change Propagation. *International Journal of Information Processing Systems*, *6*(4), 597–608. doi:10.3745/JIPS.2010.6.4.597

Gupta, C., Singh, Y., & Chauhan, D. S. (2010). Dependency based Process Model for Impact Analysis: A Requirement Engineering Perspective. *International Journal of Computers and Applications*, *6*(6), 28–33. doi:10.5120/1079-1410

Gupta, C., Singh, Y., Chauhan, D. S., & Srivastav, M. (2010). DU-Regs: Online Dynamic Approach to Visualize Impact Analysis for Regression Testing. *International Journal of Computers and Applications*, *1*(19), 8–11. doi:10.5120/411-609

Gupta, C., Srivastav, M., & Gupta, V. (2015). Software Change Impact Analysis: An Approach to Differentiate Type of Change to Minimize Regression Test Selection. *Special Issue on Software Architecture, Evaluation and Testing for Emerging Paradigms. International Journal of Computer Applications in Technology*, *51*(4), 1–10. doi:10.1504/IJCAT.2015.070498

Horwitz, S., Reps, T., & Binkley, D. (1990). Interprocedural Slicing Using Dependence Graphs. *ACM Transactions on Programming Languages and Systems*, *12*(1), 27–60. doi:10.1145/77606.77608

Huang, L., & Song, Y. T. (2006). Dynamic Impact Analysis Using Execution Profile Tracing. In *Proceedings of the Fourth International Conference on Software Engineering Research, Management and Applications* (pp. 237-243). Academic Press. 10.1109/SERA.2006.30

ISO/IEC. (2006). *International Standard - ISO/IEC 14764, IEEE Std 14764-2006, IEEE Standard for Software Maintenance*. IEEE Computer Society.

Kamkar, M. (1995). An Overview and Comparative Classification of Program Slicing Techniques. *Journal of Systems and Software*, *31*(3), 197–214. doi:10.1016/0164-1212(94)00099-9

Kilpinen, M., Eckert, C., & Clarkson, P. (2009). Assessing impact analysis practice to improve change management capability. *Proceeding of 17th International Conference Engineering Discipline*, 205–216.

Korel, B., & Laski, J. (1990). Dynamic slicing in computer programs. *Journal of Systems and Software, 13*(3), 187–195. doi:10.1016/0164-1212(90)90094-3

Larus, J. (1999). Whole Program Paths. In Proceedings of Programming Language Design and Implementation SIGPLAN (pp. 1-11). ACM.

Law, J., & Rothermel, G. (2003). Incremental dynamic impact analysis for evolving software systems. In *Proceedings of the International Symposium on Software Reliability Engineering* (pp. 430-441). Academic Press. 10.1109/ISSRE.2003.1251064

Law, J., & Rothermel, G. (2003). Whole program path-based dynamic impact analysis. In *Proceedings of the International Conference on Software Engineering* (pp. 308-318). Academic Press.

Lee, M., Offutt, A. J., & Alexander, R. T. (2000). Algorithmic Analysis of the Impacts of Changes to Object oriented Software. In Proceedings of The Technology of Object-Oriented Languages and Systems (pp.61-70). Academic Press.

Li, L., & Offutt, A. J. (1996). Algorithmic analysis of the impact of changes to object-oriented software. In *Proceedings of IEEE International Conference on Software Maintenance* (pp.171-184). IEEE. 10.1109/ICSM.1996.565004

Loyall, J. P., Mathisen, S. A., & Satterthwaite, C. P. (1997). Impact analysis and change management for avionics software. In *Proceedings of IEEE National Aerospace and Electronics Conference* (pp.740-747). IEEE. 10.1109/NAECON.1997.622723

Moreton, R. (1990). A process model for software maintenance. *Journal of Information Technology, 5*(2), 100–104. doi:10.1057/jit.1990.19

Myres, G. J. (1979). *Art of Software Testing*. New York: John Wiley & Sons.

Orso, A., Apiwattanapong, T., Law, J., Rothermel, G., & Harrold, M. J. (2004). An empirical comparison of dynamic impact analysis algorithms. In *Proceedings of the International Conference on Software Engineering* (pp. 491-500). Academic Press. 10.1109/ICSE.2004.1317471

Pfleeger, S. L. (1998). *Software Engineering: Theory and Practice*. Englewood Cliffs, NJ: Prentice Hall.

Podgurski, A., & Clarke, L. A. (1990). A formal model of program dependences and its implications for software testing, debugging, and maintenance. *IEEE Transactions on Software Engineering, 16*(9), 965–979. doi:10.1109/32.58784

Ryder, B. G., & Tip, F. (2001). Change impact analysis for object oriented programs. In *Proceedings of the ACM Workshop on Program Analysis for Software Tools and Engineering* (pp. 46-53). ACM. 10.1145/379605.379661

Seacord, R. C., Plakosh, D., & Lewis, G. A. (2003). *Modernizing Legacy Systems: Modernizing Legacy Systems: Software Technologies, Engineering Processes, and Business Practices*. Addison-Wesley.

Sherriff, M., & Williams, L. (2008). Empirical Software Change Impact Analysis using Singular Value Decomposition. In *Proceedings of 1st IEEE International Conference on Software Testing, Verification, and Validation* (pp. 268-277). IEEE. 10.1109/ICST.2008.25

Turver, R. J., & Munro, M. (1994). Early impact analysis technique for software maintenance. *Journal of Software Maintenance: Research and Practice*, *6*(1), 35–52. doi:10.1002mr.4360060104

Weiser, M. (1981). Program slicing. In *Proceedings of 5th IEEE International Conference on Software Engineering* (pp.439-49). IEEE.

Chapter 9
A Hybrid Approach of Regression-Testing-Based Requirement Prioritization of Web Applications

Varun Gupta
Amity University, India

ABSTRACT

The chapter gives a hybrid approach for prioritization of software requirements by considering regression testing efforts as one of the criteria. The prioritization is performed against the prioritized decision aspects. The chapter gives software engineers a technique for prioritizing both requirements and aspects. This technique considers regression testing effort as one of important parameters during requirement prioritization thereby reducing the regression testing effort. This technique is applied on live web application of "virtual classroom" and the results were promising. The technique resulted in decision aspect prioritization thereby leading to effective requirement prioritization and lower regression testing efforts.

1. INTRODUCTION

Requirement prioritization is an essential activity to be carried out by any software firm to deliver software in increments, with each increment implementing highest priority requirements. Priority of requirement is calculated against different conflicting prioritization dimensions by diverse stakeholders. One of the problem as reported by (L. Lehtola, M. Kauppinen, S. Kujala, 2004) was that the requirement priorities allotted by developers were on the basis of tactic knowledge or feelings about requirement priorities. This guess work about priorities allotments is done by involving few developers. There is no method for involving diverse stakeholders, no means for selection of decision aspects and no technique for negotiations for resolving conflicts. Decision aspects need to be prioritized and employed for ranking of large requirements by involving diverse stakeholders. Decision aspect prioritization and requirement prioritization are similar to the maximum extend since both require stakeholders to give their preferences that needs to be effectively mapped to single priority value. Involving a large number of stakeholders

DOI: 10.4018/978-1-5225-5951-1.ch009

would mean success but on the other hand, requires decision makers to resolve conflicts that arise due to different viewpoints. Requirement prioritization also suffers from the scalability problem which means that with the increase in the number of requirements, a prioritization process goes more complex and requires large effort. Scalability problem is less in the case of the decision aspect prioritization process.

The diverse stakeholders differ in their viewpoints, expertise, knowledge, locations, cultures etc. Managing these stakeholders separated by global distances is quite challenging. Involving a large number of stakeholders during the prioritization process is no doubt essential for project success but the ranking of aspects and thousands of requirements by involving these diverse stakeholders suffers from various problems like need for an effective method for mapping values given by "N" stakeholders into single priority value for each requirement, the need for effective resolution of conflicts due to conflicting viewpoints etc. Above mentioned requirements requires effective communication and collaboration between globally distributed stakeholders which is very difficult to maintain due to the existence of global distances. (Gupta & Srivastav, 2011) proposed a web based framework that overcomes global distances that exists in globally distributed environments, thereby making global developments easier.

During the release of next increment, software engineer's implements delayed requirements along with changed ones. This means that source code will be altered in terms of addition of new statements or modules and changes in existing ones. This task of alteration of source code will result in ripple effects in other already tested parts of source code. New test cases executing newly added code along with existing ones executing already tested code needs to be executed for this new version. In earlier work (V. Gupta, D.S Chauhan, K. Dutta, 2012a) it was found that decision aspect prioritization effects requirement prioritization process which further affects regression testing activity to the maximum extend. Improvements in the requirement prioritization process can lower regression testing activity to the maximum possible extend and improves software success rate. The requirement prioritization process can be modified in such a manner that requirement priorities are computed by analysing the effect of regression testing. This happens by implementing sets of dependent requirements independent of delayed set of dependent lower priority requirements. This gives stakeholders highest priority requirements along with extra dependent ones. This lowers regression testing effort since the addition of new requirement will have no ripple effect since it is independent of already implemented ones.

Advancements in the area of web technology had credited users and business firms with complex web applications that are complete replacement for desktop applications. This had been credited to transition from web 1.0 to web 2.0 technologies. Emerging of languages like Hypertext pre-processor (PHP), Active server pages (ASP), AJAX, XML etc had made it possible to have read/write/execute web that can solve almost any complex possible as was previously solved using desktop applications implemented in Java, C, C++ etc. Web 2.0 applications provide effective communication and collaboration. Social media like Facebook, twitter, web services like yahoo, gmail and tools like wiki etc are web 2.0 applications.

Since Web 2.0 Application exhibits similar functionality and complexity as traditional desktop applications, their incremental development will require effective decision aspect selection and sound requirement prioritization. Delivery of next increments will require effective regression testing since the addition of new requirements or changing existing ones will require changes in web pages or addition of new pages which may affect other pages and hence web application functionality.

One of the major issues with the development of Web 2.0 Application is a security issue as web applications will be assessed worldwide through the internet and hence susceptible to various types of illegal activities like hacking, data theft, denial of service attacks, malware attacks etc. For example Aurora attack against Google (http://en.wikipedia.org/wiki/Operation_Aurora), and Koobface malware attacks on Facebook and twitter like social networking websites (http://en.wikipedia.org/wiki/Koobface) are fine examples of the need for effective implementation of security requirements.

Software developments are distributed these days, with a multicultural and national sites collaborating with each other. In earlier work it was suggested to employ web application to overcome global distances that exists between development sites. Web 2.0 tools had been employed in software engineering and it had been found that Web 2.0 tools improve quality of software. Such tools provide the best platform for collaboration which is the backbone of nay collocated or global software developments. This security becomes a key issue while prioritizing requirements of Web 2.0 Applications. Since such tools have to be developed for mass markets, has to be accessed worldwide 24x7 and affects software success rate.

Web 2.0 Application incremental delivery is very complex as compared to that of desktop application due to the challenges involved in aspect selection, requirement prioritization, effective selection of security requirements etc. Web 2.0 tools employed in any software development activity that is developed due to incorrect prioritization process will degrade development quality of other software's also.

The proposed approach overcomes various problems related to prioritization process. It employs effective negotiation based voting system that aims to resolve conflicts between the views of diverse stakeholders during aspect selection and requirement prioritization. This technique is well automated to maximum extend with the help of Web 2.0 tools implemented in PHP, JavaScript and HTML with MYSQL as back end.

In this paper, methods and negotiation are considered separate activities. Further, decision aspect prioritization is considered as one of the activities involved in process of requirement prioritization and hence is not considered as separate distinct process.

2. RELATED WORK

Web 2.0 tools had reported to be employed in software development industries. (Black, 2010) has reported that use of social media in globally distributed software development had improved communication thereby improving global software developments. (Black & Jacobs, 2010) had reported that use of social media tools i.e. Web 2.0 tools improve quality of software. Decker et. al, (2007) suggested a way of supporting the active participation of software stakeholders in requirement engineering using asynchronous collaborative platform provided by wiki. Such proposal is applied to various projects and experienced problems are reported and solutions are suggested.

As per best of authors knowledge there is no paper related to requirement prioritization that prioritizes requirement of web applications and automated it using Web 2.0 Applications.

Various researchers had proposed numerous techniques that prioritize requirements against predefined prioritization dimensions. (Saaty, 1980) proposed Analytical Hierarchical Process (AHP) that compares requirements pair wise against aspect under consideration. This technique uses preference scale of 1 to 9, where 1 indicates equal value and 9 indicates the extreme value. This technique allows stakeholders to give any value of 1 to 9 for each requirement, when compared pair wise in with each other. One of advantage of this approach is ability to analyse the accuracy of pairwise comparisons since it allows calculation of accuracy metric i.e. Consistency ratio whose value should be less than 0.10 for accurate comparison. Main disadvantage related to this approach is scalability problems i.e. With an increase in the number of requirements; number of comparisons grows in large number. (Karlsson & Ryan, 1997) proposed Cost-Value Approach for Prioritizing Requirements that performs prioritization according to its value to customers and implementation cost. In this method customer allots preference for customer value aspect while developer's gives preferences against development cost aspect. AHP is employed as the prioritization method for prioritization of every requirement against each aspect separately. Results of the two prioritization were analyzed graphically, where the x-axis represented the cost to develop the requirement and the y-axis represented the customer value that the requirement is expected to yield to its customers. Analysis of this cost-value graph will allow decision maker to select those requirements that have the best value-cost ratio. Azar et. al, (2007) proposed Value-Oriented requirements Prioritization techniques where requirements are prioritized by giving preferences against business values and different risks as prioritization aspects. Preferences allotted against these aspects were filled in prioritization matrix and simply added to get priority scores. In this technique, risk is measured on negative scale. This technique is evaluated on a case study of online banking system by (Danesh, S. Mortazavi, S. Danesh, 2009). Wiegers (1999a; 1999b) proposed another requirement prioritization technique and employed it on Chemical Tracking System. This method is based on the idea that the priority of a requirement can be calculated by dividing the value of a requirement by the sum of the costs and technical risks associated with its implementation. The value of a requirement is dependent both on the value the requirement provides to the customer(s) and the penalty incurred if the requirement is absent. Machine learning based techniques for performing requirement prioritization had been proposed by Avesani et. al, (2004; 2005). (Gupta & Srivastav, 2011a) proposed requirement prioritization method that is based on multiple stakeholder preferences. This method categorized cost and penalty as reducing factors while Benefits, value, Risk, Time to Market and dependencies as prioritization required factors. Prioritization matrix is created, stakeholders give their preference for each requirement against each aspect. For each requirement, for each aspect, each value allotted is multiplied by the weight of stakeholder and added with product of weight of next stakeholder with value allotted by this stakeholder in order to map "N" values of weighted stakeholder into a single value. Finally to get the priority value of each requirement, prioritization required factors are divided by reducing factors. This technique is well supported by a web based framework that makes this technique automated as well as useful in globally distributed environments.

Some researchers had employed proposed prioritization techniques on various case studies/projects to illustrate working of prioritization techniques. Value oriented prioritization technique was evaluated on a case study of online banking system by ((Danesh, S. Mortazavi, S. Danesh, 2009).

Requirement prioritization by analyzing the effects of regression testing has not been reported by any researcher till date.

Various researchers even had evaluated some of the prioritization techniques against some criteria or parameters of prime interest in order to compare these techniques and to find out a better one (M. Hasan, A. Mahmood, M. Alam, M. Hasan, F. Rahman, 2010; M. Iqbal, A. Zaidi, S. Murtaza, 2009; Ma 2009; Ahl 2005; J. Karlsson, C. Wohlin and B. Regnell, 1998; L. Karlsson, P. Berander, B. Regnell, C. Wohlin, 2004; Perini, A. Susi, F. Ricca, C. Bazzanella, 2007; M. Aasem, M. R., A. Jaffar, 2010; Z. Racheva, M. D., L. Buglione, 2008). But unfortunately the criteria are not well established or pre-defined and agreed by all so that the results of the comparison will be different when evaluated against different criteria's. (J. Karlsson, C. Wohlin and B. Regnell, 1998) founded the AHP technique to be most valuable than Hierarchical AHP, Spanning tree, more valuable than Binary search and priority groups.

Some researchers had reported various problems that arise in requirement prioritization. (L. Lehtola, M. Kauppinen, S. Kujala, 2004) reported various problems related to the requirement prioritization activity. The authors reported the challenges on the basis of the prioritization practices carried out by two software development organizations. Main findings include: (1) Organizations does not have agreement on a single meaning of term priority and meanings of prioritization scales; (2) Few developers carry out the prioritization process by making a guess by employing tacit knowledge or feelings about requirement importance; (3) Requirements are prioritized in multiple phases, a process that is an iterative process; (4) Developers are not aware of the fact that which requirement is most important to its customer; (5) Requirement prioritization must employ multiple factors or aspects, a process that doesn't have any acceptable practice in the software industry.

As per knowledge of authors, very few authors had presented methods related to decision aspect prioritization and hence requirement prioritization against prioritized aspects. Ribeiro et. al, (2010) proposed Hybrid assessment method for requirement prioritization. This method is composed of total 5 activities divided into two phases. First phase deals with prioritization of decision aspects while the second phase generates ranking of each alternative or requirements against each criteria.

Effects of decision aspect on requirement prioritization and hence on regression testing is reported by Gupta et. al, (2012b). The authors employed two versions of the same regression testing technique proposed in (Gupta & Chauhan, 2011c), on an unsuccessful and successful project of the payroll management system. Decision aspect prioritization technique as proposed in Gupta et. al, (2012a) and requirement prioritization process as proposed in (V. Gupta, M. Srivastav, 2011a) was employed, as a result of which regression testing technique as proposed in (Gupta & Chauhan, 2011c) was executed on evolving software with minimum test cases and high fault detection capability. One of the main findings of the authors was that improvement of decision aspect prioritization will improve requirement prioritization processes that will further reduce regression testing effort by reducing the number of test cases without reducing fault detection capability (V. Gupta, M. Srivastav, 2012b).

Some researchers had reported that use of web 2.0 applications improves software quality.

The summary of various aspect and requirement prioritization techniques are given in Table 1.

3. REQUIREMENT PRIORITIZATION OF WEB 2.0 APPLICATIONS

3.1 Prioritization Dimensions/Aspects

Requirements are prioritized according to various aspects. Aspects are the properties or attributes of requirements that are used for prioritization (P. Berander, A. Andrews, 2005). During prioritization priorities are assigned to each requirement. A priority reflects the importance of requirements to various stakeholders along with several dimensions like business value, personal preferences etc. Once allotted, the result is the proper ranking of requirements and hence the implementations order of the requirements in incremental software development. While doing prioritization it is important to take into account multiple aspects. Single aspects are no doubt easier to handle but multiple aspects increases the success chances of the software if chosen correctly. These multiple aspects are often dependent on each other so it may be possible that if the analyst is interested in prioritization of requirements against one aspect, they have to compromise the second dependent one. For example if "Time to Market" is taken as one aspect and "Cost" as the other then it may be possible that high costs are associated with requirements that needs to be timely delivered in the market due to investing of more resource. Thus the aspects are not independent but interact with each other (G. Ruhe, A. Eberlein, D. Pfahl, 2003). So if the first aspect needs to be chosen, the second one needs to be abolished or compromised and vice versa. But if one is able to chose maximum aspects and do prioritization against them the probability of success of the project will increase.

Table 1. Idea of Techniques of Decision aspect and requirement prioritization.

S.No	Technique Name	Presented Approach	Strengths	Limitations
1.	Hybrid assessment method	This method is composed of total 5 activities divided into two phases. First phase deals with prioritization of decision aspects while the second phase generates ranking of each alternative or requirements against each criteria.	**1.** Performs Decision Aspect Prioritization. **2.** Employs one of the most accurate technique called AHP. **3.** Automated with help of a prototype tool.	**1.** Nothing can be commented regarding prioritization of web 2.0 application requirements. **2.** Does not support prioritization by employing diverse stakeholders during prioritization. No method is given for mapping multiple values into single preference value. **3.** Does not support negotiations for conflict resolution. **4.** Does not handle the security issue of web 2.0 applications. **5.** Since the tool is not web based, so the technique will not allow stakeholders to be distributed. **6.** Suffers from scalability problem.
2.	Web based tool supported requirement prioritization: based on multiple stakeholder preferences	This method combines the preference given by diverse stakeholders in single priority value. For such mapping, weights of stakeholders and preferences given are combined and priority is computed by dividing prioritization required factors by reducing factors. This technique is well supported with web based framework.	**1.** It takes into consideration diverse stakeholders. **2.** It can be employed in globally distributed environments. **3.** Do not suffer due to scalability problem.	**1.** Provides support for negotiation for minimum extend. **2.** Suffers from voting problems i.e. It is possible that some important requirements like security ones are given less rank due to ignorance my many stakeholders. **3.** Less suitable for web 2.0 applications. **4.** Do not provide a method for decision aspect prioritization. **5.** It employs predefined aspects.
3.	Value-Oriented requirements Prioritization.	Requirements are prioritized by giving preferences against business values and different risks as prioritization aspects. Preferences allotted against these aspects were filled in prioritization matrix and simply added to get priority scores. In this technique, risk is measured on negative scale.	**1.** Easier to automate and implement. **2.** The technique is automated since the author had given excel sheet that automatically computes priority. **3.** Results are based on a ratio scale. **4.** Less influenced to scalability problems.	**1.** Nothing can be commented regarding prioritization of web 2.0 application requirements. **2.** Does not support prioritization by employing diverse stakeholders during prioritization. No method is given for mapping multiple values into single preference value. **3.** Does not support negotiations for conflict resolution. **4.** Does not handle the security issue of web 2.0 applications. **5.** This technique will not allow stakeholders to be distributed. **6.** Suffers from voting problems. **7.** It employs predefined aspects.
4.	Wiegers Method	This method calculates a priority by dividing requirement value by the sum of the costs and technical risks associated with its implementation. The value of a requirement is computed by combining two values i.e. The value that the requirement provides to the customer(s) and the penalty incurred if the requirement is not implemented.	**1.** Easier to automate and implement. **2.** The technique is automated since the author had given excel sheet that automatically computes priority. **3.** Results are based on a ratio scale.	Same as above
5.	Cost-Value Approach	Performs prioritization according to its value to customers and implementation cost. In this method customer allots preference for customer value aspect while developer's gives preferences against development cost aspect. After AHP is employed, cost-value graph for each requirement is analyzed to select requirements with best ratio of cost-value.	**1.** It is effective since employs AHP for prioritization. **2.** Analysis of graphs makes selections of requirements against these two objectives quite simple. **3.** Easier to automate and implement for small organizations. **4.** Results are based on a ratio scale.	Same as above.
6.	Analytical Hierarchical Process (AHP)	This method compares requirements pair wise against aspect under consideration. It allows to analyze the accuracy of pairwise comparisons since it allows calculation of accuracy metric i.e. Consistency ratio whose value should be less than 0.10 for accurate comparison.	**1.** It is a most trustworthy technique. **2.** Very suitable for software with small set of requirements. **3.** Pairwise comparison reduces the judgmental errors. **4.** Results are based on the ratio scale which makes it possible to know that how much one requirement is more important than other. **5.** Technique Supports prioritization against multiple aspects.	Same as above except the last one. This technique suffers very badly from scalability problem.

3.2 Requirement Prioritization

Advancements in web technology had added complexity to web applications. For example Web 1.0 Application were simple static sites displaying the same content every time the page is visited until and unless it is modified by changing the HTML code by the site administrator. As discussed earlier that Web 2.0 Applications exhibits similar functionality and complexity as traditional desktop applications thus their incremental development will require effective decision aspect selection and sound requirement prioritization. Delivery of next increments will require effective regression testing since the addition of new requirements or changing existing ones will require changes in web pages or addition of new pages which may affect other pages and hence web application functionality. World wise accessibility requires a requirement prioritization process to be more sensitive towards security requirements. Desktop applications are secured at various levels like guarding of rooms where computers running application is placed (Human level), password protection done by the operating system (OS level), password protection security offered by the application itself and data protection offered by database management system or via some encryption technique etc. On other hand web applications accessible world wise 24x7 requires data protection at the database level or by using some encryption technique, password protection offered by application through login web pages etc. Others two levels like OS level and human level are missing in such applications. Thus requirement prioritization should take care of various security requirements to be implemented at two levels so as to make them more secure as compared to desktop applications.

The software implements thousands of requirements of diverse stakeholders. Software development takes place in resource constrained globally distributed environments. Implementation of thousands of requirements will be categorized by large delivery time. Delivery of complex software in "one go" is not a good option since it will keep its customers waiting for years to use this software and in the meantime their requirements might evolve. A good solution to this problem is the delivery of software in small increments with each increment implementing some of the requirements from evolving large set of requirements. The requirements of each increment should be of high priority. Implementation of the highest priority requirements earlier will credit customers with the most favored functionality earlier. Requirement selection is implemented with the help of complex activity called Requirement Prioritization. Requirement priorities are decided by considering a combination of various and perhaps conflicting aspects or prioritization dimensions.

Numerous prioritization techniques are there proposed in the literature like Analytical Hierarchical Process (AHP) (Saaty 1980), Software Engineering Risk Understanding and Management (SERUM) (Geer 1999), Value Oriented Prioritization (VOP) (Jim Azar 2007), Hundred dollar method, Wieger's method (Wiegers 1999b) etc. Major limitations of these techniques lies in using predefined aspects, scalability problem, and working in globally distributed environments, inability to provide a method for performing prioritization of decision aspects, no support for negotiation. These techniques suffer from problems related to voting by employing diverse stakeholders during prioritization. Without negotiation, voting or allotment of preferences leads to win-lose situation which gets translated to lose-lose situation in the near future (In, 1998). These techniques do not handle prioritization of web applications effectively.

During the prioritization main challenge is to make a balance between trade-off that exists between the various aspects and to map the interests of various stakeholders into a single value that represents the priority of individual requirement (Gupta & Srivastav, 2011a). While performing this mapping operation we must take into account the different priorities of various stakeholders (Gupta & Srivastav, 2011a). Ruhe et al., (2002) have defined the requirement prioritization problem as:

The challenge is to select the 'right' requirements out of a given superset of candidate requirements so that all the different key interests, technical constraints and preferences of the critical stakeholders are fulfilled and the overall business value of the product is maximized.

Karlsson et al., (1997) has cited requirement prioritization in following words, How do you select a subset of the customer's requirements and still produce a system that meets their needs.

3.3 Prioritization Based on Regression Testing

In earlier work it was found that effectiveness of requirement prioritization method results in a reduction in the regression testing effort of evolving software systems. Effectiveness of requirement prioritization is directly related to the effectiveness of decision aspects prioritization since the requirements are prioritized against multiple conflicting aspects. Faulty decision aspect selection results in faulty implementation order and thereby resulting in poor software quality and enhanced change requests. Due to these change requests, lots of changes have to be made in existing code which might result in ripple effects. This leads to the activity of selection of test cases to execute the modified code along with those parts that might have been affected as a result of induced changes (regression testing). The situation is more complicated when a software testing team has to test newly implemented requirements and again take into consideration ripple effects. Requirements prioritization techniques can be augmented with activities aiming at selection of those requirements that reduces regression testing effort. To achieve this objective, all the highest priority ones and those dependent on them can be part of implementation set. Implementation of dependent ones will ensure that requirements implemented in the next release are almost independent of earlier implemented ones, thereby relieving testers of testing those code parts that are affected by the changes. Thus only newly added requirements need to be tested. The requirement prioritization process must take into consideration the impact of the requirement on regression testing by considering dependencies with highest priority requirements.

3.4 Model of Requirement Prioritization

Requirement prioritization is not a simple activity but composed of several complex and risky activities. If any of such activity goes wrong then project success might be at stake. A software project is developed for a mass market i.e. Development is market driven, due to this it is essential that diverse stakeholders be involved in the process of deciding implementation order. Involving a large diverse group of stakeholders is quite challenging since it calls for proper negotiations between the conflicting viewpoints and method needs to be devised that maps values given by these stakeholders for individual requirements against aspect under consideration. The requirement prioritization method should be flexible enough to support scalability i.e. Large number of requirements and ability to perform requirement prioritization against any number of prioritized aspects rather than predefined ones. Major of prioritization techniques performs prioritization by pre-declaring decision aspects thereby leading to the situation where decision maker does not have the option to perform prioritization against any new aspect. Decision aspects do not remain same for every project. In fact they vary with feedback from previously implemented projects, targeted audience, objectives of the organization etc. Thus the requirement prioritization process must give ability to software decision makers to prioritize aspects and then rank requirements against such prioritized prioritization dimensions. Software success depends on accurate ranking of requirements against accurately ranked dimensions. Accuracy of ranking comes from the various contributors like:

1. Accurate identification of stakeholders with sound domain knowledge, expertise etc.
2. Accurate identification of decision aspects since aspects will be selected from this set only.
3. Effective rank computation formulae (mapping formulae) that distinguishes the value or importance of votes of diverse stakeholders. This happens because votes of all stakeholders are not equally important with respect to aspect under consideration.
4. Effective negotiation to resolve conflicts among diverse stakeholders and those that exists between conflicting aspects. This happens because it is possible that rank computed might create win-lose situation where some stakeholders are simply neglected. Further it is required for the requirement to score well against some aspects which happens to be conflicting i.e. One can be achieved at the cost of another (trade off).

Thus the requirement prioritization process is composed of five activities divided into two phases, with a first phase resulting in prioritization of decision aspects and second phase resulting in requirement prioritization against aspects selected in the first phase. The model of requirement prioritization is basically an iterative process with some of the activities being repeated until some consensus is built and conflicts are resolved. Higher level model of requirement prioritization is given in Figure 1.

In fact, each of the activities involved in requirement prioritization is very complex. Decision aspect involves several iterative steps involving allotment of preference values, negotiations to resolve conflicts, re allotment of preferences and final priority calculation. Requirement prioritization involves iterative methods of allotment of preferences to each requirement against each prioritized aspect, mapping values into single preference values and repeating this process in case of conflict resolution by using negotiations.

Figure 1. Model of Requirement Prioritization.

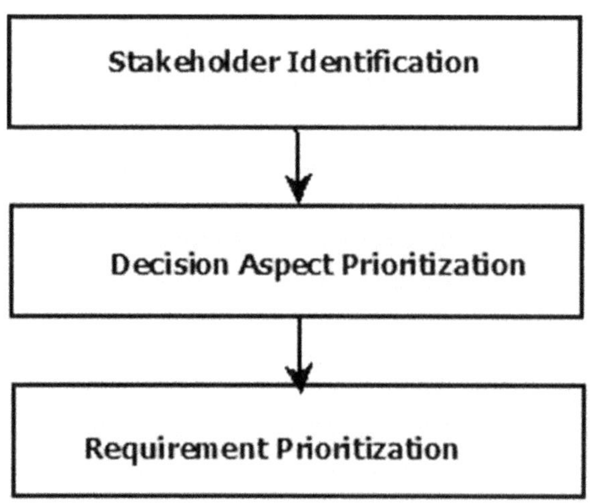

4. PROPOSED APPROACH

The proposed requirement prioritization approach builds the implementation order of Web 2.0 Application by first prioritizing decision aspects against which requirements are allotted rankings by diverse stakeholders and taking into consideration impacts on regression testing. Efforts are to select all requirements dependent on highest priority ones such that implementation of new requirements in later increments does not have any impacts on already implemented and tested code. This enables a software tester to test newly implemented requirements without considering ripple effects since new requirements will be independent of already implemented ones. Requirements of secured Web 2.0 Applications must be prioritized using the sound prioritization technique (both requirement and decision aspect), ranking being agreed by all stakeholders.

The proposed technique is hybrid approach i.e. Combination of negotiations and methods. Negotiations are the activities that result in prioritization by resolving conflicts while methods are prioritization techniques that generate ranks by allotment of preferences (Perini, A. Susi, F. Ricca, C. Bazzanella, 2009; Berander, 2005).

Since implementation of large security requirements is not feasible, so software decision makers must perform security requirement prioritization independent of prioritization of other requirements. Stakeholders involved during the security requirement prioritization could be security analysts of organization, hackers, developers, requirement analysts, some other stakeholders with experience in security requirement prioritization of similar applications. There is no need for categorization of stakeholders involved in the security requirement prioritization into groups. Security requirements can be organized hierarchically with each level of the tree representing the implementation level of requirement. For example "adding a login page" requirement can be considered as lower level as compared with *requirement for entering password for performing any activity on web tool*. Later requirement provided security at higher levels as compared with former requirement. It's good to have both requirements but under constrained environment, implementation of second one will implement the lower level requirements. Table 2 gives method for security requirement prioritization.

Since major of requirement prioritization techniques employs predefined aspects against which requirements are prioritized, thus are not flexible enough to perform prioritization against any number of applications specific decision aspects. They lack in giving a method of selection of decision aspects. The proposed requirement prioritization technique is flexible enough to handle scalability of both aspects as well as of the requirements. It provides software decision makers a method for prioritization of decision aspects and finally to prioritize requirements by employing negotiations between conflicting stakeholders to build agreements. Proposed decision aspect technique employs one of the most trustworthy method of requirement prioritization i.e. Analytical Hierarchical Process (AHP). Although AHP suffers from

Table 2. Algorithm: Security Requirement Prioritization

Input: Security Requirements, Stakeholders. **Output:** Prioritized set security requirements.
1: Compare the set of security requirements pair wise to organize security requirements at different levels of hierarchy. Pairwise comparison will determine the influence level of each requirement. Implementation of requirement at higher level will provide better security to web application especially in resource constrained environment. **2:** Implement security requirements belonging to higher levels of hierarchical i.e. tree organization of software requirements.

scalability problem but for decision aspects (which is fewer in number as compared to thousands of requirements of complex software), AHP is best suited along with proposed negotiation technique. AHP will help the software engineer to draw conclusion related to accuracy of decision methods they made during aspect prioritization by calculating Consistency Ratio (CR). This accuracy check will improve quality of aspect selection and finally of requirement ranking. AHP was also employed for aspect prioritization in (R. Ribeiro, A. Moreira, P. Broek, A. Pimentel, 2011).

Table 3 gives the algorithm for decision aspect prioritization. The requirement prioritization algorithm requires that stakeholders be organized into groups in accordance with their roles and allot preference values to each requirement against each aspect. Aspects are classified as either additive or reducing. Additive aspects are those whose higher value is required like customer value while reducing are those whose lower value is required like development cost. Dependent requirements are organized in groups called buckets. The priority of each requirement is calculated for each stakeholder and finally for each group. If the difference between the priority calculated for each requirement for each stakeholder group is greater than tolerance limit i.e. Threshold value, there is a need for negotiation between those groups that vary by huge amount in terms of allotted priority. This paper terms such groups as conflicting groups. Negotiations are done aspect wise. This happens because for some aspects, preference values given by some lower priority groups might wrongly alter requirement priority and vice versa. Such negotiations are required to create win-win situation and hence increasing acceptability of the project among its intended audience. After negotiation, conflicting preferences are altered by re allotment of values against conflicting aspects by conflicting groups and priority re computed. Finally priority given by each group for each requirement is simply added. Computed requirement priority is added with the priorities of dependent requirements and normalized by dividing the sum by the number of such dependent requirements to compute the priority of each bucket. The highest priority bucket is implemented till available resources permits. This Requirement prioritization algorithm is given in Table 4.

Table 3. Algorithm: Decision Aspects Prioritization

Input: Decision aspects, Stakeholder groups. **Output:** Prioritized set of decision aspects. Let $Priority_{(aspect(i)(j))}$ denotes the priority of aspect "i" calculated by applying AHP process on values allotted by stakeholder group "j".
1: Identify decision aspects set that need to be prioritized. For aspects identification experience from previously implemented similar projects can be quite helpful. **2:** For each stakeholder group: For each pair of aspects: Allot preference values. This fills cell of the 2-D array, the 2-D array is simply a decision matrix. Execute Analytical Hierarchical Process on 2-D decision matrix. AHP process ends after computation of Consistency Index (CI). Value of CI should be less than 0.10 since this value determines the error free state of the prioritization process. **[End level 2 inner for loop]** **[End level 1 outer for loop]** **3.** Combine the priorities given by each stakeholder group into single priority value for each aspect to determine final ranking of decision aspects. For each aspect: Compare the priority value allotted by each stakeholder group. If the difference between values allotted by stakeholder groups is above threshold, perform negotiation using the algorithm given in table 3 otherwise use equation 1 to compute final priority value of each aspect. $Priority_{(aspect(i)(j))} = \sum Priority_{(aspect(i)(j))}$ *(1)* for "i" aspects and "j" stakeholder groups. **[End level 1 outer for loop]** **4.** Select highest priority aspects. Execute requirement prioritization algorithm using the prioritized aspects list.

Table 4. Algorithm: Requirement Prioritization

Input: Requirements, Preference values, prioritized aspects.
Output: Prioritized set of requirements.

1: Arrange dependent requirements into bucket, each bucket houses dependent requirements. Two buckets houses an independent set of requirements.
2: Categorize prioritized decision aspects as either "Additive" or "Reducing" factors.
3: For each requirement:
    For each stakeholder group:
        For each aspect under consideration:
            Allot preference values.
        [End level 3 inner for loop]
        Add preference values allotted to set of additive aspect and set of reducing aspects.
        Calculate the priority of requirement for a particular group under consideration using equation 2.
        $\text{Priority}_{(group(i))}$ = (Additive value) / (Reducing value) *(2)*
        Normalize this value in accordance with a number of stakeholders in each group using equation 2.1.
        $\text{Priority}_{(group(i))} = (\text{Priority}_{(group(i))}) / N_{(i)}$ *(2.1)*
        Where $N_{(i)}$ is number of stakeholders belonging to the group "i".
    [End inner level 2 for loop]
    Compute difference of priority calculated for each stakeholder group using equation 3 and compare the value computed with a threshold value.
    $\text{Difference} = \text{Priority}_{(group(1))} - \text{Priority}_{(group(2))} - \ldots\ldots - \text{Priority}_{(group(n))}.$ *(3)*
    If (difference>=threshold) then perform negotiation using algorithm given in table 3 else compute final priority of requirement using equation 4.
    The priority $\text{requirement}_{(i)} = \sum \text{Priority}_{(group\ (i))}$ for all stakeholder groups. *(4)*
    [End level 1 outer for loop]
    The idea of threshold value computation for making a decision regarding the negotiations is also employed in (Gupta et. al, 2012a) but however the computational steps differ considerably.
4: Compute the priority of each bucket by using equation 5 and normalize it using equation 6.
    $\text{Priority}_{(bucket(i)} = \sum \text{Priority}_{\ requirement(i)}$ *(5)*
    for every requirement belonging to bucket "i".
    $\text{Final\_priority}_{(bucket(i)} = \text{Priority}_{(bucket(i)} / n$ *(6)*
    Where "n" is the total number of requirements belonging to bucket "I".
5: Implement the set of dependent requirements by selecting buckets of highest priority until resources are available.

The above algorithm aims to prioritize the buckets housing dependent requirements such that highest priority dependent requirements are implemented first. Data dependency is the criteria for the creation of the bucket. Suppose that a requirement R1 requires the requirement R2 to be executed first either for data or performing some operation of its part then it will be termed as data dependency. For example, to issue a book (R1), another code implementing the requirement of authenticating the member (R2), checking the member account details (R3) is to be executed. Thus R1 will be chosen for implementation if R2 and R3 will be selected. During the implementation of delayed requirements (which is independent of existing ones), the code will be simply added with existing one that will call the module implementing this requirement. This independent requirement will not affect the existing requirements code workings. Thus, no doubt, there will be control dependencies between all the requirements, considering data dependencies will ensure that during the implementation of new requirements, these new ones will be simply glued with existing code and there will not be any effect on existing code. Thus there will not be any ripple effects. The reachability of the existing code will be checked along with the test cases executing new requirements. This will ensure lots of savings in test cases since there is no need to test individual requirements and other affected code.

The idea of prioritization of buckets and requirements within the buckets is also employed in (M. Iqbal, A. Zaidi, S. Murtaza, 2010) but the process varies considerably. Iqbal et. al, (2010) had created a large number of bins containing few requirements. Each requirement is prioritized each bucket at a time. Finally each bucket is finally prioritized. Thus each prioritized bucket houses prioritized requirements. Iqbal et. al, (2010) has assumed that there are no dependencies between the requirements and no conflicts arise in pairwise comparisons by diverse stakeholders, the two things that are unlikely to occur in the requirement prioritization process. Key differences between the two approaches, one as proposed in Table 4 and another in (Iqbal, 2010) includes the following: (1) Different prioritization method. For example in letter technique AHP is used while in proposed method ratio scale normalization based method is proposed; (2) The proposed technique is employing negotiations while another doesn't stresses on it; (3) Criteria for creation of buckets are different. The proposed technique suggests implementing the requirements as agreed by all stakeholders after conflict resolution negotiations and all others dependent on them.

Table 5 gives an algorithm for performing conflict resolution that emerges between two stakeholder groups for each requirement. There are some decision aspects whose allotted preference value should be largely influenced by some stakeholder groups rather than all. This means that the votes of all stakeholders are not equally important for every aspect. For example, in case of customer value aspect, values allotted by software customers should get preference rather than all stakeholder groups. If the values of all stakeholders are considered equally important then the final ranking of requirements might be incorrect. A good solution for this conflict resolution is performing negotiation between these conflicting groups, reaching agreements and re allotment of values to conflicting aspects. This alters ranking of requirements and makes them correct. Table 5 gives negotiation algorithm.

5. WEB 2.0 TOOL

This technique is supported with prototype software web 2.0 tool implemented using PHP as a server side scripting language, JavaScript as client side scripting language, MYSQL as database technology. Static pages are implemented using HTML. PHP will make the page dynamic since it will enable the page to receive inputs from stakeholders, and depending on inputs received will display the new contents on a web page. These new contents are ranks computed for requirements or decision aspects. JavaScript performs some authentication at client side. In this software tool, JavaScript checks that whether the user has entered some preference or had entered nothing. It gives error message for "Nothing" and for entered value, it checks that whether it is drawn from a preference table or not. If it is not drawn from preference table, it issues an error message. This authentication is required for accuracy of the priori-

Table 5. Algorithm: Negotiation for Conflict Resolution

Input: Requirements, Preference values, prioritized aspects. **Output:** Prioritized set of requirements.
1: The system finds a conflicting group by combining those groups whose allotted preference values are almost similar, into similar group. Then from similar groups (one with highest value and others below threshold) it selects that group that has the highest value in a highest value group and other groups that has a lower one in lowest value groups. 2: Conflicting groups prepare points of agendas that represent their wishes, disagreements, alternatives or options etc. Thus they prepare points related to their feelings that why they thought some aspect or requirement as important or less important. 3: To resolve conflicts, Nominal Group Technique (NGT) can be employed. This technique is well suited for small group discussions and is free from one person dominating whole discussions since it supports discussions as a collaborative activity. NGT combines ideas generation followed by prioritization of the ideas using simple ranking methods like hundred dollar test etc.

tization process. Preference values allotted by stakeholders belonging to different groups against each aspect for each requirement are stored in the database. Also it stores decision aspect and preference values given to each aspect.

Step 1 and 2 of requirement prioritization needs to needs to be performed manually i.e. Arranging set of requirements in buckets and categorization of aspects as either additive or reducing. The working of the system is given with the help of an algorithm (Table 6).

6. CASE STUDY AND RESULTS

To illustrate the working or proposed technique, case study of "Virtual Classroom" is employed. This project was implemented as B. Tech final year major project. This project aims to provide e-learning, providing teachers, students and parents with online lecture notes, submission of assignments, evaluation of assignments, seminars etc, online uploading of papers, marks, grades, synchronous as well as synchronous face to face and textual chats. Face to face chats were no doubt an important requirement and got highest by all stakeholders but due to limited technical resources available, negotiations were made with stakeholders to lower its priority to delay its implementation. Software was implemented using PHP as a server-side scripting language; JavaScript as a client side scripting language; HTML for static pages and MYSQL as database. The deadline for the project was simply one semester i.e. Roughly 6 months. Various gathered requirements are given in Table 7.

Table 6. Algorithm: Working Algorithm of System

Input: Requirement set, Aspect set, and stakeholder group's details, Bucket details. **Output:** Prioritized set of requirements. System prioritizes decision aspects and then generates set of ranked requirements, ranked and collected under similar name or bucket on the basis of dependencies.
1: A developer enters Requirement set, Aspect set, and stakeholder group's details, bucket details. The developer is responsible for issuing usernames and passwords for each stakeholder. 2: Stakeholders give preferences for each pair of decision aspects. Such preference values are authenticated by JavaScript. 3: The system executes AHP process and calculates Consistency ratio and displays the message related to accuracy of prioritization techniques. Value of 0.10 or less is acceptable for consistency ratio. 4: The system adds the value computed using an AHP process for each stakeholder group for each aspect. For each aspect, values computed for each stakeholder group is compared with those given by other group for same aspect. 5: If the difference between the values is greater than threshold, the system allows stakeholders to perform negotiation and re-enter the preferences for conflicting aspect and conflicting groups. To do this, the system then combines those groups whose allotted values are almost similar. Then from similar groups (one with highest value and others below threshold) it selects that group that has the highest value in a highest value group and other groups that has a lower one in lowest value groups. Steps 2 and onwards steps are re executed. 6: Final priority values of each aspect are computed by using equation 1 i.e. Adding the values of each stakeholder group for each aspect. This completes decision aspect prioritization. 7: Each stakeholder gives his preference values against each aspect for each requirement. 8: Preferences for each aspect are added up group wise to generate single value. The system then computes priority value of requirement group wise. In other words if there are three groups then the requirement priority value is computed as set of three values. 9: These values are then compared with threshold value and for difference above threshold; system asks each conflicting groups to perform negotiation. The system finds a conflicting group by combining those groups whose allotted values are almost similar, into similar group. Then from similar groups (one with highest value and others below threshold) it selects that group that has the highest value in a highest value group and other groups that has a lower one in lowest value groups. The system then repeats all steps from 7 and onwards. 10: The system computes priority of requirement by adding each value belonging to the same requirement. 11: Final priority values of requirements are then used to compute the priority of each bucket, which then are stored and displayed as when required.

Table 7. Gathered Requirements of Virtual Classroom organized with Buckets.

S.No	Requirements (Bucket wise)
1. 2. 3. 4. 5.	**Bucket_1** The online test facility should be provided. Teachers should be able to specify negative marking for online tests. Teachers should be able to add questions and decide on objective or subjective nature of paper. Length for subjective and type single option or multiple options answer for objective questions be decided by the teacher. **Bucket_2** Software must allow students to submit assignments, seminar reports for evaluation. The teacher should be able evaluate submissions of students and store marks allotted to each submission for each student. The teacher should be able to upload lecture notes. **Bucket_3** Users must be able to post their problems to the systems that other users will respond by posting the solution. Users must be able to delete the problem posted by him earlier. Users be presented with a list of problems with last option be "others" so that problems can also be mailed to experts who had earlier registered in the system and solution in the form of suggested contact person be mailed. This helps to find solution in terms of contact details of experts if no one is able to find a solution. The system must be able to search specified problems posted in whole system by using the search box. Searching should be done using full string or partial tag string. Searching should allow the search process using different input parameters like problem name, domain, data of posting etc. **Bucket_4** Users must be able to chat with online users to communicate with them. Users must be able to send emails for communicating with other users using asynchronous communication mode. The system must be able to support web cam chat. The system must be able to support voice chat. **Bucket_5 (Security Requirements)** The teacher should be able to allot username and password for his students. Users must be allowed to register to the system without being waited for weeks for authentication details. Users must have to go through the authentication process so as to enhance security. The system must allow biometric authentication of the user. The system should ask for a password for each task to be performed using it.

Bucket numbered 5 will be prioritized separately for this web application while buckets numbered 1 to 4 will be subjected to negotiation based ranking prioritization by executing proposed technique. Three groups identified are Customers, Developers and Testers & Security personnel's. Testers and Security personnels were involved since prioritization is more concerned towards implementing highest priority requirements and simultaneously reducing regression testing effort and having no compromise with security. There were two customers, three developers and three belong to the last category.

Two Security requirements were selected for implementation since they belong to a higher level of hierarchy. Two requirements were *System should ask for a password for each task to be performed using it* and *Teacher should be able to allot username and password for his students.*

From the decision aspects selected from minor project of "E-booking system for tour 'N' Travels", stakeholder groups performed a selection of aspects. Two additional aspects were added to set to be prioritized i.e. Risks and Development complexity. Additions of these two aspects were because implementation of chats, submitting of lecture notes in various formats including mpeg etc, various searching operations were new to developing team. Finally selected aspects were Cost, Time, Business values, Risks and Development complexity

Each stakeholder group members allotted their preferences individually. Preferences are added for additive and reducing aspects respectively. The priority of each requirement is calculated by dividing the additive aspect total value by that of reducing aspects for each group. Thus for identified three groups, each requirement priority is partially computed into three values, each for one group. Finally for some requirements, value as computed for each group is compared with a threshold value of 10. This value is not fixed but varies in accordance with experience of organization and number of stakeholders involved. As a result of experience in the development of numerous web applications, the authors suggest to use lower value of 10 as threshold value. During such comparison need for negotiation was felt for various requirements especially web cam chat and voice chat. These requirements were given high values by customers while developers given them low values due to resource constraints. Normalized values for these requirements are given in Table 8.

Since fewer stakeholders were involved, so a threshold value of 05 was employed. After proper negotiations using NGT technique, conflicting stakeholder group's discussion made each group aware of each other views and developers were made aware of the importance of such requirements. Discussions lead to one important conclusion that work of customers can be carried out for time being without these requirements. They reached an agreement of implementing them in the next release as a result of which customer's re-allotted lower value thereby delaying the implementation of these requirements.

Finally priority values of each group for each requirement is added to compute the priority value that is added with other requirements priority and normalized by diving by total number of requirements belonging to particular bucket. Bucket 2 was of highest priority followed by Bucket 1, Bucket 3 and Bucket 4. Although Bucket 4 would be of highest priority than that Bucket 3 but after negotiation order was reversed. In first release bucket numbered 2 and 1 were implemented. This work was further carried out in next increment with the second increment implementing remaining buckets and delayed security requirements.

Addition of new requirements in second increment resulted in testing of newly added requirements without the need to take into consideration ripple effects. During second increment, developers performed only testing of new requirements rather than testing already tested code. One of the main findings of execution of the proposed technique on live system is that new decision aspects like risks and implementation complexity were selected and leads to successful delivery of first increment. More over effective negotiation during prioritization resulted in delaying of some of requirements that would have been impossible to get implemented although highest ranked due to wrong customer group preferences.

Table 8. Conflicts between allotted values

Requirement	Stakeholder Groups	Normalized Values (Before Negotiation)	Normalized Values (After Negotiation)
The system must be able to support web cam chat.	Customer	18	02
	Developers	03	03
	Testers & Security personnel's	03	03
The system must be able to support voice chat.	Customer	16	02
	Developers	03	03
	Testers & Security personnel's	03	03

7. CONCLUSION AND FUTURE WORK

The paper gives hybrid approach for prioritization of software requirements by considering regression testing efforts as one of the criteria. The prioritization is performed against the prioritized decision aspects. The paper gives software engineers a technique for prioritizing both requirements and aspects. This technique is applied on live web application of "Virtual Classroom". Some of the project success critical aspects like risks and development complexity were identified for this project that was not used in previous project of Tour "N" travel. Identification of these aspects might not have been possible if prioritization technique uses predefined aspects. Negotiations during prioritization made it possible to delay requirements related to web cam face to face chat and voice chat. These requirements were given high priority by customer groups but categorized by high implementation complexity and not possible to be implemented in short delivery time. Negotiations also made each group aware of the views of other groups. Implementation of security requirement ay highest hierarchical level leads to the delivery of a secured web application under resource constrained environment. This technique is suitable for application with thousands of requirements and suitable for geographically distributed environments. The main advantages of this technique are that it allows stakeholders to be distributed across cultural or organizational boundaries, overcomes problems related to involvement of diverse stakeholders, do not suffer from scalability problem, applicable for both Web 2.0 and desktop applications, provides a method for decision aspect selection and is automated to maximum extend.

As a future work, it is expected that this technique be applied in software development industries on complex software's and some manual activities like categorization of decision aspects, populating of buckets being done automatically. Case based reasoning can be employed to identify and categorize aspects, populating of buckets etc. This work is under progress. We do hope that the technique will be compared with existing methods to further improve the deficiencies.

REFERENCES

Aaqib Iqbal, F. M., Khan, S. & Khan, A. (2009). A Critical Analysis of Techniques for Requirement Prioritization and Open Research Issues. *International Journal of Reviews in Computing, 1,* 8-18.

Aasem, M. (2010). *Analysis and optimization of software requirements prioritization techniques.* IEEE.

Ahl, V. (2005). An experimental comparison of five prioritization methods - Investigating ease of use, accuracy and scalability. Department of Systems and Software Engineering, School of Engineering, Sweden, Blekinge Institute of Technology.

Berander, P., & Andrews, A. (2005). *Requirements prioritization. Engineering and Managing Software Requirements* (A. Aurum & C. Wohlin, Eds.). Berlin, Germany: Springer Verlag.

Danesh. (2009). Requirements prioritization in on-line banking systems: using value-oriented framework. *2009 International Conference on Computer Technology and Development.* 10.1109/ICCTD.2009.41

Geer, D., Bustard, D., & Sunazuka, T. (1999). Prioritization of system changes using cost-benefit and risks assessments. *Proceedings of Fourth IEEE International Symposium on Requirements Engineering*. 10.1109/ISRE.1999.777998

Gupta, V., & Chauhan, D. S. (2011c). Hybrid Regression Testing Technique: A Multi Layered Approach. In *IEEE annual Conference "INDICON"*. Hyderabad, India: IEEE. 10.1109/INDCON.2011.6139363

Gupta, V., Chauhan, D. S., & Dutta, K. (2012a). *Decision Aspect Prioritization Technique for Globally Distributed Developments: A Hybrid Approach*. Elsevier.

Gupta, V., Chauhan, D. S., & Dutta, K. (2012b). Impact Analysis of Requirement Prioritization on Regression Testing. *AWER Procedia Information Technology & Computer Science*, *2*, 379–383.

Gupta, V., & Srivastav, M. (2011a). Web based tool supported requirement prioritization: Based on multiple stakeholder preferences. *International Journal on Computer Engineering and Information Technology*, *25*(1), 12–19.

Gupta, V., & Srivastav, M. (2011b). *Overcoming Global Distance in Globally Distributed Developments by Web Based Framework*. CiiT International Journal of Software Engineering and Technology.

Hasan, M., Mahmood, A., Alam, M., Hasan, M., & Rahman, F. (2010). An Evaluation of Software Requirement Prioritization Techniques. *International Journal of Computer Science and Information Security*, *8*(9), 83–94.

In. (1998). *Conflict Identification and Resolution For Software Attribute Requirements*. Faculty of The Graduate School, University of Southern California.

Iqbal, M., Zaidi, A., & Murtaza, S. (2010). A New Requirement Prioritization Model for Market Driven Products Using Analytical Hierarchical Process. *2010 International Conference on Data Storage and Data Engineering*. 10.1109/DSDE.2010.49

Jim, A., Randy, S., & David, C. (2007). Value-Oriented requirements Prioritization in a Small Development Organization. *IEEE Software*, 32–37.

Karlsson, J., & Ryan, K. (1997). A Cost-Value Approach for Prioritizing Requirements. *IEEE Software*, *14*(Sept), 67–74. doi:10.1109/52.605933

Karlsson, J., Wohlin, C., & Regnell, B. (1998). An Evaluation of Methods for Prioritizing Software Requirements. *Journal of Information and Software Technology*, *39*(14-15), 939–947. doi:10.1016/S0950-5849(97)00053-0

Karlsson, L., Berander, P., Regnell, B., & Wohlin, C. (2004). Requirements Prioritisation: An Experiment on Exhaustive Pair-Wise Comparisons versus Planning Game Partitioning. *Proceedings of the 8th International Conference on Empirical Assessment in Software Engineering (EASE 2004)*. 10.1049/ic:20040407

Lehtola, Kauppinen, & Kujala. (2004). Requirements prioritization challenges in practice. *Product focused software process improvement*, 497-508.

Ma, Q. (2009). The Effectiveness of Requirements Prioritization Techniques for a Medium to Large Number of Requirements: A Systematic Literature Review. School of Computing and Mathematical Sciences, Auckland University of Technology.

Perini, Ricca, & Bazzanella. (2007). An empirical study to compare the accuracy of AHP and CBRanking techniques for requirements prioritization. *Fifth International Workshop on Comparative Evaluation in Requirements Engineering*, 23-35.

Perini, A., Ricca, F. F., & Susi, A. (2009). Tool Supported requirement prioritization: Comparing the AHP and CBRank methods. *Information and Software Technology*, *51*(6), 1021–1032. doi:10.1016/j.infsof.2008.12.001

Racheva & Buglione. (2008). *Supporting the Dynamic Reprioritization of Requirements in Agile Development of Software Products. 2nd International Workshop on Software Product Management (IWSPM)*, Barcelona, Spain.

Ribeiro, R., Moreira, A., Broek, P., & Pimentel, A. (2011). Hybrid assessment method for software engineering decisions. *Decision Support Systems*, *51*(1), 208–219. doi:10.1016/j.dss.2010.12.009

Ruhe, G., Eberlein, A., & Pfahl, D. (2003). Trade-off Analysis for Requirements Selection. *International Journal of Software Engineering and Knowledge Engineering*, *13*(4), 345–366. doi:10.1142/S0218194003001378

Ruhe, G., Eberlein, A., & Pfahl, D. (2002). Quantitative WinWin – A New Method for Decision Support in Requirements Negotiation. *Proceedings of the 14th International Conference on Software Engineering and Knowledge Engineering.*

Saaty, T. (1980). *The Analytic Hierarchy Process: Planning, Priority Setting, Resource Allocation.* McGraw-Hill.

Wiegers. (1999b). First Things First: Prioritizing Requirements. *Software Development, 7*(9).

Wiegers, K. (1999a). *Software Requirements.* Redmond, WA: Microsoft Press.

Chapter 10
Intelligent Semantics Approaches for Adaptive Web

Anu Sharma
IASRI, India

Aarti Singh
Guru Nanak Girls College, India

ABSTRACT

Intelligent semantic approaches (i.e., semantic web and software agents) are very useful technologies for adding meaning to the web. Adaptive web is a new era of web targeting to provide customized and personalized view of contents and services to its users. Integration of these two technologies can further add to reasoning and intelligence in recommendation process. This chapter explores the existing work done in the area of applying intelligent approaches to web personalization and highlighting ample scope for application of intelligent agents in this domain for solving many existing issues like personalized content management, user profile learning, modelling, and adaptive interactions with users.

1. INTRODUCTION

The quantum of information accessible from World Wide Web (WWW) is increasing exponentially over time. Retrieval of desired information from WWW has become a tedious and difficult task like finding a needle in the hay stack. So, there is a continually growing demand for more sophisticated techniques in web information retrieval. Further, current web has taken up a face and shape of a knowledge provider rather than just an information dissemination medium with the introduction of technologies like Data Mining (DM), Semantic Web (SW) and Multi-Agent Systems (MAS) (Singh, 2012). As the volume of data collected by various enterprises increase, the need for analysing the data efficiently rises sharply. DM has emerged as a cutting-edge technology with the related advancements in database technology for business intelligence through on-line analytical processing tools on data warehouses. These techniques are also applied on relational, transactional, object-oriented, spatial, temporal and text databases. This technique is found very useful in business viz. telecommunication industry, banking and finance, bio-medical and DNA analysis, agriculture and retail industry. Further, many other types of data like web content, images, video, social networking data and blogs came into picture. This change has led to the

DOI: 10.4018/978-1-5225-5951-1.ch010

emergence of Web Mining (WM) and Web Personalization (WP). WM may be defined as the use of DM methods to find patterns from the Web (Kosala & Blockee, 2000). Based on types of web data, researchers have categorized the WM into three type namely web content, structure and usage mining. These methods were found very useful in the development of adaptive/personalized web. Further, there has been tremendous growth and evolution in e-business, e-commerce, e-banking, scientific research, e-learning, social networking, web communities, blogs and other sectors which have necessitated the need for personalized information delivery from the web based on user interest and context. So, WP is emerging as the most cutting-edge technology in future web applications. The four main phases (Anand & Mobasher, 2005) of a WP system are user profile extraction, pattern discovery, recommendations and evaluation which are described in detail in section 2. WP phases can be enhanced by using SW and intelligent Software Agents (SA) for the retrieval of better knowledge oriented recommendations. Many researchers have reviewed the existing work done in this area. Kobsa (2001) has given a detailed study on generic user modeling system. Carmagnola, Cena & Jena (2011) have reviewed the work done on solving the user interoperability issue across multiple websites. Malik & Fyfe (2012) have studied in detail the contributions made in all the phases of WP. Ghorab, Zhou,O'Connor and Wade (2013) have surveyed the work done for various stages of personalized information systems. Chen, Wu and Cudré-Mauroux (2012) have surveyed the state of the art on applying computational approaches into SW applications. Singh and Sharma (2017) have reviewed of existing intelligent technologies for personalized information retrieval from web and proposed a framework for semantic web information retrieval. Singh and Sharma (2015) have explored the scope of applying SW and SA technologies for WP. This chapter extends their work by: (1) describing in details the each of WP phases; (2) including privacy and security concerns; and (3) updating the contents with latest research work. This chapter provides an insight into the various aspects of semantic and intelligent technologies along with latest research work done in this area. The next section describes the Adaptive/Personalized Web in brief.

2. ADAPTIVE WEB: AN INTELLIGENT FACE AND MIND OF WEB

Adaptive/Personalized Web may be defined as the ability of a system to provide a customized view of the Web by performing various actions for a single or group of users (Anand & Mobasher, 2003). It can deliver a wide variety of facilities to the users in the form of greetings, bookmarking, granting personalized rights, modifying web site structure, tailored offers & services and adapted web search results by applying statistical methods, DM, text mining, WM, SW and ontologies etc. (Figure 1).

WP relies on the collection, analysis and modelling of various types of web data including web access log files, client-side information in the form of cookies, browser cache, proxy server logs, click-stream data, site structure, domain ontology, explicit user profiles and location specific information. WP is achieved through the implementation of all the phases of a typical data mining cycle. Mobasher et al., (2001). There are four phases of a WP system namely learning, pattern discovery, recommendations and evaluation as shown in Figure 2.

Learning phase is mainly concerned with the creation of a user profile through modelling. User Modelling (UM) is referred to as gathering and exploiting the information about preferences, interests and behaviour of the individual user for creating user adaptive websites. Some important characteristics parameter considered for user profiling are background information, interests and preferences, goals, emotional state, context etc. A user profile may be constructed by explicitly collecting the information

Figure 1. A Depiction of Adaptive Web

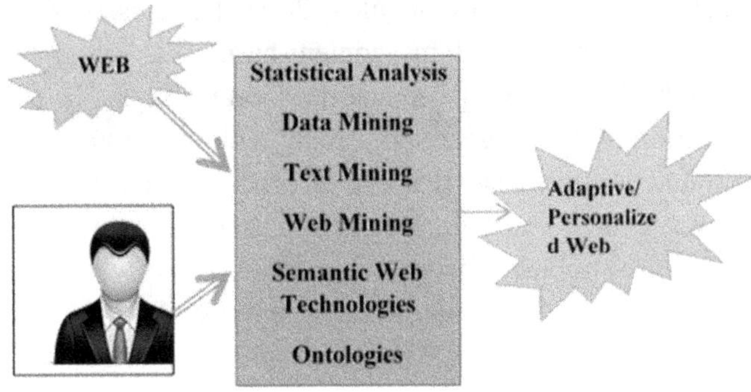

Figure 2. Web Personalization Process

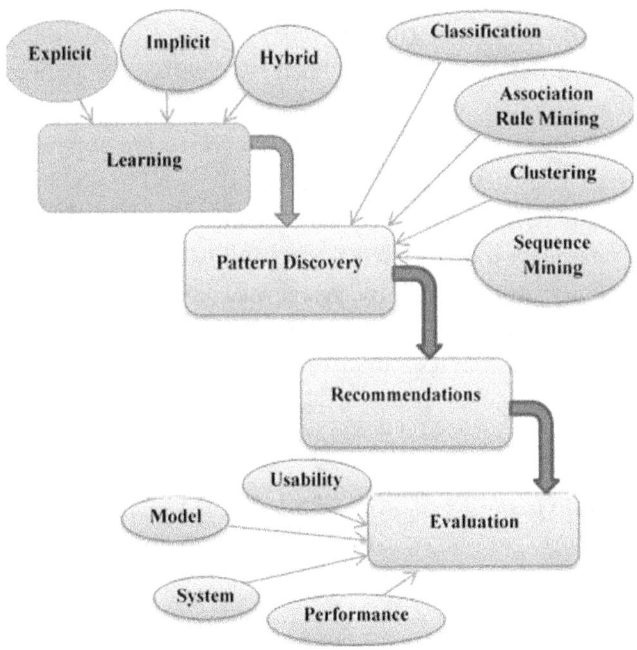

with user intervention or `by observing the user behaviour implicitly (Carmagnola, Cena & Jena, 2011). The collected information is used to represent user profile. Some commonly used techniques for representing user profiles are: (1) overlay modelling; (2) stereotype modelling; (3) keyword based profiles; (4) semantic networks and; (5) concepts based profile. Major issues related to web profiling are accuracy, inferencing/reasoning of user interest, scalability, privacy and user model interoperability.

Pattern discovery relates to applying machine learning and DM approaches for identifying unique patterns from the collected data. These methods usually consist of association rule mining, clustering, classification, sequence mining etc. (Eirinaki & Vazirgiannis, 2003). This model is used in real time for comparing the current interaction of the user with the stored model and then generating recommendations.

Recommendations phase consist of utilizing various techniques for generating recommendations. Pierrakos, Paliouras, Papatheodorou and Spyropoulos (2003) has classified WP techniques into four generic approaches manual decision rule systems, content-based filtering systems, social or collaborative filtering systems and hybrid approaches. Manual decision rule systems personalized a web-based service via manual intervention of its designer and usually with the cooperation of the user. Typically, static user models are obtained through a user registration procedure and a number of rules are specified manually concerning the web contents that are provided to users with different models. Content based techniques aims at studying the similarity between user profile and web contents and then applying machine learning methods to predict the similar items of interest. This approach has been found very suitable for a specific area. But its effectiveness decreases with the changes in the user's behaviour over time. Also, this requires the application of natural language processing and text mining techniques which are difficult to apply as compared to other approaches.

Collaborative filtering approach is to study the preferences and activities of larger number of similar people and then making suggestions using collective features. This approach is very useful as it is able to make recommendation without considering the web contents. But on other hand, recommendations are difficult to make for a new item which has not been rated earlier by anyone. Scalability becomes an important issue to take care when the number of user increases. Hybrid approaches combine two or more of above approaches for recommendations. Some other categorization of personalization approaches are reactive vs. proactive, user versus item information, memory based versus model based, and client side vs. server side phases (Anand & Mobasher, 2005). Evaluation phase deals with the evaluation of techniques utilized in earlier phases for their efficiency and validity using various parameters like prediction accuracy, confidence, trust, novelty, serendipity, privacy, scalability etc.

WP approaches and phases can be enhanced by using SW and intelligent SA for the retrieval of useful knowledge oriented recommendations (Figure 3). Next section describes the various domains where WP is applied aggressively.

Figure 3. Web Personalization in the Framework of Intelligent Semantic Technologies

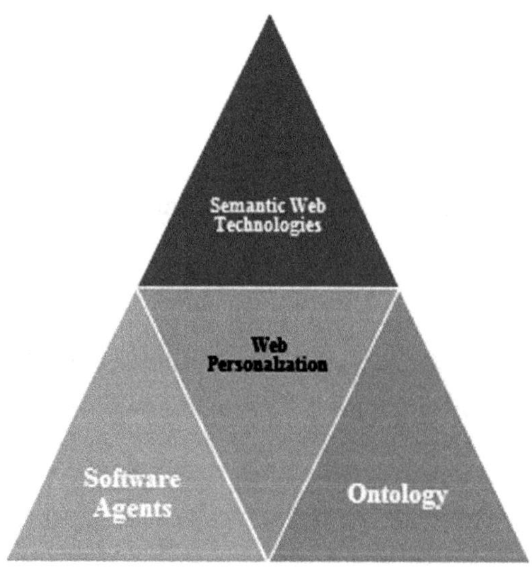

3. DOMAINS OF WEB PERSONALIZATION

Some of the major domains where WP has been found useful are:

- **Recommender Systems (RS):** RS have been widely used for suggesting an item of interest to a user. Suggestions might relate to which upcoming movies to watch to, books to read, places to visit, sports to enjoy and many more. A researcher might get suggestions on new workshops, conferences and training programs organized on his/her areas of interest along with latest research from other scholars. Amazon.com, flipkart.com, Makemytrip, Trivago, Netflix, MyYahoo, MyGoogle, Snapdeal, Tripadvisor, and IMDb are some of the important web sites based on RS. RS are found very useful in expanding the business by increasing the number of items sold, increased user satisfaction, better understanding of customer needs and selling of more diversified products. Various Recommendation Techniques (RT) have been proposed in literature to suggest new items to the users.
- **Personalized Information Filtering:** This refers to the process of eliminating the information which might not be of interest to the user. It involves the usage of user modelling to remove unwanted item during post processing. It helps to cut down the information overload by elimination (Foltz, 1992).
- **Personalized Web Search:** Personalized Web Search (PWS) refers to arranging the web search results based on user preferences. A search engine is a tool used to retrieve the desired information from web and is based on the link structure of the web to arrange the Web Search Results (WSR) for a given query. Current search engines are not able to resolve ambiguous queries and identify the user preferences automatically. This leads to retrieval of web pages which may be irrelevant to the user and wasting of time at user end in sifting through these pages. PWS takes into consideration the preferences and interests of the user and re-arrange the WSR. This leads to most interesting and useful links to come on top of ranked list and thus relieving the user from being drown into the ocean of information.
- **Personalized Query Expansion:** Personalized Query Expansion (PQE) is another domain where methods of WP has been applied to automatically enhance the web search query. This leads to retrieval of more meaningful web search results. For example, there is web surfer who searched for keyword "tourist places" at a time and in the same web session just after this web search, he searched for keyword "java". So, it is most likely that he/she might be interested in place Java Island rather than computer programming language. Using PQE, the query may be automatically extended by adding term "island". PQE also leads to better WSR. The various domain of WP are shown in Figure 4.

So far, a discussion on various aspects of WP is given and before studying the application of SW and SA technologies in this area, a brief description of SW technologies is given in next section.

4. SEMANTICS BASED INTELLIGENT TECHNIQUES

Semantics based intelligent techniques are used to add meaning to the content available on the web. This section outline the involved important paradigms in brief.

Figure 4. Domains of Web Personalization

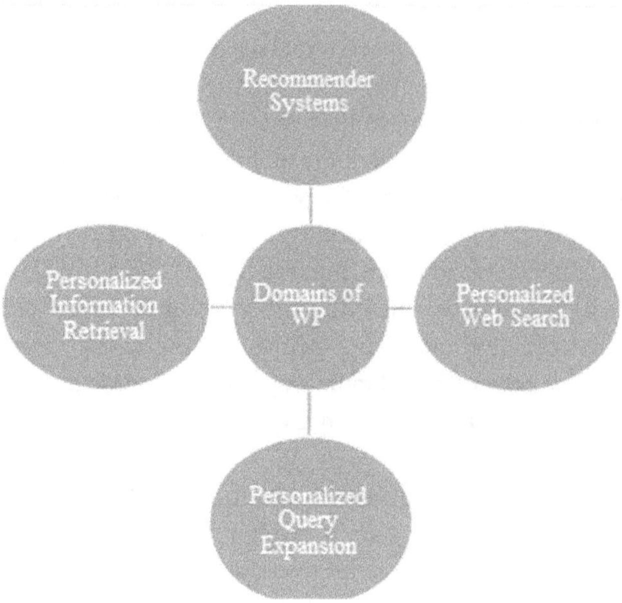

4.1 Semantic Web

Semantic Web, a concept coined by (Lee, T.B., Hendler, J. & Lassila, O., 2000), aims at changing the web from blind information display mechanism to knowledge provider. This type of web is beneficial in relating, extracting and drawing inference from web pages. It deploys semantic technologies like Resource Description Framework Schema, ontology languages for describing concepts and relationships, query processing languages, rules description language and languages for marking up data inside Web pages.

4.2 Software Agents

Intelligent SA provides an important paradigm for use in internet applications (Ehlert, 2003). Wooldridge and Jennings (Wooldridge & Jennings, 1995) defined an agent as a computer component that has been built with well design objectives for a particular environment, and is capable of operating independently. They have well defined states and properties and have complete control over their internal states and their behaviour. They are capable of existing without intervention of other systems or humans. They may react to perform their job and exhibit key attributes like:

- **Learning/Reasoning**: This is the ability of an agent to learn from past experiences and to successively modify its own behaviour in the environment.
- **Reactivity:** An agent must be able to sense and act suitably to outside environmental influences or information.
- **Autonomy:** An agent must have both control over its actions and internal states. The degree of the agent's autonomy can be specified. They may need intervention from the user only for important decisions.

- **Goal-Oriented:** Agents has well-defined goals and gradually influences its environment and so achieve its own goals.

Figure 5 shows some attributes of SA diagrammatically. An agent can act intelligently by supplementing it with ontology, machine learning techniques, user profiles and search pattern. SA can act as background processes that works on behalf of the user. SA can be of following types as shown in Table 1.

4.3 Ontology

Ontology plays a significant role in increasing the effectiveness of WP based system. According to Gruber ontology may be defined as *an explicit specification of a conceptualization* (Gruber, 1993). Some other researcher has defined it in another way as *Ontology is a formal description and specification of knowledge*. It provides a communication pathway between SA and other system by stating the

Figure 5. Some Attributes of SA

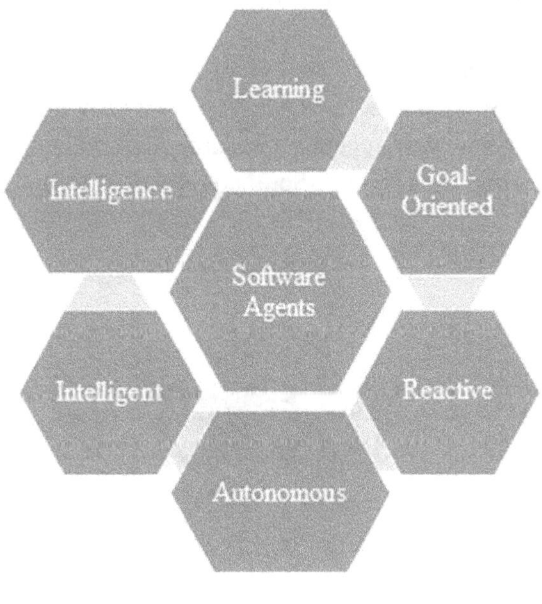

Table 1. Types of SAs

Agent Type	Purpose
Table-Driven Agents	These agents use a lookup table. A lookup table is stored in the memory with the description of causes and action.
Simple Reflex Agents	These are stateless components acting on the condition-action rules with no memory of past world states.
Agents with Memory	These agents store the historical data about its states.
Agents with Goals	These are a kind of pro-active agents which takes into consideration the possible occurrence of future events besides having the state information.
Utility-Based Agents	For a given goal, if there exist many solutions. Then utility based agent takes decision on the basis of efficiency of each solution.

clear structure of the information. Reuse of constructed domain knowledge is another benefit of using ontology. It allows to separate domain knowledge from operational part of system.

Ontology development starts with the identification and description of entities which are nothing but the any object of interest in real world. These entities are defined as a class in ontology. This step is followed by the identification of relationship between them. The identified relationships are used to arrange the classes in a hierarchy. It is followed by the specification of slots and allowable values for each slot. Finally, the slots are filled with the actual values as shown in Figure 6. Ontology are structured way of storing information.

Ontology is a widely accepted technology in MAS because it allows all the participating agents in a system to agree on the use of shared understanding of the world. It also allows inferring new knowledge through semantic reasoner and this capability is moved from agent code to ontology. This facilitate in reusing the knowledge across many applications. Further, standardization of communication language leads to clear understanding of items in a domain. It emerging as the most sought after technology for developing MAS. Next section deals with the semantics based intelligent approaches for adaptive web.

5. SEMANTICS BASED INTELLIGENT APPROACHES FOR ADAPTIVE WEB

Major challenges involved in effective personalization are personalized content management, user profile learning, modelling and adaptive interface. Personalization can be realized by using independent intelligent agents. A number of agents work together to enable personalization by recognizing individual interests and then recommending the contents. So, there is a need to have multi-agent platform using common vocabulary for web personalization. This technology can provide solution to many existing

Figure 6. Ontology Creation Process

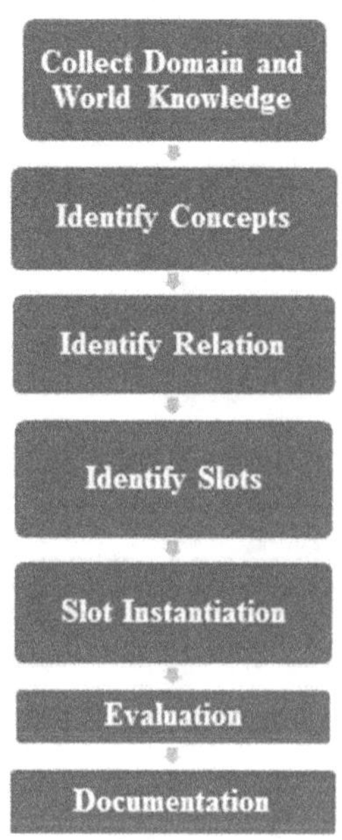

issues like scalability, robustness, efficiency, privacy, and security and user profile generation in the area. Ontology and software agents have been used in all the phases of web personalization. Some of the contributions are given in this section.

5.1 Intelligent Approaches User Profiling

Traditionally the user profiles are represented using keywords, semantic networks and concept based hierarchies. Major problem with keyword based representation is its limitation in understanding the actual interests of the user and ambiguity regarding the meaning the keywords. This is problem is addressed by using semantic networks but it can include only limited type of semantic relation. Ontolgy based user profiling using SW technologies and intelligent agents has been successfully applied to solve many problems inherent in this areas. Ontology based user profiling is powerful representational medium and provides associated inference mechanism. They provide various mechanisms for inferencing and thus allows to discover the preferences which are otherwise not visible from direct user behaviour. They are also found useful in solving the cold start problem in recommendation systems. Middleton, Shadbolt and De Roure (2004) has proposed an ontology based user profile creation by unobtrusively monitoring user behaviour and relevance feedback. Authors have also addressed the cold start problem by using external ontology to bootstrap the recommendation system. Trajkova and Gauch (2004) has investigated the techniques for the improvement of ontological user profile. They have investigated that ranking of the concepts in the profiles by number of documents assigned to them rather than by accumulated weights provides better profile accuracy. Ontological user profiles to narrow down web search space returned by search engines had been utilized by Sieg, Mobasher and Burke (2007). In this approach Open Directory Project (ODP) is used as reference to build ontological user profile. User profile is represented as hierarchical concepts from ODP along with interest score. Interest scores are automatically updated with user's browsing behaviour. User's web search results are classified or clustered using ontological user profile. The limitation of this work is manual updating of ODP ontology to accommodate changes in the ontology. A solution to this has been proposed by Safarkhani, Mohsenzadeh and Mojde, (2009). They have proposed an approach for the generation of semantically enhanced clusters using Wikipedia as reference ontology. This approach overcomes the problem of automatic creation and updating of domain knowledge along with coverage. An approach for personalizing the web using user profile ontologies has been proposed by Vigneshwari and Aramudhan, (2012). These ontologies are created dynamically using the server log file through classification and clustering techniques. A global ontology is also created by identifying the word pairs that have occurred in most of the documents. This approach considers parent-child relationship for global and user ontology construction. A multi-agent approach for building an active user profile that is able to learn and adapt itself to user activities had been proposed by Hawalah and Fasli (2011). In the proposed architecture agents are used to reduce the complexity of recommendation process by delegating the individual tasks to separate agent. Each layer in the architecture is assisted by various agents. The study uses the existing ontology for user profile construction. In another work, Hoppe (2013) have proposed a framework for automatically constructing and populating profile ontology for each user identified by the system from heterogeneous resources in big data environment. This approach is useful in providing recommendations in real-time environment. Aghabozorgi and Wah (2009) has proposed a dynamic model for user modelling by applying clustering techniques on web usage data off-line. This model is adjusted periodically through new transactions and thus solves the problem of maintaining the accuracy of user model. Chamiel and Pagnucco (2009) have used existing ontology constructed with the help of experts to specify the user's preferences in an interactive and dynamic way. Bhowmick, Sarkar, and Basu (2010) have combined static and dynamic user profiling strategies. Static user profile specifies the user's interest in a much focused manner and dynamic user profiling adds the feature of adaptability into it. Sosnovsky and Dicheva (2010) had combined together research from two different fields – user modelling and web ontologies to demonstrate how recent semantic trends in web development can be combined with the modern technologies of user modelling. Abel, Gao, Houben and Tao (2011) have used twitter posts to model the user profile semantically to contextualize Twitter activities. Aminia, Ibrahima, Othmana and Nematbakhshb (2015) have proposed a method for integrating of several domain taxonomies to build a reference ontology, and exploits this reference ontology for profiling scholars' background knowledge. Hawalah et al. (2015) have proposed ontology based user profile model created by learning the web pages visited by the user. This method can store both the long and short-term interests of the users dynamically. This modelling methodology is then used in web search by leveraging the benefits of MAS.

5.2 Intelligent Recommendation Techniques

Several researchers have contributed to the development of pattern recognition techniques and recommendation approaches towards adaptive web. This section describes the use of semantics and agents in these tasks. Semantic annotation of web pages from their contents has been given by Dai and Mobasher (2003) using text mining techniques and meta-data information. They have also integrated the domain ontologies to capture the complex relationships existing in web page objects for effective recommendation. Eirinaki, Vazirgiannis and Varlamis, (2003) have extended the normal usage logs to C-logs by incorporating the web link semantics into them. These semantically enhanced C-logs are then used as input for usage mining and thus generating better recommendations. An algorithm for improving the pattern quality by using ontologies had been developed by Senkul and Salin, (2011). In this work, authors used ontologies for annotating web pages during pattern recognition through sequence association rule mining. The annotation is done manually using existing ontologies. User based collaborative filtering techniques assumes items to be independent of each other while calculating the user to user similarity. Solution to this problem was given by Anand, Kearney and Shapcott, (2007) that use item ontology for creating semantically enriched user profiles. A measure named generalized similarity max was given which consider the items information for calculating user similarity. Fuzzy cognitive personalized agents have been proposed by (Miao, C., Yang, Q., Fang, H., Goh, A., 2007) to assist the customers in selecting the items of their interest. The proposed agent is capable of learning the user's preferences and making inferences automatically. The agent also learns from past user behaviour and generates recommendations by combining, user's and group preferences with expert opinion. But this approach still needs to be evaluated and implemented.

Context based personalization was found very useful in providing better recommendation as it considers demographic, spatial and social factors affecting the user choices. An agent based framework for providing the personalized services using context history on context-aware computing had been proposed by Hong, Suh, Kim, and Kim, 2009. A multi-agent approach for ontology based recommendation system has been given by (Pan, P., Wang, C., Horng, G. & Cheng, S., 2010). This approach aims at solving many problems prevailing in RS such as SP, NUP, OP and LCA by using agents for handling each of these. This approach may be further extended by developing more refined algorithms for filtering agents. An intelligent technique for recommendation by using information from social networking web sites like Facebook and Twitter with ontological user profiles has been proposed by (Vuljanic D, Rovan L & Baranovic M., 2010). Friend of a Friend (FOAF) profile is automatically generated from Facebook and Twitter and used in combination with CB and CF techniques to generate recommendations. This approach aims to solve the CS problem as well as generates more accurate recommendations. But the implementation of the proposed work is still not done in real environment. Fernández-Tobías et al., (2011) have proposed a generic semantic network based cross domain recommendation system. This framework uses places of interests and music together by linking the concepts between these two domains by using weighted acyclic graphs. An intelligent RT using semantics for more accurate and diverse suggestions and by using spreading activation and semantic associations has been proposed by (Blanco-Fernández Y., López-Nores M., Gil-Solla A., Ramos-Cabrer M. & Pazos-Arias J.J., 2011). Tiwari, Husain, Srivastava and Agrawal (2011) have proposed novel techniques that combines the web usage data with content, semantics and the structural properties using an ontology to compute semantically enhanced navigational patterns and recommendations. Frikha, Mhiri and Gargouri (2014) have used data from social networks sites to improve traditional recommender systems by including user preferences and influences from friends. A user interest ontology is developed to make personalized recommendations out of such information. This approach may be extended by employing SA for automating the various tasks. Most of the CF based RS do not incorporate domain ontology during recommendations. One important reason for not including domain ontology is the automatic construction of ontology. A new approach for creating the domain knowledge automatically and the utilizing this ontology to cluster the similar users from web usage data for recommending web pages to the users have been proposed by (Gopalachari & Sammulal, 2014). Gim, Jung and Jeong (2015) have proposed a modified association rule mining algorithm to improve the accuracy of the exiting Onto-Apriori algorithm. XOnto-Apriori can be used to generate association rules for suggesting more valuable candidate item sets by applying ontology reasoning function. Singh and Sharma (2017) have proposed an intelligent and adaptive mechanism deploying intelligent agents for solving new user and overspecialization problems that exist in content based recommendation systems.

5.3 Intelligent Semantic Approaches for Security and Privacy in Web Personalization

Privacy may be defined as the individual concern regarding the exposing of their private information to the outside world and is a complex issue (Berry & Linoff, 1997). Now a day, diversified form of user's information in the form personal data, banking data and other interactions using social networking is available on web. This has made a user vulnerable to many types of attacks. This section briefly describes the work in area of preserving the privacy in this information filled world. Many authors have contributed to the survey of privacy issues in RS (Jeckmans, A.J.P., Beye, M.R.T., Erkin, Z., Hartel, P.H., Lagendijk, R.L. & Tang, 2013), CF (Berkovsky S., Eytani Y., Kuflik T. & Ricci F., 2005) and user's perception on personalization efficiency (Chellappa & Sin, 2005; Toch, E., Wang, Y. & Cranor, L.F., 2012, Kobsa, A., Knijnenburg, B.P. & Livshits, B., 2014). One of the important threats to privacy is to identify the user back from the available privately held collection on personal data. A solution to this problem has been given by (Sweeny, 2002) named k-anonymity whereby a person is not distinguished from at least k-1 individual in a database.

Shen et al., (2007) have distinguished and defined the four levels of privacy protection and have analysed various architectures for personalized search. The simplest level 1 method is the use of pseudo identity which replaces the identity with single or group of users. But it still leaves the other information about user on server. In level two of privacy protection search engine builds a group profile rather than the individual profile. A proxy server may be used for interactions with search engine by many users or obfuscation of query term may be used to achieve this level of privacy protection. Third level of protection considers no information on sever side regarding the users interaction and personalization is done at client-side. This can be achieved by using anonymous network. But some information like user's original query is still stored on server. Best level of privacy is when no information about user is available on server. It involves the use of cryptographic techniques.

Privacy threats in centralized server based CF systems have been discussed by (Berkovsky S., Eytani Y., Kuflik T. & Ricci F., 2005). Based on their assumption, there are two ways by which the personal information about the user is made available to others over a probably insecure network. First way is when someone wants to receive recommendation from others and second one is when some other person wants recommendation from this person. This study proposes a method by which user has the discretion to expose their personal information to a certain degree or obfuscate a part of it. Any personalization is done only at client side. It was found that the obfuscation offers a better solution without affecting the accuracy of the system.

Xu et al. (2007) have proposed a scalable way to create a hierarchical user profile. Implicit interests of the user are stored in a hierarchical user profile with upper layer presenting the generalized interests and lower layers representing the specific interests. They have also proposed to two parameters for specifying the content and degree of details of profile information that is to be exposed to the search engine. Search engine results are personalized with public user profile. But incorporation of privacy preservation leads to compromise in quality of search results. So, the better approach would be to expose the information related to the specific query.

This aspect has been considered by Shou et al., (2014) who have proposed a framework that can adaptively generalize profiles by queries while respecting user-specified requirement. Also, the personalization is done using the user profile created at run time. This work can be further strengthened by adding background knowledge and richer relationships among topics. (Parra-Arnau, J., Rebollo-Monedero, D. & Forné, J., 2014) have made an attempt to quantify the privacy of user profiles by using KL divergence as a measure. (Jaison & Varghese, 2015) have proposed the concept of Homomorphic encryption to encrypt the server in such a way that neither an eavesdropper nor an untrusty admin could access the search words and the profile.

Ciss´ee and Albayrak, (2007) have made an attempt to use multi-agent approach for privacy preserving in RS by using important features of agents such as autonomy, adaptability and the capability to communicate. The proposed protocol could be modified so that the recommendations themselves are kept as private. Peyravi and Latif (2014) have proposed Secure Multi Agent Information Filtering (SMAIF) system based on agents to overcome information filtering challenges. Sharma and Singh (2016) have suggested an efficient and scalable method for protecting user's privacy using Elliptical Curve Cryptography (EEC) technique and also blocking unauthorized third-party access to user information on server through intelligent software agents.

6. INTELLIGENT ADAPTIVE WEB APPLICATIONS

SW and SA have been applied in many application areas like e-commerce, e-learning, e-banking (Favre-Bonte, Elodie, & Thevenard-Puthod, 2013), social networking, blogs, personalized web search etc. The major focus is to provide services (Slimani, 2013). to the users based on their interests. Adaptive web revolves around user interests and preferences, so as to attract and retain them on a website. Some applications of web personalization based on these aspects are listed below:

- **News Services:** A web based personalized news service named, *myPlanet*, had been developed by (Kalfoglou, Domingue, Vargas-Vera & Shum, 2001). This service provides easy access of new stories to readers by identifying user interests. The ontology is used to automatically deduce the user's interest by reasoning. It also supports the annotation of existing stories using ontologies. It also provides the semi-automatic methods using information extraction and domain specific rule templates. Benefits of recommending news items using domain ontology instead of using a term-based approach had been proposed by (IJntema, Frasincar, Goossen & Hogenboom, 2010).
- **Tourism:** (Gavalas & Kenteris, 2011) have used temporal, spatial and weather conditions for representing context along with the collaborative filtering techniques for deriving improved recommendations in pervasive environments. The exact location of the tourists has been identified by using wireless sensor network (WSN) and mobiles are used to gather the tourist's interests and ratings. But this approach was not based on the use of semantic web and SA technologies. However, (Batet, Moreno, Sánchez, Isern, & Valls, 2012) have developed a recommender system for tourist activities through agents.
- **E-Commerce:** (Lee, Hendler, & Lassila, 2002) has developed two recommender systems using intelligent agents. The first one focuses on learning the user preferences for product recommendations and the capacity to adapt to its changing preferences. The other one assisted the customers to find the optimal products based on his personal preferences. Various agents were used to accomplish tasks in these two systems. (Rosaci & Sarnè, 2014) have developed a multi-agent system, named MAST, for facilitating various e-activities like promotional campains and expenditure in e-commerce. MAST uses personalized approaches for helping both customers and merchants. The system also considers the privacy and trust issues between merchants and customers in e-payment model.
- **E-Learning:** Chellatamilan and Suresh (2011) have proposed a framework for adaptation of MOODLE interface, course contents organization and student learning path by using software agents, ontologies and learners preference knowledge base. Analyzing the user behavior is a challenging area considered by (Razmerita, 2011) for an ontology-based framework for modeling user behavior in knowledge management to bridge the gap towards personalized knowledge management. The user behavior is characterized in terms of level of information sharing, type of activity and level of activity. The framework is divided into three layers user front end layer, intelligent service layer and data layer. Intelligent service layer may provide various functionalities (Acharjee, Chakraborty, Karaa, Azar & Dey, 2014) by using software agents for identifying the characteristics of user, interest of user, yellow page agent and conflict-detection agents. Use of ontology and agents will enable the efficient and automatic inferences from facts towards knowledge.
- **Movie and Music Recommendation:** Emotions play an important role in buying and learning behaviors of consumers in the areas of music and movie recommendations. This aspect is studied

by (Fong, Zhou, Hui Siu, Tang, & Hong, 2012) for automatic generation of semantically enabled personalized web usage ontology. They considered the emotions reported by user after accessing each web page for ontology creation. The ontology generated may be used by software agents for web personalization and semantic search. The work may be further extended by using software agents to create personalized pages for users. Recommending music based on text queries is considered by (Hariri, Mobasher & Burke, 2013). Since collection of meta-data over large collection of songs is a tedious task, so they used music annotations retrieved from social tagging websites such as last.fm and use them as textual descriptions of songs. A probabilistic model has been proposed that takes into account the user's preference history as well as tag co-occurrences in songs.

Software Engineering

Castro-Herrera, Duan, Cleland-Huang, and Mobasher (2009) has proposed a framework for requirements elicitation. This framework uses data-mining techniques and personalization to specify the requirements in a very big and distributed project. The proposed recommender model consist of a number of discussion forums working collaboratively with defined purpose. Pakdeetrakulwong and Wongthongtham (2013) has applied the web personalization and multi-agent systems in the area of software engineering. They have proposed a conceptual framework of multi-agent based recommender approach for active software engineering ontology. This work is very useful in situations where different phase of software life cycle are executed at distributed places. Software ontology helps in sharing of knowledge across the team members whereas agents help by working autonomously towards the realization of various SE tasks.

- **Event Recommendation:** Neves, Carvalho and Ralha (2014) proposed an agent oriented architecture using ontology and the spreading algorithm for context-aware and adaptive event recommendation. The use of ontology allows defining the domain knowledge model, while the spreading activation algorithm learns user patterns by discovering user interests.
- **Food Recommendation**: A framework for diet and food recommendations using SW and SA technologies had been proposed by Prakash and Sivakumar, 2013. A semantic framework that uses the personalization techniques based on integrated domain ontologies, pre-constructed by domain experts, to recommend the relevant food that is consistent with people's needs had been presented by Al-Nazer, Helmy and Al-Mulhem, (2014).

7. CONCLUSION

This chapter describes in brief the concepts, approaches and techniques used in WP. Further, a thorough review of research in applying SW and SA in this area is presented. Ontologies and intelligent agents have been successfully applied in user profiling, recommendation and user's privacy and security concerns. Similarly, they have also been applied at pattern generation stage. Although, the literature survey has highlighted the use of taxonomies and ontologies at various phase but only a few implementations have been found for the use of intelligent agents. So, there exist an ample scope in applying these technologies in this upcoming area. This chapter will be highly useful for researchers engaged in applying intelligent techniques for WP.

REFERENCES

Abel, F., Gao, Q., Houben, G., & Tao, K. (2011). Semantic enrichment of twitter posts for user profile construction on the social web. In G. Antoniou, M. Grobelnik, E. Simperl, B. Parsia, D. Plexousakis, P. D. Leenheer, & J. Pan (Eds.), *The Semanic Web: Research and Applications* (pp. 375–389). Berlin: Springer. doi:10.1007/978-3-642-21064-8_26

Acharjee, S., Chakraborty, S., Karaa, W. B. A., Azar, A. T., & Dey, N. (2014). Performance evaluation of different cost functions in motion vector estimation. *International Journal of Service Science, Management, Engineering, and Technology*, *5*(1), 45–65. doi:10.4018/ijssmet.2014010103

Aghabozorgi, S. R., & Wah, T. Y. (2009). Dynamic modeling by usage data for personalization systems. *Proceedings of 13th International Conference on Information Visualisation*, 450-455, 10.1109/IV.2009.111

Al-Nazer, A., Helmy, T., & Al-Mulhem, M. (2014). User's profile ontology-based semantic framework for personalized food and nutrition recommendation. *Procedia Computer Science*, *32*(1), 101–108. doi:10.1016/j.procs.2014.05.403

Aminia, B., Ibrahima, R., Othmana, M. S., & Nematbakhshb, M. A. (2015). A reference ontology for profiling scholar's background knowledge in recommender systems. *Expert Systems with Applications*, *42*(2), 913–928. doi:10.1016/j.eswa.2014.08.031

Anand, S. S., Kearney, P., & Shapcott, M. (2007). Generating semantically enriched user profiles for web personalization. *ACM Transactions on Internet Technology*, *7*(4), 22, es. doi:10.1145/1278366.1278371

Anand, S. S., & Mobasher, B. (2005). Intelligent Techniques for Web Personalization. In B. Mobasher & S. S. Anand (Eds.), Lecture Notes on Computer Science: Vol. 3169. *Intelligent Techniques for Web Personalization* (pp. 1–36). Berlin: Springer-Verlag. doi:10.1007/11577935_1

Batet, M., Moreno, A., Sánchez, D., Isern, D., & Valls, A. (2012). Turist@: Agent-based personalised recommendation of tourist activities. *Expert Systems with Applications*, *39*(8), 7319–7329. doi:10.1016/j.eswa.2012.01.086

Berkovsky, S., Eytani, Y., Kuflik, T., & Ricci, F. (2005). Privacy-enhanced collaborative filtering. *Proceedings of PEP05 UM05 Workshop on Privacy-Enhanced Personalization*, 75-83.

Berry, M., & Linoff, G. (1997). *Data Mining Technique for Marketing, Sales and Customer Support*. New York: John Wiley.

Bhowmick, P. K., Sarkar, S., & Basu, A. (2010). Ontology based user modeling for personalized information access. *International Journal of Computer Science and Applications*, *7*(1), 1–22.

Blanco-Fernández, Y., López-Nores, M., Gil-Solla, A., Ramos-Cabrer, M., & Pazos-Arias, J. J. (2011). Exploring synergies between content-based filtering and spreading activation techniques in knowledge-based recommender systems. *Information Sciences*, *181*(21), 4823–4846. doi:10.1016/j.ins.2011.06.016

Carmagnola, F., Cena, F., & Gena, C. (2011). User model interoperability: A survey. *User Modeling and User-Adapted Interaction*, *21*(3), 285–331. doi:10.100711257-011-9097-5

Castro-Herrera, C., Duan, C., Cleland-Huang, J., & Mobasher, B. (2009). A recommender system for requirements elicitation in large-scale software projects. *Proceedings of the 2009 ACM symposium on Applied Computing,* 1419-1426, 10.1145/1529282.1529601

Chamiel, G., & Pagnucco, M. (2009).Ontology Guided Dynamic Preference Elicitation, *Proceedings of 3rd ACM Conference on Recommender Systems & the Social Web*, 41-48.

Chellappa, R. K., & Sin, R. G. (2005). Personalization versus privacy: An empirical examination of the online consumer's dilemma. *Information Technology Management*, *6*(2-3), 181–202. doi:10.100710799-005-5879-y

Chellatamilan, T., & Suresh, R. M. (2011). Intelligent agents for the semantic adaptive e-learning system. *International Journal of Computer Applications ICACT*, *1*(1), 1–5.

Chen, H., Wu, Z., & Cudré-Mauroux, P. (2012). Semantic web meets computational intelligence: State of the art and perspectives. *IEEE Computational Intelligence Magazine*, *7*(2), 67–74. doi:10.1109/MCI.2012.2188592

Ciss'ee, R., & Albayrak, S. (2007). An agent-based approach for privacy-preserving recommender systems. *Proceedings of the 6th international joint conference on autonomous agents and multi-agent systems ACM New York USA*. 10.1145/1329125.1329345

Dai, H., & Mobasher, B. (2003). A road map to more effective web personalization: integrating domain knowledge with web usage mining. *Proceedings of the International Conference on Internet Computing*, 1-8.

Ehlert, P. (2003). *Intelligent user interfaces: introduction and survey.* Research Report DKS03-01/ICE 01. Faculty of information technology and systems, Delft University of Technology.

Eirinaki, M., & Vazirgiannis, M. (2003). Web mining for web personalization. *ACM T Internet Tech.*, *3*(1), 1–27. doi:10.1145/643477.643478

Eirinaki, M., Vazirgiannis, M., & Varlamis, I. (2003). SEWeP: Using Site Semantics and a Taxonomy to Enhance the Web Personalization Process. *Proceeding of SIGKDD '03*, 99-108. 10.1145/956750.956765

Favre-Bonte, V., Elodie G. & Thevenard-Puthod, C. (2013). Identifying different forms of innovation in retail banking. *International Journal of Service Science, Management, Engineering, and Technology,* *4*(4), 43-57, doi: 10.4018/ijssmet.2013100103

Fernández-Tobías, I., Kaminskas, M., Cantador, I., & Ricci, F. (2011). A generic semantic-based framework for cross-domain recommendation. *Proceedings of HetRec '11 Chicago USA*, 25-32. 10.1145/2039320.2039324

Foltz, P. W., & Dumais, S. T. (1992). Personalized information delivery: An analysis of information filtering methods. *Communications of the ACM*, *35*(12), 51–60. doi:10.1145/138859.138866

Fong, A. C. M., Zhou, B., Hui Siu, C., Tang, J., & Hong, G. Y. (2012). Generation of personalized ontology based on consumer emotion and behavior analysis. *IEEE Transactions on Effective Computing*, *3*(1), 152–164. doi:10.1109/T-AFFC.2011.22

Frikha, M., Mhiri, M., & Gargouri, F. (2014). Toward a User Interest Ontology to Improve Social Network-Based Recommender System. In J. Sobecki, V. Boonjing, & S. Chittayasothorn (Eds.), *Advanced Approaches to Intelligent Information and Database Systems* (Vol. 551, pp. 255–264). Springer International Publishing. doi:10.1007/978-3-319-05503-9_25

Gavalas, D., & Kenteris, M. (2011). A web-based pervasive recommendation system for mobile tourist guides. *Personal and Ubiquitous Computing*, *15*(7), 759–770. doi:10.100700779-011-0389-x

Ghorab, M. R., Zhou, D., O'Connor, A., & Wade, V. (2013). Personalised information retrieval: Survey and classification. *User Modeling and User-Adapted Interaction*, *23*(4), 381–443. doi:10.100711257-012-9124-1

Gim, J., Jung, H., & Jeong, D. (2015). XOnto-Apriori: An Effective Association Rule Mining Algorithm for Personalized Recommendation Systems. In Computer Science and its Applications(pp 1131-1138). Berlin: Springer-Verlag.

Gopalachari, M. V., & Sammulal, P. (2014). Personalized collaborative filtering recommender system using domain knowledge. *Proceedings of International Conference on Computer and Communications Technologies (ICCCT)*, 1066-1071. 10.1109/ICCCT2.2014.7066693

Gruber, R. T. (1993). Toward principles for the design of ontologies used for knowledge sharing. *International Journal of Human-Computer Studies*, *43*(5-6), 907–928. doi:10.1006/ijhc.1995.1081

Hariri, N., Mobasher, B., & Burke, R. (2013). Personalized text-based music retrieval. *Intelligent Techniques for Web Personalization and Recommendation: Papers from the AAAI 2013 Workshop*, 24-30.

Hawalah, A., & Fasli, M. (2011). A multi-agent system using ontological user profiles for dynamic user modeling. *IEEE/WIC/ACM International Conferences on Web Intelligence and Intelligent Agent Technology*, 430-437. doi:10.1109/WI-IAT.2011

Hawalah, A., & Fasli, M. (2015). Dynamic user profiles for web personalisation. *Expert Systems with Applications*, *42*(5), 2547–2569. doi:10.1016/j.eswa.2014.10.032

Hong, J., Suh, E., Kim, J., & Kim, S. (2009). Context-aware system for proactive personalized service based on context history. *Expert Systems with Applications*, *36*(1), 7448–7457. doi:10.1016/j.eswa.2008.09.002

Hoppe, A., Nicolle, C., & Roxin, A. (2013). Automatic ontology based user profile learning from heterogeneous web resources in a big data context. *Proceedings of the VLDB Endowment International Conference on Very Large Data Bases*, *6*(12), 1428–1433. doi:10.14778/2536274.2536330

IJntema, W., Frasincar, F., Goossen, F., & Hogenboom, F. (2010). Ontology-Based News Recommendation. *Proceedings of EDBT '10, New York, USA.*

Jaison, G. V., & Varghese, C. M. (2015). Privacy protection in personalized web search using homomorphic encryption, *Int. J. of Scientific and Research Publications*, *5*(9), 1–5.

Jeckmans, A.J.P., Beye, M.R.T., Erkin, Z., Hartel, P.H., Lagendijk, R.L. & Tang, Q. (2013). Privacy in recommender systems. *Proceedings of Social Media Retrieval, Computer Communications and Networks*, 263-281.

Kalfoglou, Y., Domingue, J., Vargas-Vera, E. M. M., & Shum, S. B. (2001). *myPlanet: an ontology-driven Web-based personalized news service*. Retrieved from www.kmi.open.ac.uk/tr/papers/kmi-tr-102.pdf

Kobsa, A. (2001). Generic User Modeling Systems. *User Modeling and User-Adapted Interaction, 11*(1/2), 49–63. doi:10.1023/A:1011187500863

Kobsa, A., Knijnenburg, B. P., & Livshits, B. (2014). Let's do it at my place instead?: attitudinal and behavioural study of privacy in client-side personalization. *Proceedings of the SIGCHI Conference on Human Factors in Computing Systems (CHI '14) ACM New York USA*, 81–90. 10.1145/2556288.2557102

Kosala, R., & Blockee, H. (2000). Web mining research: A survey. *SIGKDD Explorations, 2*(1), 1–15. doi:10.1145/360402.360406

Lee, T. B., Hendler, J., & Lassila, O. (2000). The semantic web. *The Scientific Americans, 15*(1), 36. Retrieved from http://www.scientificamerican.com/2001/0501issue/0501berbers-lee.html

Lee, W., Liu, C., & Lu, C. (2002). Intelligent agent-based systems for personalized recommendations in Internet commerce. *Expert Systems with Applications, 22*(4), 275–284. doi:10.1016/S0957-4174(02)00015-5

Malik, Z. K., & Fyfe, C. (2012). Review of web personalization. *Journal of Emerging Technologies in Web Intelligence, 4*(3), 285–296. doi:10.4304/jetwi.4.3.285-296

Miao, C., Yang, Q., Fang, H., & Goh, A. (2007). A cognitive approach for agent-based personalized recommendation. *Knowledge-Based Systems, 20*(4), 397–405. doi:10.1016/j.knosys.2006.06.006

Middleton, S. E., Shadbolt, N. R., & De Roure, D. C. (2004). Ontological User Profiling in Recommender Systems. *ACM Transactions on Information Systems, 22*(1), 54–88. doi:10.1145/963770.963773

Mobasher, B., Honghua, D., Tao, L., & Miki, N. (2001). Effective personalization based on association rule discovery from web usage data. *Proceeding of 3rd International Workshop on Web Information and Data Management*, 9–15. 10.1145/502932.502935

Neves, A. R. M., Carvalho, Á. M. G., & Ralha, C. G. (2014). Agent-based architecture for context-aware and personalized event recommendation. *Expert Systems with Applications, 41*(2), 563–573. doi:10.1016/j.eswa.2013.07.081

Pakdeetrakulwong, U., & Wongthongtham, P. (2013). State of the art of a multi-agent based recommender system for active software engineering ontology. *International Journal of Digital Information and Wireless Communications, 3*(4), 29–42.

Pan, P., Wang, C., Horng, G., & Cheng, S. (2010). The development of an ontology-based adaptive personalized recommender system. *Proceedings of International Conference on Electronics and Information Engineering (ICEIE 2010), 1*, V1-76-V1-80. 10.1109/ICEIE.2010.5559842

Parra-Arnau, J., Rebollo-Monedero, D., & Forné, J. (2014). Measuring the privacy of user profiles in personalized information systems. *Future Generation Computer Systems, 33*, 53–63. doi:10.1016/j.future.2013.01.001

Peyravi, F., & Latif, A. (2014). Secure multi agent information filtering. *International Journal of Computer Theory and Engineering, 6*(3), 240–246. doi:10.7763/IJCTE.2014.V6.869

Pierrakos, D., Paliouras, G., Papatheodorou, C., & Spyropoulos, C. D. (2003). Web Usage Mining as a Tool for Personalization: A Survey. *User Modeling and User-Adapted Interaction*, *13*(4), 311–372. doi:10.1023/A:1026238916441

Prakash, K., & Sivakumar, R. (2013). Ontology-based intelligent multi-agent for diet food recommendation. *International Journal of Scientific & Engineering Research*, *4*(1), 1–8.

Razmerita, L. (2011). An ontology-based framework for modeling user behavior—A case study in knowledge management. *IEEE Transactions on Systems, Man, and Cybernetics. Part A, Systems and Humans*, *41*(4), 772–782. doi:10.1109/TSMCA.2011.2132712

Rosaci, D., & Sarnè, G. M. L. (2014). Multi-agent technology and ontologies to support personalization in B2C E-Commerce. *Electronic Commerce Research and Applications*, *13*(1), 13–23. doi:10.1016/j.elerap.2013.07.003

Safarkhani, B., Mohsenzadeh, T. M., & Mojde, M. M. R. (2009). Deriving semantic sessions from semantic clusters. *International Conference on Information Management and Engineering*, 523-528. doi:10.1109/ICIME.2009.131

Senkul, P., & Salin, S. (2011). Improving pattern quality in web usage mining by using semantic information. *Knowledge and Information Systems*, *30*(3), 527–541. doi:10.100710115-011-0386-4

Sharma, A., Al-khwaldeh, A. T., & Singh, A. (2016). ABSEP3S- An Agent Based Security Engine for Privacy Preserving in Personalized Search. *International Journal of Computing Academic Research*, *5*(3), 170–176.

Shen, X., Tan, B., & Zhai, C. (2007). Privacy protection in personalized search. *Proceedings of ACM SIGIR Forum*, *41*(1), 4-17.

Shou, L., Bai, H., Chen, K., & Chen, G. (2014). Supporting privacy protection in personalized web search. *IEEE Transactions on Knowledge and Data Engineering*, *26*(2), 453–467. doi:10.1109/TKDE.2012.201

Sieg, A., Mobasher, B., & Burke, R. (2007). Ontological user profiles for personalized web search. American Association for Artificial Intelligence, 84-91.

Singh, A. (2012). Wisdom Web: The WWW generation next. *International Journal of Advancements in Technology*, *3*(3), 123–126.

Singh, A., & Sharma, A. (2017). *Web Semantics for Personalized Information Retrieval. In Web Semantics for Textual and Visual Information Retrieval* (pp. 166–186). IGI Global.

Singh, A., & Sharma, A. (2017). MAICBR: A Multi-agent Intelligent Content-Based Recommendation System. *Springer Lecture Notes in Networks and Systems*, *10*, 399–411. doi:10.1007/978-981-10-3920-1_41

Singh, A., Sharma, A., & Dey, N. (2015). Semantics and agents oriented web personalization: State of the art. *International Journal of Service Science, Management, Engineering, and Technology*, *6*(2), 35–49. doi:10.4018/ijssmet.2015040103

Slimani, T. (2013). Approaches for semantic web service: An overview. *International Journal of Service Science, Management, Engineering, and Technology*, *4*(3), 18–36. doi:10.4018/ijssmet.2013070102

Sosnovsky, S., & Dicheva, D. (2010). Ontological technologies for user modeling. *International Journal of Metadata, Semantics and Ontologies*, *5*(1), 32–71. doi:10.1504/IJMSO.2010.032649

Sweeny, L. (2002). k-anonymity: A model for protecting privacy. *Int. J. Uncertain. Fuzz.*, *10*(5), 557–570. doi:10.1142/S0218488502001648

Tiwari, R. G., Husain, M., Srivastava, V., & Agrawal, A. (2011). Web personalization by assimilating usage data and semantics expressed in ontology terms. *Proceedings of International Conference and Workshop on Emerging Trends in Technology (ICWET 2011) – TCET 13-15 December Mumbai India*, 516-521. 10.1145/1980022.1980133

Toch, E., Wang, Y., & Cranor, L. F. (2012). Personalization and privacy: A survey of privacy risks and remedies in personalization-based systems. *User Modeling and User-Adapted Interaction*, *22*(1), 203–220. doi:10.100711257-011-9110-z

Trajkova, J., & Gauch, S. (2004). Improving ontology-based user profiles. *Proceedings of RIAO 2004, University of Avignon (Vaucluse), France, April 26-28*, 380-389.

Vigneshwari, S., & Aramudhan, M. (2012). A novel approach for personalizing the web using user profiling ontologies. *IEEE- Fourth International Conference on Advanced Computing (ICoAC), 13-15 Dec., Chennai*, 1-4. doi:10.1109/ICoAC.2012.6416815

Vuljanic, D., Rovan, L., & Baranovic, M. (2010). Semantically enhanced web personalization approaches and techniques. *Proceedings of the ITI 2010 32nd Int. Conf. on Information Technology Interfaces Croatia*, 217-222.

Wooldridge, M., & Jennings, N. (1995). Intelligent agents: Theory and practice. *The Knowledge Engineering Review*, *10*(2), 115–152. doi:10.1017/S0269888900008122

Xu, Y., Zhang, B., Chen, Z., & Wang, K. (2007). Privacy-enhancing personalized web search. *Proceeding of WWW 2007 Banff Canada*, 591-600.

Chapter 11
Applying Formal Methods to Verify Web Services Orchestration and Choreography

Zohra Sbai
Prince Sattam Bin Abdulaziz University, Saudi Arabia & Tunis El Manar University, Tunisia

ABSTRACT

The contribution presented in this chapter is to provide a formal framework ensuring the model checking based verification of the Composition of Web Services (WSC). For this, the authors propose first to model the web services composition by an interaction of open workflow nets: a special class of Petri nets. Then, they detail how to check behavioral properties specified in temporal logic using the model checker NuSMV. A WSC is with added value only if the involved services are compatible. So, in this context, across the translation proposed, the authors develop a verification layer of the WSC compatibility. This work is developed in a framework named D&A4WSC which allows to model the WSC by oWF-nets and to check their compatibility by invoking the model checker NuSMV. Furthermore, the authors extended D&A4WSC so that it permits a web services choreography described in a WS-CDL specification. For this they developed a translation from WS-CDL to a composition of oWF-nets, so that one can verify this choreography by the aforementioned approach.

INTRODUCTION

Composition of reusable services often provided by different service providers is the key of service oriented software development. This involves the composition of software systems from reusable components and the ability to dynamically locate necessary services at run-time. These services may be provided by other parts as Web services, but the essential element is the dynamic nature of the connection between the service users and the service providers. Web service technology provides a flexible and universal approach to the interoperability of heterogeneous systems. We distinguish two categories of coordination

DOI: 10.4018/978-1-5225-5951-1.ch011

processes: processes which implement the choreography of services and processes implementing their orchestration. While the orchestration is centralized around a composition engine and there is a static service composition which follows a predefined pattern, the choreography form a composition approach based on decentralized cooperation between a collection of services. Dealing with the one or the other, the formal verification of such a composition is a very interested research subject.

In the SOA framework, process automation is a major focus, with the concepts of orchestration and composition of services. This is to centralize the logic of a process in a dedicated component, which supports the Web and business rules associated. This approach tends to reduce the impacts related to fluctuations in the process.

The generalization of the orchestration approach leads to the choreography which consists in designing a decentralized coordination of applications, in which there is no privileged machine but a network of interconnected machines that exchange messages.

Several languages have been proposed to ensure the composition of Web services namely BPEL (Ma, C., Xu, Q., and J.W. Sanders, 2009) and WS-CDL (Hongli, Y., Xiangpeng, Z., Zongyan, Q., Geguang, P., & Shuling, W., 2006) which are based on the XML syntax. BPEL is used to define the abstract and executable business processes as a set of Web services interaction coordinated recursively. WS-CDL is a language that describes peer-to-peer collaborations of participants by defining, from a global viewpoint, their common and complementary observable behavior; where ordered message exchanges result in accomplishing a common business goal.

For the best achievement of this common goal, the composed services should be compatible. This property ensures to obtain a composite service that does not suffer from problems of deadlock. It is therefore necessary to make sure that these services will be able to interact properly, which requires a study, a clear understanding and a verification of the concept of compatibility. In this context, we propose to use formal methods to model a given composition and thus to be able to verify it. In particular, we propose to use open workflow nets (oWF-nets), a subclass of Petri nets, which are characterized by their interface places that ensure the communication between the different services. In addition to this, given that the choreography refers to the description of coordinated interactions between two or more parties, we propose to translate a WS-CDL description into a composition of oWF-nets so that we will be able to formally verify the given choreography description.

To implement our approach, we propose D&A4WSC as a framework having as a main goal the verification of WSC using formal methods. The verified WSC can be generated to a valid BPEL process so that it can be executed by an orchestration engine. More precisely, D&A4WSC allows first the modeling of a WSC using a composition of oWF-nets or the generation of oWF-nets composition from a WS-CDL description. Secondly, it allows the compatibility verification. For this, we first check the compatibility at the level of exchanged messages between the different services, that we call syntactical verification. Then, we move to the semantic verification based on the principle of model checking, and more precisely, on the model checker NuSMV (Cimatti, A., Clarke, E.M., Giunchiglia, F., & Roveri, M., 1999). In fact, model checking is one of the most used formal verification methods because of its effectiveness, its ease of use and the generation of a counter example in case of error. If the compatibility is valid, then, we generate an owfn file of the composition and we invoke the compiler oWFN2BPEL (Lohmann, 2008) which takes as input an owfn file and generates a BPEL file as output.

This chapter is organized as follows: Section 2 presents preliminaries on oWFnets, compatibility, BPEL, WS-CDL and the model checking technique used in this proposal. Section 3 describes the approach of formal checking of the compatibility of the composition of Web services and explains the translation of oWF-nets to BPEL process. In Section 4, we present the framework D&A4WSC that implements the proposed approach and a case study illustrating our approach. In section 5, we discuss the related works. Finally, section 6 concludes the chapter and outlines the future work.

PRELIMINARIES

We present in this section the main notions used in the rest of this chapter.

Web Services Composition

The Web services composition allows multiple services to communicate and collaborate by exchanging messages and to implement a composite service that can perform consumer complex tasks. With increasing attention given to WSC, it becomes more and more important to standardize the used terminology. Orchestration refers to automated execution of a workflow, i.e., a workflow is defined using an execution language such as BPEL, and an orchestration engine is used to execute the workflow at runtime. An orchestrated process is typically exposed as a service and does not describe a coordinated set of interactions between two or more parties. However, choreography refers to a description of coordinated interactions between two or more parties. In the rest of this section, we describe BPEL as an orchestration language and WS-CDL as a choreography language.

BPEL Language

Given the multiple applications and partners to integrate, as well as the strong need to parallelize business processes which are becoming increasingly complex. This necessitated a change in processes. It is in this way that several languages for business process modeling appeared. They are based on similar concepts namely the business, operations, exchange messages, error handling, etc.

Among these languages, we focus on BPEL (Business Process Execution Language). It is a language, written in XML, which describes business processes and interacts with Web services. It is a part of the description of orchestration of Web service. It allows, in an automated way the integration of a large number of applications published in the form of services, without dependence on neither platform nor language.

To implement a BPEL process, there are two different tools that can be used which are the graphical editor of BPEL, usually integrated into an IDE such as Eclipse and BPEL engine integrated with application servers.

BPEL processes interact in two ways with Web services:

- Invoke operations from other services.
- Receive invocations from clients.

These two relationships between Web services and BPEL processes are presented by a partner link called "PartnerLink". This link transmits messages between the different partners and is stored in the variables. These variables allow the BPEL process to have an internal state by manipulating the different exchanged messages.

The behavior of BPEL processes is described by a set of activities that handle exceptions, manipulate XML data and ensure communication BPEL process with other Web services.

The activities of BPEL processes are classified into two types:

- **Basic Activities:** They allow to build the business logic of the BPEL process and are: 1) Reply (to send messages); 2) Invoke (call a service); 3) Receive (receive a message); 4) Terminate (end a process); 5) Throw (throw an exception); 6) Assign (Manipulate data) and; 7) Wait (suspend the process for a specified period).
- Structured activities, such as sequential, parallel, conditional and loops activities.

WS-CDL Language

A WS-CDL description is a specification based on an XML document which aims to describe the collaborative observable behavior of multiple services which interact to achieve a common goal.

A WS-CDL specification is composed of two parts (Hongli, Y., Xiangpeng, Z., Zongyan, Q., Geguang, P., & Shuling, W., 2006) as described in Figure 1.

- **Static Part**: Specifies the entities that cooperate in WS-CDL description such as:
    - **RoleType:** Specifies the implementation process given by a participant with a specific role.
    - **ParticipantType:** An entity having a particular set of roles in the choreography.
    - **RelationshipType:** Defines the relation between the roles and declares that two role Types have a need to interact to achieve some set of Web services.
    - **ChannelType:** The medium used to communicate between participant Type with specifying the where and how information is exchanged.

Figure 1. WS-CDL description

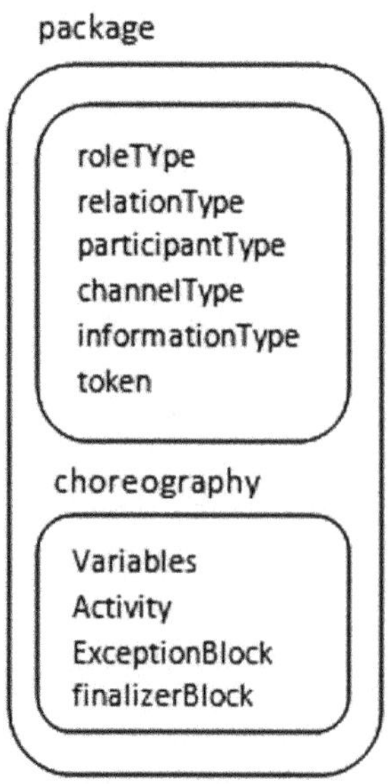

- **Dynamic Part:** Describes interactions between the participants and their temporal interdependencies. It includes a definition of the set of variables used to realize the data dependent behavior of the choreography. The choreography activity construct presents the basic building block of WS-CDL. It describes the behavior of the choreography, while an exception block defines the handling of exceptional events during the execution of the choreography activity, and a finalize block describes the actions to be taken upon termination of the choreography. There are three classes of activities:
  - **Basic Activity:** Interaction is the principal basic activity. It describes the communication between two roles with exchanging information.
  - **Structural Activity:** Refers to an activity that present the scheduling rules in a choreography.
  - **Work-Unit Activity:** It is a conditional and looping construct combined in an enclosing activity.

Open Workflow Nets

A Petri net is a formal language used to model the functionality behavior of concurrent systems. This modeling language is represented as an oriented graph which makes it readable and ensures an easy communication with different designers. Petri nets can be considered as a tool for verifying business processes since they have strong mathematical foundations. A Petri net is represented by a graph containing: a set of places P, a set of transitions T and a set of arcs: $F \subseteq (P \times T) \cup (T \times P)$. Thus, a Petri net (PN) is defined by a quadruplet N= (P, T, F, W) with W is the set of weight of the arcs (x,y).

A marking of a PN at time t is $M(t) = \{ m_1(t); m_2(t);..; m_n(t)\}$ with $m_i(t)$ is the number of tokens in place i at time t. If W(u)=1 ∀ u∈ F then N is said to be an ordinary net and it is denoted by N = (P,T,F).

We denote by $[M_i >$ the set of all accessible marking from M_i. The previous *set* of a node x ∈ P ∪T is given by •x= {y ∈ P U T / (y,x) ∈ F } and the *next set* is given by x•= {y ∈ p U T / (x,y) ∈ F }.

A transition t ∈ T is enabled in a marking M (denoted by M [t>) if and only if ∀ p ∈ •t, M(p)≥W(p,t).

Petri nets are drawn as follows: the places are represented by circles and the transitions by rectangles. For the flow relation, an arc is drawn between x and y for each (x,y) in the relation. The weight function labels the arcs in case their weights are greater than 1. For the marking M of a Petri net, we draw M(p) tokens into the circle representing the place p.

Many subclasses of Petri nets have been proposed in the literature. For WSC, we propose to use the subclass of open workflow nets (oWF-net). An oWF-net is a PN which has an input place indicating the start of process execution, and an output place indicating its end. Besides, an oWF-net has interface places serving for communication with other oWF-nets. They are classified into two types: Input interface places that receive tokens from other oWF-net, and Output interface places that send tokens to other oWF-net. By sending and receiving tokens, a communication mechanism is processed between different oWF-nets. This mechanism can be with benefit when modeling Web services composition.

Consequently, an oWF-net (Sbaï & Barkaoui (2013)) N= (P, T, F, I, O, m_0, m_f) consists of:

- A Petri net N = (P; T; F; m_0; m_f) with m_0 the initial marking and m_f the final marking.
- Interface places (I U O) ⊂ P knowing that:

$$\forall p \in \left(I \cup O\right), m_o\left(p\right) = m_f\left(p\right) = 0$$

$$\forall p \in I, p^{\cdot} = \Phi\left(\text{Input interface places}\right)$$

$$\forall p \in O, p^{\cdot} = \Phi\left(\text{Output interface places}\right).$$

In this work, we model a WSC by a composition of oWF-nets. Hence, n Web services are modeled by n oWF-nets N_1, .., N_n communicating together via the places in I_i and O_i, i=1..n.

Compatibility

Compatibility verification in a WSC is an important step. In Fact, a WSC is compatible only if the two following points are ensured:

- **The Compatibility at the Level of the Exchanged Messages Between Services:** Since we use oWF-nets, we need to check the compatibility of their input and output interface places that model the exchange messages.
- **The Compatibility of the Composite Service:** In fact, here we need to verify that after the composition of autonomous services, the composite service does not suffer from any problem that could damage its performance such as deadlock problem.

Both verification steps have already been addressed in many previous works.

Below is the verification of interface places.

Definition 1 (Henzinger, T.A., Nicollin, X., Sifakis, J., & Yovine, S., 1994):

Let N1 and N2 be two oWF-nets with pair wise disjoint components except for the interface places. If input interface places of an oWF-net have intersected with output interface places of another oWF-net output, that is to say: I1\I2 = Φ and O1\O2= Φ, then N1 and N2 are interface compatible.

Below is the definition of the compatibility verification of the composite service which consists in verifying the absence of deadlock.

Definition 2 (Hongli, Y., Xiangpeng, Z., Zongyan, Q., Geguang, P., & Shuling, W., 2006): *Given a Petri Net N, N is deadlock-free if and only If* $\forall M \in [M_o >, \exists t \in T$, *M[t>.*

Unlike (Bordeaux, L., Salaun, G., Berardi, D., & Mecella,M., 2005; Foster, H., Uchitel, S., Magee, J., & Kramer, J., 2004) Martens, A., (2003) who just verified the absence of deadlock to prove compatibility, Barkaoui et al. (2010) have classified this type of checking as a verification of weak compatibility and proposed two other different types of compatibility: compatibility and strong compatibility. Below are the definitions of these three compatibility subclasses:

- **Weak Compatibility:** A composition is weak compatible if it is deadlock-free.

- **Compatibility:** A composition is compatible if it is already weak compatible and furthermore it verifies the proper termination property.
- **Strong Compatibility:** A composition is strong compatible if it is compatible and quasi-live.

The proper termination and quasi-liveness are explained later in the chapter.

In Barkaoui et al. (2010), the authors theoretically verified these three types of compatibility while we propose to formally verify them using the principle of model checking.

Model Checking

Formal methods and tools have been proved useful to give high level and precise descriptions of computer systems, and to analyze exhaustively these systems at early phases of the system development process. They are a particular kind of mathematically based techniques for the specification, development and verification of software and hardware systems.

Model checking is a formal verification method which is based on its ability to explore exhaustively all the possible states of the model and then to verify if the desired requirement is guaranteed in every possible execution. In addition, this formal method permits to produce a counter example if the checked property is violated and hence it facilitates the process of error characterization as well as its correction. For this reason, we choose to use the model checking formal method.

Tools automating the model checking (named model checkers) are as well available. Among many model checkers, we choose to use one of the main model checkers, NuSMV, which is based on SMV (Symbolic Model Verifier) (Henzinger, T.A., Nicollin, X., Sifakis, J., & Yovine, S., 1994). NuSMV is an extension and re-implementation of SMV symbolic model checker, the first model checking tool which is based on Binary Decision Diagrams (BDDs) (Antonik, A., & Huth, M., 2006). The tool has been designed to be an open architecture for model checking. It has been developed in a joint project between ITC-IRST, the University of Genoa, the University of Trento and Carnegie Mellon University.

NuSMV offers many features including essentially the navigation of a counter example. In fact, all the model checkers can generate a counter example, the particularity of NuSMV is that it offers the possibility to the user to move from one state to another and even to assess the temporal logic expression of the counter example status (Cimatti, A., Clarke, E.M., Giunchiglia, F., & Roveri, M., 1999).

NuSMV verifies temporal logic properties over finite state systems. It uses a symbolic representation of the specification in order to check if a model satisfies a property. NuSMV is able to verify properties specified in LTL (Linear Temporal Logic) as well as in CTL (Computation Tree Logic) (Marc, F., Benoit, T., Romain, C., Raphael, C.Y.F., & Mohammed, O., 2010).

We chose to adopt in this work the temporal logic CTL since it permits to use universal as well as existential quantifiers which allow us to reason on a state and many future states possible from it.

CTL formulas are inductively defined as follows:

$$\begin{aligned}\varphi ::= \top \,\big|\, \bot \,\big|\, p \,\big|\, \neg\varphi \,\big|\, \varphi \wedge \varphi \,\big|\, \varphi \vee \varphi \,\big|\, \varphi \rightarrow \varphi \,\big|\, AX\varphi \,\big|\, EX\varphi \,\big|\, AF\varphi \,\big| \\ \big|\, EF\varphi \,\big|\, AG\varphi \,\big|\, EG\varphi \,\big|\, A\big[\varphi \sqcup \varphi\big] \big|\, E\big[\varphi \sqcup \varphi\big]\end{aligned}$$

where A models all possible paths from the current moment, E models at least one path from the current moment. A and E are usually followed by one of the following prefixes: X, U, G and F.

A CTL formula f is built from atomic propositions corresponding to variables, boolean operators such as: ¬, ∨, ∧, → and temporal operators. Each temporal operator consists of two parts. A path quantifier (A or G) followed by a temporal modality (F, G, X, U). The temporal operators are interpreted relatively to a current state s. The path quantifier indicates if the temporal operator expresses a property that should hold on all paths starting at s (denoted by the universal path quantifier A), or at least on one of these paths (denoted by the existential path quantifier E). The temporal modalities are interpreted as follows:

- Xp (neXt time p) is true if the formula p is true in the state reached immediately after the current state.
- Fp (Future p) is true if there exists a state in the path where the formula p is true.
- Gp (Globally p) is true if p is true at every state in the path.
- pUq (p Until q) is true if there is a state in the path where q is true and p is true in all preceding states (if any). This definition is called strong until. Weak until would also be true when q holds forever.

FORMAL MODELING AND VERIFICATION OF WSC

Our aim is to propose a generic framework called D&A4WSC (Design and Analysis For Web Services Composition) dedicated to the formal modeling and analysis of WSC. We propose to part from a formal, simple and expressive model which is oWF-net and then to build a solid approach of checking the compatibility of the different services involved in the WSC. The verification is based on the principle of model checking which returns either a confirmation of the validity of the property, or the invalidity of the property with a counter example.

In addition to the departure from a composition of oWF-nets to model and verify a WSC in D&A4WSC, one can enter a WS-CDL specification to D&A4WSC and benefit from the formal verification of the compatibility of the given Web services chorography ensured by this framework.

Once the Web services compatibility is checked, we aim to generate the correct WSC into common languages used to deal with Web services composition such as BPEL.

So, the approach we propose is concretized in the framework D&A4WSC which architecture is described in Figure 2.

This framework is keen on five modules summarized as follows:

- Modeling of a composition of Web services using the composition of oWF-nets.
- Translation of a given WS-CDL specification of Web services choreography into a composition of oWF-nets.
- Translation of a composition of oWF-nets to SMV in order to be checked with NuSMV model checker.
- Formal verification of the compatibility of Web services given by oWF-nets composition.
- Translation of a WSC verified compatible into a BPEL process.

Figure 2. Architecture illustrating the proposed framework

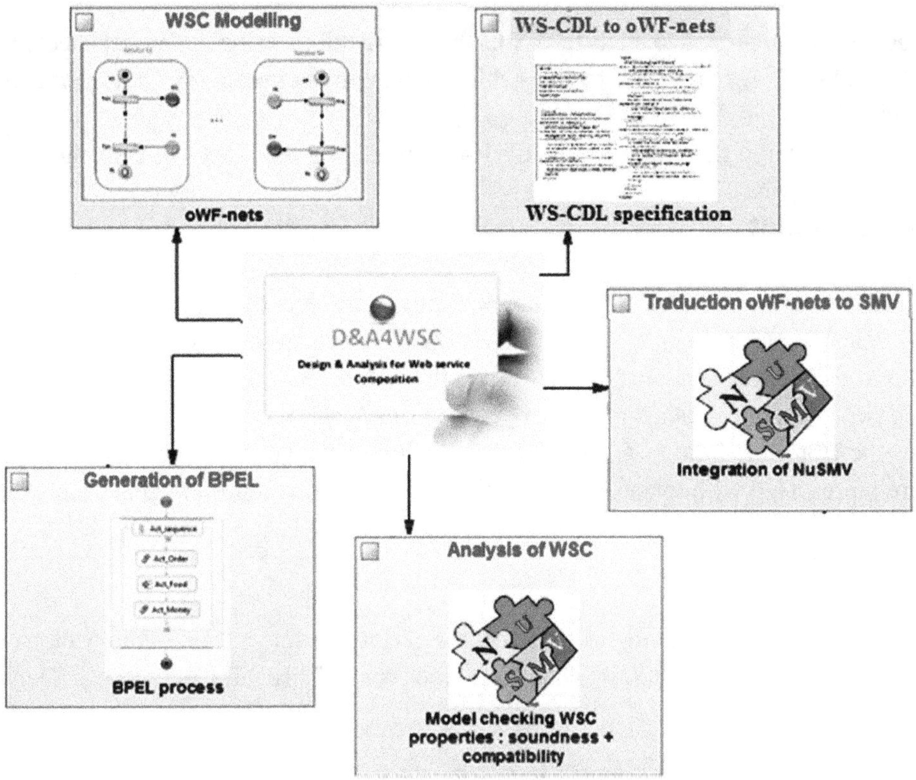

We explain in the following each of these modules.

Modeling of the WSC With a Composition of oWF-Nets

To model Web services, we used the oWF-nets. Indeed, as we noted in section I.1, oWF-nets are characterized by their input and output interface places providing communication between each other. Thus, we model a composition of Web services by oWF-nets which communicate by exchanging messages. The exchanged messages are represented in interface places of the oWF-nets. Consequently, each Web service is modeled by an oWF-net with:

- **An Input Place:** Containing a token indicating the start of execution of the service,
- **An Output Place:** The place that mentions the end of execution of the service,
- Places that represent the preconditions of the service,
- Transitions that represent the tasks that a service can perform,
- Interface places which connect a service to one or more service(s). The interface places are of two types: input interface places (which receive messages from other services) and output interface places (which send messages to other services).

WS-CDL Translation to oWF-Nets

In order to develop an approach to analyze WS-CDL specifications using the model checking technique assured by D&A4WSC, we propose first to establish a mapping between WS-CDL elements and oWF-nets constructs.

We briefly represent in the following the oWF-nets modeling for WS-CDL elements.

Interaction Element

Interaction element is the basic block in a choreography. It reflects the information exchange between roleTypes (Decker, G., Overdick, & H., Zaha, J.M., 2006) which are represented by places and referred to as the *fromRoleType* (the sender of message) and the *toRoleType* (the receiver of message).

There are two types of exchange messages between *roleTypes*: request or response. Request is made always from the fromRoleType to *toRoleType* and response is made in the opposite direction. These messages are represented with tokens. An example of interaction is represented in Figure 3.

Activity Element

Activity element defines the ordering rules of records performed into a choreography element (Decker, G., Overdick, & H., Zaha, J.M., 2006). There are three types of structural activities namely Sequence, Parallel and Choice.

- Sequence Activity represents one or more activities performed in the same order as their definition (Figure 4a).
- Parallel Activity represents one or more activities performed at the same time (Figure 4b).
- Choice Activity represents the performing of one activity chosen among a set of activities (Figure 4c).

In the choreography, each Web service is modeled by an oWF-net describing the control flow between the tasks ensured by this service and containing interface places used to communicate with other Web services. The overall obtained model is a Petri net with a set of input places, a set of output places and a set of interface places. Before processing to the generated of:

Figure 3. oWF-net Model for request-response exchange messages

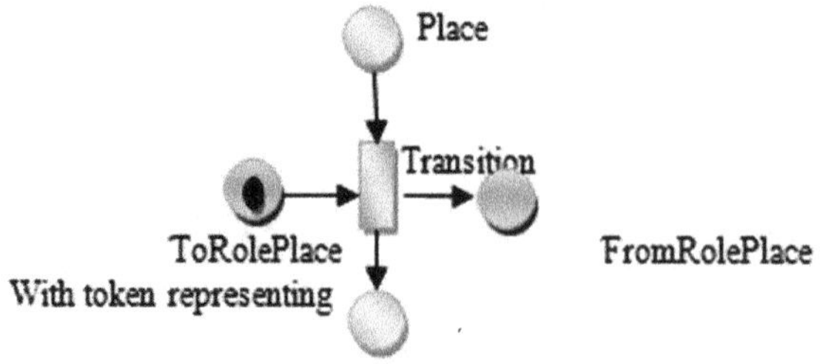

Figure 4. Sequence, Parallel and Choice activities

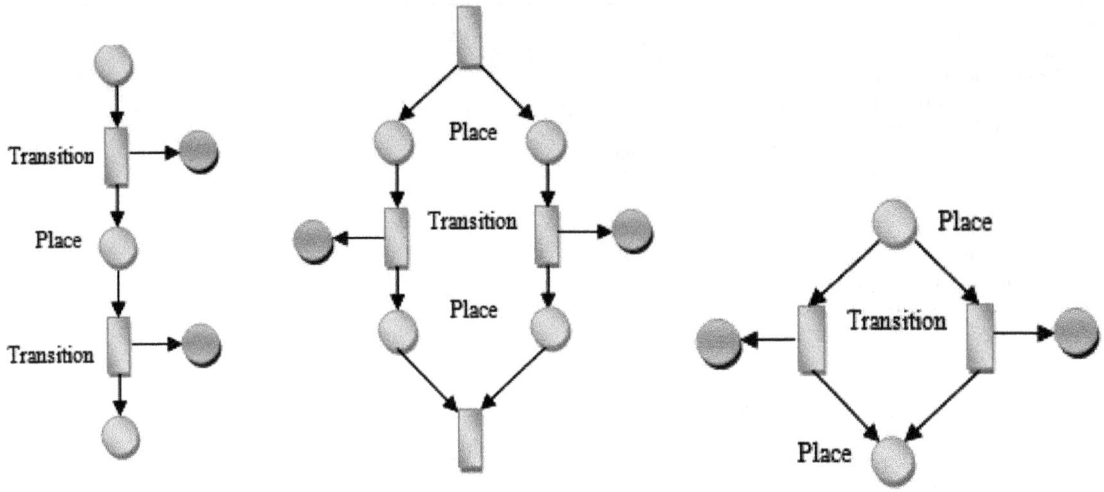

This generation is done according to algorithm 1, knowing that PT represents an array of services and S represents the array of interactions for each service. In this algorithm, we assume the following definitions:

- **Int:** A transition will be drawn with its input interface if the service is the receiver or output if the service is the sender.
- **SeqInt:** A series of transitions will be displayed with corresponding interfaces according to the role of the service. A sequence has at least two interactions.
- **ChInt:** If the service is involved in two different interactions with two different services and has to choose between them, so two arrows will be drawn each one is followed by a transition and its appropriate interface.
- **ChSeqInt:** The same principle of "ChInt" except that it is a sequence of transitions to draw.
- **ParaInt:** If the service is involved in two interactions that take place at the same time. So two arrows will be drawn each one is followed by a place and a transition with appropriate interface.
- **ParaSeqInt:** The same principle of "ParaInt" except that it is a sequence of transitions to draw.

Compatibility Checking

Having modeled the set of Web services, we can move to the phase of verification of the compatibility of these services. This is an important step and must be accomplished before composing services. To do this, we first check the syntactic compatibility of Web services that mainly focuses on the conformance of the interface places. Then, if the syntactic compatibility is checked, we move to verify the semantic compatibility by checking the three classes of compatibility mentioned in the preliminaries section using the principle of model checking. We detail in the following the syntactic compatibility as well as the three types of semantic compatibility.

Algorithm 1. oWF-nets generation from WS-CDL

```
Input: pt,i,k,n: integer
1: Begin
        for pt=1 to PT.length do
                draw(StartPlace);
                for i=1 to n do
                        if PT[pt]=M[i][2] or PT[pt]=M[i][3] then
                                /* M[i][2]="FromRole",M[i][3]="ToRole" */
                                fill(S);
                        end if
                end for
                for k=1 to S.lenght do
                        if S[k][0]="Int" then
                                draw(Transition);
                                draw(InterfaceO/I);
                                draw(place);
                        else if S[k][0]="SeqInt" then
                                draw(SequenceTransition);
                                draw(InterfaceO/I);
                        else if S[k][0]="ChInt" then
                                draw(place);
                                draw(ChoiceTransition);
                                draw(InterfaceO/I);
                        else if S[k][0]="ChSeqInt" then
                                draw(ChoiceSequenceTransition);
                                draw(InterfaceO/I);
                        else if S[k][0]="ParaInt" then
                                draw(ParallelTransition);
                                draw(InterfaceO/I);
                        else if S[k][0]="ParaSeqInt" then
                                draw(ParallelSequenceTransition);
                                draw(InterfaceO/I);
                        else /* "FinalizerBlock/ExceptionBlock" */
                                draw(Transition);
                                draw(InterfaceO/I);
                                draw(EndPlace);
                        end if
                end for
        end for
end
```

Syntactic Verification of Compatibility

The syntactic verification of compatibility consists in:

- Checking the number of interface places: Here we must verify that each service has at least one interface place, whether input or output place, so that it can communicate with other services. In fact, if a service has no interface place, then, it cannot send nor receive messages from other services, so it can't even exchange messages with other services.
- Verification of compliance of the names of interface places: Here we must verify that for each input interface place to a service, named "i", there must be at least one output interface place from other service named "i" and vice versa (i.e., we must also ensure that each output interface place of S1, having "o" as a name, must have at least one input interface place to another service having as name "o"). Thus, each message sent by a service is surely received by another one.

To implement these two steps, we are based on two matrices:

- **Matrix I:** For which the number of rows corresponds to the number of services, and the number of columns corresponds to the number of input interface places of each service. The matrix O contains all input interface places of the WSC.
- **Matrix O:** For which the number of rows corresponds to the number of services and the number of columns corresponds to the number of output interface places of each service. The matrix O contains all output interface places of the WSC.

To check the first condition of syntactic compatibility, it is necessary to check that each row of one of the matrix O and I has at least one non-empty element, i.e. each service has at least one (input or output) interface place.

Once the number of interface places is checked, we move on to verify the conformity of these places. Indeed, we rely here on the matrices O and I to implement the algorithm 2 given as follows, knowing that:

- n: is the number of rows of the matrices I and O.
- m: is the number of columns of the matrix I.
- m1: is the number of columns of the matrix O.

Where Compare-Noun(I,O,n,m) compares the name of the element I[i][j] with all elements of matrix O except those of line I and Assign-Value(I;O; i; j) assigns a value 11 to an element of the array test if I[i][j] has no name that conforms in the matrix O.

Suppose that:

Algorithm 2. Checking the conformity of names of interface places

```
For i = 1 to n do
    for j = 1 to m do
        Compare-Noun(I,O,n,m)
        test ← Assign-Value(I;O; i; j)
    end for
end for
for i=1 to n1 do
    for j=1 to m1 do
        Compare-Noun(O,I,n1,m1)
        test ← Assign-V alue(O; I; i; j)
    end for
end for
for i=1 to indice do
    if test[i] = 11 then
        y y + 1
    end if
end for
if y = 0 then
    Name of interface places does not conform.
else
    The composition is syntactically compatible
end if
```

- T_1 (n;m): is the number of elementary operations of the function Compare-Noun(I,O,n,m).
- T_2 (n1;m1): is the number of elementary operations of the function Compare-Noun(I,O,n1,m1).
- T_3 (l): is the number of elementary operations of the function Assign-Value(I,O, i, j).

Then, the complexity of this algorithm is given as follow:

$$(T_1(n;m) + T_2(n1;m1) + T_3(l))\theta(n^* m) + (T_2(n1;m1) + T_3(l))\theta(n1^* m1)$$

Semantic Verification of Compatibility

Semantic verification consists in checking three types of compatibility namely: weak compatibility, compatibility and strong compatibility. These properties are verified using the principle of model checking. In particular, we use the symbolic model checker NuSMV which takes the SMV language as input language. So, we first translate the modeling of the composition of Web services into SMV code. Then, we express the compatibility properties with CTL and send them to the model checker to be verified. Here, NuSMV will either return a positive response or a negative one with a counter example to help the user to diagnose the error. We summarize this step in the following two points:

- Translation of the modeling to SMV code.
- Expression of the compatibility properties in CTL temporal logic.

These steps will be detailed in the following.

Translation of the Modeling into SMV Code

To generate the SMV code, we defined three tables in which we inserted the various elements of oWFnets. Places, input places and output ones are all included in a table named **PLS**. Each table **PLS** represents a Web service modeled by an oWF-net. The number of columns in this table is the number of places, input places and output ones of the service that it represents. As for transitions, they are introduced in a table named **TRS**. The number of **TRS** tables is the number of Web services. Note that the number of columns of each TRS table corresponds to the number of transitions of the service it represents. Finally, input and output interface places of all services included in the composition are inserted into an array named **I**, i.e. there is only one array **I** for the whole interface places of the composition. We note that two or more interface places having the same name will be considered as a single interface place. We rely on these three tables to define the different parts of the structure of SMV code. In fact, this code consists of four main parts which are:

- **Var**: This part is dedicated to the declaration of tables. Here, we declare the three types of tables: PLS, TRS and I. Recall that the number of tables TRS and PLS is equal to the number of services included in the composition.

- **INIT**: This section is devoted to the initialization of declared elements. We initialize all the elements of the tables PLS, TRS and I to zero except for the input places, that is to say $PLS_i[0]$.will be initialized to 1 since they have a token.
- **TRANS**: This part expresses the evolution of transitions. In fact, every transition has at least one input place and one output place. To make this transition allowed, its input place must send a token to its output place. Note that during the evolution of this transition, all other transitions and places don't evolve. We define a variable in this part which we call "deadlock" and which is verified only if all places and transitions do not change.
- **DEFINE**: This part is devoted to defining the condition of crossing of each transition. We verify if each input place at each transition is strictly positive, which means that it has at least one token. The variable "Deadlock" is not verified if all transitions are allowed.

In order to implement this structure of SMV code, we rely on the matrix M in which we saved the different types of objects. In fact, every object has an integer type unique to him as follows:

- The type of place is 2,
- The type of input place is 0,
- The output place has 3 as type,
- The transition is of type 1,
- The types of input and output interface place are respectively 4 and 5.

The number of rows of this matrix is equal to the number of services included in the composition, and the number of columns is equal to the number of objects that composes a service.

To better understand the translation that we proposed, we express a simple example of service, which contains an input place, an output interface place, a transition and an output place, with the following SMV specification.

For example, the matrix of the previous example is giving as follows:

M	0	4	1	3

Relying on matrix M, we implemented an algorithm to generate the SMV code. The SMV code of the previous example is giving in the Figure 5.

CTL Formulas of the Compatibility Properties

Checking the semantic compatibility amounts to verify the three properties of compatibility by using the symbolic and open source model checker NuSMV (Cimatti, A., Clarke, E.M., Giunchiglia, F., & Roveri, M., 1999). It takes as input an SMV specification and a formula expressed in CTL or LTL. In this work, we use CTL since it can reason about the non-deterministic systems taking into account the temporal operators and quantization operators on the execution paths. Grammar CTL contains four types of prefixes: X (next), F (future), G (globally) and U (Until), which are always preceded by A (which means all accessible paths from the current time) or E (i.e. at least one accessible path from the current time).

Figure 5. Example of SMV code

```
MODULE main
VAR

    PLS1: array 0..2 of boolean;
   TRS1: array 0..1 of boolean;
    I: array 0..1 of boolean;

 INIT
    (PLS1[0]=1 & PLS1[1]=0 & TRS1[0]=0 &
I[0]=1)

TRANS
    -TRS1[0]=T1
TRS1[0].enabled & next(TRS1[0])=TRS1[0]+1 &
next(PLS1[0])=PLS1[0]-1 & next(I[0])=I[0]-1 &
next(PLS1[1])=PLS1[1]+1
|
Deadlock & next(TRS1[0])=TRS1[0] &
next(PLS1[0])=PLS1[0] & next(I[0])=I[0] &
next(PLS1[1])=PLS1[1]

 DEFINE
    TRS1[0].enabled:= PLS1[0]>0 & I[0]>0;
    Deadlock: =! (TRS1[0].enabled);
```

Below are the expressions of the compatibility properties.

- Weak Compatibility

A composition is weakly compatible if it is deadlock-free. Thus, we express the property of deadlock-freeness in CTL which consists in verifying the existence of a transition allowed for any marking.

(N,M0) is deadlock-free $\Leftrightarrow \forall M \in [M_o >, \exists t \in T$. *M[t>*. The expression of the weak compatibility in CTL is given as follows:

$$EG(EF(\&_{s=1}^{nbS}(PLS_s\left[f_s\right] > 0))).$$

- Compatibility

A WSC is compatible only if it is weak compatible and verifies the proper termination property. Indeed, the proper termination property allows the process to finish its execution at any time but requires that at the time of termination, all places should be empty except the output ones which must contain at least one token. Therefore, we should check whether there is a marking M for which all places are empty except output places. The expression of proper termination is given as follows:

$$\forall M \in [M_0 >, \left(M \geq M_f \Rightarrow M = M_f\right).$$

The expression of this property in CTL is given as follows:

$$AG\left(EF\left(\&_{s=1}^{nbS}\left(\left(\&_{p=1}^{nbp}(PLS_s[p] = 0 \ \& PLS_s[f_s] > 0)\right)\&\left(\&_{i=1}^{nbi} I[i] = 0\right)\right)\right)\right).$$

- Strong Compatibility

We focus here on the property of quasi-liveness. Indeed, a transition is said to be quasi-live if, from the initial marking, it is crossed at least once. A PN is quasi-live if all its transitions are quasi-live.

$$t \in T \, is quasi - live \Leftrightarrow \exists M \in [M_o > such as M[t >.$$

The expression of the strong compatibility in CTL is as follows:

$$\&_{s=1}^{nbS}\left(\&_{t=1}^{nbt} EG\left(EF\left(TRS_s[t] > 0\right)\right)\right)$$

Translation Into BPEL

Having modeled the composition of Web services using the composition of oWFnets, and having checked the syntactic and semantic compatibility of this composition, we can generate the BPEL process corresponding to the composite service. To do this we should generate the corresponding XML file. Foe this, we use the oWFN2BPEL compiler which allows the generation of BPEL process from a description called owfn. Indeed, the description owfn is easier to generate than the XML file. So, we translate, first, the composition of the oWF-nets into an owfn file, then, we invoke the oWFN2BPEL to convert this file to BPEL.

The translation approach we propose is described in Figure 6.

FRAMEWORK D&A4WSC

The approach proposed above is implemented in our framework D&A4WSC developed in java. D&A4WSC is characterized by its simple interface to use, allowing the user to model the composition of Web services

Figure 6. Steps of BPEL process generation

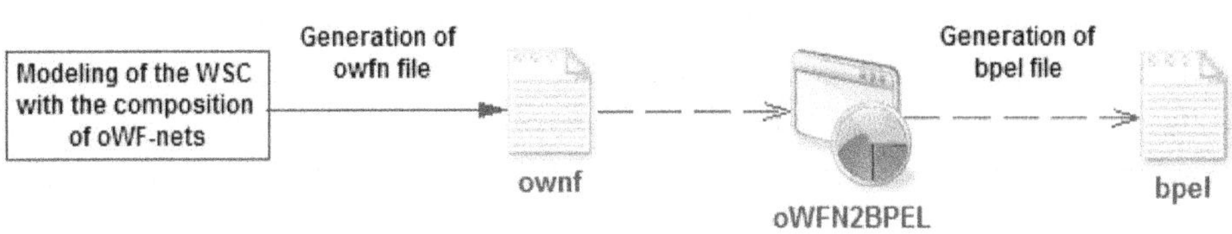

by a composition of oWF-nets. It allows in special, the compatibility checking and the generation of a BPEL process of the composite service. Mainly, D&A4WSC user can:

- Model a WSC by a direct edition of oWF-nets,
- Request for a translation of a given WS-CDL description to oWF-nets,
- Request the verification of the WSC composition,
- Request the syntactic and semantic verification of the compatibility of the composition,
- Request the generation of a BPEL process describing the composition.

We describe in the following the main features of D&A4WSC, and then we study the process of orders treatment and selling share.

D&A4WSC Features

First of all, the user may model the involved services by oWF-nets. In this step, he has to put different objects (places, transitions, interface places, etc.) in the graphical editor. Then, he compiles this modeling. Here, D&A4WSC can return two types of responses: either the composition is compiled with errors, or a composition is compiled without errors. In the case of the validity of the composition, the user may request directly the verification of the services compatibility. He may also request the verification of the compatibility of this composition. Here, he may check the syntactically and semantically compatibility in its three classes defined above. The WS-CDL friendly user can enter a description of a choreography in WS-CDL to D&A4WSC which will be translated to oWF-nets so that the user could process to the formal verification of this services compatibility.

In case of validity of the strong compatibility property, the user may request the generation of the BPEL process of the composite service. We describe the generation steps of the BPEL process with the sequence diagram given in Figure 7.

Case Study

To illustrate the scenario of WSC modeling with oWF-nets and the generation of BPEL process in case of compatible services with D&A4WSC, we propose to study the case of *Processing Orders and Selling Shares*.

The case study introduces the composition of five autonomous services which are: Customer service, Broker service, Analytics service, Exchange service and Supervision service. The scenario, given by Figure 8, is as follows. First of all, the customer sends a sell order to the broker. Then, the broker transmits it to the analytics Service with the request parameters. The analytics service analyzes the request, generates a sale plan and sends it to the customer for confirmation. In this case, the customer checks the plan and can either approve it or reject it.

If the plan is approved, the broker submits the order to the exchange service, according to the sale plan established. Finally, the supervision service oversees each order and transactions to detect possible illegal actions such as insider trading. The modeling of this scenario with D&A4WSC is captured in Figure 8.

Since each service has an input and output interface place, and for each input interface place, we have an output interface place with the same name, then the syntactic compatibility is valid in this model.

Figure 7. Steps of BPEL process generation

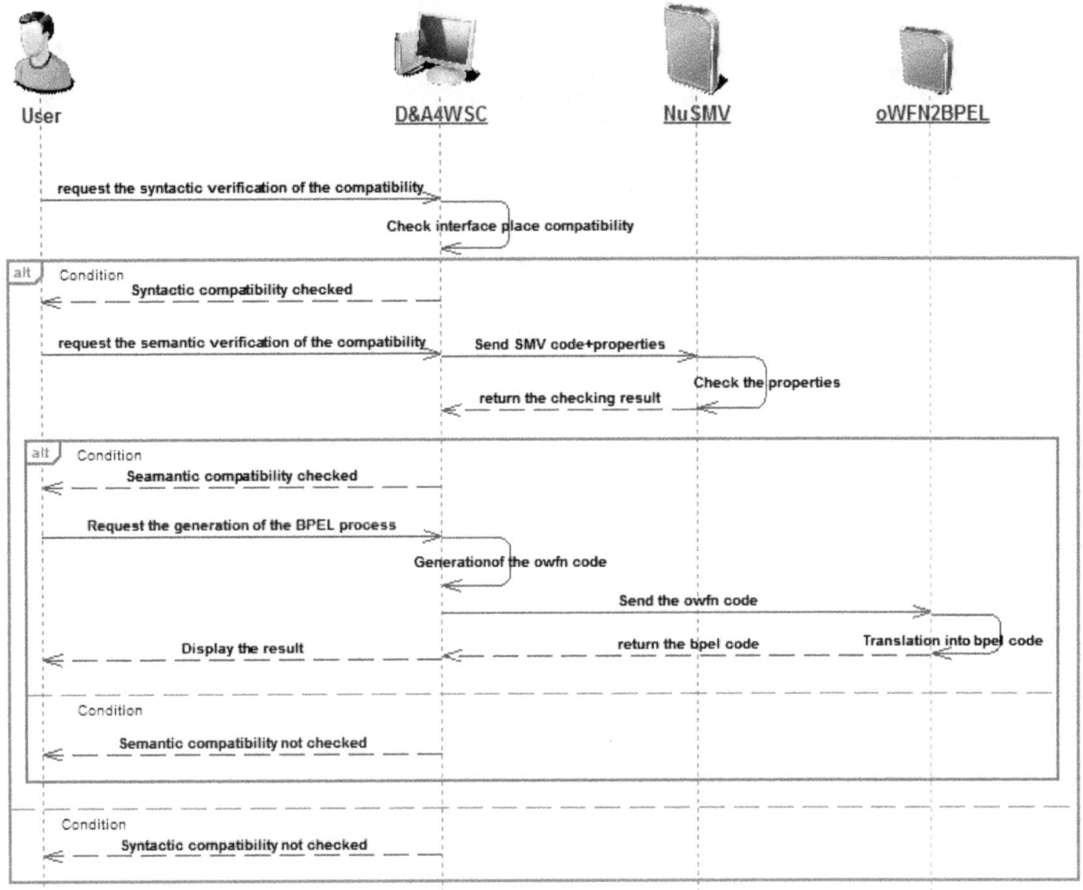

Besides, all the output places of the composition can receive a token when the net evolves, therefore, there will not be any blocking situation and hence the composition is deadlock-free. Hence, the composition is weakly compatible. In addition, at the time of termination process, all places are empty except final output ones. Thus, the property of proper termination is valid and the composition is compatible. Last but not least, all the transitions are allowed, i.e. there is no dead transitions. Then, the WSC is quasi-live and therefore the strong compatibility is ensured. In this case, the user can generate the BPEL process corresponding to this composition. To do so, he just requests this generation and D&A4WSC translates the modeling into owfn file and sends it to the compiler for BPEL process generation. We capture in figure 8 the graphical interface of this process which is represented using the graphical editor BPEL included in the Eclipse environment.

RELATED WORK

As it is shown in the above sections, we dealt in this chapter with the two aspects of Web services composition which are orchestration and choreography for a same objective consisting of the formal analysis

Figure 8. Generation of BPEL process for "Processing orders and selling shares"

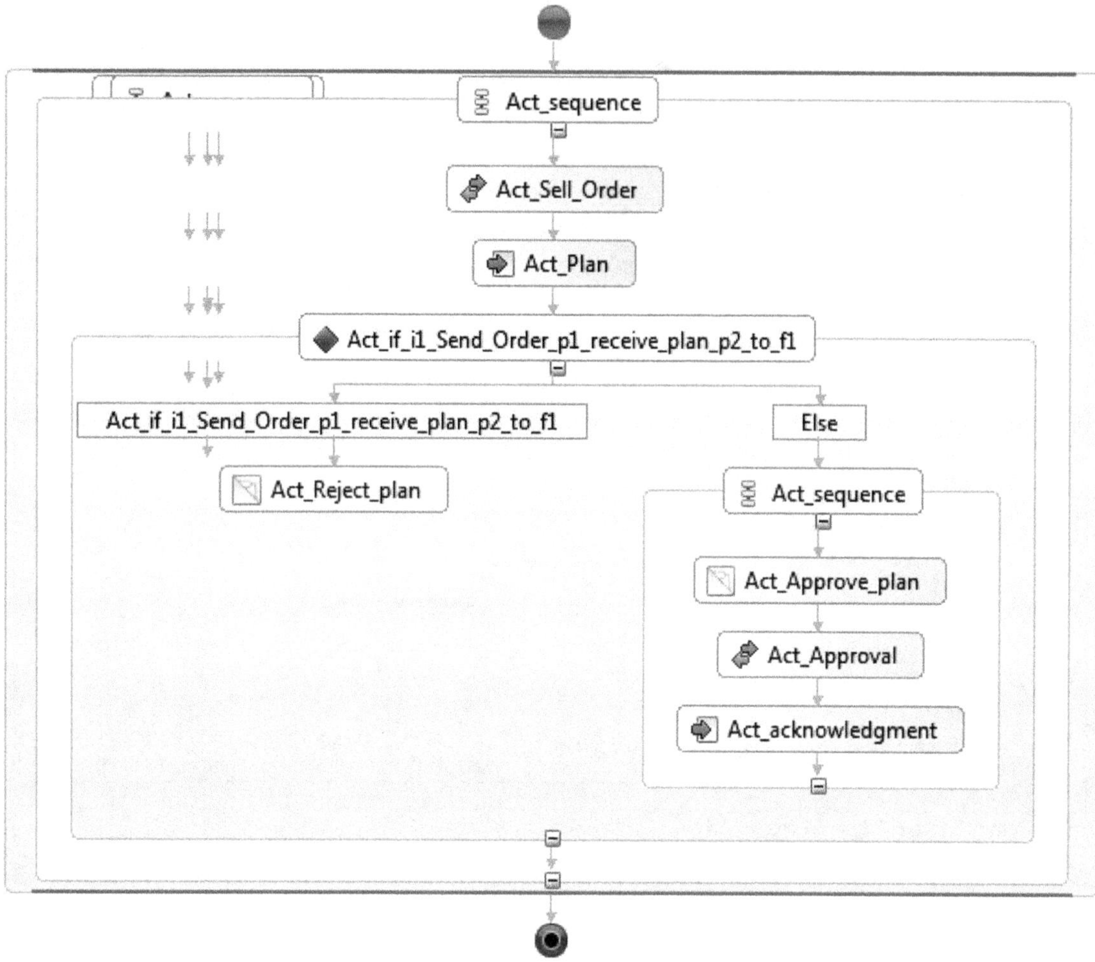

of Web services composition. This analysis is mainly oriented to verify the services compatibility but the approach proposed can be easily generalized to verify any property of the composition.

Several works have been proposed to verify the compatibility of Web services composition. We cite as an example the work of (Bordeaux, L., Salaun, G., Berardi, D., & Mecella, M., 2005), in which the authors modeled Web services by Labeled Transition Systems (LTS) and defined three types of compatibility: opposite behavior, unspecified reception and absence of deadlock.

We cite also the work of Guermouche et al. (2008) who proposed to extend the Web Service Timed Transition System (WSTTS) for presenting services. In this work, they checked the compatibility at the level of exchanged messages taking into account the time constraints.

Tan et al. (2009) proposed an approach for verifying the compatibility of Web services interfaces described with BPEL, and correcting them if they are not compatible. To do this, they modeled the composition using SWF-nets, a subclass of Colored Petri Nets (CPN). These approaches have focused on verifying the compatibility of the WSC, while it is important to represent to the user the composite service after having checked it.

In this sense, Lassen et al. (2006) proposed a WorkowNet2BPEL4WS tool to translate a composition modeled by colored Petri nets using CPN to a BPEL code. This approach is limited to the simple translation of the composition while we propose a verification of the compatibility of services before the translation of their composition.

Lohmann et al. (2008) proposed also the framework, Tools4BPEL for the translation of Petri nets in BPEL using the tool oWFN2BPEL and the translation of BPEL processes into Petri nets using BPEL2oWFN compiler. This framework also included the model checker LOLA to check the properties of deadlock and livelock of a WSC. Our approach does not stop at these two properties, but also verifies proper termination and syntactic compatibility of the WSC.

Focusing on choreography especially with WS-CDL, (Geguang, P., Jianqi, S., Zheng, W., Lu, J., Jing, L., & Jifeng, H., 2007) defined an approach that uses Java as a modeling language for WS-CDL specification. Then, they applied the Java Path Finder model checker (JPF) to check the deadlock property expressed in LTL. However, the defined Java model cannot provide a clear representation of the interactions in a choreography. In addition, it requires user effort by adding the keyword Assert to facilitate the verification process, which presents a barrier for its use.

Finally, the authors of (Diaz, G., Cambronero, M.E., Pardo, J.J., Valero, V., & Cuartero, O., 2007) defined an approach that translates a given WS-CDL specification to a timed automata as an intermediate model to verify the accuracy of Web services generated with the model checker UPPAAL. However, timed automata cannot handle errors and exceptions of a composite Web service and the objective of the authors was to generate WS-BPEL documents and not necessarily to analyze and verify WS-CDL specifications.

CONCLUSION

We proposed in this work a formal approach to model and verify Web services composition. Furthermore, we fully implemented this approach in a framework that we named D&A4WSC. This framework ensures the modeling of WSC by oWF-nets or the generation of oWF-nets from a WS-CDL specification as well as the model checking of compatibility classes of this WSC. Once verified, D&A4WSC generates a BPEL process for the WSC using the oWFN2BPEL compiler. For this purpose, it generates an owfn file of the given composition and sends it to oWFN2BPEL. Note that the reverse process is possible, in fact, we can part from a BPEL description of a WSC and translate it to oWF-nets composition and thus we can benefit from the formal analysis ensured by D&A4WSC.

Note that the generation of a BPEL process should be accomplished only if the composition is strong compatible. Therefore, we propose, as a perspective, to extend D&A4WSC with an adapter to edit a composition in case it is not strong compatible. Furthermore, we propose to rely on the formal verification of the temporal compatibility (Sbaï, Z., & Barkaoui, K., 2014; Sbaï, Z., & Barkaoui, K., 2015) and other advanced properties (Jbeli N., Sbaï, Z., & Ben Ayed, R., 2016) that are required in many cases.

REFERENCES

Antonik, A., & Huth, M. (2006). Effcient patterns for model checking partial state spaces in CTL intersection ltl. *Electronic Notes in Theoretical Computer Science*, *158*, 41–57. doi:10.1016/j.entcs.2006.04.004

Barkaoui, K., Eslamichalandar, M., & Kaabachi, M. (2010). A structural verification of web services composition compatibility. *Proceedings of the 6th International Workshop on Enterprise & Organizational Modeling and Simulation, EOMAS '10*, 30-41.

Ben Azaiez, A., & Sbaï, Z. (2015). Model Checking Web Services Choreography. *Lecture Notes in Business Information Processing, 231*, 171-186.

Bordeaux, L., Salaun, G., Berardi, D., & Mecella, M. (2005). When are two web services compatible? *Proceedings of the 5th International Conference on Technologies for E-Services, TES'04*, 15-28. 10.1007/978-3-540-31811-8_2

Cimatti, A., Clarke, E. M., Giunchiglia, F., & Roveri, M. (1999). Nusmv: A new symbolic model verifier. *Proceedings of the 11th International Conference on Computer Aided Verification, CAV '99*, 495-499.

Decker, G., Overdick, H., & Zaha, J.M. (2006). On the suitability of ws-cdl for choreography modeling. In Proceedings of methoden, Konzepte Und Technologien Fur Die Entwicklung Von Dienstebasierten Informationssystemen.

Diaz, G., Cambronero, M. E., Pardo, J. J., Valero, V., & Cuartero, O. (2007). *Model Checking Techniques applied to the design of Web Services*. Academic Press.

Foster, H., Uchitel, S., Magee, J., & Kramer, J. (2004). Compatibility verification for web service choreography. *Proceedings of the IEEE International Conference on Web Services, ICWS '04*, 738. 10.1109/ICWS.2004.1314806

Geguang, P., Jianqi, S., Zheng, W., Lu, J., Jing, L., & Jifeng, H. (2007). The Validation and Verification of WSCDL. *Asia-Pacific Software Engineering Conference*, 81-88.

Guermouche, N., Perrin, O., & Ringeissen, C. (2008). Timed Specification For Web Services Compatibility Analysis. *Electronic Notes in Theoretical Computer Science, 200*(3), 155–170. doi:10.1016/j.entcs.2008.04.098

Henzinger, T.A., Nicollin, X., Sifakis, J., & Yovine, S. (1994). Symbolic Model Checking for Real-Time Systems. *Information and Computation, 111*(2), 193–244.

Hongli, Y., Xiangpeng, Z., Zongyan, Q., Geguang, P., & Shuling, W. (2006). A formal model of web service choreography description language (WS-CDL). *International Conference on Web Services*.

Jbeli, N., Sbaï, Z., & Ben Ayed, R. (2016). On expressiveness of TCTLhdelta for Model Checking Distributed Systems. *International Conference on Computational Collective Intelligence*.

Lassen, K. B., & van der Aalst, W. M. P. (2006). Workownet2bpel4ws: A tool for translating unstructured workflow processes to readable bpel. *Proceedings of the 2006 Confederated International Conference on On the Move to Meaningful Internet Systems: CoopIS, DOA, GADA, and ODBASE*, 127-144.

Lohmann, N. (2008). *oWFN2BPEL: translates a Petri net model into an abstract BPEL process*. GNU Free Documentation License.

Lohmann, N., & Kleine, J. (2008). Fully-automatic translation of open workflow net models into simple abstract bpel processes. Modellierung 2008, 57-72.

Ma, C., Xu, Q., & Sanders, J. W. (2009). *A survey of business process execution language (bpel)*. UNU-IIST Report No. 425.

Marc, F., Benoit, T., Romain, C., Raphael, C. Y. F., & Mohammed, O. (2010). Comparison of Model Checking Tools for Information Systems. *Proceedings of the 12th International Conference on Formal Engineering Methods and Software Engineering*, 581-596.

Martens, A. (2003). On compatibility of web services. *Petri Net Newsletter,* (65), 12-20.

Rao, J., & Su, X. (2005). A survey of automated web service composition methods. *Proceedings of the First International Conference on Semantic Web Services and Web Process Composition, SWSWPC'04*, 43-54. 10.1007/978-3-540-30581-1_5

Sbaï, Z., & Barkaoui, K. (2013). Vérification formelle des processus workflow - Extension aux workflows inter-organisationnels. *Revue Ingénierie des Systèmes d'Information, 18*(5), 33-57.

Sbaï, Z., & Barkaoui, K. (2015). On quantitative Analysis of Time Open Workflow Nets and Parametric Extension. *Proceedings of the 9th Workshop on Verification and Evaluation of Computer and Communication Systems, VECoS 2015*, 97-108.

Sbaï, Z., Barkaoui, K., & Boucheneb, H. (2014). Compatibility Analysis of Time Open Workflow Nets. *Proceedings of the International Workshop on Petri Nets and Software Engineering (PNSE 2014), in conjunction with Petri nets and ACSD*, 249-268.

Sbaï, Z., & Guerfel, R. (2016). CTL Model Checking of Web Services Composition based on Open Workflow Nets Modeling *International Journal of Service Science, Management, Engineering, and Technology*, *7*(1), 27–42. doi:10.4018/IJSSMET.2016010102

Tan, W., Fan, Y., & Zhou, M. C. (2009). A petri net-based method for compatibility analysis and composition of web services in business process execution language. *IEEE T. Automation Science and Engineering*, 94-106.

Chapter 12
Periodic Patterns in Dynamic Network:
Mining and Parametric Analysis

Hardeo Kumar Thakur
Manav Rachna University, India

Anand Gupta
Netaji Subhas Institute of Technology, India

Anshul Garg
Netaji Subhas Institute of Technology, India

Disha Garg
Netaji Subhas Institute of Technology, India

ABSTRACT

The periodic interactions represented as a dynamic network possess two aspects: structure and weight. This chapter introduces and explores the use of a third aspect that is associated with the periodic interactions, namely the directional aspect. Moreover, the authors have showcased how some applications require mining of patterns on both aspects—1) on direction and 2) on weight of directed interactions—for a better understanding of their behavior. With the aim of overcoming the limitation of existing frameworks, which only mine periodic patterns individually, a framework is proposed to mine periodic patterns on both the aspects together. Further, the patterns are analyzed to develop a better understanding of the dynamic network. A set of six parameters, defined later in the chapter, are used to conduct a microscopic study on the behavior of interactions. The framework is tested on real world and synthetic datasets. The results highlight its practical scalability and prove its efficiency.

INTRODUCTION

In recent times, an enormous amount of data is available around consumer behaviour, social interaction, weather forecast pattern, financial market, DNA pattern and many other different types of behavioral pattern. The volume of raw data available is put to better use when data visualisation and mining techniques

DOI: 10.4018/978-1-5225-5951-1.ch012

Hassanein et al., (2015) are applied on it. Graph is one such effective way to arrange and organize such passive data into a meaningful form. It helps to visualize the relationship between entities with significant clarity. As amount of data available is immense and its nature varies with time, so it becomes more meaningful when data is modelled in form of dynamic network of graphs which evolve over a period of time. After modelling data in such a form, it becomes a lot easier to study the behaviour and nature of interactions between participating entities. Not only it provides insight about existing behaviour, but also assists to obtain significant information about the future and thereby making fruitful predictions.

A considerable amount of effort has already been put in to study the nature of interactions in dynamic network(Apostolico, A., Barbares, M., & Pizzi, C., 2011; Borgwardt, K. M., Kriegel, H. P., & Wackersreuther, P., 2006; Desikan & Srivastava, 2006; Duan, D., Li, Y., Jin, Y., & Lu, Z., 2009; Gupta, A., Thakur, K. H., & Kishore, P., 2014; Halder, S., Han, Y., & Lee, Y., 2013; Inokuchi, A., Washio, T., & Motoda, H., 2000; Lahiri & Wolf, 2010; Liu, G., Wong, L., & Chua, H. N., 2009; Obulesu, O., Reddy, M. R., & Reddy, T. M., 2014; Qin, G., Yang, J., Gao, L., & Li, J., 2011; Rasheed, F., Alshalalfa, M., & Alhajj, R., 2011; Yang, J., Wang, W., & Yu, P., 2014). Among these work, a significant amount of effort has been put in to understand frequent (Borgwardt, K. M., Kriegel, H. P., & Wackersreuther, P., 2006; Inokuchi, A., Washio, T., & Motoda, H., 2000; Yang, J., Wang, W., & Yu, P., 2014) and regular (Gupta, A., Thakur, K. H., & Kishore, P., 2014; Qin, G., Yang, J., Gao, L., & Li, J., 2011) behavior of interactions. Along with such patterns, there are other type of behavior that interactions exhibit, namely periodic patterns. Periodic patterns repeat itself over a period of time.

The initial step to explore and extract periodic patterns dates back to 1996 when Srikant and Agrawal (1996) developed a model based on similarity of time sequences. Later, a model for discovery of association rules has been developed by Ozden et al.(1998) which displays cyclic variations over time. And soon after this, Han et al. (1999) constructed a model for efficient mining of partial periodic patterns in time series database. Subsequently, Ma and Hellerstein (2001) have proposed an algorithm for mining partially periodic event patterns with unknown periods. Subsequently, Berberidis et al. (2002) have devised an approach to deal with datasets in which periodicity is initially unknown. Further, Yang (2003) developed a model to mine asynchronous periodic patterns. Later in 2010, Lahiri & Wolf came up with an algorithm to mine parsimonious periodic patterns in a dynamic network. It is soon followed by efforts of Apostolico et al. (2011), where the time complexity of algorithm (Lahiri & Wolf, 2010) has been improved. Subsequently, a noise resilient suffix tree based approach to detect all periodic patterns in a time series and sequence database has been put forward by Rasheed et al. (2011). Further, Halder et al. (2013) proposed super graph based approach to mine periodic patterns. Later, Obulesu et al. (2014) have proposed a framework to find frequent and maximal periodic patterns in Spatiotemporal Databases.

The aforementioned works have contributed significantly to study periodic behaviour of interactions based on two characteristics: 1) topology or structure (Apostolico, A., Barbares, M., & Pizzi, C., 2011; Halder, S., Han, Y., & Lee, Y., 2013; Lahiri & Wolf, 2010; Obulesu, O., Reddy, M. R., & Reddy, T. M., 2014; Rasheed, F., Alshalalfa, M., & Alhajj, R., 2013); and 2) weight on edges denoting interactions (Lahiri & Wolf, 2010). But analysis on either of the two characteristic has been carried out individually. However, there exist many real world applications having characteristic of direction as well. Further, more insight on the nature of interaction is obtained when periodic patterns are extracted on two aspects together: 1) direction and; 2) weight. Here, weight aspect of interaction refers to weight on direction aspect. In the chapter, weight on direction aspect is denoted as *dw* aspect (Gupta, A., Thakur, K.H., Garg, A., Garg, D., 2016). Significant information is obtained by paying attention on both direction and *dw* aspect of interaction. Let us illustrate the meaning of *dw* aspect and significance of mining periodic patterns on both the aspects with the help of following example. Consider an email network depicting relationships between sender and receiver of an email. Let P_1 and P_2 be two vertices in a graph corresponding to the two participants. A directed edge exists from P_1 to P_2, if P_1 sends an email to P_2. The weight on the edge represents the number of emails exchanged in a week. Let us limit the attention to one particular edge of the network between P_1 and P_2 and study its behaviour over a series of 10 timesteps. Suppose the observed sequence is **a**bc**b**db0**b**ab. Here, the non-bold character represents weight on edge from P_1 to P_2 and bold from P_2 to P_1. Also, '0' denotes the absence of edge, 'a' denotes a number between 1-50, 'b' between 51-100 and so on. The periodic occurrence of edge between P_1 and P_2 based on *dw* aspect is from P_1 to P_2 starting at timestep 2 with period 4. From the sequence, it is observed that considering only direction or *dw* aspect individually results in either losing the information on the direction of communication or the number of emails exchanged. But on considering both the aspects, it is inferred that the exchange of email from P_1 to P_2 is periodic in nature which starts at timestep 2 and recurs with period 4, and weight on the directed edge gives the volume which for the above example is 'b'.

Above example clearly illustrates that analyzing patterns obtained on both, direction and *dw* aspect provide vital information which would have not been conveyed if only single aspect is considered. However, existing algorithm (Lahiri & Wolf, 2010), called PSEMINER, find patterns only on a single aspect of interaction in a single run. Therefore, in this chapter, a framework has been presented based on the work of Lahiri and Wolf (2010) to mine periodic patterns on both direction and weight on direction (*dw*) aspect. The framework is known as *Directed and DW Periodic Patterns Framework, DDWPPF.* It constitutes the following components: 1) a pairing function for assigning unique numbers; 2) a mapping procedure which associates weight component to edges exhibiting periodic pattern on direction aspects; 3) a modified version of existing algorithm (Lahiri & Wolf, 2010) for mining periodic patterns and; 4) an inverse pairing function for converting the unique numbers back to original form. The framework first extracts periodic patterns on direction aspect and then on *dw* aspects.

Further, to underline its practicality and prove its efficiency, real world datasets are used. Also, a synthetic dataset is used to highlight its scalability. Subsequently, a microscopic analysis has been carried out based on six parameters to obtain a better insight about behaviour of interactions. The following six parameters are used for further analysis of results obtained:

- **Global Purity:** Measures the likeliness of a periodic pattern on direction to exhibit periodicity only in an observed range of timesteps in proportion to its occurrence over the entire time series of interactions.
- **Reciprocity:** Is a measure of the biasness of flow of information.
- **Maturity:** Measures the likeliness of a periodic pattern to become periodic after its introduction in the dynamic network.
- **Density:** Is a measure of the connectivity of a graph in terms of the available pathways between nodes.
- **Uniformity:** Measures the variance in distribution of the weights over the edges of a weighted directed graph.
- **Weighted Reciprocity:** Measures the variance in distribution of weights over the reciprocal edges in a weighted directed graph.

The chapter is organized as follows:

The next section presents some preliminary definitions and notations. Then, data structures used in implementation of DDWPPF are presented. The section following that contains discussion on modified *PSEMINER* used to mine periodic patterns in a dynamic network. Also, a detailed description of DDWPPF is also presented here. It is followed by description of datasets used for experimental analysis. Also, results obtained from experiments on three datasets are demonstrated in a succinct manner using line plots. Then, the parameters used for analyzing periodic patterns are defined. Further, analysis results on two real world datasets is also presented. Finally, conclusion and discussion are summarized in later section.

PRELIMINARIES AND PROBLEM FORMULATION

This section provides some elementary background for a better understanding of the topic, followed by the problem definition.

Preliminaries

In this chapter, a simple weighted directed graph G is represented as (V, E, W), where V is the vertex set, $E = \{<u, v> \mid u, v \in V\}$ is the edge set, with <u, v> representing an ordered pair possessing a weight $w \in W$. All the elements of V, E,W belong to non-negative integer set.

Such a weighted directed graph represents a snapshot of the network at a particular time. The network may have varying directional and weighted edges at consecutive timesteps. The authors conducted a study on a series of such consecutive temporal networks namely weighted directed dynamic network.

Definition 1 (Weighted Directed Dynamic Network): A weighted directed dynamic network (Duan et al., 2009) $G = (G_1, G_2, ... G_T)$ is a sequence of graphs at T timesteps, where $G_i = (V_i, E_i, W_i)$ is a simple weighted directed graph with edge set E_i possessing weight w across vertex set V_i. $E = E_1 \cup E_2 \cup ...... E_T$ is the universal edge set of G. Also, an universal vertex set is $V = V_1 \cup V_2 \cup .... V_T$. A weighted directed dynamic network across 8 timesteps is shown in Figure 1.

Figure 1. Weighted Directed Dynamic Network

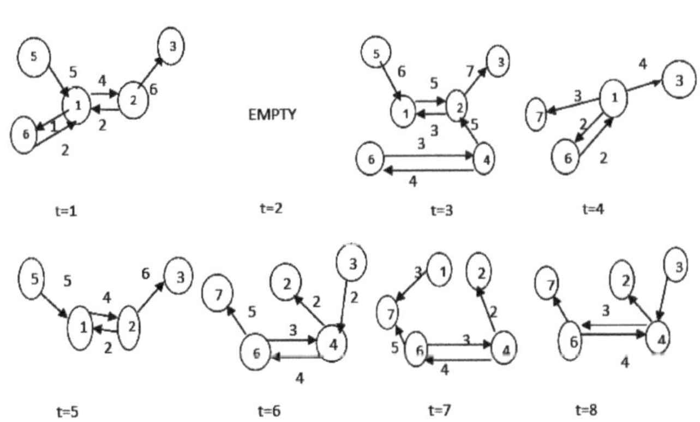

Weighted directed dynamic network may have some edges that only appear in a few timesteps. Such edges are insignificant and unreasonably consume processing time. So, the concept of reduced weighted directed dynamic network is being introduced.

Definition 2 (Reduced Weighted Directed Dynamic Network): Given a weighted directed dynamic network G, a reduced weighted directed dynamic network G' = (V, E'), where E' ⊆ E and each edge e ∈ E' occurs at-least minimum threshold times, which by default is 3. Figure 2 shows the reduced weighted directed dynamic network for the network shown in Figure 1. Here, edge <1, 3> gets removed as it occurs less than minimum threshold times.

The authors studied the periodic subgraph in reduced weighted directed dynamic network. Since the periodic subgraph is a particular type of frequent subgraph. The next definition is about support of the frequent subgraph.

Definition 3 (Support): Given a sequence of graph G_1, G_2, ... G_T and an arbitrary graph G_a, the support set S of G_a is the set of t_i at which G_a is contained in G_i, i.e. $G_a \subseteq G_i$ as in Lahiri and Wolf (2010). Also, support of G_a = | S |.

Occurrence of G_a is significant when it occurs at-least minimum threshold times. It is called minimum support threshold, denoted by σ. Directed support set is defined for an arbitrary graph $G_{ad} = (V_{ad}, E_{ad})$ and directed support of G_{ad} is the cardinality of its directed support set. σ_d denotes minimum directed support threshold. *DW* Support set is defined for an arbitrary graph $G_{adw} = (V_{adw}, E_{adw}, W_{adw})$ and *dw* support of G_{adw} is the cardinality of its *dw* support set. σ_{dw} denotes minimum *dw* support threshold.

The frequent graphs that satisfy minimum support are redundant in nature. Mining frequent closed subgraph is one of the best solutions to the redundancy of frequent subgraph. The next definition is about closed subgraph as in Lahiri and Wolf(2010).

Definition 4 (Closed Subgraph): An arbitrary graph G_a is a closed subgraph in a weighted directed dynamic network G if it is maximal for its support set as in Lahiri and Wolf(2010). Directed Closed Subgraph represents an arbitrary $G_{ad} = (V_{ad}, E_{ad})$. Adding an edge <u,v> to G_{ad} modifies its directed

Figure 2. Reduced Weighted Directed Dynamic Network

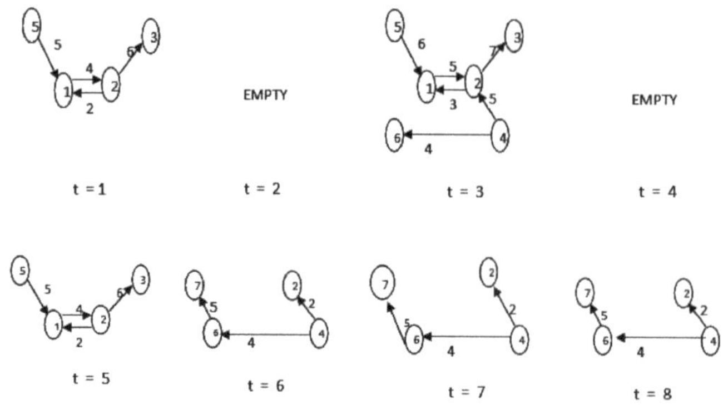

support set. *DW* Closed Subgraph represents an arbitrary graph $G_{adw} = (V_{adw}, E_{adw}, W_{adw})$. Adding an (edge, weight) pair to G_{adw} changes its *dw* support set.

In this chapter, a Pairing function and an Inverse pairing function has been applied to uniquely assign number to an edge and vice-versa. The next two definitions are about Pairing function and Inverse pairing function.

Definition 5 (Pairing Function): Given two non-negative integers '*a*' and '*b*', pairing function *ElegantPair(a,b)* assigns a unique integer to a pair <a, b> as in Szudzik(2006). The pairing function is defined as in equation (1):

$$ElegantPair(a,b) = \begin{matrix} b^2+a & if\ a \neq \max(a,b) \\ a^2+a+b & if\ a = \max(a,b) \end{matrix} \tag{1}$$

Definition 6 (Inverse Pairing Function): Given a non-negative integer *c*, the inverse pairing function *ElegantUnpair(c)* associates pair <a, b> with *c* as in Szudzik (2006). It is defined as in equation (2):

$$ElegenyUnpair(c) = \begin{matrix} \{c-\left|\sqrt{c}\right|^2, \left|\sqrt{c}\right|\} & if\ \{c-\left|\sqrt{c}\right|^2\} < \left|\sqrt{c}\right| \\ \{\left|\sqrt{c}\right|, c-\left|\sqrt{c}\right|^2-\left|\sqrt{c}\right|\} & if\ \{c-\left|\sqrt{c}\right|^2\} \geq \left|\sqrt{c}\right| \end{matrix} \tag{2}$$

Formulation of Problem Definition

The definitions above have formed the basis to define periodic subgraph in weighted directed dynamic network. A graph may have multiple periodic support sets to permit for multiple, overlapping or disjoint periodic behaviour. Thus, all periodic subgraph embeddings are significant. The next two definitions are regarding Periodic Support Set and Periodic Subgraph Embedding.

Definition 7 (Periodic Support Set): Given a weighted directed dynamic network G, the periodic support set S_p of an arbitrary graph G_a in G, denoted by <i, p, s>, is a maximal, ordered set of s timesteps starting at t_i having a period of p timesteps as in Lahiri & Wolf (2010). Directed periodic support set S_{pd} for an arbitrary graph $G_{ad} = (V_{ad}, E_{ad})$ is denoted by $<i_d, p_d, s_d>$. *DW* periodic support set S_{pdw} for an arbitrary graph $G_{adw} = (V_{adw}, E_{adw}, W_{adw})$ is denoted by $<i_{dw}, p_{dw}, s_{dw}>$.

Definition 8 (Periodic Subgraph Embedding): Given a weighted directed dynamic network G, a periodic subgraph embedding(PSE) of an arbitrary graph G_a is a pair $<G_a, S_p>$, where G_a is closed over the periodic support set S_p and $|S_p| \geq$ minimum support threshold σ as in Lahiri & Wolf (2010). Directed periodic subgraph embedding (DPSE) is defined for an arbitrary graph $G_{ad} = (V_{ad}, E_{ad})$. *DW* periodic subgraph embedding (DWPSE) is defined for an arbitrary graph $G_{adw} = (V_{adw}, E_{adw}, W_{adw})$.

The definition of DPSE and DWPSE allows some redundant patterns as well. The patterns whose periods are multiples of a base period do not provide any new information and therefore needs to be eliminated. Hence, the idea of subsumption has been formed based on the principle of parsimony in Lahiri & Wolf(2010). The next definitions convey the idea of subsumption.

Definition 9 (Parsimonious Periodic Subgraph Embedding): A periodic subgraph embedding (PSE) which is not subsumed by any other PSE is a parsimonious periodic subgraph embedding (PPSE) as in Lahiri and Wolf (2010). Directed parsimonious periodic subgraph embedding (DPPSE) represents a PPSE on direction aspect. Similarly, *DW* parsimonious periodic subgraph embedding (DWPPSE) represents a PPSE on *dw* aspect.

Here, only those periodic patterns (DPPSE or DWPPSE) having at least two edges have a significance. The edges are in the form of unique number. Since the edges are in form of unique number, the author name such periodic pattern as Directed Inverse Periodic Pattern and *DW* Inverse Periodic Pattern.

Definition 10 (Directed Inverse Periodic Pattern): A DPPSE for an arbitrary graph $G_{ad} = (V_{ad}, E_{ad})$ with directed periodic support set S_{pd} denoted by $<G_{ad}, S_{pd}>$, is a directed inverse periodic pattern (DIPP), if cardinality of edge set is greater than 1 i.e. $|E_{ad}| > 1$.

Definition 11 (DW Inverse Periodic Pattern): A DWPPSE represented by $<G_{adw}, S_{pdw}>$, where $G_{adw} = (V_{adw}, E_{adw}, W_{adw})$ is an arbitrary graph and S_{pdw} its *dw* periodic support set, is a *dw* inverse periodic pattern(DWIPP), if cardinality of edge set is greater than 1, i.e. $|E_{adw}| > 1$.

Since the DIPP and DWIPP are in the form of unique numbers assigned to an edge $<u_a, v_a>$ and to the triplet $<<u_a, v_a>, w_a>$ respectively, therefore they need to be converted into original form using the inverse pairing function.The next two definitions are about the kind of periodic patterns that the authors are interested to mine in weighted directed dynamic network.

Definition 12 (Periodic Pattern on Direction): Given a weighted dynamic network G, periodic pattern on direction(PPD) is obtained by applying inverse pairing function to each element in the DIPP, denoted by $<G_{ad}, S_{pd}>$ where $G_{ad} = (V_{ad}, E_{ad})$ is an arbitrary graph with directed periodic support set S_{pd}.

Definition 13 (Periodic Pattern on dw): Given a weighted dynamic network G, periodic pattern on *dw*(PPDW) is obtained by applying inverse pairing function to each element in the DWIPP, denoted by $<G_{adw}, S_{pdw}>$ where $G_{adw} = (V_{adw}, E_{adw}, W_{adw})$ is an arbitrary graph with *dw* periodic support set S_{pdw}.

Prearranged the above definition, the objective of this chapter is given below:

Problem Definition: Given a weighted directed dynamic network $G = \{G_1, G_2, \ldots\ldots, G_T\}$ and minimum {directed, *dw*} support threshold > 2, the problem is to mine all periodic patterns on direction(PPD) and periodic patterns on *dw* (PPDW) together which are respectively above the minimum {directed, *dw*} support threshold and analyse the patterns found.

In the following section, the data structures used in the framework are described.

DATA STRUCTURES

The primary data structures used in PSEMINER (Lahiri & Wolf, 2010) are as follows:

1. **Pattern Tree:** It consists of treenodes which satisfies the property that the graph associated with child treenode is subgraph of its parent treenode. It enables to search periodic patterns in an efficient way.
2. **Subgraph Hash Map:** It enables randomised access to treenodes of the pattern tree.
3. **Timeline List:** It maintains a list of timesteps at which a subgraph is expected to occur in future.

Along with aforementioned data structures, following data structures are also used in DDWPPF (Gupta, A., Thakur, K.H., Garg, A., Garg, D., 2016):

- **Directed Edge Table:** It is a data structure used to store the unique values of directed edges (assigned using pairing function) present in input weighted directed dynamic network. It is stored in the form of a table in which each row corresponds to a timestep and consists of unique values of edges present at that timestep.
- **DDW Numbered Map:** It is a map in which an array of unique values assigned to weighted directed edge at T timesteps is mapped to the unique value of the corresponding directed edge. Figure 3 shows the format of a DDW numbered map. Here, X_j represents the unique value assigned to a directed edge $<u_j, v_j>$ and Y_{ji} represents the unique value assigned to $<<u_j, v_j>, w_{ji}>$ at timestep t_i.
- **Directed Edge List:** It is a list consisting of edges exhibiting periodic behaviour on direction which is used together with DDW numbered map to construct a dw edge table.
- **DW Edge Table:** It is a table consisting of entries from DDW numbered map matched against the edges present in directed edge list.

DESCRIPTION OF DDWPPF

The DDWPPF framework for identifying and mining all PPDs and PPDWs in a weighted directed dynamic network is described in this section. The heart of DDWPPF is MPSEM which is based on PSEMINER (Lahiri & Wolf, 2010).

Modified PSEMINER

Modified PSEMINER (MPSEM) extracts parsimonious periodic patterns in a dynamic network. It is based on building a pattern tree that maintains list of all PSEs seen till a timestep t_i, where $i \in [1,T]$ and T denotes total number of timestep in the network. Only the PSEs which cease to exhibit periodic

Figure 3. DDW Numbered Map

$$X_1\ Y_{11}\ Y_{12} \cdots Y_{1T}$$
$$X_2\ Y_{21}\ Y_{22} \ldots\ldots Y_{2T}$$
$$X_j\ Y_{j1}\ Y_{j2} \ldots\ldots\ Y_{jT}$$

behaviour at a timestep t_i but consists of at least two edges and satisfy the minimum support threshold are considered as potential candidates for PPSEs. The additional condition on PSE to consist of at least two edges has been imposed in MPSEM so as to eliminate isolated interactions and emphasize on the entire network as a whole. The further are steps are same as in PSEMINER (Lahiri & Wolf, 2010).

Figure 4 and Figure 5 show the block diagrams of PSEMINER and MPSEM respectively.

Detailed DDWPPF

DDWPPF extracts periodic patterns on Direction and *DW* aspect in a dynamic network. Figure 6 shows the steps involved in the framework. The detailed description of the steps followed in DDWPPF is as follows:

Figure 4. Block diagram of PSEMINER

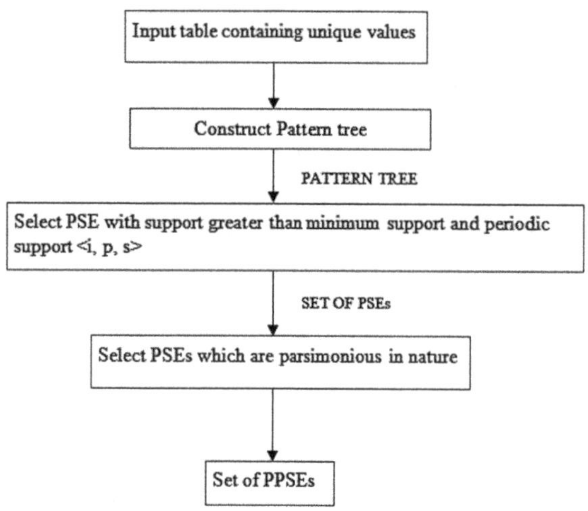

Figure 5. Block diagram of MPSEM

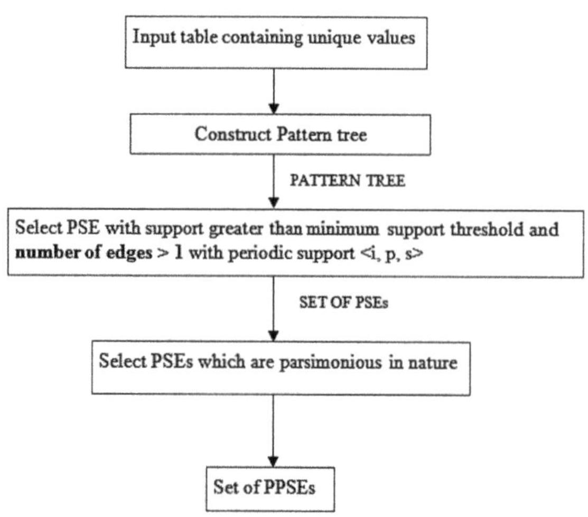

Figure 6. Block diagram of DDWPPF

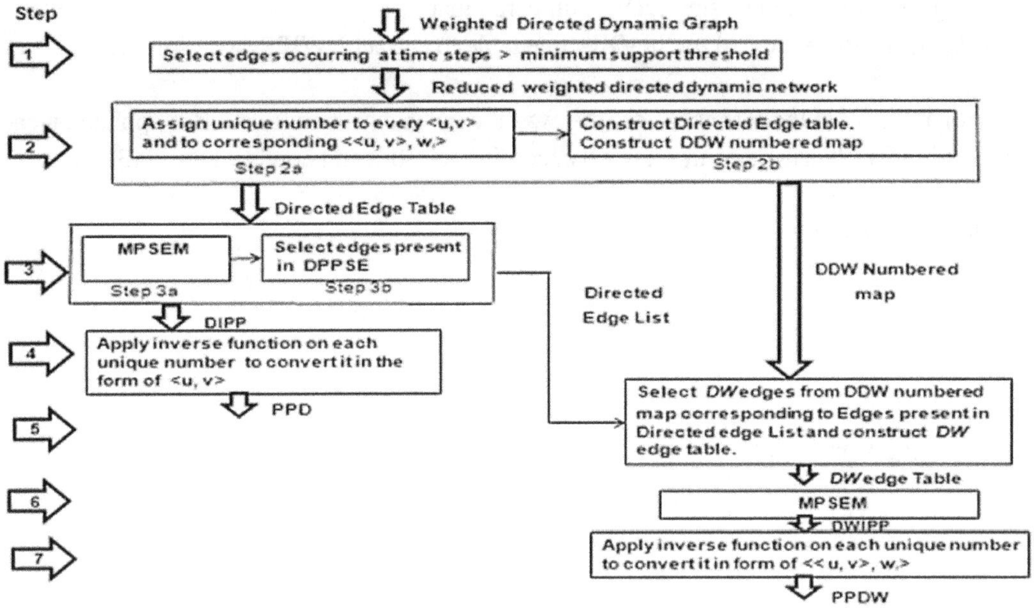

INPUT: A weighted directed dynamic graph in the form of u, v, w_1, w_2, .. w_T, where <u, v> represents a directed edge from u to v and w i for i ∈ [1,T] are the weights on the edge at T timesteps. Figure 1 shows an example of weighted directed dynamic network.

Step 1: Edges which do not satisfy minimum support threshold are eliminated from weighted directed dynamic network. It helps to reduce the processing time by removing non-essential edges. A reduced weighted directed dynamic network is obtained as a result of this step.

Step 2: A unique number is assigned to every edge <u, v> and to its corresponding weighted pair <<u, v>, w> to construct a directed edge table and DDW numbered map corresponding to reduced weighted directed dynamic network.

Step 3: MPSEM is executed using directed edge table obtained in step 2 as input. It provides DIPPs and simultaneously edges which are a part of DPPSE are collected in directed edge list.

Step 4: PPD in the form of <u, v> is obtained by applying inverse pairing function on each element present in DIPP. A PPD obtained for using weighted directed dynamic network(Figure 1) as input is shown in Figure 7.

Figure 7. PPD with Support Set <6, 1, 3>

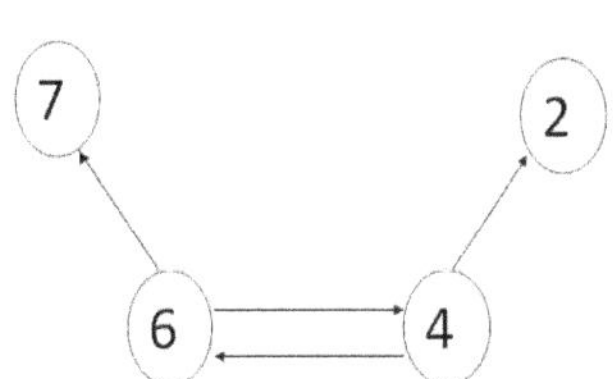

Step 5: DW edge table is constructed using the edges present in directed edge list and their corresponding weighted components using DDW numbered map.
Step 6: MPSEM is now executed on the *DW* edge table to extract DWIPPs as result set.

To obtain a PPDW in the form of <<u, v>, w>, the inverse pairing function is applied on each element present in DWIPP as shown in Figure 8 according to following condition as in (3) and (4):

(a, b) = ElegantUnpair (c)

if (a is odd)

w = (a + 1) / 2 (3)

< u, v > = ElegantUnpair (b / 2)

Else

w = (b + 1) / 2 (4)

< u, v > = ElegantUnpair (a / 2)

OUTPUT: PPD and PPDW as shown in Figures 7 and 8 are obtained as output of the above steps.

EXPERIMENTAL STUDIES

The experiments have been carried out on 2 real world data sets to establish practical applicability of DDWPPF in real world scenarios as well as on one artificial dataset to prove its scalability. The framework has been implemented in C++ programming language. Further, a 64-bit Ubuntu(v14.04 LTS) virtual machine have been used to perform the experiments. The machine is configured with Intel core processor with i5-3337U CPU running 1.80 GHz with 8.8 GB RAM.

Datasets

- **Stock Market Dataset (Historical Data for S&P 500 Stocks, 2013):** The dataset comprises of opening and closing prices of stock for 447 days. The size of dataset has been reduced by taking average price of a stock over 6 days so that each timestep represents a week. This has been done so

Figure 8. PPDW with Support Set<6,1,3>

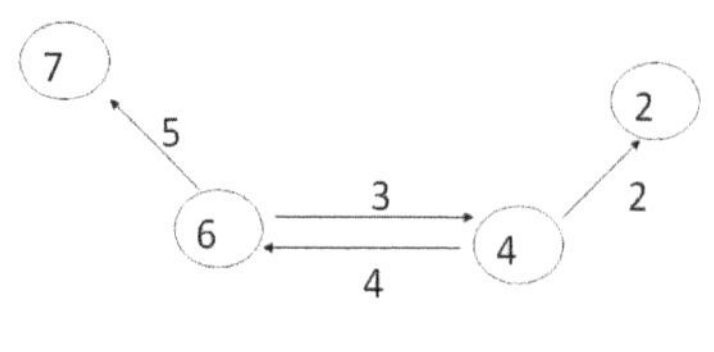

as to analyses profit or loss patterns of a stock over a week which forms basis for more meaningful analysis. The data is organized in the form of dynamic network where a stockholder may choose to buy or sell stocks based on their performance. The vertices of the network represents stocks and a directed edge from one stock to another exist only if the source vertex is at loss and target vertex is incurring profit. The weight on the edge corresponds to net gain in switching from loss incurring stock to profit making stock. The network has 500 vertices, 249221 edges spread over 74 timesteps and density of network is 0.15.

- **Enron Email Dataset:** The dataset is based on email data exchanged between employees of an organisation(Gupta et al, 2014). The vertex of the network represents an employee and a directed edge exists only if an email has been sent by an employee(source vertex) to another(target vertex). The weight on the edge represents the number of emails exchanged between two vertices. The network has 2359 vertices, 99154 edges spread over 28 timesteps and density of network is 0.0014.
- **Artificial Dataset:** The dataset consists of synthetically generated directed edges and the weights are randomly assigned to edges. The network has 2000 vertices, 80000 edges spread over 100 timesteps and density of network is 0.015.

Execution Time Comparison

A graphical comparison between execution time of DDWPPF and PSEMINER is shown in Figure 9. As observed from the plots, DDWPPF is faster than PSEMINER for Stock Market and Artificial Datasets which are dense networks. Enron dataset presents an exception as it is sparse network. On further analysis, it is observed that overhead cost for maintaining Directed Edge Table and DDW Numbered Map in case of sparse networks is higher than performance gain achieved for sparse networks. In case of dense networks, the performance gain is much higher than the overhead cost incurred.

The time complexity of DDWPPF for weighted directed dynamic network is $O((V+E)T^3 \ln T)$ and space complexity is $O((V+E)\, T^2 \ln T + T^4 \ln T)$ which is same as for PSEMINER. However, execution performance of DDWPPF is better than performance of PSEMINER as shown in Figure 9.

Figure 9. Time comparison between PSEMiner and DDWPPF

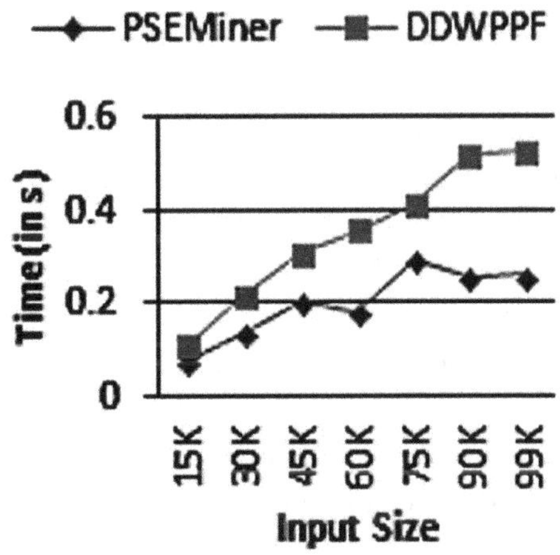

The following section contains a detailed parametric analysis performed on the patterns obtained during experiments.

ANALYSIS

This section presents the analysis done by the authors, on the periodic patterns mined from two real world dynamic networks - STOCK MARKET NETWORK and ENRON EMAIL NETWORK.

STOCK MARKET NETWORK ANALYSIS

A study on behavioral pattern of financial market was conducted with the help of proposed parameters. It has helped in answering the following questions:

1. Does the periodic behaviour of stock have any effect on market stability? If so, what is the effect?
2. How can shareholder reduce loss or maximize profit?
3. How would a stockholder and a buyer respond to occurrence behavior? What might their next steps be?
4. How is the market going to behave in long run?
5. What are the possible options a buyer has when switching from one stock to another?
6. How much is the risk and return trade off?

A total of 3935 PPDs are obtained, and 78.36% edges exhibit periodicity on direction aspect in stock market dataset. The observations based on different parameters are summarized below.

1. **Maturity:** This parameter answers the first question raised above, in the affirmative. It is a useful measure to determine the stability of the market forces. It is inversely related to the stability of performance. A higher value can be correlated to the launch of a new product which results in the temporary shifting of the stockholders towards it, followed by permanent adoption of the product after having evaluated its response .From loss point of view, the later the occurrence of a behavior, the better it is, because the investor won't get impatient and would be ready to wait and evaluate market sentiments. Also it can be seen that, higher the maturity, higher would be the stability of the market. The average maturity value is found to be 0.35, and 27.42% of the PPDs possess higher maturity level than the average. Figure 10 shows a PPD with minimum maturity value 0.02 and directed periodic support set <71, 1, 4>.
2. **Global Purity:** It has helped to evaluate the performance of a stock. If global purity is high, then the loss behaviour of a stock is because of its own performance, and a low value of global purity indicates the effect of external factors as well. Figure 11 shows a PPD with maximum global purity value 0.67 and directed periodic support set <16, 1, 14>. As seen in the figure, the stock 402 is incurring losses because of its own performance. So the buyer has two options either to go for stock 6 or 357. Similarly, there is a PPD with minimum global purity value 0.14 and directed periodic support set <47, 7, 3>.

Figure 10. PPD with minimum maturity (0.02)

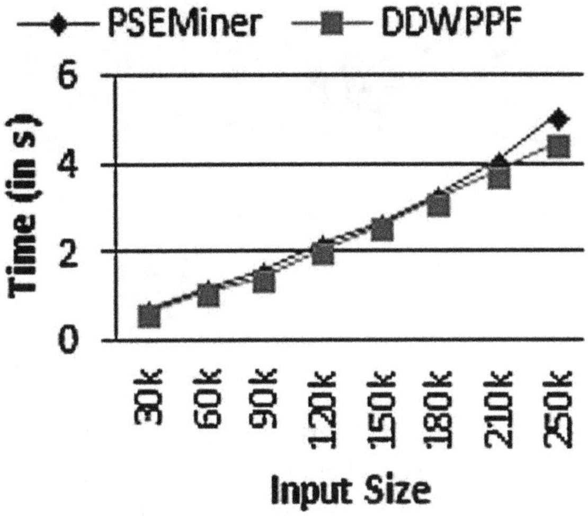

Figure 11. PPD with maximum global purity (0.67)

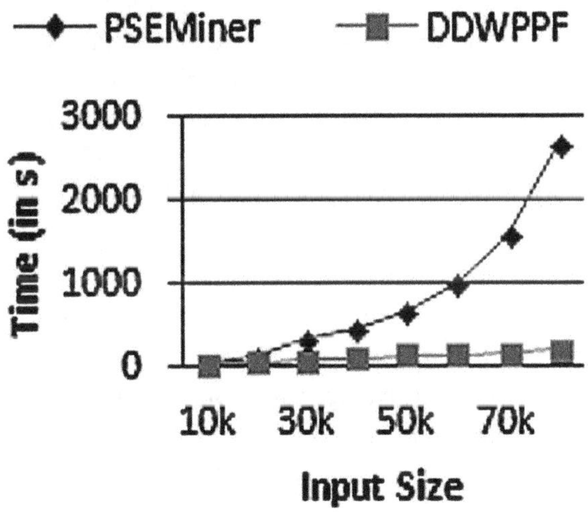

Global Purity and Maturity: In cases where two PPDs have the same value for global purity, maturity acts as a distinguishing factor.The two PPD's shown in Figure 12 possess same value of global purity i.e. 0.31 but PPD in Figure 12a has lower maturity value than that in Figure 12b. It is inferred that the PPD having lower maturity value has more chances to stabilize in long run.

3. **Reciprocity:** It helps in the prediction of response of the buyer towards periodic behaviour of occurrence patterns. From the high reciprocity value, it is inferred that a loss incurring stock would recover in the long run. This means that buyer should wait and observe further behavior before jumping at a decision. So this parameter provides an answer to the third question. There is a PPD with maximum reciprocity value 0.33 and directed periodic support set <12, 13, 4>. The behaviour of PPD with period 13 i.e. 3 months is because of the quarterly change in interest rate having a temporary effect.

Figure 12. PPD with same global purity and different maturity: a. PPD with maturity value (0.75), b. PPD with maturity value 1.31

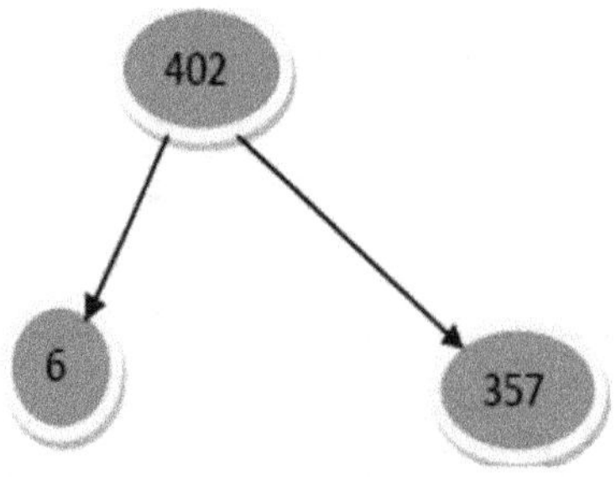

4. **Density:** It is a very useful parameter to determine the possible options a buyer has when switching from a loss incurring stock to another. There is a PPD with minimum density 0.017 and directed periodic support set <16, 5, 9>. More the number of switching options that are available to the stockholder, more are the chances of maximizing profit or minimizing the loss.
   Density and Reciprocity*:* It is found that some PPDs have identical density values and therefore cannot be distinguished. In such a scenario, reciprocity value helps to make a distinction among them and analyze their occurrence behaviour. There are two PPDs with same density i.e. 0.015 and different reciprocity values. It is inferred from their reciprocity values that PPD with higher value incurs less switching cost and hence is more stable than the other PPD.

As compared to PPDs, only 3592 PPDWs are obtained. Following are the observations made by analyzing them on the basis of parameters:

1. **Uniformity:** It is useful to determine the risk and return trade off involved in the occurrence pattern of stock behaviour. The average uniformity value is found to be 0.031 with only 21.82% of PPWDs having uniformity above the average value. Since majority of PPDWs have low uniformity, therefore there is a high risk involved and returns are high as well. There is a PPDW with minimum uniformity 0.0016 and weighted directed support set <1, 2, 4>.
2. **Weighted Reciprocity:** It has helped to determine the buyer's response towards the occurrence behaviour of patterns. The average weighted reciprocity value is evaluated to be 0.0014 and only 10.5% PPDWs have a higher value. There is a PPDW with maximum weighted reciprocity 0.08 and *dw* periodic support set <55, 3, 5>. It is inferred that the stockholder has more faith and would avoid spending more transition cost as the loss incurring stock at present may recover in future.

ENRON EMAIL NETWORK ANALYSIS

The proposed parameters helped in making some interesting observations on the occurrence behaviour of Enron email network. It has helped to determine the following:

- Whether the nature of interactions involved in the network is formal or informal.
- Whether the direction of flow of information within the network is bidirectional or unidirectional.

- Whether the level of communication in a network is horizontal or hierarchical.
- What is the connectivity of the network and the time involved in broadcasting a particular piece of information?
- Which organisation is well set up?

Observations based on different patterns in PPDs are summarized next.
The total number of PPDs obtained is 737.

1. **Global Purity:** Global Purity is a measure of the extent of periodicity exhibited by an interaction in a network. Interactions, that do not occur at time steps other than their support set, exhibit purely periodic character (gp =1). Such interactions are of particular interest in the analysis of a social network and can be characterised as formal or informal based on their periodicity. If the period is 1, then the interaction is considered to be an informal interaction happening between two people with a strong bonding on a daily basis. But for a purely periodic interaction with period greater than 1, it could be a formal interaction. One such example is the interaction that takes place between the employees of an organisation participating in a meeting which is held on a periodic basis to plan sprints.
2. **Reciprocity:** Reciprocity is a measure to characterize the flow of information within a network based on the number of mutual connections. A high reciprocity value corresponds to a bidirectional flow of information in the network whereas a low reciprocity value corresponds to a unidirectional flow of information. Another perspective that is used to view reciprocity is its use to determine the level of communication in a network. A network with a horizontal level of communication exhibits a higher value of reciprocity due to the presence of more mutual connections as compared to a network with a vertical or hierarchical level of communication. There is a PPD with minimum reciprocity value of 0.00 and directed periodic support set <4, 8, 4>. It is clearly seen that the network possesses a hierarchical structure with node 28 at the highest level, node 46 at a lower level and nodes 684, 402 and 288 at the lowest level. The flow of information is purely unidirectional thus resulting in a minimum reciprocity of the network. Experimentally, the average value of reciprocity obtained for the enron network is 0.138076 with the maximum value being 0.4. The low value of average reciprocity drives us to the conclusion that the data corresponds to an organization that exhibits a hierarchical structure.

   Reciprocity and Global Purity: Consider two networks having same values of reciprocity. Then they are distinguished on the basis of global purity values. The network having a higher global purity tends to have a higher hierarchical structure as compared to the network having a lower global purity value. It is because communications at horizontal level neither necessarily stick to the support set and nor exhibit purely periodic nature.
3. **Density:** Density is a useful measure that reflects the various pathways available for communication. In a denser network, the time involved in broadcasting information is less due to the presence of more direct connections between nodes. In other words, density gives a measure of the connectivity of a network. An average value of 0.019026 is observed, while the maximum and minimum values of density are 0.25 and 0.002417 respectively. The values and the plot clearly indicate that the average density of the network is very low hence verifying the claim made in Section *Execution Time Comparison* that the network is sparse due to which DDWPPF exhibits higher run time complexity as compared to PSEMINER.

4. **Maturity:** Maturity expresses the likelihood of an interaction to become periodic after entering the dynamic network. The sooner an interaction starts exhibiting periodic nature, the more is its maturity. This measure is used to differentiate between two organizations that enter the dynamic network at the same time. The one that starts exhibiting periodicity in its interactions sooner will be more mature as compared to the other in which interactions start showing periodicity later. Higher value of maturity implies that the organization is well set up.

There is a PPD with maximum maturity of 2.00 and directed periodic support sets <7, 2, 11>. It implies that it starts showing periodic character as soon as it enters the dynamic network.

Also, a relation observed between purity and maturity is that the subgraphs having the maximum purity of 1.00 also possess the maximum maturity. But the reverse is not necessarily true, that is, interactions having the highest maturity value of 2.00 do not necessarily have a purity of 1.00 as there might be interactions that start showing periodicity right after entering the dynamic network, but have occurrences also beyond their directed periodic support set. Figure 13 shows a PPD with minimum maturity value of 0.15 and directed periodic support set <14, 1, 15>.

It is observed that the set of nodes N = {68, 108, 128, 181, 202, 203} is also found to exhibit an interaction starting at timestep 7 and having a maximum maturity value as seen in Figure 13. But with the introduction of node 53, the maturity value of the resultant network drops to 0.15 thus having the minimum maturity.

The initial timestep *t* of occurrence of the interactions has been found using equation (5):

$$0.15 = 1/(14\text{-}t) \Rightarrow t = 7. \tag{5}$$

It is seen, that the interaction between the node set N and node 53 exists from timestep 7, but starts exhibiting periodicity at timestep 14, thus reducing the overall maturity of the new group, whereas the interaction between the nodes of set N starts exhibiting periodic character at t=7 right after entering the network. One possible reason for this observation of drop in maturity of the group N could be the shifting of node 53 to the department N at timestep 14 after its interactions also assume periodic nature. Initially when node 53 is not a part of N, the network exhibits maximum maturity. Later at timestep 14, when 53 joined the department N, it affects overall maturity.

Maturity and Density: Networks having same maturity values are distinguished on the basis of density. A denser network has more links or connections and hence the team size is predicted to be smaller. In large team, it is not practically feasible to have a direct link between each and every people and therefore the numbers of associations are less thus resulting in low density.

Figure 13. PPD with minimum maturity (0.15)

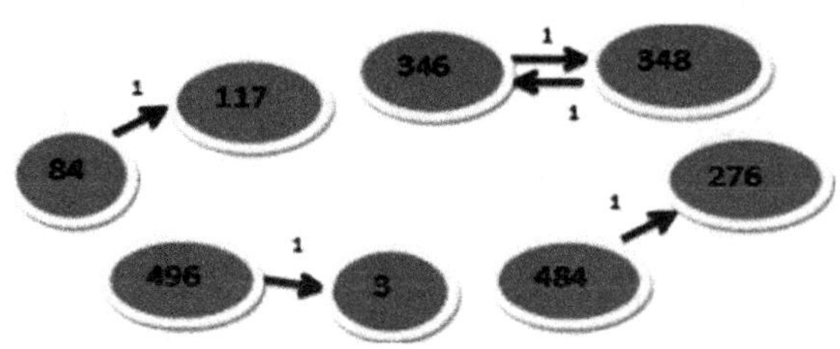

Observations based on different patterns in PPDWs are summarized below:

Total number of PPDWs mined = 304

1. **Uniformity:** Uniformity in a network is a measure of the distribution of weights in the network. More the uniformity, lesser the difference between the maximum and minimum weights on the edges present in the network. Figure 14 shows the plot of uniformity values obtained for the mined periodic patterns on *DW*. One important property of uniformity is that when weights on all edges are unity, uniformity reduces down to density of the network. The maximum value of uniformity obtained is 0.3333 and the average and minimum values obtained are 0.04910 and 0.00030 respectively. As seen from the plot and the values, the average uniformity is low leading to the conclusion that the network has a high variation in the number of emails exchanged between different pairs of interactions.
2. **Weighted Reciprocity:** Weighted reciprocity is a measure of the distribution of weights on the mutual pairs present in a network. More weighted reciprocity implies more uniformity in distribution of weights on the mutual pairs. Figure 15 shows the plot of weighted reciprocity values of the network for all the patterns mined on *DW*. It is clearly observed, that the overall weighted reciprocity of the network is very low, thus leading to the conclusion that the distribution of weights on mutual connections is not uniform i.e., the number of emails exchanged between pairs of connections is biased towards one direction thus imparting a unidirectional character to the network. This observation further strengthens the claim that the network under study has a hierarchical structure.

Figure 14. Uniformity of PPDWs

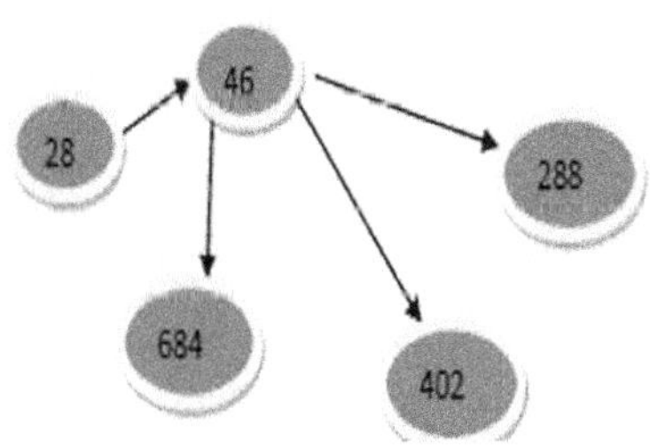

Figure 15. Weighted Reciprocity of PPDWs

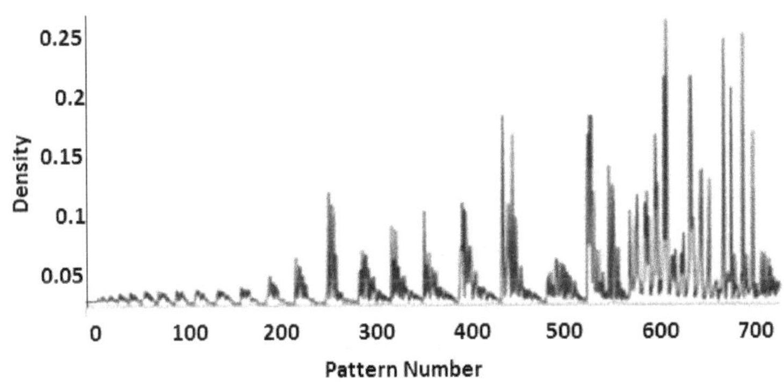

CONCLUSION AND DISCUSSION

In this chapter, the authors have discussed about a framework to efficiently mine periodic patterns on direction and *dw* aspects of an interaction together in a single run. To achieve the same, the patterns are first extracted on direction aspect followed by extracting patterns on *dw* aspect. The graphical plots of experimental results on real world and artificial datasets proves the efficiency of framework and highlights its practical applicability. Further, the patterns obtained in result set are analyzed at microscopic level using six patterns which has helped to study the behaviour of interactions in dynamic network. The parameters are categorized based on direction and dw aspects. Global purity, maturity, density and reciprocity are used to analyze patterns periodic on direction aspects. Uniformity and weighted reciprocity take into consideration the weight of interactions as well and hence perform an analysis on interactions periodic on *dw* aspect. Each parameter has the same mathematical significance, while holding a different meaning based on the nature of application. For example, global purity helps to predict the performance of stocks in a stock market network, while in enron email network, it helps to determine the extent of formality in an interaction. The detailed analysis on Enron and stock market datasets has provided useful insights of the networks and has helped to make important conclusions. Under the influence of external factors on interactions, the values of parameters may change. For instance, in stock market network, factors such as inflation and change in policies may affect parameter values such as density. This could form the basis of an intensive analysis for future work.

REFERENCES

Akoglu, L., Melo, P. O. S. V. D., & Faloutsos, C. (2012). *Quantifying Reciprocity in Large Weighted Communication Networks* (Vol. 7302). Advances in Knowledge Discovery and Data Mining Lecture Notes in Computer Science; doi:10.1007/978-3-642-30220-6_8

Apostolico, A., Barbares, M., & Pizzi, C. (2011). Speedup for a periodic subgraph miner. *Information Processing Letters, 111*(11), 521–523. doi:10.1016/j.ipl.2011.02.016

Berberidis, C., Aref, G. W., Atallah, M., Ioannis, V. I., & Elmagarmid, K. A. (2002). Multiple and Partial Periodicity Mining in Time Series Databases. *Proceedings of the 15th European. Conference on Artificial Intelligence*, 370-374.

Borgwardt, K. M., Kriegel, H. P., & Wackersreuther, P. (2006). Pattern Mining in Frequent Dynamic Subgraphs. *Proceedings of the Sixth International Conference on Data Mining (ICDM'06)*, 818 – 822. 10.1109/ICDM.2006.124

Desikan, P., & Srivastava, J. (2006). *Mining Temporally Changing Web Usage Graphs* (Vol. 3932). Advances in Web Mining and Web Usage Analysis Lecture Notes in Computer Science; doi:10.1007/11899402_1

Duan, D., Li, Y., Jin, Y., & Lu, Z. (2009). Community Mining on Dynamic Weighted Directed Graphs. *Proceeding of the 1st ACM international workshop on Complex networks meet, information & knowledge management*, 11-18. 10.1145/1651274.1651278

Faust, K. (2006). Comparing Social Networks: Size, Density, and Local Structure. *Metodološki Zvezki, 3*(2), 185–216.

Fournier-Viger, P., Lin, J. C.-W., Kiran, R. U., Koh, Y. S., & Thomas, R. (2017). A Survey of Sequential Pattern Mining. *Data Science and Pattern Recognition, 1*(1), 54–77.

Gupta, A., Thakur, K.H., Garg, A., & Garg, D. (2016). Mining and Analysis of Periodic Patterns in Weighted Directed Dynamic Network. *International Journal of Service Science, Management, Engineering, and Technology, 7*. doi: 10.4018/IJSSMET.2016010101

Gupta, A., Thakur, K. H., & Kishore, P. (2014). Mining Regular Patterns in Weighted Directed Networks. *Proceedings of 13th International Conference Of Information Technology (ICIT '14)*, 215-220. 10.1109/ICIT.2014.44

Halder, S., Han, Y., & Lee, Y. (2013). Discovering Periodic Patterns using Supergraph in Dynamic Networks. *Research Notes in Information Science, 14*, 148–151.

Han, J., Dong, G., & Yin, Y. (1999). Efficient mining of partial periodic patterns in time series database. *Proceedings 15th International Conference on Data Engineering*, 106-115.

Hassanien, A. E., Azar, A. T., Snasel, V., Kacprzyk, J., & Abawajy, J. H. (2015). *Big Data in Complex Systems: Challenges and Opportunities, Studies in Big Data* (Vol. 9). Springer-Verlag GmbH Berlin/Heidelberg.

Historical Data for S&P 500 Stocks. (n.d.). Retrieved from http://pages.swcp.com/stocks/#historical%20data

Inokuchi, A., Washio, T., & Motoda, H. (2000). An Apriori-based Algorithm for Mining Frequent Substructures from Graph Data. Principles of Data Mining and Knowledge Discovery. *Lecture Notes in Computer Science, 1910*, 13–23. doi:10.1007/3-540-45372-5_2

Lahiri, M., & Berger-Wolf, Y. T. (2010). Periodic subgraph mining in dynamic networks. *Journal of Knowledge and Information Systems, 24*(3), 467–497.

Liu, G., Wong, L., & Chua, H. N. (2009). Complex discovery from weighted PPI networks. *Journal of Bioinformatics, 25*(15), 1891–1897. doi:10.1093/bioinformatics/btp311 PMID:19435747

Ma, S., & Hellerstein, L. J. (2001). Mining partially periodic event patterns with unknown periods. *Proceedings of 17th International Conference on Data Engineering*, 205 – 214.

Obulesu, O., Reddy, M. R., & Reddy, T. M. (2014). Finding Frequent and Maximal Periodic Patterns in Spatiotemporal Databases towards Big Data. *International Journal of Emerging Technology and Advanced Engineering, 4*(1), 265–266.

Ozden, B., Ramaswamy, S., & Silberschatz, A. (1998). Cyclic Association Rules. *Proceedings of 14th international conference on Data Engineering*, 412-421. 10.1109/ICDE.1998.655804

Qin, G., Yang, J., Gao, L., & Li, J. (2011). Evolution pattern discovery in dynamic networks. *Proceedings of IEEE International Conference on Signal Processing, Communications and Computing (ICSPCC)*, 1-6. doi:10.1109/ICSPCC.2011.6061813

Rasheed, F., Alshalalfa, M., & Alhajj, R. (2011). Efficient periodic pattern mining in time series & sequence databases. *Journal of IEEE Transactions on Knowledge and Data Engineering, 23*(1), 79-94.

Srikant, R., & Agrawal, R. (1996). Mining Sequential Patterns: Generalizations and Performance Improvements. *Proceedings of the 5th International Conference on Extending Database Technology: Advances in Database Technology*, 3-17. 10.1007/BFb0014140

Szudzik, M. (2006). *An elegant Pairing Function*. Retrieved from http://szudzik.com/ElegantPairing.pdf

Yang, J., Wang, W., & Yu, P. (2003). Mining asynchronous periodic patterns in time series data. *Journal of IEEE Transactions on Knowledge and Data Engineering*, *15*(3), 613–628. doi:10.1109/TKDE.2003.1198394

Yang, Y., Yu, J. X., Gao, H., Pei, J., & Li, J. (2014). Mining most frequently changing component in evolving graphs. *Journal of World Wide Web.*, *17*(3), 351–376. doi:10.100711280-013-0204-x

Chapter 13
An Optimal Hybrid Regression Testing Approach Based on Code Path Pruning

Varun Gupta
Amity University, India

ABSTRACT

Hybrid regression testing approaches involve the combinations of test suite selections, prioritizations, and minimizations. The hybrid approaches must reduce size of test suite to minimal level and enhance fault detection rate. The chapter proposes a new hybrid regression testing approach that reduces the number of test cases by reducing the paths of source code on the basis of the dependency between the statements and the changes. The proposed technique is evaluated to be better than the existing hybrid approach in terms of percentage savings in test cases and fault detection rate.

1. INTRODUCTION

Regression testing is aimed at the execution of new versions of software, in order to find errors in newly added or modified parts of software and also to ensure that no new errors have been introduced in previously tested code. Software testers execute software against the test cases and match the output obtained with the expected outcome. The number of test cases could be so large that the testing activity may take very long time and effort, making this task itself infeasible. Also, due to the limited resources available, it is not feasible to execute all the test cases. The tester must select most influential test cases which are representatives of all the errors in the software.

In software industry, various approaches are adopted to conduct regression testing, such as test cases selection, prioritization and minimization. Selections results in reduction in size of test suites by selecting test cases that executes the modified parts of the code (S. Biswas, R. M., M. Satpathy & S. Sukumaran, 2011). Thus, regression testing by test cases selection is a modification aware testing that selects only those test cases that are associated with changed parts (Yoo & Harman, 2007). Regression prioritization creates the order of test cases being arranged in order of decreasing priorities, while minimization

DOI: 10.4018/978-1-5225-5951-1.ch013

deletes non-influential test cases from larger test case array. Research had shown that the test case prioritizations also improves fault detection rate. Rothermel et al., (1999) had measured efficiency of nine test case prioritization techniques in terms of fault detection rate. As fault detection metric, average of the percentage of faults detected *(APFD)* was employed. Value of APFD was found high for every prioritization technique although relative APFD for various techniques varied.

In recent years, hybrid approaches (Aggarwal, Singh & Kaur, 2004; Gupta & Chauhan 2011a), with a combination of selection and prioritization methods have been proposed. Selected test cases are prioritized and executed for testing purpose. Aggarwal, Singh and Kaur (2004) presented hybrid regression testing approach which prioritizes test cases on the basis of number of lines of code covered by each test case. Authors have hypothesized that; larger the coverage of test case, greater is its priority. Process of prioritization continues until all lines of code are covered by selected test cases. The work was extended by Malhotra, Kaur and Singh (2010). Proposed technique selected test cases from test suite containing prioritized test cases. Selected test cases were further prioritized. According to the authors, test case of higher priority is used first till resources are available or confidence about correctness is build up to a considerable good extend.

Singh, Kaur and Suri (2010) presented hybrid regression testing technique that employs various levels of selections and prioritization. This technique first selects the test cases that execute the changed variables or variables computed from these changed variables. Prioritization is again performed depending on the value of dependency count of variables.

All these techniques suffer from several drawbacks like non-coverage of unaffected parts and incomplete handling of control dependencies. Further, the techniques have not been tested on complex live system with larger data set of test cases.

Path pruning can be used as one of the steps toward reducing test suite size. Path pruning involves identifying the path that involves statements containing variables that are directly or indirectly influenced by the evolution of software and thus ignoring the path not involving such variables. This leads to the rejection of test cases associated with the ignored paths. Heuristics can be applied on test cases identified for pruned paths by considering some parameters like number of times the statement is used. Further improvements in terms of reduction in testing efforts can be attributable to parallelization of testing technique.

The paper is organised as follows: Section 1 presents an introduction to regression testing techniques and the suitability of hybrid approaches for regression testing. Section 2 discusses various phases of regression testing. Section 3 presents the related work in the area of regression testing that emphasises on the previous work done by various researchers that laid the foundation for this paper. Section 4 gives problem statement of regression testing for which this paper tries to find appropriate solution. Section 5 gives the proposed approach/algorithm for the problem in the area of regression testing as given in previous section. The proposed algorithm is version specific hybrid approach, comprising of different layers of selections and prioritizations of test cases. Automatization and parallelization of the proposed technique is suggested by taking advantage of web technology. Next section presents case study of library management software system on which this technique is applied. The results are presented in section 7 where comparison is made between the results obtained by executing the algorithm proposed in this paper and one presented in Singh, Kaur and Suri (2010). Section 8 analyses the reasons for difference in results obtained for two techniques. Paper is finally concluded and directions for future work are given in section 9.

2. REGRESSION TESTING AND ITS PHASES

Regression testing is performed on the modified code so as to validate modified software and detect whether new faults have been introduced into the previously tested code (Malhotra, Kaur & Singh, 2010; Singh, Kaur & Suri, 2010). Regression testing involves three phases:

- **Phase 1:** The objective of this phase is to retest the modified code. Implementation of changes in the source code might have adverse effects (ripple effects) on the other parts of the code. In phase
    - Following issues have to be addressed:
        - Testing the modified code
        - Testing of those sections of the code that might have been affected by the implemented change.
        - Execution of those paths that were unaffected by the given variable usage. Considering only variable usage and definition does not guarantees that unaffected statements will continue to function/execute properly after modification without introducing any new or unexpected error.
        - Making a decision about which test case will be executed i.e. will be the part of test case array. Since there are thousands of test cases in the test suite to be executed, re-running of all the test cases will result in increase in term of cost, testing time as well as efforts.
        - Updation of test case array with minimum number of test cases. The updation is based on some optimization measures like total statement coverage, early fault detection etc. Decision must be made about selection of optimization criteria's and selection of testing approaches like minimization, selections and prioritization or their combinations.
- **Phase 2:** In this phase, new test cases are created to test the newly added codes that represent new requirements representing new functionality or requests for fixing of bugs or replacement of misinterpreted requirements.
- **Phase 3:** In this phase, the test case array is executed as updated in phase 1 and phase 2. Software is executed by supplying the inputs as given in test cases and comparing the expected output with the actual received output. In case, there is a mismatch between the actual received and expected output, the tester must locate the error in the code. After having corrected the code, the test case that had failed should be re-executed.

However, the main problem lies in phase 1. It is difficult for the testing team to come up with test case array with minimum number of test cases. The effect of executing of these test cases has to be equivalent to the execution of whole test case array in terms of fault detection capability. The minimum number of test cases used to perform regression testing is termed as chromatic number of the software increment. Effective technique needs to result in low chromatic number of test cases, which is a challenge for the testing team.

3. RELATED WORK

Yoo and Harman (2007) surveyed regression testing approaches that include minimization, selections and prioritization. Biswas et al. (2011) surveyed only regression testing selection techniques. This survey attempt to give an overview and critical analysis of various regression test selection techniques for variety of programming paradigms like procedural, object-oriented, database driven, component-based, web applications etc. Such surveys are helpful in obtaining exhaustive list of regression testing approaches and nurturing future directions.

Vipindeep and Jalote (2005) presented regression testing technique based on static path pruning of source code. This approach relies on control flow graph (CFG) of source code and prunes the graph by removing those parts from the graph that had already been tested thoroughly and hence, tester has faith in them. Reduction in parts of CFG implies reduction in test cases to be executed. This approach relies on the availability of coverage information to prune those paths that are tested with great confidence. But this technique cannot be applied to regression testing resulting as a result of changes like addition, modification and/or deletion of statements.

Dynamic Impact analysis technique (Gupta, Singh & Chauhan, 2010a, 2010b) classifies the changes arising due to modification in the software. Definition and usage information pertaining to the statement variables is used for performing impact analysis. Impact analysis helps in selection of those test cases that executes those code fragments that are impacted by changes in other parts of code.

Few researchers (Aggarwal, Singh & Kaur, 2004; Malhotra, Kaur & Singh, 2010; Singh, Kaur & Suri, 2010; Gupta & Chauhan, 2011a) have recently focused on hybridization of regression testing. Main limitation of technique presented in Aggarwal, Singh and Kaur (2004) is inability to take into consideration ripple effects. The work was extended in Malhotra, Kaur and Singh (2010), but it does not handle control dependencies properly. A hybrid regression testing technique employing various levels of selections and prioritizations has been proposed by Singh, Kaur and Suri (2010). This technique assigns values for priority depending on variable usage and definition, and dependency count between these variables. Neglecting unaffected parts and prioritization on the basis of just variable usage and definition is a major limitation of this approach. The approach is unable to handle special cases of control structures. For example, deletion of a variable may remove control dependency between two fragments of code. Thus some parts of the code will become un-reachable from other part of code due to deletion of variable that connects the two fragments. Recently, genetic algorithm approach for regression testing has also been employed (Malhotra & Garg, 2011). The technique relies on employing adequacy based testing criteria using mutation analysis.

Gupta and Chauhan (2011a) proposed a multi layered hybrid regression testing approach. The approach clusters the test cases and prioritizes the clusters on the basis of priorities of requirements represented by the clusters. Clustered test cases are subjected to sequences of selections and prioritizations at test case level. Execution of these test cases results in the testing of highest priority requirements. This is achieved by focussing on those statements that had resulted in large failures and had highest number of dependencies with other statements.

One of the problem associated with the approach presented in the paper is non-handling of unaffected parts. Another problem is the incapability of the technique to handle those control dependencies that exist between various statements, not associated with the variable usage and definition. The technique was employed on a hypothetical example and required to be verified on some real project.

(Dennis Silva, Pedro Oliveira, Ricardo Rabelo, Matheus Campanhã, Pedro Neto, & Ricardo Britto, 2016) proposed a hybrid regression testing technique that works in five stages i.e. mapping of features into classes, Calculating class criticality, Computing tests criticality, Selecting and prioritizing test cases. The proposed technique is evaluated and compared with two approaches i.e. greedy approach and exhaustive approach, where the proposed technique out performs the other two approaches.

`The summary of various hybrid regression techniques are available in Table 1.

Many researchers have now shifted their attention from co located software development to multisite software development called global software development (GSD). Some researchers and practitioners have explores the functioning of multisite companies involved in GSD so as to find out various challenges in such development (Damian & Zowghi, 2003; Eoin O. Conchuir, H. H., Par J. Agerfalk, Brian Fitzgerald, 2006; Conchúir, Ågerfalk, Holmström & Fitzgeral, 2009; Prikladnicki, Audy & Evaristo, 2003) while others have highlighted the benefits that can be reaped out of GSD (Damian, 2006; Eoin O. Conchuir, H. H., Par J. Agerfalk, Brian Fitzgerald, 2006; Conchúir, Ågerfalk, Holmström & Fitzgeral, 2009). Research on problems has suggested that in GSD main task is to overcome problems as in col-located development and also as posed by distance between developments sites. Gupta and Srivastav (2011b) have proposed a web based frameworks that overcome the barriers of global distances in GSD oriented project. For regression testing in GSD environment, a preliminary work has been reported by Gupta and Chauhan (2011a).

Table 1. Idea of Techniques of Hybrid Regression Testing

S.No	Research Paper	Presented Approach	Data Set Used and Results	Limitations
1.	Regression Testing Technique: A Multi Layered Approach. (Gupta and Chauhan, 2011a)	Regression testing comprises several layers. First layer selects test cases associated with requirements of high priorities. These test cases are further selected on basis of dependency of the statement on large number of other statements (and vice versa). These selected test cases are prioritized; with statement causing failures frequently getting higher priority.	Hypothetical example of software that is composed of three requirements. This Paper had claimed 80% saving in test cases and hence testing effort.	**1.** Neglecting of the unaffected parts. **2.** Inability to handle those dependencies that do not arises as a result of variable usage and definition. **3.** Technique was employed on very simple hypothetical example. This made it impossible to comment on the effectiveness of technique. It needs to be employed on live system.
2.	A Hybrid Approach for Regression Testing in Interprocedural Programs. (Singh, Kaur & Suri, 2010)	This technique assigns values for priority depending on variable usage and definition and dependency count between these variables. Several data structures are employed for this job.	Technique was employed on various small programs like triangle classification, k-maps, time-date, calendar etc and compared with that proposed by Rothermel's technique. Technique resulted in selection of few test cases and higher faults detected i.e. higher APFD metric value.	Similar problems as with above paper. Although results of technique proved to be very effective but nothing can be said about how effective technique really is. Since technique was not employed on live systems with bigger lines of code and thousands test cases.
3.	A Regression Test Selection and Prioritization Technique. (Malhotra, Kaur & Singh, 2010)	Technique is extension of paper given in Sr. no 4. Test cases are prioritized, selected and then finally prioritized. Thus the sequences of selections followed by prioritizations of test cases leads to creation of test suite for regression testing.	Technique was employed on very small program of 42 lines of code and 13 test cases. Program solves the triangle classification problem. Paper claims to have 75% savings in number of test cases.	**1.** It does not consider ripple effects. **2.** Technique applied over very simple example of triangle classification problem. **3.** It does not handles control dependencies properly. **4.** Neglection of unaffected parts of code.
4.	Code Coverage Based Technique For Prioritizing Test Cases For Regression Testing. (Aggarwal, Singh & Kaur, 2004)	Technique selects only those test cases that are of higher priority. Higher priority test cases are those that execute large number of changed statements.	Technique was employed on hypothetical example. Paper claimed to have 60% savings in number of test cases.	**1.** Technique does not consider ripple effects. **2.** Technique effectiveness applied on hypothetical example.

Gupta, Chauhan and Dutta (2012b) have studied the impact of decision aspect prioritization on requirement prioritization and regression testing activities of iterative software development. Authors concluded that in order to reduce regression testing effort and to enhance fault detection rate, requirement priorities can be used as an initial step towards pruning of test cases. Accuracy of such pruning and hence regression testing effectiveness will depend on accuracy of decision aspect prioritization and hence on requirement prioritization process. Results are reported by applying two slightly different versions of regression testing technique as proposed in Gupta and Chauhan (2011a) and decision aspect prioritization technique as given in Gupta, Chauhan and Dutta (2012a) on payroll management software system.

The present paper is based on the work presented in Gupta and Chauhan (2011a).

4. PROBLEM STATEMENT

The software tester has to update test case array with minimum number of test cases. The effect of the execution of these test cases has to be equivalent in terms of fault detection capability, to the execution of the whole test suite. Minimum number of test cases used to perform regression testing of the software can be defined using chromatic number. Thus, tester must solve two problems. First problem is to find the chromatic number of test cases. Second problem is to employ "early and complete fault detection", as one of the optimization criteria. The outcome of the second problem deals with the generation of serial schedule of test cases. Selected prioritized order of test cases is generated whose execution is equivalent to the execution of whole test suite. To find such an order, tester must take into consideration ripple effects that might arise due to changes in code. Whenever some changes are made in one part of code, other parts get affected. Thus changes in one part of code may have adverse effects on other parts of code resulting in ripple effect (Aggarwal & Singh, 2004). Concept of serial schedule is same as used in transaction management of data base management system. Thus problem for which this paper tries to find solution is:

How do the software testing teams will find the chromatic number of test cases, execution of which will lead to early fault detection as one would find after execution of whole test suite, test cases of which lies in thousands?

5. PROPOSED APPROACH

The proposed Hybrid approach tries to minimize the total number of test cases by employing selections of test cases followed by test case prioritizations (number and order of selections and prioritizations may vary) as in (Malhotra, Kaur & Singh, 2010; Singh, Kaur & Suri, 2010). The layered model focuses on the following functionalities:

- **Layer 1 (Top Most Layer):** This layer performs path pruning by reducing the paths to be tested to a manageable number of test cases. The criterion used for path pruning is that path should involve only those statements that affect or are affected by each change in the code/statement. In other words, for taking ripple effect into consideration, one must do impact analysis either static or dynamic. Various approaches are there to perform impact analysis as in (Gupta, Singh & Chauhan, 2010b; Gupta, Singh & Chauhan, 2010a). One of the simplest method on which our algorithm is based is just to take into consideration the definition and usage of variables used in statements as proposed in (Gupta, Singh & Chauhan, 2010b; Gupta, Singh & Chauhan, 2010a; Singh, Kaur & Suri, 2010). Also the variable(s) part of control dependency with changed statement are considered. This helps to take into account the ripple effects generated as a result of modifications.

- **Layer 2 (Second Layer):** This layer minimizes the size of original test suite by selecting some of the test cases while neglecting others. The criterion used for selection is that test case should be executing the paths selected by layer 1.
- **Layer 3 (Third Layer):** This layer tries to further minimize the test array as created by Layer 2. The objective of this layer is to find chromatic number of test cases without sacrificing the criteria of fault detection. To find chromatic number of test cases, paths are prioritized on the basis of *number of dependencies between the changed statements and other statements belonging to same/other paths*. To take into consideration unaffected parts, tester must trace back from output to source code, generating list of variables that are responsible for generation of such variables and those dependent on these variables. Test cases executing these variables are also added to test case array.
- **Layer 4:** This layer results in execution of minimized test case array. The decision is taken on whether to do debugging or deliver the software as determined by the results of execution of the software.

Proposed approach involves various layers as shown in Figure 1.

Figure 1. Proposed Technique

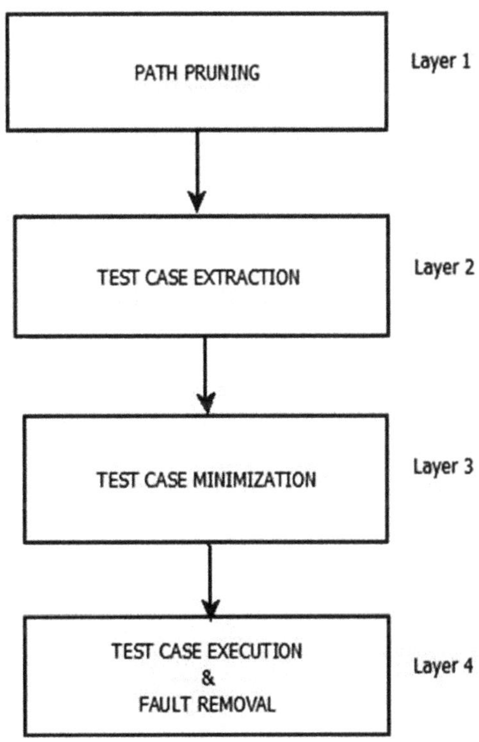

The algorithmic implementation of the proposed approach is presented below:

Algorithm: Regression_testing_increment (Test_suite, Source_code, changed_variables, added_variables, deleted_variables, unaffected_variables, N[i])

Input: Test_suite, Source_code, changed_variables, added_variables, deleted_variables, unaffected_variables.

Output: Test case array i.e. $\mathbf{T_{array}[i]}$. This variable houses the test cases numbers.

/* **Test_suite** is the collection of test cases that includes the test cases of old increment of the software, Source_code is the modified source code that had gone evolution as a result of addition of functionality of new increment, deletion of non-required one or modification of the code. Correspondingly tester have four sets of variable arrays as given as changed_variables, added_variables, deleted_variables, where first one is set of those variables that are modified as a result of statement modification, second one are newly added one as a result of new functionality implementation, third one is set of deleted variables while latter one defined those variables that are not affected by any change. N[i] is 1-D array that stores number of times statement is executed during the use of the software*/

Step 1: For each variable "v" in changed_variable array

{

Fill statement_in and statement_out, 1-D arrays. Former array stores the statement numbers that affects while latter are affected by the changes in variable.

Generate path details that exist between the changed variable and statements in statement_in and statement_out arrays. Store result in Path_changed array.

} // End of loop

Step 2: For each variable "v" in added_variable array

{

Fill statement_in and statement_out, 1-D arrays. Former array stores the statement numbers that affects while latter are affected by the addition of variable.

Generate path details that exist between the added variable and statements in statement_in and statement_out arrays. Store result in Path_added array.

} // End of loop

Step 3: For each variable "v" in deleted_variable array

{

Fill statement_in and statement_out, 1-D arrays. Former array stores the statement numbers that affects while latter are affected by the deletion of variable.

Generate path details that exist between the deleted variable and statements in statement_in and statement_out arrays. Store result in Path_deleted array.

} // End of loop

Step 4: Generate test cases that execute each statement in Path_changed array. Store results in Test_changed_case array by using Test_suit and Path_changed array.

Step 5: Generate test cases that execute each statement in Path_added array. Store results in Test_added_case array by using Test_suit and Path_added array.

Step 6: Generate test cases that execute each statement in Path_deleted array. Store results in Test_deleted_case array by using Test_suit and Path_deleted array.

Step 7: Now trace back from output to program code and identify the "output" variables that are responsible for generation of software outputs. Also identify variables that are dependent (data or control) on identified output variables. Denote such variables as "un-affected" variables.

Step 8: For each variable "v" in unaffected_variable array

{

Fill statement_in and statement_out, 1-D arrays. Former array stores the statement numbers that affects while latter are affected by such unaffected variables.

Generate path details that exist between the unaffected variable and statements in statement_in and statement_out arrays. Store result in Path_unaffected array.

} // End of loop

Step 9: Generate test cases that execute each statement in Path_unaffected array. Further these minimal test cases are prioritized by selecting boundary values i.e. Minimum, Maximum, and average from test suite. Store results in Test_unaffected_case array by using Test_suite and Path_unaffected array.

Step 10: Combine the test cases by removing redundant ones. T_{array} will store all the test cases that are housed in Test_changed_case, Test_added_case, Test_deleted_case and Test_unaffected_case by removing redundant ones since it may be possible that same statement may be in more paths of different variables, thus redundant test cases are simply ignored and what gets stored in test case array is one copy of redundant test cases. A test case is redundant if it covers same statements as covered by other test case.

Step 11: Generate the value of "N_changed" for each path in Path_changed array by adding the values of "n" of each statement belonging to the path. Value of "n" represents that how many times the statement is used when user uses the software.

Step 12: Generate the value of "N_added" for each path in Path_added array by adding the values of "n" of each statement belonging to the path. Value of "n" represents that how many times the statement is used when user uses the software.

Step 13: Generate the value of "N_deleted" for each path in Path_deleted array by adding the values of "n" of each statement belonging to the path. Value of "n" represents that how many times the statement is used when user uses the software.

Step 14: Generate the value of "N_unaffected" for each path in Path_unaffected array by adding the values of "n" of each statement belonging to the path. Value of "n" represents that how many times the statement is used when user uses the software.

Step 15: Now for each path in Path_changed, Path_added, Path_deleted and Path_unaffected array, rate each path according to dependency with other paths. To accomplish this task work at statement level (statements part of path _changed, Path_added, Path_deleted and Path_unaffected array) and look at dependencies (considering case 1, 2, 3 and 4) and add up number of dependencies that exist between various variables in such statements to give total dependency count to the path.

Step 16: Starting with Test cases with highest value of dependencies as found in step 15, those with highest value of N i.e. "N_changed", "N_added", "N_deleted","N_unaffected" are one of highest priority.

Step 17: Thus as far as our resources permits, test cases are executed of most frequently used dependent statements.

Execution of algorithm performs series of selections and prioritization of test cases to reduce bigger test suite to chromatic number of test cases. The proposed approach considers every form of software evolution. For example, software evolved due to the change in the customer's requirements, addition of new functionality planned for current software release, or deletion of obsolete ones. Every form of software evolution involves working with source code at highest level of abstraction, with modules at somewhat lower level and thereafter at statement level.

Tester working at statement level can reduce test cases by determining the statements that are either control dependent or data dependent and highlighted the path connecting the statement involving the change to the statements affected by change. Various cases may arise for the statement involving a change, as given under:

Case 1: The statement is dependent on some other statement. This might be the case where one statement is generating value or defining some variables that the changed statement is using as operand. For example, changed statement c=a+b; is data dependent on statement a=9.

Case 2: Other statements are dependent on the statement involving a change. Thus changed statement is generating some value that is being consumed in other statements as operands. Consider as an example statement c=a+b; which is data dependent on changed statement a=9.

Case 3: One statement is housing some variable that defines control dependency with the changed statement. For example consider the following "Case" control structure:

```
          switch(choice)
{
Case1:
          if (n==p)
          {
printf (" Matrices can be Multiplied\n");
          for (i=0;i<p; i++)
              {
                    for (j=0; j<q; j++)
                    {
                                                    for (k=1;k<=n;k++)
                                                        {
C[i][j] = C[i][j] + (A[i][k] * B[k][j]);
                                                         }
                         }
                    }
printf ("The Multiplication of two Matrics is:\n);
for (i=0;i<m;i++)
     {
          for (j=0; j<n; j++)
          {
          printf ("%d", c[i][j]);
          printf ("\n");
          }
        }
}
```

```
        else
        {
printf (" Matrices cannot be Multiplied\n");
                }
        break;
default:
printf ("Please enter the valid choice");
}
```

Statement like **int choice** or choice=choice1+choice2; is dependent on case control structure. If testing of the code for matrix multiplication is of prime interest, then it should be reachable, so that black box testing is applied over the code. Suppose that choice=choice1+choice2 is altered/deleted, then testing of matrix multiplication would never had been possible.

Case 4: Changed statement defines control dependency with some other statement(s). This also covers the statements, whose execution either determines execution of changed statement or is dependent on it. Consider for example, following code fragment:

```
        printf("Please enter two numbers");
        scanf(("%d %d", &a, &b);
if(a>b)
        {
        printf("A is larger than B, perform Binary search");
        binary(a[ ]);          //Function call
        }
        else
        {
        printf("B is larger than A, so perform Bubble sort first followed by
Binary search");
        Bubble_binary(a[ ]);
        }
        Bubble_binary() module houses code to perform bubble sort and calls
binary() module. Now binary() and Bubble_binary() modules are control depen-
dent on first two value assignment statements. Infact binary() module is also
control dependent on Bubble_binary().
```

Two variables are dependent if one determines another in terms of data dependency or control dependency. In both the types of dependencies, tester highlights the path(s) connecting the statements that are dependent on changed variable(s). Only those test cases that execute the highlighted paths are selected. It is also believed that unaffected parts now malfunction. Thus tester must execute the unaffected statements at least once so that confidence can be generated about quality of unaffected parts. To accomplish this, tester must trace back from output to source code, generating list of variables that are responsible for generation of such output and other variables dependent on them. Test cases obtained can be prioritized further by prioritizing the paths. The path that has highest dependencies value i.e. highest number of dependencies between changed statements and other statements belonging to same path, and with other paths, are first to be selected for execution. Thus one can neglect path of lowest dependency since the dependency covered by this path will definitely be covered by higher priority paths. The prioritization of dependency paths leads to path pruning where low priority paths as well as paths connecting unaffected statements are simply ignored, resulting in neglection of test cases executing pruned paths. Result of sequence of selections and prioritizations is the chromatic number of test cases with high fault detection capability. Test cases obtained results in the execution of those statements that are executed less frequently and have lower value of dependency. Reduction in test cases as a result of path pruning was also presented in (Vipindeep and Jalote, 2005) where source code paths were pruned by doing static coverage analysis. Figure 2 gives Control Flow Graph for hypothetical application.

Figure 2 gives control flow graph for hypothetical application. Node numbered 4, 5, 12, 14, 16, and 21 are decision nodes. Let us consider that changes are to be made to node numbered 2 and 3. As per the dependencies mentioned in above four cases, Let us consider data dependency between node 2, 3, 7 and 10. Assume that node 2 and 3 assigns value to variable, consumed in decision node 5, thus relating them through control dependency. Thus test cases executing statements represented by nodes 5 to 11 through both 6 and 9 are selected for execution. Test cases executing other nodes are simply ignored. To take into consideration unaffected parts, neglected nodes must be executed with minimal number of test cases. To select minimal test cases, only those nodes are selected, that result in software output and are dependent on other variables. Further these minimal test cases are prioritized by selecting boundary values i.e. Minimum, Maximum, and average from test suite. Thus path pruning resulted in reduction in number of test cases to be executed on modified source code.

Objective of the proposed regression testing technique is to reduce the testing effort by reducing the number of paths to be tested thereby reducing the number of test cases to be executed. This technique can be made more advantageous if following features are provided with the technique:

- The process should be automated in such a way that different testing teams can use the same framework. When supplied with the change information, source code and test suite, and the framework will automatically generate different paths. These paths can be tested and chromatic number of test cases to be executed can be obtained.
- Since software development is a multisite affair these days and might involve the testing activities being distributed at various sites (Global Software Development), the technique must be accompanied with a framework that overcomes the barriers as provided by global distances (Damian

Figure 2. Control Flow Graph (CFG)

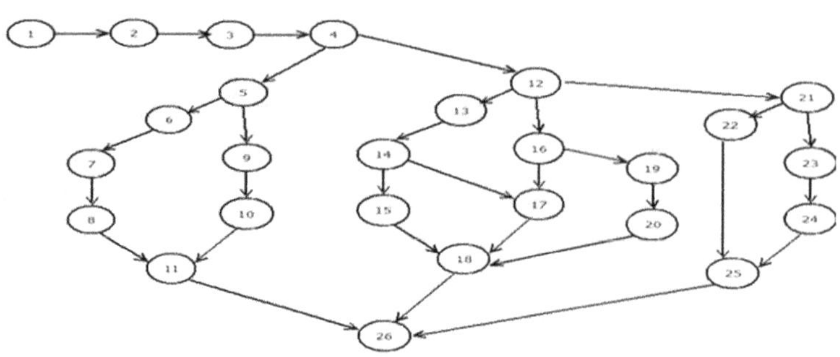

& Zowghi, 2003; Damian, 2006; Eoin Conchuir, H. H., Par J. Agerfalk, Brian Fitzgerald, 2006; Eoin O. Conchuir, P. J. A., Holmstrom, H., Brian Fitzgerald, 2009; Prikladnicki, Audy & Evaristo, 2003).

Web based frameworks was proposed by (Gupta & Srivastav, 2011b). The framework was found to be suitable for overcoming the barriers of global distances allowing testing activities to be performed in parallel. Thus, if testers are able to distribute the testing work to different globally distributed sites, the overall testing time will be divided and lowered.

In order to provide above features with this technique, paper proposes a web based framework that will automate proposed technique in globally distributed environments such that the activities are performed in parallel. Web application development languages will be employed to make both static as well as dynamic web pages, static is for information delivery while dynamic will be updated in accordance with the inputs supplied. Different dynamic web pages will be displayed for different persons who are logging in. For example, administrator page displayed will have options like create parallel slices, distribute slices and combine distributed results. Similarly, for a developer team member, the web page displayed will only have details of new assignments given, source codes, test suites etc with the option to employ the testing approach to generate chromatic number of test cases and option to execute these test cases. Now, one question that comes in mind is that why to distribute the work to different sites, since only thing one has to do is "click on" generate chromatic number of test cases or execute test cases. Reason is that today's software's are highly complex in nature, categorized by thousands of test cases and thousands of lines of code. After the execution of test cases, received output will be compared with the expected one and mismatches in comparison process will be generated. Mismatched represents errors in code that calls for performing debugging. Software will automatically trace back to the statements that are dependent on the variables represented by inputs parameters. These statements might be faulty and needs to be manually analyzed by tester. Thus it will make it difficult for a single person to handle the whole testing process. The implementation of web based framework employs the following:

- **Languages:** Hyper Text Markup Language (HTML), JavaScript, Hyper text Preprocessor (PHP)
- **Development framework:** Dreamweaver, or Microsoft front page.
- **Database:** MySQL.

Since PHP pages are platform independent, developing the pages on Linux or windows or even UNIX makes no issues. Similarly, tester can run the web based framework on any of these operating systems running on server. HTML makes static pages; JavaScript will be employed to have client side scripting i.e. implementing some checking of input fields at client side like filling of mandatory fields in correct format expected etc, PHP for dynamic pages will have code to link to database offering server side scripting language.

Web based application will be composed of various web pages with some web pages implementing some functionality while others implementing some other. Web application can be visualized as a collection of modules in accordance with the functionality implemented by them. Following will be modules that comprise the application:

- **Create_member:** This module is responsible for creating the authorized users of this web framework. Potential users includes administrator (responsible for creating new users, parallelization and distribution and combination of testing reports), testers (who are distributed by global distances and responsible for doing regression testing).

- **Authentication:** This module is responsible for authenticating the user by the combination of triplets i.e. user name, password and type of user. Different types of users have different privileges. For example, administrator has administration powers but testers don't have such. For successful login, users are taken to the main page, while for unsuccessful attempt users are given option to re-enter the triplet's combinations.
- **Parallelization:** This module performs the most important function of converting the testing task into parallel assignments that are to be distributed among distributed testing teams. One of the biggest requirements for such parallelization to occur is that the source code needs to be divided into slices that are independent of each other. This independence implies that the code that is part of the slice do not have any data or control dependency with other slices. To compensate for control dependencies, fragment of code that triggers the modules or makes the modules control to exhaust, are added by the framework itself. Finally the test cases executing the modules or statements in slices are extracted from the test suite and made part of the slice to be distributed. Thus, bigger software is decomposed into smaller independent ones that are tested individually at several testing nodes. At such nodes, errors are corrected and main node that houses testing information is informed about the corrections. It is possible that multiple nodes may process slices of codes with some similar statements. The sub modules of parallelization module interact with each other and with various databases is shown in Figure 3.

After the static analysis of code, static analyzer will generate the virtual ordering of the statements that represents individual slice along with test cases associated with it. Fragmenter will take this information as an input and generate fragments in the form suitable for allocation to different nodes along with the associated test cases. Such generated fragments will be assessed by the distribution module. After considering various allocation parameters like current load on particular node, resources available at that node, processing capability of node etc, slices along with test cases are distributed to testing nodes.

Figure 3. Sub modules and databases of parallelization module

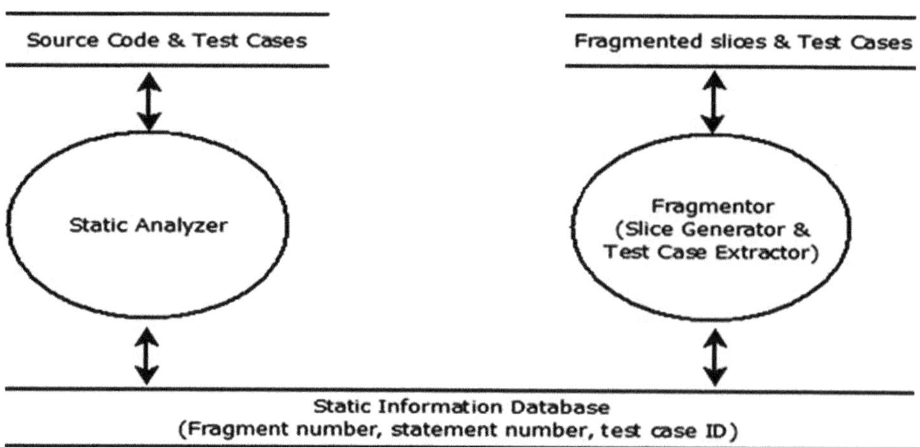

Appendix A gives simple software of calculator as an example and illustrated how it will be parallelized and distributed to two testing nodes.

- **Distribution:** This module distributes the slices to various testing nodes for execution. Each testing team at distributed node executes the testing algorithm, locates and removes errors (if any) in the program. Distribution of slices is done by taking into consideration the load at various computing nodes since each node may have old assignments that are to be solved.
- **Testing:** This module implements the proposed algorithm. The module takes the slice as an input i.e. both subset of source code statements along with the test cases implementing them and finds chromatic number of test cases by finding the paths that connects the variables with the changed, deleted or added variables. Finally, the test cases are executed to find faults in each path. For any mismatch between the received output and expected output; the framework highlights the statements that are related to the variables that are computed wrongly. This is achieved with the help of following modules:
    - **Chromatic_testcases:** This module is responsible for the generation of chromatic number of test cases. After analyzing the change impact, test cases executing the path connecting statement containing changed variable with other statements belonging to same path, starting with path of higher dependency are selected for execution.
    - **Execution:** This module execute the test cases selected by the Chromatic_testcases module and match the expected output with the actual output received. Here, match signifies no error while non matching condition represents error or fault.
    - **Error_visualization:** If an error is encountered then it must be traced in the code (Perform debugging). Set of Paths between the variables that are dependent on variable responsible for the error, are identified by this module.
- **Report Generation:** This module generates the outcome of the testing process in the form of a report that is submitted by each testing node. If report is "successful" from every node, the testing is assumed to be successful.
- **Combination:** This module combines and analyses the report as submitted by each node. If report is "successful" from every node, testing is assumed to be successful.

The proposed web based framework has numerous advantageous that will truly improve quality of the software product. These advantages are given as under:

- It employs parallelism among testing activities. Thus testing effort, time is divided among "N" testing nodes.
- It overcomes barriers as imposed by global distances that makes globally distributed software development a big problem.
- It automated the proposed technique to maximum extend ranging from creation of parallel slices and their distribution, to illustration of path of statements that has created a mismatch between the expected output and actually received output i.e. faults, thus reduces the testing time, costs as well as effort.

6. CASE STUDY

The effectiveness of the proposed technique is tested on the prototype model of "Library Management System", project with 486 lines of code, 13 functions and 125 test cases. As a scenario of software evolution, following modifications were induced in the software:

- New functionality was added to the software as a result of implementation of pending requirement of "calculating fine of students and blocking their accounts if they have not paid their fine within 14 days". This case involves calculating fine by fetching student details like books issued, data of issue etc and multiplying the late submissions by 5 (for Rs. 5 per day). After 14 days, system must block the account so that no new book can be issued to the member and student is allowed to return books.

After inducing new functionality, paths connecting the variables affecting the variable "fine" and other variables affected by this variable "fine" are identified. The statements in Issue() functions are affected by this variable "fine" while Book(), Member_details() are the functions affecting "fine" variable. The distribution of test cases is given in Table 2:

Analysis of Table 2 makes it very clear that if tester adds up the number of test cases required to test these three modules, they result in just 32 test cases. Thus tester has to run only 32 out of 125, which results in saving of approximately 93 test cases or equivalent 74% savings. These 32 test cases are further prioritized, resulting in total 21 test cases. Execution of the program with 21 test cases results in 20 faults thus leaving 2 undetected. But, such undetected faults will have no effect on quality of software since they are being used less frequently. The test cases selected as per data and control dependency are further prioritized in accordance with the *number of times paths are executed* criteria. This lead to the detection of all faults that were seeded in the software. Finally only 21 cases have to be tested, leaving 104 un-executed leading to a total savings of 83%.

Table 2. Distribution of Test Cases for Library Management System

S.No.	Segment of Code	Type of Dependency	Number of Test Cases
1	Path between variables in function Book() and added variable "fine".	Data Dependency	04
		Control Dependency	06
2	Path between variables in function Issue() and added variable "fine".	Data Dependency	02
		Control Dependency	06
3	Path between variables in function Member_details() and added variable "fine".	Data Dependency	04
		Control Dependency	08
4	Path between variables generating output (unaffected statements) and variables dependent on it.	Data Dependency	24
		Control Dependency	22
Redundancy in Test Cases			44
Total Number of Test Cases			32

- In order to analyse fault detection capability of the technique, the source code was transformed to a faulty version by seeding some known faults. Maximum of these faults were seeded in statements having control dependencies with changes statements. Higher fault detection capability would be attributable to higher number of faults detected. Faults are introduced as a result of changes (due to ripple effect) and other that was due to the unexpected error in the unaffected parts. Most of the errors (22 in number) were seeded in that part of source code which is control dependent on changed variables statements but do not houses its usage. Some of the faults were there in Data dependent statements, while some were seeded in the code that was unaffected by the changes in the variables. After test suite was minimized to a manageable number, all test cases were executed and output was analyzed to see whether all seeded faults would be detected. After the run of software against the minimized test suite test cases, software faulted for these test cases for all seeded faults. This means software has 100% fault detection rate. This happened since proposed approach was able to handle ripple effects as well as errors that can be introduced in unaffected parts.

7. EXPERIMENTAL SETUP

The technique was applied on Library Management Software System. The experimental analysis is performed as follows: For the given software with 125 test cases, and 486 lines of code, variables contained in statements that are affected or affect the added new variable "fine" were identified. Once all the statements are identified, set of statements are visualized as a path. Once such path is established, test cases that execute the path were identified from test suite. Such a path will contain dependent statements that are either data or control dependent. Control dependency is automatically there when one variable in a statement is used in some other statement. However, it may happen that changed variable is not there in the control structure but its modification affects the other statements to be tested.

The savings in test cases are due to the fact that the test cases executing other paths are simply ignored and only those executing it are selected. Such test cases are executed after taking into consideration two factors namely dependencies between paths and number of times path or statement is executed. Improved savings are the result of the fact that since one is interested in only those paths that contain dependen cies between the changed/deleted or added variables and statements that affects or are affected by them. This result in total number of paths required to be tested gets decreased. Some more paths related to unaffected statements can be prioritized and selected for execution so as to make sure that no new error are introduced in such statements. Decreased number of paths implies decreased number of test cases to be executed and hence less testing time, efforts and costs are to be invested.

Thus, more such savings will be seen if this technique is employed on a code that is categorized by thousands of lines of code and thousands of test cases. To analyse the effectiveness of proposed technique, it is compared with the technique presented in (Singh, Kaur & Suri, 2010) with respect to following parameters i.e. similarity in working of techniques, difference in working, difference in results obtained after the two techniques was applied on Library Management System. The parameters of comparison are as under:

- **Similarity in Proposed Techniques:** There are various similarities in the two techniques that includes the following:
    - Both are hybrid based techniques for regression testing.

- Both are multi layered techniques and perform test cases reduction for regression testing by carrying out selections followed by prioritizations. The sequence can vary in terms of number or layer of application of selections or prioritizations.
- Both can handle software of real world that are interprocedural in nature i.e. for software with many modules.
- Both are based on variable usage and definition but such is taken as only base in the proposed technique as presented in this technique.

- **Differences in Techniques:** Since the two techniques seems to be same from above section but differs in terms of following aspects:
    - The approach proposed by Singh, Kaur and Suri,(2010) takes into account the variable usage and definition but does not consider the need for executing those paths that were unaffected by the given variable usage. In our work we hypothesize that considering only variable usage and definition does not guarantees that unaffected statements will function/executes properly after modification without introducing any new or unexpected error. Thus such statements needs to be executes at least once.
    - Proposed technique presented in this paper is based on finding data and control dependencies between changed statements and other parts of code. Highlighted paths are prioritized on the basis of the dependencies and the count of *how many times a statement and hence the path is executed when user executes the software*. This makes sure to execute not only affected parts but also unaffected parts of the software *starting with most dependent statements that are executed very often.* The technique in (Singh, Kaur & Suri, 2010) only prioritizes the test cases that execute changed variables contained in statements on the basis of dependency count i.e. number of times such variable is used as operand which neglects the control structures defined in the program code. For example, deletion of a variable will make a control structure "Non-control dependent" and hence non-reachable on/from changed statement (example use of goto statements where there is control dependency but no variable usage or definition). This makes black box testing of such "non-reachable" code impossible.
- **Differences in Results Obtained:** The two techniques were employed on Library Management System for which source code and test cases were already available. Test case reduction (in number) and fault detection is taken as a main parameter for discussion. Table 3 illustrates the comparison of results obtained after the application of two techniques on Library Management System against two performance parameters:
    - Test case reduction and,
    - Fault detection.

Regression testing technique is better if it selects smaller set of test cases that had 100% fault detection rate. There can be some compromise with number of test cases if difference is being limited to single or double digit but not with fault detection capabilities.

Where **D.D** refers to Data Dependency while **C.D** refers to Control Dependency. P1 refers to proposed technique as presented in this paper while P2 refers to that proposed in (Singh, Kaur & Suri, 2010). The results obtained for two techniques for same project of Library Management System, executed against similar test suite but against different test cases obtained, can be represented graphically as shown in Figure 4.

Table 3. Detailed description of results obtained

S.No.	Segment of Code	Type of Dependency	No. of Test Cases (a)		Fault Detected (Yes/No) (b)		Reason for (a) (P1)	Reason for (b) (P2)
			P1	P2	P1	P2		
1	Path between variables in function Book() and added variable "fine".	D.D	04	02	Yes	No	Different Prioritization schemes.	Two faults out of 5 were identified since one is related to control structure and another was in unaffected parts.
		C.D	06	04			Control Dependency exists in form of variable usage in control struct.	
2	Path between zvariables in function Issue() and added variable "fine".	D.D	02	02	Yes	No	Yielded same set of test cases.	Three out of five went unnoticed.
		C.D	06	00			Control Dependency not considered	
3	Path between variables in the function i.e. Member_details() and added variable "fine".	D.D	04	03	Yes	Yes	Different Prioritization schemes.	All two seeded fault were identified since it was in variable usage.
		C.D	08	04			Control Dependency considered to same extend due to some variable usage in control struct.	
4	Path between variables generating output and variables dependent on it.	D.D	24	00	Yes	No	Unaffected statements are not considered.	Unaffected statements are not considered. So all five faults were not identified.
		C.D	22	00			Unaffected statements are not considered	
Redundancy in Test Cases			44	00	Redundancy only considered for deleted variables that has no usage.			
Total Number of Test Cases			32	17	Basically 32 are what obtained after selecting test cases on basis of dependencies. Actual are 21 obtained after considering number of times path is executed.			
Number of Faults Detected			Technique P1 detected 20 faults while P2 detected 10 out off 22.				P2 do not handles unaffected statements and control dependent statements. Only those control dependencies are handled that has variable usage.	

Figure 4. Comparison results for two techniques

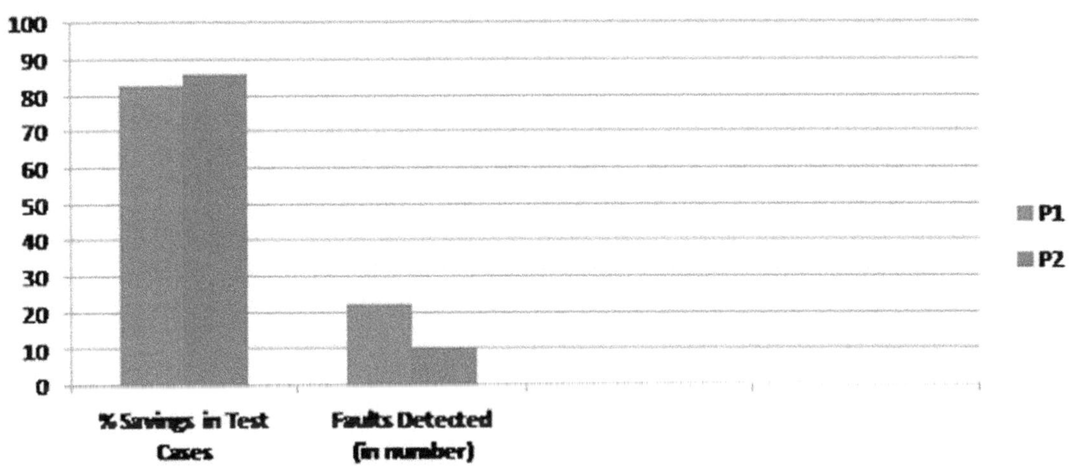

8. RESULT ANALYSIS

Figure 4 and Table 3 gives the details of results obtained after applying two regression testing techniques on prototype project of Library Management system. The proposed technique, P1 is better than P2 technique (Singh, Kaur & Suri, 2010), Reason for the difference in reduction in test case is that P2 technique selects only those test cases that were executing the changed variable and variables that are computed from these variables. Only these selected test cases were prioritized while in P1 technique, all the test cases that were part of the highlighted path were selected. Highlighted path is one that exists between the changed variable and other statements that is data or control dependent on it. Moreover, P2 technique does not consider the fact that unaffected statements might not execute properly. Due to this fact, P1 technique has some large test cases but difference in being limited to few tens (since only few prioritized test cases executing control dependent paths and few executing unaffected statements are to be added).

Only test case reduction had no role to play in regression testing if it is not making a faulty program to fault when executed against selected test cases. Thus, scope of P2 technique is limited as it is not able to make program to generate faults. Thus, no doubt if proposed technique P1 is utilized, software testing team had to execute some extra test cases (number being in single to double digit) but ensures better fault detection rate. The P1 technique finds chromatic number of test cases that had good likelihood of finding all errors housed in a program. Here, 32 test cases have not been further prioritized based on values of dependency and number of times path (and hence statements) are executed. Thus, if further steps are employed, program will be executed with 21 test cases. Execution with 21 test cases, executes 20 faults, thus leaving 2 undetected. But, such undetected faults will have no effect on quality of software since they are being used very less. This leaves difference between the number of test cases between the two techniques P1 and P2 as 5 but with huge difference in fault detection rate.

9. CONCLUSION AND FUTURE WORK

Goal of this paper is to find out chromatic number of test cases with high fault detection capability. Paper proposes to automate hybrid regression testing approach by performing activities in parallel using web framework. The proposed hybrid regression testing technique leads to reduction in number of paths to be tested thereby reducing number of test cases to be executed. Proposed technique will be useful in software industry dealing with maintenance of "aged" software in resource constrained environments. The proposed hybrid regression testing technique overcomes various drawbacks of existing techniques like non-coverage of unaffected parts and incomplete handling of control dependencies.

Regression testing technique proved to be effective for a program of size 486 lines of code and test cases in range of few hundreds.

It is expected that in near future this technique will be employed on those software whose sizes and test cases are in few thousands. It is also expected that web based framework will be implemented in near future and some software development firm will employ the tool and take advantage of it. Work on automation and parallelization of regression testing is currently in progress.

REFERENCES

Aggarwal, Y. S. (2004a). *Software Engineering*. New Age International.

Aggarwal & Kaur. (2004b). Code Coverage Based Technique For Prioritizing Test Cases For Regression Testing. *ACM SIGSOFT Software Engineering Notes, 25*, 1-4.

Biswas, Satpathy, & Sukumaran. (2011). Regression Test Selection Techniques: A Survey. *Informatica, 35*, 289–321.

Chetna Gupta & Chauhan. (2010a). DU-Regs: Online Dynamic Approach to Visualize Impact Analysis for Regression Testing. *International Journal of Computer Applications, 1*, 8-11.

Chetna Gupta & Chauhan. (2010b). A Dynamic Approach to Estimate Change Impact using Type of Change Propagation. *Journal of Information Processing Systems, 6*, 597-608.

Conchúir, Holmström, & Fitzgerald. (2009). Global Software Development: Never Mind the Problems – Where are the Benefits. *Communications of the ACM,* 52.

Conchúir. Ågerfalk, & Fitzgerald. (2006). Global Software Development: Never Mind the Problems – Are There Really Any Benefits? *29th Information Systems Research Seminar in Scandinavia (IRIS).*

Damian, D. (2006). Requirements Engineering in Distributed Projects. *IEEE International Conference on Global Software Engineering (ICGSE'06).* 10.1109/ICGSE.2006.261218

Malhotra & Singh. (2010). A Regression Test Selection and Prioritization Technique. *Journal of Information Processing Systems, 6*, 235-252.

prikladnicki & Evaristo. (2003). Global Software Development in Practice Lessons Learned. *Software Process Improvement and Practice, 8*, 267–281.

Damian, D. Z. (2003). Requirements Engineering challenges in multi-site software development organizations. *Requirements Engineering Journal, 8*, 149–160. doi:10.100700766-003-0173-1

Rothermel, Chu, & Harrold. (1999). Test Case Prioritization: An Empirical Study. *Proceedings of the International Conference on Software Maintenance.*

Ruchika Malhotra, M. G. (2011). An Adequacy Based Test Data Generation Technique Using Genetic Algorithms. *Journal of Information Processing Systems*, *7*(2), 363–384. doi:10.3745/JIPS.2011.7.2.363

Silva, D., Oliveira, P., Rabelo, R., Campanhã, M., Neto, P., & Britto, R. (2016). A Hybrid Approach for Test Case Prioritization and Selection. *IEEE Congress on Evolutionary Computation.* 10.1109/CEC.2016.7744363

Varun Gupta, . (2011b). *Overcoming Global Distance in Globally Distributed Developments by Web Based Framework.* CiiT International Journal of Software Engineering and Technology.

Varun Gupta & Dutta. (2012a). Decision Aspect Prioritization Technique for Globally Distributed Developments: A Hybrid Approach. Elsevier.

Varun Gupta & Dutta. (2012b). Impact Analysis of Requirement Prioritization on Regression Testing. *2nd World Conference on Innovation and Computer sciences (INSODE).*

Varun Gupta. (2011a). Hybrid Regression Testing Technique: A Multi Layered Approach. I*EEE annual Conference "INDICON"*.

Vipindeep, V. (2005). *Efficient Static Analysis with Path Pruning using Coverage Data. International Worskhop on Dynamic Analysis*. WODA.

Yogesh Singh & Suri. (2010). A Hybrid Approach for Regression Testing in Interprocedural Programs. *Journal of Information Processing Systems, 6*, 21-32.

Yoo. (2007). Regression Testing Minimization, Selection and Prioritization: A Survey. *Software Testing, Verification & Reliability*.

Chapter 14
Impact Analysis of Intelligent Agents in Automatic Fault-Prone Components Prediction and Testing:
Impact Analysis of Intelligent Agents in Test Automation

Jeya Mala Dharmalingam
Thiagarajar College of Engineering, India

ABSTRACT

Software quality is imperative for industrial strength software. This quality will be often determined by a few components present in the software which decides the entire functionality. If any of these components are not rigorously tested, the quality will be highly affected. Without knowing which of these components are really critical, it will not be possible to perform high level testing. Hence, to predict such fault-prone or critical components from the software prior to testing and prioritizing them during the testing process, an agent-based approach is proposed in this chapter. The framework developed as part of this work will certainly reduce the field failures and thus will improve the software quality. Further, this approach has also utilized important metrics to predict such components and also prioritized the components based on their critical value. Also, the work proposed in this research has also been compared with some of the existing approaches and the results reveal that, this work is a novel one and can both predict and test the components from the software.

INTRODUCTION

As per the study of National Institute of Standards and Technology, the cost for an inadequate infrastructure for software testing is estimated to be from \$22.2 to \$59.5 billion (Tassey, 2002). As exhaustive testing (testing 100%) is not feasible (David C., Jinlin Yang, Sarfraz Khurshid, Wei Le and Kevin

DOI: 10.4018/978-1-5225-5951-1.ch014

Sullivan, 2005; Myers,1979), the industries are forced to stop the testing process at one point of time and deliver the software to the customers. This leads them to compromise the quality of the software due to customers' need for quick delivery of quality software, reduced software development lifecycle, changing markets with global competition and rapid development of new processes and technologies.

The surveys have indicated that, many of the complex systems' failures are due to insufficient testing of software before they are deployed to the customer side (Bernardi, 2011; Schneidewind, 1978). After the analysis, it has been identified that, the highly critical components are not being properly tested or simply ignored without knowing their critical level due to time and cost compromises.

The identification of such critical components from the software prior to testing is still a research area, since the automated testing tools available in the market doesn't address the said problem. Based on our field surveys conducted in several organizations during the past months, it has been identified that, most of the defects reported by the customers after delivery are present in these higher critical components. This gives us the insight on the importance of critical components identification and their verification prior to the delivery of the software. But, the identification of the criticality level of a component involves the evaluation of various metrics and measures associated with them.

Hence, an automated software testing framework that can identify and prioritize the critical components based on the various metrics and measures associated with each of the components and can also provide an optimized critical paths list which can reduce the time and cost needed in the testing process without compromising the testing of these critical components is the need of the hour.

From the literature survey, it has been identified that only a very few works have been conducted in the said research area, and that too have been limited by the type and number of metrics used by them. If the software under test is small and simple then the identification can be done manually. As the real time complex systems have huge functionalities, there is a need for an approach that embeds both intelligence and automation as a tool.

Since software testing is NP-hard (Non-Polynomial hard) (Nagappan, 2006), and as manual testing is costly and error prone, several existing research works on structural testing have employed computationally intelligent techniques, such as artificial intelligence and evolutionary computation methods, to achieve optimization in the testing process (Alok Singh, 2009; Basturk & Karaboga, 2006; Baykasolu A, Lale Özbakır & Pınar Tapkan, 2009; Dorigo M., Maniezzo, V., Colorni, A., 1996, Teodorović & Dell, 2006; Fathian M, Babak Amiri & Ali Maroosi., 2007; Karaboga, 2008, 2009; Pham D.T., Ghanbarzadeh A., Koc E., Otri S., Rahim S., & Zaidi M., 2006; R. Srinivasa Rao, S.V.L. Narasimham, M. Ramalingaraju., 2008; Wong Li-Pei, Chi Yung Puan, Malcolm Yoke Hean Low, Chin Soon Chong., 2008).

As several existing knowledge based approaches are population based approaches with sequential execution behavior, from the literature survey it has been identified that, agents based approach gives promising results when compared to the other solutions. However, this research work applied the multi agents based approach in which each agent is responsible for doing its activities in a parallel manner.

Hence, this research work proposed, an automated critical components identification and verification framework using multi-agents based approach to identify the critical components and to generate and optimize the number of critical paths needed for their verification.

PROBLEM FORMULATION

The proposed research problem has its origin from various industrial surveys taken during the recent years. Even though achieving zero-defect quality software is the ambition, it is not possible in reality (Pressman, 2007). Software testing can be very costly and it typically consumes at least 50% of the total cost involved in software development (Bernardi S, Javier Campos, & José Merseguer., 2011; Li and Lam, 2004). There should be some stopping criterion to be followed to stop testing at one point of time and release the software to the customer side. Generally, the industries have different stopping points based on cost, time, resource utilization or some acceptable level of quality.

Due to customers' need for quick delivery of quality software; reduced software development lifecycle; changing markets with global competition and rapid development of new processes and technologies, lead the industries to skip off some of the components during testing. If these skipped components are highly critical, then their impact on the quality of the software will be huge. Hence, an approach is the need of the hour to prioritize the testing of critical components and perform the testing process in an optimized way without compromising quality.

This insight is the motivation behind this research work to propose a novel framework to identify critical components early in the life cycle and generate the critical paths that will cover all the critical components in minimum number of test executions. This will in turn help to reduce the cost involved in testing and also the huge amount of loss incurred in terms time, resources, human life etc. because of the lack of testing of critical components.

This research work, proposed an multi-agents based framework, motivated by the intelligent behavior of the agents to act, react and work autonomously to automate the critical paths generation and optimization process. In the proposed approach, the SUT is converted into a graph based representation in which each node represents a component in the SUT. Each critical path is represented as a possible solution in the optimization problem and critical value - a heuristic introduced to each component corresponds to the quality or fitness of the associated component to be added to the critical path. The nodes are selected by means of an intelligent search through the Software under Test (SUT) based on the fitness value associated with each node which is calculated using the complexity and dependability metrics associated with each node in the SUT. The intelligent agents modify the search paths with time and the agent's aim is to discover the places of nodes with higher critical value and finally the one with the highest critical value.

The objective function is to maximize the critical value of each path by finding the sum of the critical values associated with each node along a test path based on constraint satisfaction. Now, a few efficient test paths that can cover all the critical components in the system model in less time are generated and stored in the optimal critical path repository.

Since the intelligent agents exhibit autonomy, social ability and interactivity with other agents (Russel & Norvig, 1995), they are well suited for applications that need intelligent decision making process. The performance of the agents in software test optimization was illustrated in a previous work (Mala & Mohan, 2009), for test sequence optimization. Based on the results of the previous work, the proposed approach implemented the agents to perform the critical path optimization activities seamlessly. In the proposed approach, since the local search methods and global search methods are managed by three types of agents, the proposed approach attains near global optimal solution.

The test adequacy criterion applied in this work ensures component coverage. This test adequacy criterion has been applied by researchers as a potential measure to evaluate the coverage during system testing (Horgan, J., London, S., & Lyu, M., 1994). Hence, in the proposed approach component coverage measure is applied as a way to optimize the test paths from the search space.

Further this paper is organized as follows: Section II discusses the related work in the proposed research area; Section III deals with the proposed work in which the proposed algorithm and the Critical Path Optimizer and Section IV gives the experimentation and result analysis.

RELATED WORK

Kaur (2015) has investigated various metrics that that have been used for fault prediction and have classified the existing metrics into three broad categories. Since no type of metric could accurately identify the fault prone modules, an integrated metric has been proposed to predict the fault prone modules.

Jin et.al. (2012), proposed a reduction dimensionality phase, which can be generally implemented in any software fault-prone prediction model. In this study, the authors presented applications of artificial neural network (ANN) and support vector machine in software fault-prone prediction using metrics. A new evaluation function for computing the contribution of each metric is also proposed in order to adapt to the characteristics of software data. The vital characteristic of this approach is the automatic determination of ANN architecture during metrics selection.

Bernardi et al.(2011) applied UML, model driven transformation and formal techniques for assessing the risk of timing failure by evaluating the software design.

Shin et al. (2011) showed that the metrics available from development history are stronger indicators of vulnerabilities than code complexity metrics. In their study they investigated three categories of metrics such as complexity, code churn, and developer activity metrics to identify vulnerable files.

According to Zimmerman et al. (2011), factors such as complexity measures, computed metrics, churns are related to field failures and dependencies. They observed that metrics such as size, churn metrics and code metrics like fan in, fan out, and cyclomatic complexity are related to field failures. Their study showed that depending on a component that has failures does not have an effect on failures of the dependent component.

Mizuno and Hata (2010), have applied two apporaches namely a text feature metrics based approach using naive Bayes classifier and a complexity metrics based approach using logistic regression to identify the fault-prone components from the given software. Their approach is an integrated one in which the text feature based approach's high recall, and complexity metrics based approach's high precision were applied.

Surl and Kumar (2010) designed a simulator to identify critical components. In their study they have used Component Execution Graph (CEG) for developing simulator. They have used only the reusability metric of the component to find the criticality index of each component.

Gegick et.al. (2009) have created a predictive model that identifies the software components that pose the highest security risk in order to prioritize security fortification efforts. The input variables to their model are available early in the software life cycle and included security-related static analysis tool warnings, code churn and size, and faults identified by manual inspections. These metrics have been validated against vulnerabilities reported by testing and those found in the field. But all these are dynamic metrics and can be obtained only after field test.

Andrea et al (2008) focused on the difficulties involved in testing the container classes with nature inspired algorithms. They applied input space reductions and a novel testability transformation to aid the search algorithms.

Xiao (2007) analyzed the evaluation of optimization algorithms when used in goal oriented automated test data generation. They identified that goal-oriented approach is a promising approach to devise automated test data generators to cover the software.

Li and Sun (2007) proposed an algorithm based on reversed binary tree for automated test data generation. They claimed that their algorithm can automatically find out all of the feasible paths in the program from the source node to the base node and can automatically generate test data for each founded feasible path thus covers all the components in the software.

Ramon and Jose (2006) described two approaches which employ Estimation of Distribution Algorithm as a Meta-Heuristic technique for test data generation. The approach first extracts a region in the initial search space that incorporates static information from the SUT. If this method is not enough, then a grid search method is applied.

Wes (2007) empirically evaluated several test case filtering techniques based on coverage and profile distribution based techniques. Also, they compared the performance of random sampling against filtering techniques in exercising various program elements or components. The observations indicated that, distribution based filtering techniques did not perform well when compared to coverage based filtering techniques.

Dennis et al (2006) described an approach for checking the methods of a class against a full specification. It shares with traditional model checking - the idea of exhausting the entire space of executions within some finite bounds.

Zhou and Leung (2006) in their study showed that design metrics are statistically related to fault-proneness of classes across fault severity and the prediction capabilities of the investigated metrics greatly depend on the severity of faults.

Nagappan et al. (2006) built a regression model to predict the likelihood of post-release defects on code metrics. According to their study metrics such as number of classes in a module, inheritance depth, class coupling measures have significant influence on post-release defects.

David et al (2005) have provided a bounded exhaustive testing technique for software quality assurance.

The research works (Basili,1996; Subramanyam & Krishnan, 2003; Zhou & Leung, 2006) have applied CK suite of OO design metrics for the analysis of design complexity. They have applied metrics such as Depth of Inheritance Tree, Response for a class, and Number of children, lines of code etc., to find out design complexity. The design metrics used in all their research works addressed only the static aspect of OO design complexity.

Martina and Antonia (2003) exploited entity subsumption and spanning trees to determine the reduced set of coverage entities or components such that coverage of reduced set implies the coverage of unreduced set.

Beydeda and Gruhn (2003) proposed a binary search based test case generation approach. The primary use of the BINTEST algorithm is in testing the methods of a class or component. In their paper, the proposed binary search algorithm which is used to determine the test cases, requires certain assumptions but allows efficient test case generation.

Katerina Goseva et al (2003) and McGregor (1997) have proposed risk analysis technique for critical component identification. Kemmerer (1985) demonstrated the need for testing formal specifications during design phase. The formal specification language that is used in this paper is a variant of Ina Jo®, which is a nonprocedural assertion language that is an extension of first-order predicate calculus. The language assumes that the system is modeled as a state machine. The key elements of the language are types, constants, variables, definitions, initial conditions, criteria, and transforms.

Kim and Boldyreff (2006) suggested four categories of metrics such as class, message, use case and methods applicable to UML model. They proposed a tool called UML metrics producer which is used to calculate complexity metrics value for each class in a program.

Ebert (2006) evaluated classification techniques such as pareto classification, classification trees, factor-based discriminate analysis, fuzzy classification, neural networks for identifying critical components. Their study showed that the fuzzy classification is the best technique among five techniques for complexity-based criticality prediction.

Genero et al. (2000) classified metrics based on the elements of UML class diagram such as packages, classes, attributes and operations of each class and their relationships such as association, aggregation, generalization and dependencies.

Agrawal (1994) used the notion of dominators and super blocks to derive coverage implications among the basic blocks with the goal of reducing coverage requirements for testing a program.

Schneidewind (1979) represented a matrix form of directed graph to identify structural complexity and error properties. According to them the matrix forms of the directed graph identify program properties which are useful for software program development and testing.

PROPOSED MULTI-AGENTS BASED FRAME WORK

Intelligent Agents: An Introduction

Intelligent Agents are pieces of software that are designed to make computing and other tasks easier by assisting and acting on behalf of the user. The user can interact with the agent through the user interface while the agent can sense and act according to the condition of the external environment. The agent performs its tasks by taking information from the environment in which it is working (Russel & Norvig, 1995).

Agents can be constructed with a wide range of capabilities. In agent-based approach a complex processing function can be broken into several smaller and simpler ones. An autonomous agent is a system situated within and a part of an environment that senses that environment and acts on it, over time, in pursuit of its own agenda and so as to effect, what it senses in the future (Russel & Norvig, 1995).

As in the earlier work on the application of Intelligent Agents in Software Testing (Dhavachelvan & Uma, 2003), agents can be implemented with sophisticated intellectual capabilities such as the ability to reason, learn, or plan. In addition, intelligent software agents can utilize extensive amounts of knowledge about their problem domain.

Among many type of Intelligent Agents, this approach has focused on the design of Task-Specific Software Agents to perform software testing task. Since testing activity consists of a number of uncertainties, the test agent which is used in this research work checks up all the conditions in the SUT in generating optimal test sequences.

In the proposed framework, the functionalities are extended to three agents namely Selector Agent, Exploration Agent and Random Selection Agent to generate the critical paths with priority being given for the coverage of critical components from the number of components in the SUT.

The three agents work independently as per their assigned task and communicate with other agents whenever they have to exchange information. Because of the parallel behavior of these agents, the solution generation becomes faster and makes the approach an efficient one.

Since the basic test adequacy criteria used is coverage based; the quality of the software is improved by not missing any of the critical components as the fitness function to cover the components is based on their critical value.

SOLUTION METHODOLOGY IN THE PROPOSED APPROACH

The proposed frame work shown in Figure 1 has the following modules:

- The SUT is given as input to the critical path optimizer module which consist of two sub modules namely
  - Critical component identification module and,
  - Execution of test cases along Critical test paths module.
- In critical component identification module, the critical components are identified by the modules such as component extraction, criticality calculation and criticality analysis. As a result, the critical component list will be prepared and will be given as input to the critical path identification module.
- In critical test path execution module, the critical test paths are identified and tested using this module.

Figure 1. Critical Path Optimizer Framework

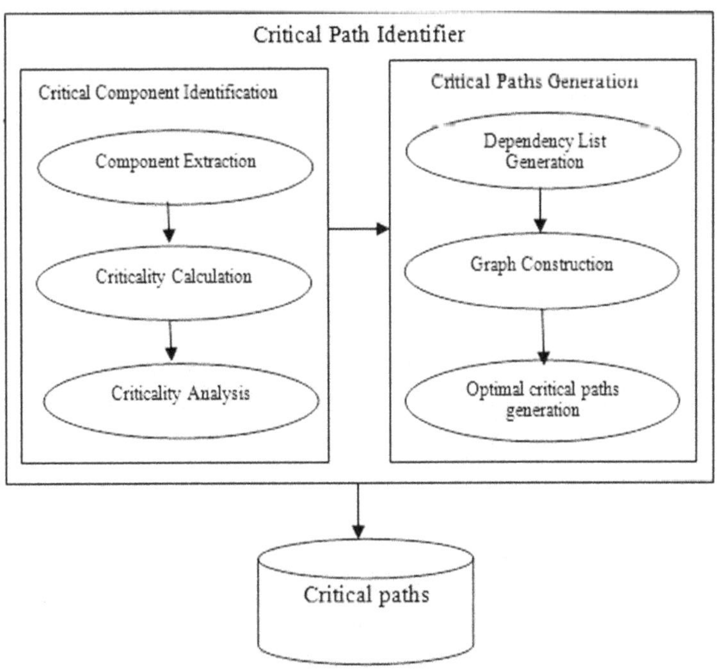

MATHEMATICAL MODEL FOR THE PROPOSED APPROACH

The proposed solution methodology is given by means of a mathematical model with multi-objective functions as given below:

Min.

$$Count\ (Cpath) \quad (1)$$

$$TTime\ (SUT) \quad (2)$$

Where, CPath – Critical paths;
Ttime(SUT) – Testing time required for testing SUT

Max.

$$CCover\ (CComp_i) \quad (3)$$

Where, $CComp_i$ – Critical Component I;CCover ($CComp_i$) - Component Coverage of Critical Component i.

Sub. to

$$Cover\ (CComp_i) = \begin{cases} false & \text{if } CComp_i \text{ is uncovered} \\ true & \text{otherwise} \end{cases} \quad (4)$$

$$Status\ (CComp_i) = \begin{cases} 1 & \text{if } Critical\_Value = MAX \text{ and } Cover\ (CComp_i) = false \\ 0 & \text{if } Critical\_Value = 0 \\ -1 & \text{if } Dep\_Metric\_Value = 0 (\text{Individual Component}) \end{cases} \quad (5)$$

$$Critical\ Value\ (CComp_i) = \sum_k (C_k(CComp_i)) \quad (6)$$

Where, Ck = complexity and dependability metric values associated with node $CComp_i$

$$CPath_j = \begin{cases} CPath_j + CComp_i & \text{if } Status(CComp_i) = 1 \\ CPath_j & \text{otherwise} \end{cases} \quad (7)$$

$$Continue\text{-}execute\text{-}flag\ (SUT) = \begin{cases} 1 & \text{if } Cover(x) = false \\ 0 & \text{otherwise} \end{cases} \quad (8)$$

The objective function in Equation 1 is to minimize the number of critical paths. The next objective function in Equation 2 is to minimize the total testing time required to complete the testing activity. The objective function in Equation 3 is to maximize the coverage of critical components. Constraint in Equation 4 indicates the continuation of generation of critical paths till no more frontier nodes to explore. Constraint in Equation 5 shows the value of status flag associated with each node that guides the proposed search algorithm about the presence / absence of critical components. Its value is set as 1 if the component is associated with MAX value; it is set as 0 if the component is not the node with highest critical value; it is set as -1 if the component is an independent component or an empty or a dummy component that does not have any associated metrics. Constraint in Equation 6 indicates the C_k metric value associated with $CComp_i$. Constraint in Equation 7 is used to construct the critical path based on the Status flag associated with each node. Constraint in Equation 8 indicates whether the exploration has to be continued or not based on the result of constraint in Equation 4.

PROPOSED ALGORITHM

Phase I: Critical Components Identification and Critical Test Paths Execution using Multi-Agents

- **Step 1:** Read the SUT as input.
- **Step 2:** Convert it into a Component Relationship Graph.
- **Step 3:** Identify the critical components by the following steps from 3.1 to 3.4 using Selector Agent.
  - **Step 3.1:** Extract the components from the SUT.
  - **Step 3.2:** Calculate the criticality value for all the components based on the following dependability and complexity metrics.

    Fan-In: Fan-In is the number of other functions calling a given function in a module.

    Fan-Out: Fan-Out is the number of other functions being called from a given function in a module.

    Number of Children: Number of classes inherited from a given class.

    Inheritance Depth: Inheritance depth is the maximum depth of inheritance for a given class.

    Coupling: This metric signifies coupling to other classes through:

    a. Class member variables;
    b. Function parameters;
    c. Classes defined locally in class member function bodies;
    d. mmediate base classes; and
    e. Return type.

    Number of Methods: Number of methods in a class including public, private and protected methods

    Number of Classes: This is the count of the number of classes in the component

    Number of Interfaces: This is the count of the number of interfaces in the component

    Number of Parameters: This is the count of the total number of parameters in the component

    Number of Fields: This is the count of the total number of fields in the component

    Number of Static Methods and Fields: This is the count of the total number of static methods and static fields in the component. The two metrics were counted separately

    Cyclomatic Complexity: This metric is used to show the complexity of a given component by counting the number of independent paths in the given component.
  - **Step 3.3:** Analyze the criticality of each component based on the critical value and generate a priority list.
- **Step 4:** Initialize the components list with extracted components with their critical values calculated in Phase I. Set j=0. Cycle = 1;
- **Step 5:** Apply **Selector Agent** to do the following:
  - **Step 5.1:** If this is the initial run/cycle, then search through the CRG using the components list to find the node with the highest critical value. Set ***CPath[j] = node*** $(max(\sum_k h(C_k(X_i))))$ ***for all i=1 to m and k=1 to p.***

- **Step 5.2:** If the node is selected by the Random Selection Agent, then set ***CPath[j] = X_i where X_i is the node selected by the Random Selection Agent.***
- **Step 5.3:** For both the cases, mark the node as covered after executing the test cases for it and place it in the components list and color the node in the graph for visualization.
- **Step 5.4:** Then, explore the frontier nodes from the current node to indicate a test sequence (A test sequence is composed of a sequence of executable components/nodes).
- **Step 5.5:** Communicate the selection information to the Exploration Agent.

- **Step 6:** Apply **Exploration Agent** to perform the following tasks:
  - **Step 6.1:** Evaluate the Critical value of each uncovered frontier node based on the information given by the Selector Agent.
  - **Step 6.2:** If a node with next higher critical value is found, then remember the node by appending this with the previously selected node to indicate a part of the critical path.
  - **Step 6.3:** Update ***CPath[j] = CPath[j] + node (max ($\sum_k h(C_k(X_r))$)) where r = 1 to no. of currently explored nodes.*** Mark the node as covered after executing the test cases for it.
  - **Step 6.4:** Abandon the remaining nodes in the current exploration and store them in Temporary Node list. The temporary node list indicates the uncovered node information associated with the previous explored node list and is used to ensure the coverage of all nodes and is refreshed during every cycle.
  - **Step 6.5:** Then, explore the frontier nodes from the currently selected node using Steps 5.4 and 5.5.
  - **Step 6.6:** If no node is found with expected level of critical value, store the current test sequence as a critical path and store it in the Critical Path Repository for further exploration in the subsequent stages.
  - **Step 6.7:** Send the nodes inefficiency information to Random Selection Agent.
- **Step 7:** Apply **Random Selection Agent** to perform the following tasks:
  - **Step 4.1:** The Random Selection Agent identifies a new node from the search space with higher critical value among all the uncovered nodes in the search space, on its own.
  - **Step 4.2:** Set j=j+1 to indicate a new exploration.
  - **Step 4.3:** Communicate the new selection to the Selection Agent for further exploration process.
- **Step 5:** Set Cycle=Cycle+1;
- **Step 6:** The steps from 2 to 5 are repeated until all the components are covered at least once or the number of cycles reached the Maximum Number of Cycles (MCN).

EXPERIMENTATION AND EVALUATION

Tested Programs

A range of case studies is taken to find the efficiency of the proposed approach against the approaches based on Random Testing and Sequential search. The tested programs are listed in Table 1 and the experimental setup is done as in the Table 2. The tested problems are classified as industrial and academic.

The industrial problems are taken from various industrial surveys and through the online project libraries, from some software industries and also some of them are given by the industries such as IBM

Table 1. Tested Software

Case Study #	Object Oriented Systems – in Java		
	Case Study	**Type**	**#Classes**
1.	Patient Monitoring System	Industrial	22
2.	Banking Management System	Industrial	19
3.	Library Management System	Industrial	20
4.	Shopping Cart	Industrial	12
5.	Hospital Management System	Industrial	28
6.	Network Monitor	Industrial	23
7.	Workflow Management system	Industrial	35
8.	Dealership Management System	Industrial	17
9.	Insurance Management System	Industrial	27
10.	Stock Management System	Industrial	14
11.	Credit Card Validation	Industrial	22
12.	Vendor Management System	Industrial	16
13.	Anti Money Laundering System	Industrial	56
14.	Microfinance Management System	Industrial	12
15.	Online Bus Reservation Management System	Industrial	15

Table 2. Experimental Setup – Sequential, Proposed Parallel, ACO and Random Search

Parameter Setup	Proposed Multi-Agent Based Parallel Search	Linear Search	Ant Colony Optimization (ACO)	Random Search
Type of Algorithm	Parallel Search Based	Sequential Search Based	Sequential Search Based	Random Search Based
Fitness Function	Critical value based on Component Coverage	Critical value based on Component Coverage	Pheromone Level/ Critical value of each Component	Component Coverage
No. of Cycles	MCN (Prescribed by the tester) - Maximum Cycle Number	MCN (Prescribed by the tester) - Maximum Cycle Number	No. of Cycles	No. of Iterations
Termination Criteria	Max. no. of cycles / Acceptable Component coverage measure	Max. no. of cycles / Acceptable Component coverage measure	Max. no. of cycles / Acceptable Component coverage measure	Max. no. of Iterations
Pheromone / Non – Pheromone Based	Non-Pheromone Based	Non-Pheromone Based	Pheromone Based	Non-Pheromone Based
Communication About Selection	Status flag	Status flag	Pheromone Trace/ Status Flag	Status flag
Working Nature	Parallel	Sequential	Sequential	Random
Hardware Used for Experimentation	CPU Speed: 3GHZ Processor: Core i3 RAM: 2 GB Temporary Storage: 320GB	CPU Speed: 3GHZ Processor: Core i3 RAM: 2 GB Temporary Storage: 320GB	CPU Speed: 3GHZ Processor: Core i3 RAM: 2 GB Temporary Storage: 320GB	CPU Speed: 3GHZ Processor: Core i3 RAM: 2 GB Temporary Storage:320GB

and Microsoft to compete in the National Project Competition such as 'The Great Mind Challenge' contest and 'Student Research Project' respectively. The specifications and technology are given by the industries and the code is sent to them for their evaluation. After the evaluations, the software is observed by the industries. Hence, these problems are also taken for testing the proposed approach.

The experimental setup for each of the approach is shown in the table 2.

EXPERIMENTATION: CASE STUDIES

Case Study: Analysis of Banking Management System

The specification document has been delivered by constructing a rigorous analysis of several real time applications related to Banking Management System (BMS). It consists of 19 components.

From the Table 3, it has been identified that 'Admin' component has the highest critical value. The calculation of the criticality values and the critical path generation are done using the tool developed in this research work.

Once, the components criticality value is identified, a parallel selection, exploration and random selection processes are employed using multi-agents. The selection of components based on their critical values and the construction of critical paths based on their order of exploration and selection is shown in Table 4.

Table 3. Complexity Metrics Calculation using the proposed tool for BMS

Class_ ID	Class_Name	fin	Fout	Child	Methods	Inhdep	Coupling	cc	Interfaces	par	Fields	Class	Staticmf	Total
C1	Account	4	0	4	1	0	0	11	0	2	4	0	0	26
C2	**Admin**	**0**	**0**	**0**	**4**	**1**	**0**	**35**	**0**	**16**	**14**	**0**	**0**	**70**
C3	CheckBalance	2	0	0	2	0	2	26	0	5	8	0	0	45
C4	ChequeBook	0	1	0	3	1	1	28	0	6	9	0	0	49
C5	CurrentAcc	1	1	0	1	1	0	5	0	2	5	0	0	16
C6	Customer	0	0	0	3	1	0	19	0	5	5	0	0	33
C7	Deposit	2	2	0	2	1	4	30	0	7	12	0	0	60
C8	FixedAcc	1	1	0	1	1	2	5	0	2	4	0	0	17
C9	FundTransfer	0	2	0	1	1	2	5	0	5	10	0	0	26
C10	Month	1	1	0	1	1	2	1	0	2	2	0	0	11
C11	OnlineServices	2	0	2	1	0	2	11	0	2	4	0	0	24
C12	ProfileUpdation	2	1	0	2	1	3	4	0	7	6	0	0	26
C13	RecurringAcc	1	1	0	1	1	2	5	0	2	4	0	0	17
C14	SavingAcc	1	1	0	1	1	2	5	0	2	4	0	0	17
C15	Statements	1	0	1	1	0	1	13	0	2	4	0	0	23
C16	Transaction	0	1	0	1	0	0	1	0	2	3	0	0	8
C17	User	0	0	0	2	0	0	7	0	4	4	0	0	17
C18	Withdraw	1	0	0	2	1	0	31	0	7	12	0	0	54
C19	Year	1	1	0	0	0	0	0	0	0	0	0	0	2

Table 4. Critical Paths Generation based on Critical Value of the Components

C.No.	Cpath1	Cpath2	Cpath3	Cpath4	Cpath5	Cpath6	Cpath7	Cpath8	Cpath9
C1					(*)	(*)	(*)	(*)	
C2	(*) (*)								
C3		(*)		(*) (*)					
C4			(*) (*)						
C5					(*)	(*)	(*)	(*)	
C6	(*)								
C7		(*) (*)							
C8								(*)	
C9		(*)		(*)					
C10									(*)
C11			(*)						
C12			(*)						
C13							(*)		
C14						(*)			
C15									(*) (*)
C16		(*)		(*)					
C17	(*)								
C18		(*)(*) (*)							
C19									(*)

In Table 4, each 'Cpath' represents a critical path that has higher priority on the execution of critical components. The components selection is done by analyzing the order of exploration from the currently selected component. If none of the explored components are having higher critical value than some other component which is not yet covered, then, that component will be selected by means of random selection by the Random Selection Agent. Hence, the component selection, exploration and random selection are done using the proposed approach within less time because of the parallel execution of the agents.

From the generated critical paths, the optimization is applied by identifying repeated components coverage. This is showed in Table 5 with information on the components that needs to be executed and the components that have been already covered.

From Table 5, the optimized list of critical paths are generated using the proposed framework and the final list of paths required to cover all the components at least once with priority being given for higher critical components is less. The final list of optimal number of critical paths is shown in Table 6.

From Table 7, it has also been inferred that, the critical components are executed with highest priority by creating execution sequences. Each execution sequence is a path that guides the testing process to do continuous execution using the test suit meant for that path. Some paths have single components, indicating that their connected components have been already tested and these components can be individually tested which thus reduce the total time spent if all the components in that path are executed again.

Table 5. Critical Paths with Coverage Information

S.No.	Critical Paths With Coverage Identification	Components Executed
1.	C02-C17-C06-C02(Covered)	C02,C17,C06
2.	C18-C07-C03-C09-C07(Covered) -C16-C18(Covered)	C18,C07,C03,C09,C16
3.	C04-C12-C11-C04 (Covered)	C04,C12,C11
4.	C03 (Covered) -C09(Covered)-C16(Covered)-C03 (Covered)	For this path all the components are already covered
5.	C01-C15	C01,C15
6.	C01 (Covered) -C14-C05	C14,C05
7.	C01 (Covered) -C13-C05(Covered)	C13
8.	C01(Covered) -C08-C05 (Covered)	C08
9.	C15-C19-C10-C15(Covered)	C15,C19,C10

Table 6. Optimized List of Critical Paths

S.No.	Optimal No.of Critical Paths
1	C2-C17-C06
2	C18-C07-C03-C09-C16
3	C04-C12-11
4	C01-C05
5	C15-C19-C10
6	C13-C08
7	C14 (individual component)

Table 7. Total No. of Test runs taken by different approaches for Sample Case Studies

S.No.	Search Technique	Components Coverage %		Critical Components Coverage %		Number of Test Runs Required		Time Taken for Paths Generation (Nano Sec.)	
		PMS	BMS	PMS	BMS	PMS	BMS	PMS	BMS
1.	Random Search	62%	65%	30%	45%	>350	>350	840528866182	820423835201
2.	Linear Search	92%	94%	85%	82%	300	286	810725386562	634267763328
3.	ACO	100%	100%	100%	100%	125	78	77689005172	36455823855
4.	Proposed Multi-Agents Based Parallel Search	100%	100%	100%	100%	31	19	23388843385	23202043568

Testing of the Critical Components

Unit Testing

In this work, unit testing is done by generating test cases using Genetic Algorithm based approach based on our previous work (Jeya Mala et.al. 2013). To ensure the test effectiveness, branch coverage based test adequacy is used. This has been done by means of the automated testing tool developed as part of this research work. The screenshot is shown in the figure 2.

Figure 2. Unit testing of each identified component

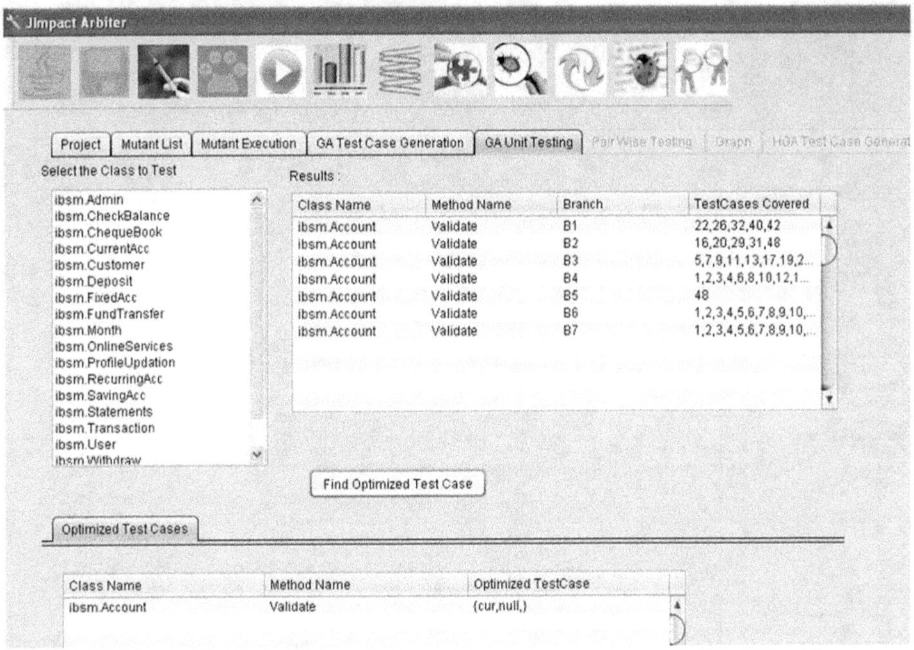

Integration Testing

The integration testing is done by means of pair-wise testing of components based on their connectivity. The identification of connected components is done by means of fan-in and fan-out measures calculated. The screenshot of the integration testing being done using the developed tool is shown in Figure 3.

METHOD OF ESTIMATION

To do the estimation of test runs taken by the proposed work and the other existing searching techniques, the algorithms have been coded in Java. The time taken is estimated in nano seconds and the number of execution cycles is estimated by counting the number of executions done by the three agents to achieve optimization. Based on that, the different approaches are evaluated.

RESULT ANALYSIS

To conduct the comparative analysis in covering all the critical components within less amount of time and increasing the coverage of all the components, the existing graph searching approaches ranging from traditional searching techniques such as Random Search and Linear Search to intelligence based searching technique like ACO and the proposed heuristic guided intelligent searching technique are applied.

Under the test bed having Windows 7 operating system, Core I3 processor with 2 GB RAM and NetBeans IDE application development environment all the techniques have been executed for the two

Figure 3. Integration Testing of Components

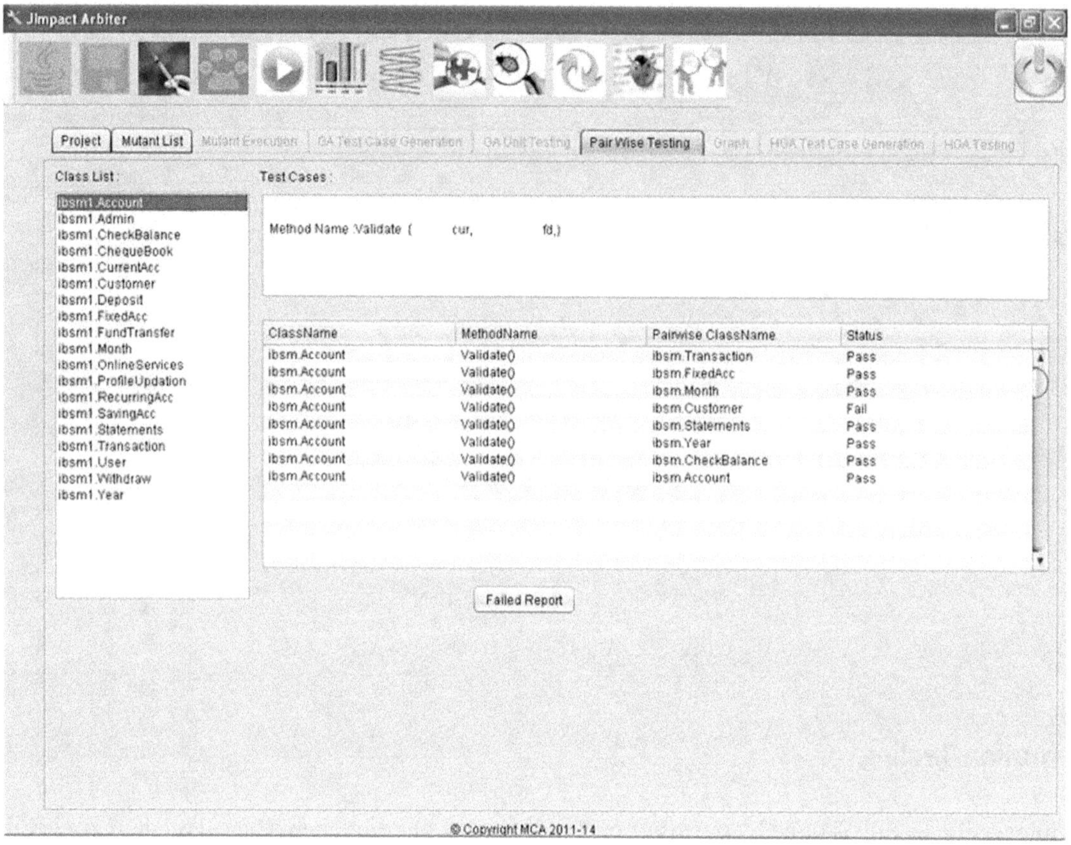

given case studies to identify the best technique that can generate optimal number of critical paths that can cover all the critical components within less time.

Table 7 provides the number of test runs required by existing search approaches and the proposed search approach to cover all the critical components. The maximum number of test runs is initialized as 350. For random and linear search techniques the number of test runs exceeds 300, but has not provided the optimal solution. It simply generated the paths without consideration on optimality.

When the results of ACO and multi-agents based approach are analyzed, it has been identified that both ACO and proposed critical path optimizer approach have provided the optimal number of critical paths with coverage of all the critical components with less number of test runs. But still when compared to the number of test runs and total time taken to generate the critical paths, the proposed 'critical path optimizer' performs better when compared to ACO. The Problems Identified in ACO are (i) as the tracking of nodes is based on the pheromone values at each node, the values of them must be updated frequently to keep its current level. This process presents substantial overhead in the optimization process; (ii) In ACO, the final optimal solution can be obtained only by examining all of the solution candidates created by ant exploration.

Since, the process is a sequential one in which the solution selection is done only at the end, it leads to computational overhead and memory limitation problems and (iii) Suppose a group of ten ants have been deployed for the optimal solution generation, and if this group of ants fails, then a new group of ten other ants have to be deployed. The time spent for the initial process will be a mere waste and leads to substantial time overhead.

The inferences from the table have been depicted by graph based visualizations as shown in Figures 4, 5 and 6. Among them, Figure 4 shows the component coverage% of each of the approaches. This shows that, ACO and the proposed Critical Path Optimizer have outperformed and they achieved 100% of coverage.

Based on the critical components coverage to ensure quality of the software product, a graph as shown in Figure 5 is constructed and it shows that, ACO and the proposed approach have achieved 100% coverage of critical components.

Figure 4. Components Coverage% Vs. Search Approaches

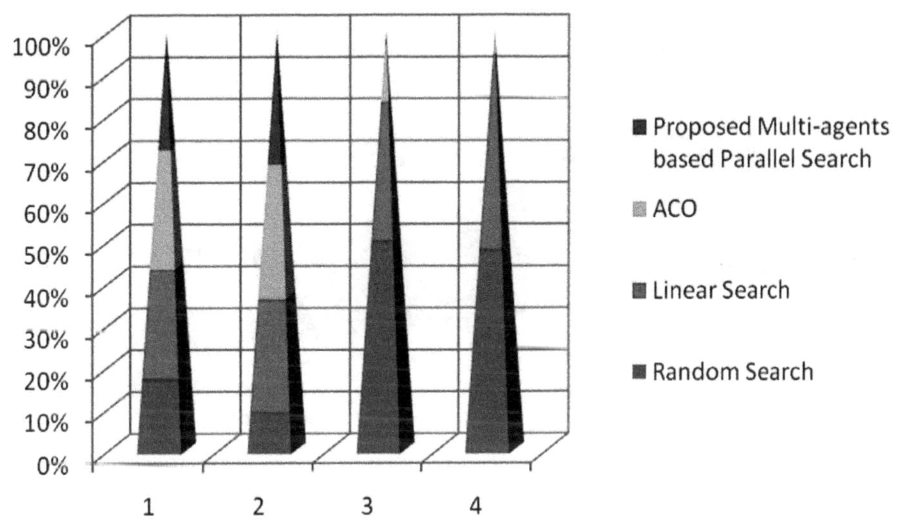

Figure 5. Critical Components Coverage% Vs. Search Approaches

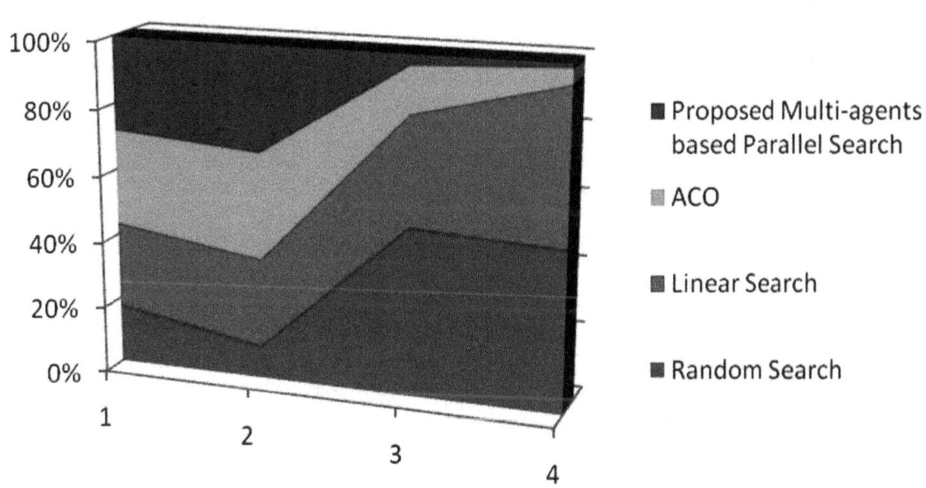

Eventhough, ACO and proposed approach provided 100% coverage in terms of components and critical components, the test runs required to identify the critical components and the critical paths which can ensure the coverage of critical components during testing shows that the proposed approach outperforms all the other approaches. This has been depicted in Figure 6.

Based on the sample case studies, the time required to generate the critical paths in order to cover all the critical components is analyzed and is depicted in Figure 7. It has been inferred that, the total time needed for critical paths generation and optimization is reduced using the proposed novel critical path optimization approach based on multi-agents based approach. Hence, it has been concluded that the proposed tool is a suitable one for the identification and verification of critical components within less amount of time without compromising quality.

Figure 6. Test runs required to cover all the critical components by different approaches

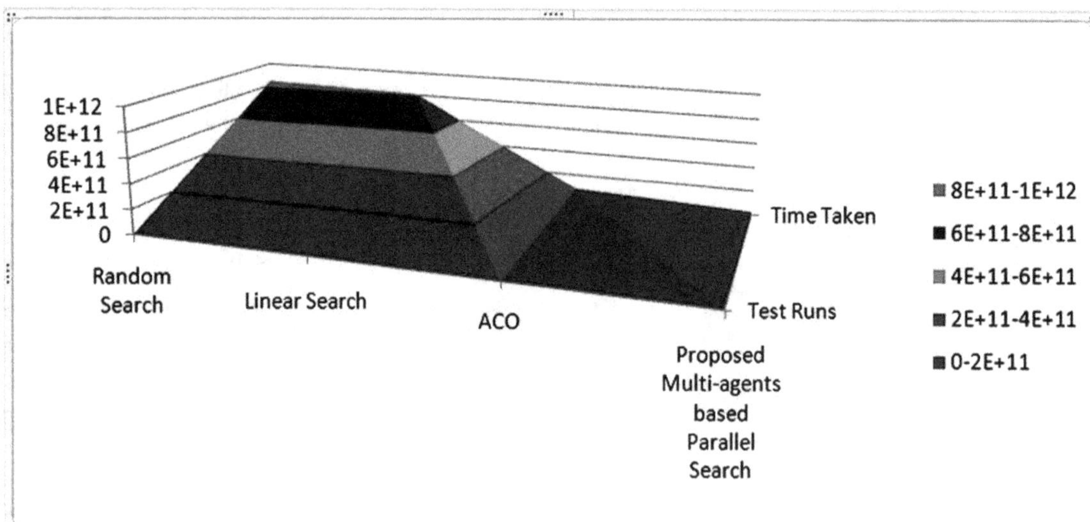

Figure 7. Time taken by various apporaches in (Nano Seconds)

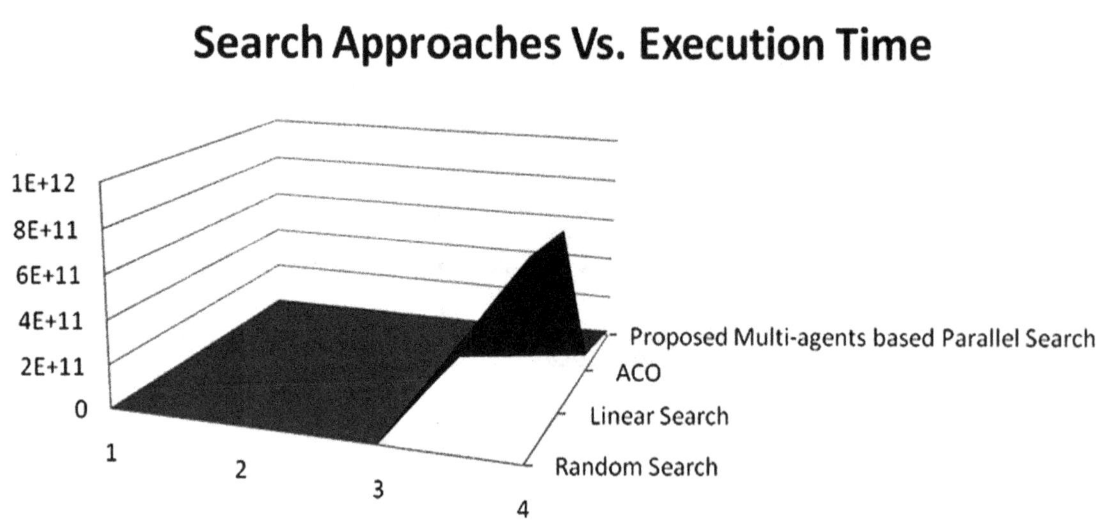

From Table 8, one can infer that, the number of false positives in the proposed approach is 1 among the total of 19 components. The false negative, true negative are identified accurately using the proposed approach.

Existing Work Based Qualitative Analysis

The comparison of the performance of the proposed approach with existing research works in critical components identification based on various parameters is given in Table 9.

Table 8. Predicted vs. Actual fault prone components

Actual	Predicted	
	No fault Components (True)	Fault Components (False)
No Fault Components (Positive)	15 (True Positive)	1 (False Positive)
Fault Components (Negative)	3 (True Negative)	0 (False Negative)

Table 9. Comparison of Existing and Proposed Research works based on their working

S.No.	Research Work	Inference Based on	Metrics Used for Critical Components Identification	Traditional / Intelligence Applied	Components Verification / Critical Test Paths Generation and Tool Development
1.	Ruchika et.al.	Source Code Analysis	WMC, NPM, LOC, DAM, MOA, AMC, RFC, CBO, EC, IC,CBM, LCOM,CAM,CC	Traditional	No
2.	Janes et.al.	Source Code Analysis	RFC	Traditional	No
3.	Shin et al. (2011)	Development history	Complexity, Code Churn, And Developer Activity Metrics	Traditional	No
4.	Zimmerman et al. (2011)	Source Code Analysis	Size, Churn Metrics And Code Metrics Like Fan In, Fan Out, And Cyclomatic Complexity	Traditional	No
5.	Suri and Kumar(2010)	Simulator	Reusability Metric	Traditional	No
6.	Zhou and Leung (2006)	Design Documents	Design Metrics	Traditional	No
7.	Nagappan et al. (2006)	Source Code Analysis	Number Of Classes In A Module, Inheritance Depth, Class Coupling Measures	Traditional	No
8.	TiborGyimo´ thy et.al (2005), Subramanyam and Krishnan (2003), and Basili et.al (1996)	Analysis of design documents	CK Suite Of OO Design Metrics Such As Depth Of Inheritance Tree, Response For A Class, And Number Of Children, Lines Of Code	Traditional	No
9	Proposed Multi Agents based Approach	Analysis of UML based design documents, OCL based formal specification documents, Source Code Analysis and Java Byte Code Analysis	Complexity and Dependability Metrics	Intelligence Based	Yes – Tool Name: Critical test path Optimizer

Other Observations

When compared to the other existing approaches on critical components identification and dynamic verification, the proposed approach provides an intelligent search process to do the testing of the components rigorously.

Also, the other observations related to the proposed approach:

- As per 'No Free Lunch' procedure, the proposed approach too has some tradeoffs. If the number of components in the given software is less, then time taken for completing the process is higher when compared to linear or random search processes because of the switching time between the agents.
- Also, prior to applying ABC, the algorithm itself takes some amount of time to complete its critical value calculation prior to testing. The time taken to complete the Phase – I of identification of critical components by calculating complexity, severity and Phase – II of risk exposure calculation, it requires some more time when compared to the design metrics alone based approach.

But, when considering the criticality and quality level of the problem concerned, the time it requires for the proposed approach is very less and is acceptable.

IMPACT ANALYSIS OF PROPOSED APPROACH USING STATISTICAL ANALYSIS

A correlation analysis is conducted to find the relationship between the number of classes in an application against the time taken and test runs required to complete the fault-prone components identification is conducted. For all the cases, the critical components coverage is 100%.

To identify the impact of the proposed multi agents approach in terms of total time taken to identify the fault-prone components and the number of test runs required to test them is statistically analyzed. Even though the proposed approach has taken only less time when compared to the existing approaches, it has been observed that, the total number of classes or class components present in an application will heavily impact on the time taken and test runs requirement. These analytical results are provided here for further research in this field.

```
Initial Set-up:
z for 95% CI= 1.96
Invalid: 0
Cases-N: 15
r (#classes.Timetaken)= 0.9691
p= 0
r (#classes.TestRuns)= 0.9813
p= 0
r (Timetaken.TestRuns)= 0.9242
p= 0
```

Table 10. Fault Prone Components Identification and Testing Results Analysis

Case Study #	Application	No.of Classes	Time Taken (NanoSec)	Time Taken (Sec)	Test Runs
1.	Patient Monitoring System	22	2.34E+10	23.3888434	31
2.	Banking Management System	19	2.32E+10	23.2020436	19
3.	Library Management System	20	2.32E+10	23.2150553	28
4.	Shopping Cart	12	2.17E+10	21.7154112	12
5.	Hospital Management System	28	3.2E+10	32.0147525	38
6.	Network Monitor	23	2.35E+10	23.5123874	31
7.	Workflow Management system	35	4.15E+10	41.520478	46
8.	Dealership Management System	17	2.22E+10	22.1542797	17
9.	Insurance Management System	27	3.15E+10	31.4715427	32
10.	Stock Management System	14	2.15E+10	21.4789875	11
11.	Credit Card Validation	22	2.32E+10	23.2478875	31
12.	Vendor Management System	16	2.19E+10	21.9471548	14
13.	Anti Money Laundering System	56	5.91E+10	59.1248643	72
14.	Microfinance Management System	12	2.17E+10	21.7154112	12
15.	Online Bus Reservation Management System	15	2.11E+10	21.1247885	16

The correlation coefficient (r) can range in value from −1 to +1. The larger the absolute value of the coefficient, the stronger the relationship between the variables. For the Pearson correlation, an absolute value of 1 indicates a perfect linear relationship. A correlation close to 0 indicates no linear relationship between the variables.

From our experimental results, the Pearson correlation values (r), r (#classes.Timetaken), r (#classes. TestRuns) and r (Timetaken.TestRuns) are highly correlated. They have exhibited a large positive relationship as the values are nearer to 1.

In the regression table shown in Table 11, corresponds to E(Time Taken) the s.e value which represents the standard error value shows the average distance that the observed values fall from the regression line. Conveniently, it tells how wrong the regression model is on average using the units of the response variable. Smaller values are better because it indicates that the observations are closer to the fitted line. From the regression table of E(Time Taken), the s.e value is 0.063 which is <=2.5 which produces a sufficiently narrow 95% prediction interval.

Table 11. Regression Table for E(Timetaken) against #classes

Regression Table for E(Timetaken)				
	B	s.e.	t	p
#classes	0.899	**0.063**	14.168	0
Intercept	7.127			

Now, the scatter plot is drawn between #classes and Time Taken and the relationship curve indicates that it is linear which means that, they are highly correlated. The points fall close to the line, which indicates that there is a strong linear relationship between the variables. The relationship is positive because as one variable increases, the other variable also increases.

By using the box plot as shown in Figure 9, one can infer that, the relationship between #classes and Time taken are highly correlated and shows positive correlation.

From the regression Table 12, the s.e value is 0.078 which is <=2.5 which produces a sufficiently narrow 95% prediction interval.

Also, a scatter plot is drawn between #classes and Tesr Runs as shown in Figure 10. It has been observed that, the plot is a linear regression which means that, they are highly correlated and are providing a positive correlation.

From the regression Table 13, the s.e value is 0.166 which is <=2.5 which produces a sufficiently narrow 95% prediction interval. A scatter plot between Time Taken and Test Runs as shown in Figure 11, indicates that the two series of values show a positive correlation and shows a linear regression line.

Figure 8. Scatter plot for correlation analysis between mean differences of classes and time taken

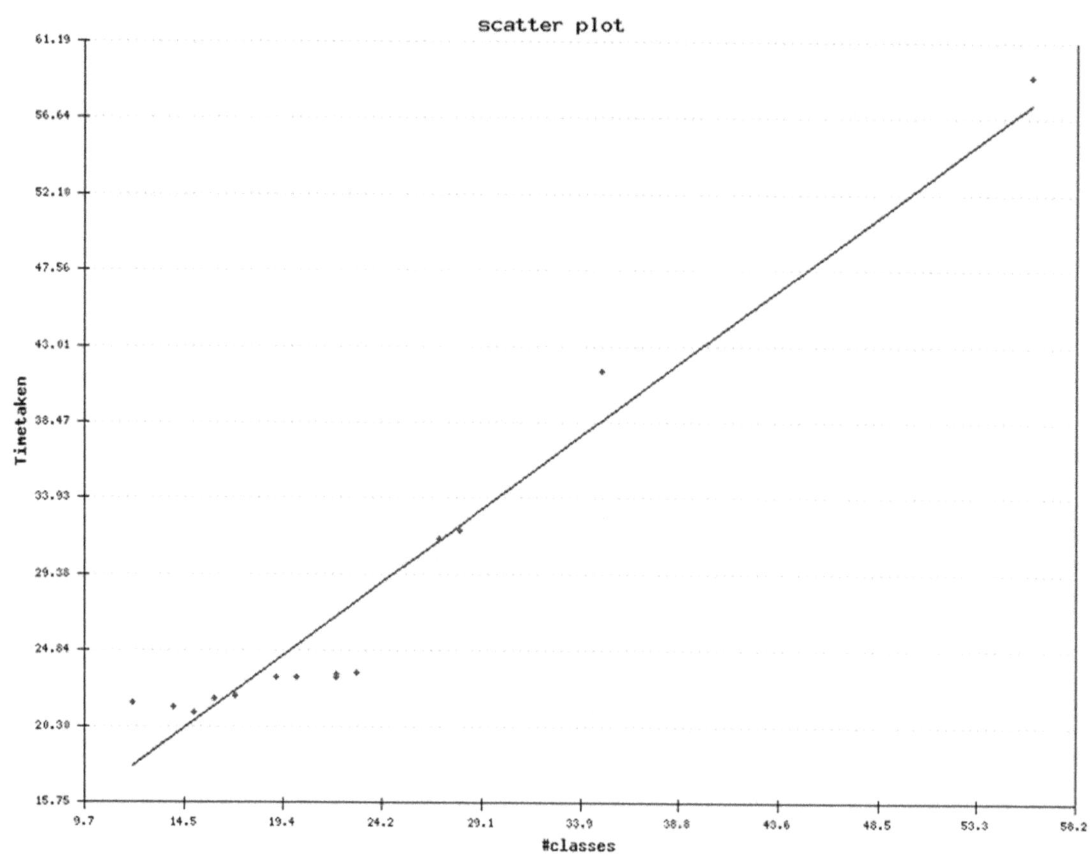

Figure 9. Box Plot for correlation analysis between #classes and Time Taken

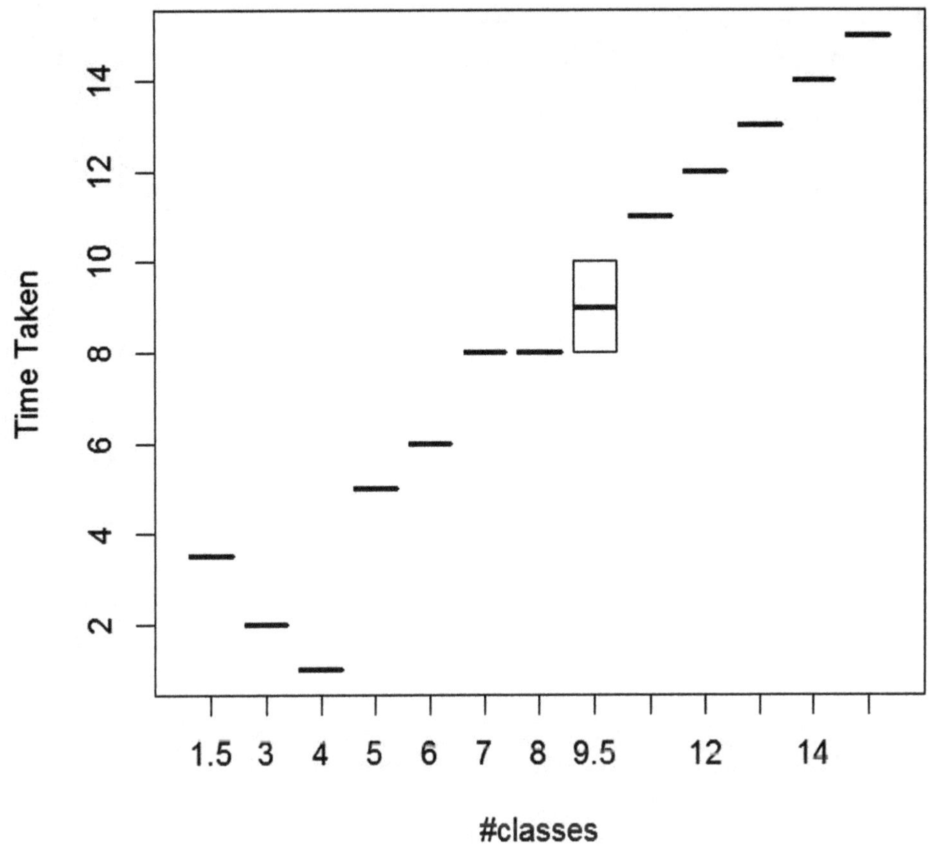

Table 12. Regression Table for E(TestRuns) against #classes

Regression Table for E(TestRuns)				
	B	s.e.	t	p
#classes	1.426	**0.078**	18.374	0
Intercept	-4.808			

Variance Analysis Using ANOVA and the F-Test

ANOVA is used to compare differences of means among more than 2 groups. It does this by looking at variation in the data and where that variation is found. Specifically, ANOVA compares the amount of variation between groups with the amount of variation within groups. It can be used for both observational and experimental studies. Mathematically, ANOVA can be written as:

$$x_{ij} = \mu_I + \varepsilon_{ij} \quad (1)$$

Figure 10. Scatter plot for correlation analysis between mean differences of classes and time taken

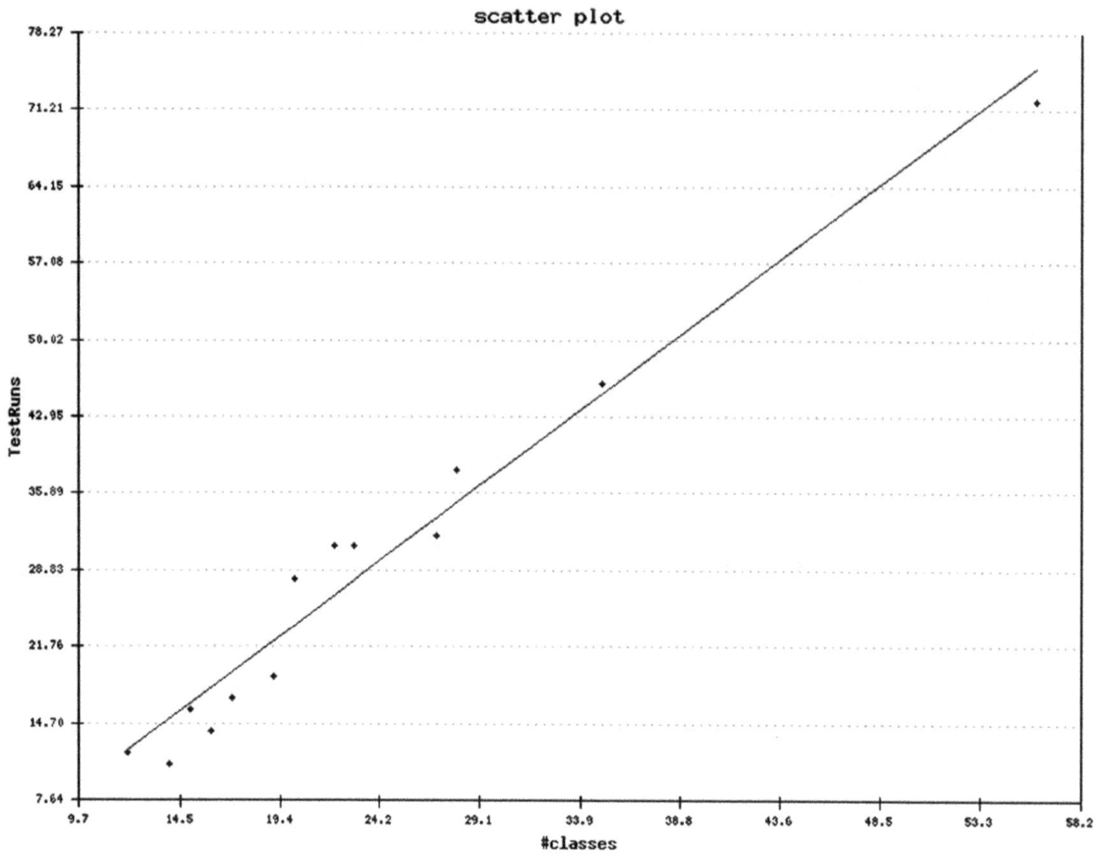

where x are the individual data points (I and j denote the group and the individual observation), ε is the unexplained variation and the parameters of the model (μ) are the population means of each group. Thus, each data point (x_{ij}) is its group mean plus error. To use the F-test to determine whether group means are equal, it's just a matter of including the correct variances in the ratio. In one-way ANOVA, the F-statistic is this ratio:

F = variation between sample means / variation within the samples (2)

In general, an F-statistic is a ratio of two quantities that are expected to be roughly equal under the null hypothesis, which produces an F-statistic of approximately 1. The low F-value shows a case where the group means are close together (low variability) relative to the variability within each group. The high F-value shows a case where the variability of group means is large relative to the within group variability. In order to reject the null hypothesis that the group means are equal, we need a high F-value.

In the ANOVA table shown in Figure 12, one can conclude that, the F-Value is 0.7 which indicates that the group has low variability. That is the group means are close together relative to the variability within each group.

Figure 11. Scatter plot for correlation analysis between time taken and Test Runs

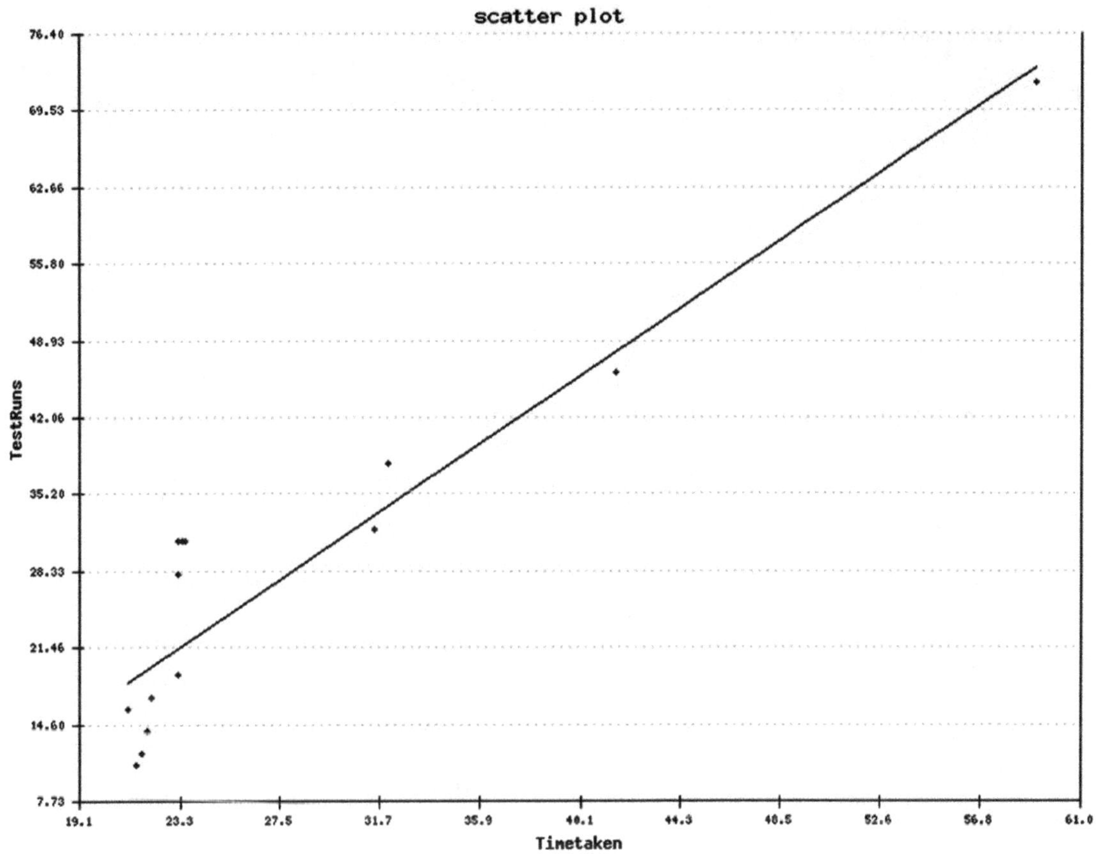

Table 13. Regression Table for E(TestRuns) against Time Taken

Regression Table for E(TestRuns)				
	B	s.e.	t	p
Timetaken	1.448	**0.166**	8.7237	0
Intercept	-12.321			

CONCLUSION

The intelligent agents based fault-prone components identification and testing framework was successfully implemented as a tool using Java. Also, the proposed framework has been compared with the existing approaches and found to be the better candidate to identify the fault-prone components within less time and cost. The merits of this research work include: (1) Critical Components Identification during early development process; (2) a multi-agent based Critical Components Coverage; (3) Critical Paths execution without redundancy of covering the same components to reduce total time and cost involved in the testing process without compromising quality.

Figure 12. Anova Table based on regression analysis

Data Summary						
	Samples					
	1	2	3	4	5	Total
N	15	15	15			45
ΣX	338	410.8339	410			1158.8339
Mean	22.5333	27.3889	27.3333			25.7519
ΣX^2	9386	12775.921	14946			37107.921
Variance	126.4095	108.8302	267.0952			165.1316
Std.Dev.	11.2432	10.4322	16.343			12.8504
Std.Err.	2.903	2.6936	4.2198			1.9156

standard weighted-means analysis					
ANOVA Summary	1				
Source	SS	df	MS	F	P
Treatment [between groups]	233.0993	2	116.5497	0.7	0.502286
Error	7032.6896	42	167.445		
Ss/Bl					
Total	7265.7889	44			

Ss/Bl = Subjects or Blocks depending on the design.
Applicable only to correlated-samples ANOVA.

Moreover, the proposed work has also statistically evaluated the significance of the proposed approach in identifying the fault-prone components from a given software. The correlation analysis results indicate that, the total number of classes/class components in the software has positive correlation with the total time taken and the number of test runs needed by the proposed approach. This means that, the traditional approaches which are not having the parallel behavior and the automation in terms of intelligent decision making will be heavily affected if the complexity of the software increases.

Hence, the proposed work has higher level of impact on the identification of fault-prone components in spite of the complexity of the software. As a future work, the components which are identified as fault-prone are going to be prioritized based on their test execution sequence so that, the total test runs needed to complete the testing task will be optimized.

REFERENCES

Agrawal, H. (1994). Dominators, Super Blocks, and Program Coverage. *Proceedings of the 21st ACM SIGPLAN-SIGACT Symposium on Principles of Programming Languages*, 25-34. 10.1145/174675.175935

Andrea, A., & Xin, Y. (2008). A Novel Co-Evolutionary Approach to Automatic Software Bug Fixing. In *IEEE World Congress on Computational Intelligence*. IEEE Press.

Basili V. R., Briand L. C., & Melo W. L. (1996). A validation of object-oriented design metrics as quality indicators. *IEEE Transaction on Software Engineering, 22*(10), 751–761.

Basturk & Karaboga. (2006). An Artificial Bee Colony (ABC) Algorithm for Numeric function Optimization. *IEEE Swarm Intelligence Symposium.*

Baykasolu, Özbakır, & Tapkan. (2009). *Artificial Bee Colony Algorithm and Its Application to Generalized Assignment Problem.* Academic Press.

Bernardi, S., Campos, J., & Merseguer, J. (2011, February). Timing-Failure Risk Assessment of UML Design Using Time Petri Net Bound Techniques. *IEEE Transactions on Industrial Informatics*, *7*(1), 90–104. doi:10.1109/TII.2010.2098415

Beydeda, S., & Gruhn, V. (2003). Test Case Generation According To The Binary Search Strategy. *Lecture Notes in Computer Science*, *2869*, 1000–1007. doi:10.1007/978-3-540-39737-3_124

David, C., & Yang, J. (2005). Software Assurance by Bounded Exhaustive Testing. *IEEE Transactions on Software Engineering*, *31*(4), 328–339. doi:10.1109/TSE.2005.52

Dennis, F. S., & Chang, D. J. (2006). Modular Verification of Code with Sat. *ISSTA-2006*, 1-11.

Dhavachelvan, P., & Uma, G. V. (2003). Multi-agent based integrated framework for intra-class testing of object-oriented software. *LNCS, 2869.*

Dorigo, M., Maniezzo, V., & Colorni, A. (1996). The Ant System: Optimization by a Colony of Cooperating Agents. *IEEE Transactions on Systems, Man, and Cybernetics. Part B, Cybernetics*, *26*(1), 29–41. doi:10.1109/3477.484436 PMID:18263004

Ebert, C. (2001). Metrics for identifying critical components in software projects. In *Handbook of Software Engineering and Knowledge Engineering* (Vol. 1). World Scientific Publishers. doi:10.1142/9789812389718_0017

Fathian, M., Amiri, B., & Maroosi, A. (2007). Babak Amiri and Ali Maroosi, "Application of honey-bee mating optimization algorithm on clustering. *Applied Mathematics and Computation*, *190*(2), 1502–1513. doi:10.1016/j.amc.2007.02.029

Gegick, M., Rotella, P., & Williams, L. (2009). Predicting Attack-prone Components. *2009 International Conference on Software Testing Verification and Validation*, 181-190. 10.1109/ICST.2009.36

Genero, L. M., Piattini, M., & Calero, C. (2000). Early measures for UML class diagrams. *L'Objet*, *6*, 2000.

Gyimothy, T., Ferenc, R., & Siket, I. (2005, October). TiborGyimo´thy, Rudolf Ferenc, and Istva'nSiket (2005), "Empirical Validation of Object-Oriented Metrics on Open Source Software for Fault Prediction. *IEEE Transactions on Software Engineering*, *31*(10), 897–910. doi:10.1109/TSE.2005.112

Horgan, J., London, S., & Lyu, M. (1994). Achieving Software Quality with Testing Coverage Measures. *IEEE Computer*, *27*(9), 60–69. doi:10.1109/2.312032

Jeya Mala, D., & Iswarya, R. (2014). Multi Agents Based Approach for Critical Components Identification and Verification. *International Journal of Systems and Service-Oriented Engineering*, *5*(1), 21–39. doi:10.4018/ijssoe.2014010102

Jeya Mala, D., Kamalapriya, S., & Mohan, V. (2010). Automated Software Test Optimization Framework – an Artificial Bee Colony Optimization based Approach. *IET Software*, *4*(5), 334–348. doi:10.1049/iet-sen.2009.0079

Jeya Mala, D., & Mohan, V. (2009). On the Use of Intelligent Agents in Test Sequence Selection and Optimization. *International Journal of Computational Intelligence and Applications*, *8*(2), 155–179. doi:10.1142/S1469026809002515

Jin, C., Jin, S.-W., & Ye, J.-M. (2012). Artificial neural network-based metric selection for software fault-prone prediction model. *IET Software, Vol*, *6*(6), 479–487. doi:10.1049/iet-sen.2011.0138

Karaboga, D., & Basturk, B. (2008). On The Performance Of Artificial Bee Colony (ABC) Algorithm. Applied Soft Computing, 8(1), 687-697.

Karaboga, N. (2009). A new design method based on artificial bee colony algorithm for digital IIR filters. *Journal of the Franklin Institute*, *346*(4), 328–348. doi:10.1016/j.jfranklin.2008.11.003

KaterinaGoseva, P. (2003, October). Architectural-Level Risk Analysis Using UML. *IEEE Transactions on Software Engineering*, *29*(10), 946–959.

Kaur, I. (2015). A Compound Metric for Identification of Fault Prone Modules. *IOSR Journal of Computer Engineering, 17*(6), 31-35.

Kemmerer, R. A. (1985, January). Testing Formal Specifications to Detect Design Errors. *IEEE Transactions on Software Engineering*, *1*(1), 32–44.

Kim, H., & Boldyreff, C. (2002). Developing Software Metrics Applicable to UML Models. *ECOOP Workshop on Quantitative Approaches in Object-Oriented Software Engineering (QAOOSE).*

Li, H., & Peng Lam, C. (2004). Software Test Data Generation using Ant Colony Optimization. *Transactions on Engineering, Computing and Technology.*

Li, J., & Sun, J. (2007). Automated Test Data Generation Algorithm Based on Reversed Binary Tree. *SNPD*, *3*, 1124–1128.

Li-Pei, W., Puan, C. Y., Malcolm, Y. H. L., & Chong, C. S. (2008). Bee colony optimization algorithm with big valley landscape exploitation for job shop scheduling problems. *Applied Soft Computing*, *8*(1), 687–697.

Martina, M., & Antonia, B. (2003). Using Spanning Sets for Coverage Testing. *IEEE Transactions on Software Engineering*, *29*(11), 974–984. doi:10.1109/TSE.2003.1245299

McGregor, J. D. (1997). Component Testing. *Journal of Object-Oriented Programming*, *10*(1), 6–9.

Mizuno, O., & Hata, H. (2010). An integrated approach to detect fault-prone modules using complexity and text feature metrics. *Proceedings of the 2010 international conference on Advances in computer science and information technology*, 457-468. 10.1007/978-3-642-13577-4_41

Myers. (1979). The Art of Software Testing. John Wiley & Sons, Inc.

Nagappan, N., Ball, T., & Zeller, A. (2006). Mining Metrics to Predict Component Failures. *28th International Conference on Software Engineering (ICSE)*.

Pham, D. T., Ghanbarzadeh, A., Koc, E., Otri, S., Rahim, S., & Zaidi, M. (2006). The Bees Algorithm, (2006), "A Novel Tool for Complex Optimization Problems. In *Proc 2nd International Virtual Conf on Intelligent Production Machines and Systems (IPROMS 2006)*. Oxford, UK: Elsevier.

Pressman, R. S. (2007). *Software Engineering – A Practitioners Approach*. McGraw Hill Publishers.

Ramon, S., & Jose, A. L. (2006). Scatter Search in Software Testing, Comparison and Collaboration with Estimation of Distribution Algorithms. *European Journal of Operational Research*, *169*(2), 392–412. doi:10.1016/j.ejor.2004.08.006

Russel, S., & Norvig, P. (1995). *Artificial Intelligence: A modern approach*. Prentice-Hall Inc.

Schneidewind, N. F. (1979, August). Application of Program Graphs and Complexity Analysis to Software Development and Testing. *IEEE Transactions on Reliability*, *R-28*(3), 192–198. doi:10.1109/TR.1979.5220563

Shin, Y., Meneely, A., & Williams, L. (2011). Evaluating Complexity, Code Churn, and Developer Activity Metrics as Indicators of Software Vulnerabilities. *IEEE Transactions on Software Engineering*, *37*(99), 1–16.

Singh, A. (2009). An artificial bee colony algorithm for the leaf-constrained minimum spanning tree problem. *Applied Soft Computing*, *9*(2), 625–631. doi:10.1016/j.asoc.2008.09.001

Srinivasa Rao, R., Narasimham, S. V. L., & Ramalingaraju, M. (2008). Optimization of Distribution Network Configuration for Loss Reduction Using Artificial Bee Colony Algorithm. *Proceedings Of World Academy Of Science, Engineering And Technology*, *35*, 709–715.

Subramanyam, R., & Krishnan, M. S. (2003, April). Empirical Analysis of CK Metrics for Object-Oriented Design Complexity: Implications for Software Defects. *IEEE Transactions on Software Engineering*, *29*(4), 297–310. doi:10.1109/TSE.2003.1191795

Surl, P. K., & Kumar, S. (2010, June). Simulator for Identifying Critical Components for Testing in a Component Based Software System. *IJCSNS International Journal of Computer Science and Network Security, VOL.*, *10*(6), 250–257.

Tassey, G. (2002). *The Economic Impacts of Inadequate Infrastructure for Software Testing. Final Report*. National Institute of Standards And Technology.

Teodorović, D., & Dell, M. (2006). *Bee Colony Optimization – A Cooperative Learning Approach To Complex Transportation Problems*. Advanced OR and AI Methods in Transportation.

Wes, M., Andy, P., & David, L. (2007). An Empirical Study of Test Case Filtering Techniques Based on Exercising Information Flows. *IEEE Transactions on Software Engineering*, *33*(7), 454–477. doi:10.1109/TSE.2007.1020

Xiao, M., Mohamed, E. A., Marek, R., & James, M. (2007). Empirical Evaluation Of Optimization Algorithms When Used in Goal-Oriented Automated Test Data Generation Techniques. *Empirical Software Engineering*, *12*(2), 183–239. doi:10.100710664-006-9026-0

Zhou, Y., & Leung, H. (2006, October). Empirical Analysis of Object-Oriented Design Metrics for Predicting High and Low Severity Faults. *IEEE Transactions on Software Engineering*, *32*(10), 771–789. doi:10.1109/TSE.2006.102

Zimmerman, T., Nagappan, N., Herzig, K., Premraj, R., & Williams, L. R. (2011). An Empirical Study on the Relation between Dependency Neighborhoods and Failures. *IEEE fourth international conference on Software Testing (ICST)*, 347-356.

Chapter 15
Towards a New Extracting and Querying Approach of Fuzzy Summaries

Ines Benali-Sougui
University of Tunis El Manar, Tunisia

Minyar Sassi Hidri
Imam Abdulrahman Bin Faisal University, Saudi Arabia

Amel Grissa-Touzi
University of Tunis El Manar, Tunisia

ABSTRACT

Diversification of DB applications highlighted the limitations of relational database management system (RDBMS) particularly on the modeling plan. In fact, in the real world, we are increasingly faced with the situation where applications need to handle imprecise data and to offer a flexible querying to their users. Several theoretical solutions have been proposed. However, the impact of this work in practice remained negligible with the exception of a few research prototypes based on the formal model GEFRED. In this chapter, the authors propose a new approach for exploitation of fuzzy relational databases (FRDB) described by the model GEFRED. This approach consists of 1) a new technique for extracting summary fuzzy data, Fuzzy SAINTETIQ, based on the classification of fuzzy data and formal concepts analysis; 2) an approach of assessing flexible queries in the context of FDB based on the set of fuzzy summaries generated by our fuzzy SAINTETIQ system; 3) an approach of repairing and substituting unanswered query.

INTRODUCTION

Techniques based on data summaries are now considered as a good way to handle large amounts of data, especially when the precise values of these data are not needed. Thus, several databases (DB) summarization model have been proposed such as SaintEtiQ model (Raschia, 2002), which is close to our

DOI: 10.4018/978-1-5225-5951-1.ch015

research tasks. This system makes it possible to generate a hierarchy of summaries making it possible to cover parts of the data base.

Unfortunately, all of these techniques are limited to the exact data and cannot be applied to fuzzy data.

We are confronted more and more with the situation where applications need to manage fuzzy data and to make profit their users from flexible querying. We speak then about flexible querying and Fuzzy Databases (FDB) (Galindo, J., M. P., & Urrutia, A., 2006; Ben Hassine, M.A., Grissa-Touzi, A., Galindo, J. & Ounelli, H., 2008).

Among the results of the query, the user is more interested in the most important or the best, called the Top-K answers. The main reason for this interest is that they avoid overloading the user with a large number of uninteresting answers.

The querying mechanism also provides the means to detect the failed queries reasons. The purpose of the repairing approaches is to provide satisfactory answers even when the query does not accept a result in the strict sense. However, existing approaches only handle Boolean queries and cannot be applied to FRDB.

In this work, we present an extension of the SaintEtiQ summarization model for modeling fuzzy data and therefore to exploit to flexible FSQL query by turning the top *k* result. Our goal is also to found a result even in the case of the absence of summaries corresponding strictly to the query.

The rest of the paper is organized as follows: Section 2 presents the basic concepts of Fuzzy Database (FDB), fuzzy attributes in GEFRED model and the basic concept of fuzzy FCA. Section 3 exposes the fuzzy queries' modeling and an overview of FCA-based summary model. Section 4 gives the limits of the existing summarization approach and of the exploitation of fuzzy data. Then it presents the architecture of our proposed approach. Section 5 describes the first step data organization of our proposed approach. Section 6 describes the second step the querying phase of our proposed approach. Section 7 gives a new kind of repaired query. Section 8 concludes the paper and gives some future work.

BACKGROUND

In this section, we present the basic concepts of FDB, Fuzzy attributes in GEFRED model and the theoretical foundations of fuzzy FCA.

Fuzzy Databases (FDB)

In this section, we present the basic concepts of FDB. A FDB is an extension of the relational database. This extension introduces fuzzy predicates under shapes of linguistic expressions that, at the time of a flexible querying, permits to have a range of answers (each one with a membership degree) in order to offer to the user all intermediate variations between the completely satisfactory answers and those completely dissatisfactory (Bosc, P., O., & Litard. L., 1998). The FRDB models are considered in a very simple shape and consist in adding a degree, usually in the interval [0,1], to every tuple. It allows maintaining the homogeneity of the data in DB. The main models are those of PradeTestemale (Prade, H., & C. T., 1987), Umano-Fukami (Umano, M., M. M. K. T., & Fukami, S., 1980), Buckles-Petry (Buckles, B. P. & F. E. P., 1982), ZemankovaKaendel (Zemankova-Leech, 1985) and GEFRED of Medina et al. (1994). This last model constitutes an eclectic synthesis of the various models published so far with the aim of dealing with the problem of representation and treatment of fuzzy information by using relational DB.

Fuzzy Attributes in GEFRED Model

The GEFRED model (Generalised model Fuzzy heart Relational Database) has been proposed in 1994 by Medina et al. (1994). One of the major advantages of this model is that it consists of a general abstraction that allows for the use of various approaches, regardless of how different they might look. In fact, it is based on the generalized fuzzy domain and the generalized fuzzy relation, which include respectively classic domains and classic relations. In order to model fuzzy attributes we distinguish between two classes of fuzzy attributes: fuzzy attributes whose fuzzy values are fuzzy sets and fuzzy attributes whose values are fuzzy degrees (Galindo, 2005; Galindo, J., M. P., & Urrutia, A., 2006).

Fuzzy Sets as Fuzzy Values

These fuzzy attributes may be classified in four data types. This classification is performed taking into account the type of referential or underlying domain. In all of them the values Unknown, Undefined, and Null are included:

- **Fuzzy Attributes Type 1 (FTYPE1):** These are attributes with "precise data", classic or crisp (traditional, with no imprecision). However, they can have linguistic labels defined over them, which allow us to make the query conditions for these attributes more flexible.
- **Fuzzy Attributes Type 2 (FTYPE2):** These attributes admit both crisp and fuzzy data, in the form of possibility distributions over an underlying ordered domain (fuzzy sets). It is an extension of the FTYPE1 that does, now, allow the storage of imprecise information.
- **Fuzzy Attributes Type 3 (FTYPE3):** They are attributes over *data of discreet non-ordered dominion with analogy*.
  In these attributes some labels are defined ("blond", "red", "brown", etc.) that are scalars with a similarity (or proximity) relationship defined over them, so that this relationship indicates to what extent each pair of labels be similar to each other.
- **Fuzzy Attributes Type 4 (FTYPE4):** These attributes are defined in the same way as Type 3 attributes, without it being necessary for a similarity relationship to exist between the labels.

Fuzzy Degrees as Fuzzy Values

The domain of these degrees can be found in the interval [0,1], although other values are also permitted, such as a possibility distribution (usually over this unit interval) (Medina, J.M., M., & Pons, O., 1994; Galindo, 2005). The meaning of these degrees is varied and depends on their use. The most important possible meanings of the degrees used by some authors are: Fulfillment degree, Uncertainty degree, Possibility degree and Importance degree.

Fuzzy FCA

In this section, we discuss the Fuzzy FCA proposed by Quan et al., (2004), which incorporates fuzzy logic into Formal Concept Analysis, to represent vague information.

Definition 1: A fuzzy formal context is a triple $K_f = (G,M, I = \varphi(G \times M))$ where G is a set of objects, M is a set of attributes, and I is a fuzzy set on domain $G \times M$. Each relation $(g,m) \in I$ has a membership value $\mu(g,m)$ in [0, 1].

A fuzzy formal context can also be represented as a cross-table as shown in Table 1. The context has three objects representing three documents, namely *D1, D2* and *D3*. In addition, it also has three attributes, *Data Mining* (*D*), *Clustering* (*C*) and *Fuzzy Logic* (*F*) representing three research topics. The relationship between an object and an attribute is represented by a membership value between 0 and 1.

A confidence threshold T can be set to eliminate relations that have low membership values (Quan, T.T., Hui, S.C., & Cao, T.H., 2004). Table II shows the cross-table of the fuzzy formal context given in Table 1 with $T = 0.5$.

Each relationship between the object and as a membership value in fuzzy formal context, then the intersection of these membership values should be the minimum of these membership values, according to fuzzy theory (Zadeh, 1965).

Definition 2: Fuzzy formal concept: Given a fuzzy formal context $K_f = (G,M, I = \varphi(G \times M))$ and a confidence threshold T, we define $A^* = \{m \in M | \forall g \in A: \mu(g, m) \geq T\}$ for $A \subseteq G$ and $B^* = \{g \in G | \forall m \in B: \mu(g,m) \geq T\}$ for $B \subseteq M$.

A fuzzy formal concept (or fuzzy concept) of a fuzzy formal context $K_f = (G,M, I = \varphi(G \times M))$ with a confidence threshold T is a pair $(A_f = \varphi(A), B)$ where $A \subseteq G$, $B \subseteq M$, $A^* = B$ and $B^* = A$. Each object $g \in \varphi(A)$ has a membership μ_g defined as $\mu_g = min(\mu(g,m))$ and $m \in B$ where $\mu(g,m)$ is the membership value between object g and attribute m, which is defined in I. Note that if $B = \{\}$ then $\mu_g = 1$ for every g.

Definition 3: Let (A_1, B_1) and (A_2, B_2) be two fuzzy concepts of a fuzzy formal context (G, M, I). $(\varphi(A_1), B_1)$ is the sub-concept of $(\varphi(A_2), B_2)$, denoted as $(\varphi(A_1), B_1) \leq (\varphi(A_2), B_2)$, if and only if $\varphi(A_1) \subseteq \varphi(A_2) (\Leftrightarrow B_2 \subseteq B_1)$.

Equivalently, (A_2, B_2) is the super-concept of (A_1, B_1).

Definition 4: A fuzzy concept lattice of a fuzzy formal context K_f with a confidence threshold T is a set $F(K_f)$ of all fuzzy concepts of K_f with the partial order $\leq$ with the confidence threshold T.

Definition 5: The similarity of a fuzzy formal concept $K_1 = (\varphi(A_1), B_1)$ and its subconcept $K_2 = (\varphi(A_2), B_2)$ is defined as

Table 1. Fuzzy Formal Context with T = 0.5

	D	C	F
D1	0.8	0.12	0.61
D2	0.9	0.85	0.13
D3	0.1	0.14	0.87

$$E(K_1,K_2) = \frac{|\ \phi(A_1) \cap \phi(A_2)|}{|\ \phi(A_1) \cup \phi(A_2)|} \tag{1}$$

Figure 1 (a) gives the traditional concept lattice generated from table 1, in which crisp values *Yes* and *No* are used instead of membership values. Figure 1 (b) gives the fuzzy concept lattice generated from the fuzzy formal context given in table 1.

RELATED WORK

In this section, we present the fuzzy queries modeling and an overview of FCA-based summary model (Sassi, M., H. O. I. A., & Grissa-Touzi.A., 2010).

Fuzzy Queries' Modeling

To model flexible queries, Medina et al. (1998) introduces an extension of the SQL language called FSQL (Fuzzy SQL). So all valid queries in SQL are also valid in FSQL. This Flexibility comes from the fact that we can introduce fuzzy attributes, Fuzzy constants, fuzzy comparators, fuzzy qualifiers and quantifiers, etc. while returning a range of responses each with a satisfaction level. Each atomic condition combines a satisfaction level μϵ [0, 1] with an attribute value. For all the attributes of an n-uplet, the semantics of degrees are the same which involve that all criteria are commensurable. FSQL query has the following syntax:

Select *< attribute >*

From *< strict relation >*

Where *< fuzzy condition >* THOLD α

Where < fuzzy condition > can incorporate blocks of queries nested or partitioned. Parameters α from *Select* clause limit the number of answers by using a qualitative calibration (the data that satisfy the query with a threshold greater than α).

Figure 1. (a) A concept lattice generated from traditional FCA. (b) A fuzzy concept lattice generated from Fuzzy FCA

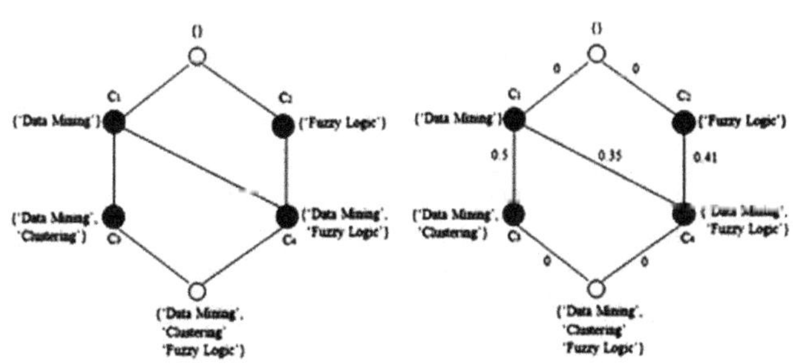

The FSQL language enriches the SQL language with:

- **Linguistic Labels:** If an attribute is accessible to fuzzy processing, then linguistic labels can be defined on it, preceded by the $ symbol to distinguish them easily. These labels are of two types, depending on the attribute domains, ordered (used for Ftype 1 and 2) or unordered (used for Ftype 1, 3, and 5).
- **Fuzzy Comparators:** In addition to the usual comparators (=,>, etc.), FSQL includes fuzzy comparators that can compare a column (or attribute) to a constant or two columns of the same type. These comparators are presented in Table 2.
- **The Fuzzy Qualifiers:** They are of two natures, absolute and relative:
  - Absolute qualifiers can answer such questions *the total number of tuples is large, small, much, a little, etc.* In this case, the qualifier depends on a single quantity.
  - Relative qualifiers can answer questions in which the truth of the qualifiers depends on two quantities.
- **Fuzzy Constants:** In addition to the usual constants, FSQL includes fuzzy constants (UNKNOWN, UNDEFINED, NULL, …).
- **Fuzzy Attributes:** The classification adopted in the GEFRED model.
- For attribute types is based on "imprecise" representation and data processing approaches.

With FSQL language, preferences are considered only as constraints and are taken into account through the expression of fuzzy predicates commensurable, modeled using fuzzy sets of values more or less satisfactory.

Besides, Top-k queries have attracted much interest in many different areas such as network and system monitoring (Babcock, B. & C. O., 2003), information retrieval (Kimelfeld,B. & Y. S., 2006), sensor networks (Silberstein, A., C. E. K. M. & Braynard, R., 2006; Wu, M., X. T. W.-C. L., & Xu, J., 2006), large databases (Grissa-Touzi, A. & H.O., 2012), multimedia databases (Chaudhuri, S., A. M., & Gravano, L., 2004), spatial data analysis (Ciaccia, P., & M. P., 2002; Hjaltason,G.R., & H. S., 2003), P2P systems (Akbarinia, R., P. V., & Pacitti, E., 2007), data stream management systems (Mouratidis, K., D. P., & Bakiras.S., 2006), etc.

The main reason for such interest is that they avoid overwhelming the user with large numbers of uninteresting answers which are resource-consuming.

Table 2. Fuzzy comparators

Id	Name	Age
FEQ	NFEQ	Fuzzy Equal
FGT	NFGT	Fuzzy Greater Than
FGEQ	NFGEQ	Fuzzy Greater or Equal
FLT	NFLT	Fuzzy Less Than
FLEQ	NFLEQ	Fuzzy Less or Equal
MGT	NMGT	Much Greater Than
MLT	NMLT	Much Less Than

The problem of answering top *k* queries can be modeled as follows (Fagin, R., M. N., & Lotem, J., 2003). Suppose we have m lists of n data items such that each data item has a local score in each list and the lists are sorted according to the local scores of their data items. And each data item has an overall score which is computed based on its local scores in all lists using a given scoring function. Then the problem is to find the *k* data items whose overall scores are the highest. This problem model is simple and general. Let us illustrate with the following examples. Suppose we want to find the top *k* tuples in a relational table according to some scoring function over its attributes. To answer this query, it is sufficient to have a sorted (indexed) list of the values of each attribute involved in the scoring function, and return the *k* tuples whose overall scores in the lists are the highest.

We plan to extend the FSQL language to support the top-k best answers. This new language is called FSQL-extended.

The FCA-Based Summary

In (Sassi, M., H. O. I. A., & Grissa-Touzi.A., 2010), they proposed to extend the SaintEtiQ summarization model (Raschia, G. & N. M., 2002) by introducing some optimization processes including minimization of the expert risks domain, building of the summary hierarchy from DB records, and cooperation with the user by giving him summaries in different hierarchy levels. In this case, the summarization process can be divided into two major steps. For the first step, the pre-processing, we have considered a fuzzy clustering that permits the generating of a membership matrix associating the DB records to generated clusters through the membership degrees. The second step, the post processing, generates the summary hierarchy. Thus, we have proposed to use fuzzy FCA in order to generate fuzzy hierarchy. In this step, we cooperate with the user by giving him summaries in different hierarchy levels.

SOLUTIONS AND RECOMMENDATIONS

In recent years, several methods of DB summarization have been proposed. Unfortunately, all these techniques cannot be applied to the large Fuzzy DB, they can only applied to simple data sets.

In this paper, we propose a new approach of linguistic summarization for fuzzy databases, called Fuzzy-SaintEtiq (BenAli-Sougui, I., Sassi-Hidri, M. & Grissa-Touzi, A., 2013; Benali-Sougui, 2016). This approach is an extension of the SaintEtiQ summarization model (Raschia et al., 2002) and the FCA-based Summary (Sassi, M., H. O. I. A., & Grissa-Touzi.A., 2010) to support the fuzzy data represented by GEFRED model.

Although our solution is based on the fuzzy model GEFRED and it can be applied to other fuzzy models.

In fact, we are confronted with the situation where applications need to manage fuzzy summaries and to profit their users from flexible querying. We have proposed a flexible FSQL query based on fuzzy linguistic summaries.

In the following we present the architecture of our proposed approach, which is divided in two major phases: fuzzy data organization phase and querying fuzzy summaries phase.

However, querying the summaries as explained above is interesting in order to rapidly get a rough idea of the properties of tuples in a relation but sometimes queries may have a null answer.

The user has to think about another query, which might fail. Thus, the idea is to repair and substitute query to provide an approximate answer.

Our goal in this paper is also to found a result even in the case of the absence of summaries corresponding strictly to the query. So, we speak about repaired query.

FUZZY DATA ORGANIZATION PHASE

Fuzzy-SaintEtiq takes the FDB records and provides knowledge. The summarization act considered like a process of knowledge discovery from database, in the sense that it is organized according to two following principal steps.

- **The Preprocessing Step:** This step organizes the database records in homogeneous clusters having common properties. This step gives a certain number of clusters for each attribute. Each tuple has values in the interval [0..1] representing these membership degrees according the formed clusters. Linguistic labels, which are fuzzy partitions, will be assigned on attributes domain.
- **The Post Treatment Step:** This step takes into account the result of the fuzzy clustering on each attribute, visualizes by using the fuzzy concepts lattices. Then, it imbricates them in a fuzzy nested lattice. Finally, it generalizes them in a fuzzy lattice associating all records in a simple and hierarchical structure. Each lattice node is a fuzzy concept which represents a concept summary.

This structure defines summaries at various hierarchical levels. This step consists in organizing the summaries within a hierarchy such that the most general concept summary is placed at the root of the fuzzy lattice, and the most specific concepts summaries are the leaves. This summary model corresponds to prototypical approaches since the intention of a concept summary present for each attribute the various possible values in the form of a fuzzy descriptors and the representativeness of these descriptors within the specified concept summary.

Fuzzy Clustering

For clustering fuzzy data, we have used an extension of Fuzzy C-Means clustering algorithm to deals with fuzzy data called Fuzzy-FCM (Grissa-Touzi, 2010; BenAli-Sougui, I., Sassi-Hidri, M. & Grissa-Touzi, A., 2013) in order to support different types of data represented by GEFRED model (Medina, J.M., M., & Pons, O., 1994).

The Fuzzy-FCM algorithm allows the user to select attributes according to which he wants to carry out classification. This treatment gives a refined intermediate matrix only formed of the codes of the selected attributes. Once the selection achieved, the FCM algorithm is applied on the refined table to get a matrix of adherence and a cut is exercised on this matrix of adherence to purify it by eliminating all values lower to the cut. The main idea of the algorithm is to define an intermediate matrix to model fuzzy data.

Let consider a relation R = (*Id, Age, Income, Experience*) from an FDB Employee table. Table 3 gives a sample of FDB Employee table. In the following we will use this example.

Each cluster of a partition is labeled by linguistic descriptor provided by a domain expert. For example, the fuzzy label Young belongs to a partition built on the domain of age attribute. Linguistic variables

Figure 2. The process of extracting and querying fuzzy summaries

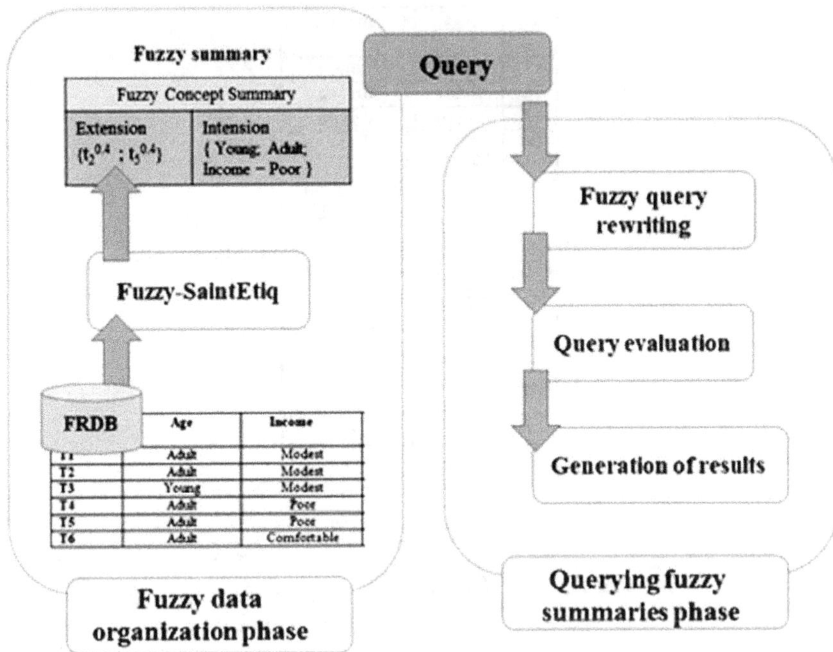

Table 3. Data sample

Id	Age	Income	ProfessionalBackground
T1	Adult	Modest	10
T2	Adult	Modest	5
T3	Young	Modest	3
T4	Adult	Poor	20
T5	Adult	Poor	7
T6	Adult	Comfortable	12

associated with the attributes of *R*. These linguistic variables constitute the new attribute domains used for the rewriting of tuples in the summarization process. For clusters generation, we carry out a fuzzy clustering (Vanisri, 2010) while benefiting from fuzzy logic.

This operation makes it possible to generate, for each attribute, a set of membership degrees.

Figure 3 presents the results of fuzzy clustering applied to *Age and Income.*

For Age, Income and Experience attributes, fuzzy clustering generates three clusters. The minimal value (resp. maximal) of each cluster corresponds to the lower (resp.higher) interval terminal of the values of this last. Each cluster of a partition is labeled with a linguistic descriptor provided by the user or a domain expert. For this, the following abbreviations are used:

- **For *Age* Attribute:** Young Age, Adult Age and Old Age.
- **For *Income* Attribute:** Low Income, Modest Income and Comfortable Income.

Figure 3. Fuzzy-FCM process applied on the Employee table

Id	Age	Income
T1	Adult	Modest
T2	Adult	Modest
T3	Young	Modest
T4	Adult	Poor
T5	Adult	Poor
T6	Adult	Comfortable

Select classification attributes
Generate intermediate matrix

Id	Age	Income
T1	1.2	2.8
T2	1.2	2.8
T3	1.1	2.8
T4	1.2	2.4
T5	1.2	2.4
T6	1.2	2.12

FCM Classification
Apply an α-cut

Id	Age		Income		
	Young	Adult	Poor	Modest	Comfortable
T1	0.5	0.5	-	0.6	0.4
T2	0.4	0.6	0.4	0.6	-
T3	0.7	0.3	-	0.8	-
T4	0.2	0.8	0.6	0.4	-
T5	0.6	0.4	0.7	-	-
T6	0.5	0.5	-	-	0.8

Fuzzy Summarization

After clustering, we generate the corresponding fuzzy concept lattice with similarity distance (Sassi, M., H. O. I. A., & Grissa-Touzi.A., 2010).

For a more formal expression of a query, let be consider:

- $A = \{A_1, .., A_k\}$ is a set of attributes.
- A_k is the kth attribute which appears is query Q,
- $R(A)$ is the relation whose tuples are summarized,
- t_i is the ith tuple, $i = 1..N$,
- $L_k = \{l_{1k}, .., l_{jk}\}$ is a set of linguistic terms of attribute A_k,
- μ_{ijk} is a membership degree of tuple t_i (record) to the linguistic term l_{kj} of attribute A_k.
- R_{zf} is the sub set of involved tuples in summary z_f,
- I_{zf} is the sub set of linguistic terms which appears in summary z_f,
- $z_f = (R_{zf}, I_{zf})$ is the concept summary, in which R_{zf} and I_{zf} are respectively the extension and the intension of the concept,
- *level* is the level of the concept summary in the concept lattice,

- $R_z = \{t_1, t_2, \ldots, t_N\}$. Each t_i is associated to one primitive DB records, i.e. an element of R.
- $|R_{zf}|$ is the number of candidates tuples in R_{zf}.

Table 4 shows the fuzzy formal concept lattice generated according the fuzzy formal context given in Figure 2.

Each concept summary z_f of the set of concept's summary Z_f is thus a description of a set of n-uplets of R which jointly form his extension and which is noted by R_{zf}. Table 4 shows concept summaries for each level extracted from fuzzy concept lattice. Thus, the root contains the summary of all the candidate records, whereas leaves represent only one combination of fuzzy linguistic labels over all the attributes. Levels 0 and 5 are both the root and leaves concept summary.

QUERYING FUZZY SUMMARIES PHASE

This phase takes place in 3 steps which will be developed later:

- Fuzzy query rewriting
- Query evaluation
- Generation of results

Fuzzy Query Rewriting

In the FSQL-extended query we have two parameter α and k. As previously said parameters α and k from Select clause limits the number of answers by using a quantitative calibration (k best responses) or qualitative calibration (the data that satisfy the query with a threshold greater than α).

Table 4. Fuzzy Concept summaries' generation

Level	Fuzzy Summaries
0	$z_0 = (\{t_1; t_2; t_3; t_4; t_5; t_6\}; \varnothing)$
1	$z_{11} = (\{t_1^{0.5}; t_2^{0.6}; t_4^{0.8}; t_5^{0.4}; t_6^{0.5}\}; \{\text{Adult}\})$ $z_{12} = (\{t_1^{0.5}; t_2^{0.4}; t_3^{0.7}; t_5^{0.6}; t_6^{0.5}\}; \{\text{Income –Modest}\})$ $z_{13} = (\{t_1^{0.5}; t_2^{0.4}; t_3^{0.7}; t_5^{0.6}; t_6^{0.5}\}; \{\text{Young}\})$
2	$z_{21} = (\{t_1^{0.5}; t_2^{0.6}; t_4^{0.4}\}; \{\text{Adult; Income –Modest}\})$ $z_{22} = (\{t1^{0.5}; t_2^{0.4}; t_3^{0.7}\}; \{\text{Young; Income –Modest}\})$ $z_{23} = (\{t_2^{0.4}; t_4^{0.6}; t_5^{0.7}\}; \{\text{Adult; Income – Poor}\})$ $z_{24} = (\{t_1^{0.5}; t^{2}_{0.6}; t_5^{0.4}; t_6^{0.5}\}; \{\text{Young; Adult}\})$
3	$z_{31} = (\{t_2^{0.4}; t_4^{0.6}\}; \{\text{Adult; Income – Poor; Income –Modest}\})$ $z_{32} = (\{t_1^{0.5}; t_2^{0.4}\}; \{\text{Young; Adult; Income –Modest}\})$ $z_{33} = (\{t_2^{0.4}; t_5^{0.4}\}; \{\text{Young; Adult; Income – Poor}\})$ $z_{34} = (\{t_1^{0.4}; t_6^{0.8}\}; \{\text{Young; Adult; Income – Comfortable}\})$
4	$z_{41} = (\{t_2^{0.4}\}; \{\text{Young; Adult; Income – Poor; Income –Modest}\})$ $z_{42} = (\{t_1^{0.4}\}; \{\text{Young; Adult; Income –Modest; Income – Comfortable}\})$
5	$z_5 = (\{\varnothing\}; \{\text{Young; Adult; Income – Poor; Income –Modest; Income – Comfortable}\})$

The parameter k is given by the user and the parameter is calculated depending on the number of clusters which α involved in condition of Select query.

$$\alpha = \frac{1}{\max\left(\mathrm{NClus}\right)}. \tag{2}$$

Where NClus is the number of clusters involved in the condition of select clause.

Let us consider the example in the Table 3. The queries are as follows:

Q_1: What is the experience and income of each young employee with a minimum degree of 0.5.
Select Income, ProfessionalBackground
From Employee
Where Age FEQ $Jeune THOLD 0.5;

And

Q_2: What is the experience of "young" employees with a "comfortable" income with a minimum degree of 0.3.
Select ProfessionalBackground
From Employee
Where Age FEQ $Young THOLD 0.3 AND Income FEQ $Comfortable THOLD 0.3;

In a query, descriptors like Young, Comfortable in Q_1 and Q_2 are called required characteristics and embody the properties that a record must consider them as an element of the answers. A query also defines the attributes for which required characteristics exist. The set of these input attributes is denoted by Inputs(A_Q). The expected answer is a description over a set of other attributes, denoted by Outputs(A_Q). It is the complement of Inputs(A_Q) relatively to A_Q (the set of attributes appears in the query Q):

$$\mathrm{Inputs}(A_Q) \cup .\mathrm{Outputs}(A_Q) = A \tag{3}$$

and

$$\mathrm{Inputs}(A_Q) \cap .\mathrm{Outputs}(A_Q) = \varnothing . \tag{4}$$

Hence a query Q defines not only a set $Inputs(A_Q)$ of input attributes but also for each attribute A_k, the set $L_{Ak}(Q)$ of its required characteristics which define the set of linguistic terms of attribute A_k appears query Q. The set of sets $L_{Ak}(Q)$ is denoted by $L(Q)$.

For example, for Q_2, this set is determined as follows:

- $Inputs(A_{Q2})$ = {Income, Age};
- $Outputs(A_{Q2})$ = { *ProfessionalBackground* };
- $L_{Age}(Q_2) = \{Young\}$, $L_{Income}(Q_2) = \{Comfortable\}$;

- $L(Q_2) = \{L_{Age}(Q_2), L_{Income}(Q_2)\}$.
- The degree of membership to this query is 0.3.

The query rewriting in a logical proposition $P_f\,(Z_f, Q)$ used to qualify the link between the fuzzy summary Z_f and the query Q. $P_f\,(Z_f, Q)$ is in a conjunctive form in which all descriptors are literals. Then, each set of descriptors yields one corresponding clause. Thereafter we will apply an α.cut on this new form with the parameters α.is calculated previously.

Consider the query Q3:

Q3: Which are the "young" and "adult" employees who have a "poor" or "modest" income with a minimum degree of 0.3.

Select ***
From *Employee*
Where *Age FEQ ($Young, $Adult) THOLD 0.3;*
And Income FEQ ($Poor, $Modest) THOLD 0.3;

We have then:

- $Inputs(A_{Q3}) = \{Age, Income\}$;
- $L_{Age} = \{Young, Adult\}$;
- $L_{Income} = \{Poor, Modest\}$;
- The degree of membership to this query is 0.3;
- P(Q3) = (0,3-*cut*(*Young*) ˅ 0,3-*cut* (*Adult*)) ^ (0,3-*cut* (*Poor*) ˅ 0,3-*cut*(*Modest*)).

Query Evaluation

In this section, we try to evaluate the proposed approach. For this, the searching procedure should take into account all fuzzy summaries in the concept lattice that correspond to the query Q.

Our approach is based on our definition of fuzzy summary α-summary and the valuation function, which are defined as follows:

Definition 6: (BenAli-Sougui, I., Sassi-Hidri, M. & Grissa-Touzi, A., 2013) An α*.summary*, denoted as Z_f, is a fuzzy summary $Z_f = (R_{Zf}, I_{Zf})$ in which Z_f is a collection of candidate records $R_{Zf} = \{t_1, t_2, \ldots, t_N\}$ which represents the extent and I_{Zf} is the intent. Each tuple t_i of R_{Zf} existing in the summary has a membership value. It can be formulated as follows:

$$Z_f = \{\forall\ t_i \in Z_f \mid \mu(t_i) \geq \alpha\} \text{ with } \mu(t_i) \in [0, 1] \tag{5}$$

Definition 7: (BenAli-Sougui, I., Sassi-Hidri, M. & Grissa-Touzi, A., 2014) Let be consider v_f the valuation function. It is obvious that the valuation of $P_f\,(Q)$ depends on the summary Z_f. Thus $v_f\,(P_f\,(Q)_{Zf})$ denotes the valuation of $P_f\,(Q)$ in the context of Z_f, $L_{Ai}\,(Z_f)$ the set of linguistic terms that appear in Z_f and $L_{Ai}(Q)$ the set of linguistic terms that appear in query Q. We can distinguish between three assumptions:

- *Coresp*(Z_f, *Q*) = *Exact*: $v_f(P_f(Q)_{Zf})$ = *true* and $L_{Ai}(Z_f) \subseteq L_{Ai}(Q)$: All tuples including in Z_f verify the query *Q*;
- *Coresp*(Z_f, *Q*) = *False*: $v_f(P_f(Q)_{Zf})$ = *false* and $L_{Ai}(Z_f) \neq .L_{Ai}(Q)$ Linguistic terms appear in Z_f do not correspond to terms in query Q;
- *Coresp*(Z_f, *Q*) = *Indecision*: $\exists$i, $L_{Ai}(Z_f) - L_{Ai}(Q) \neq \varnothing$. There are some tuples in Z_f satisfying *Q*.

These situations reflect a global view of the matching of a fuzzy summary Z_f with a query *Q*.

Figure 4 presents the steps of our flexible query approach based on fuzzy data summaries.

Algorithm 1 describes the different steps of our approach.

The procedure *Insert*($Z_{f,}$ L_{result}) lets us to insert $Z_{f,}$ in the list L_{result} of all fuzzy summary responding to the query *Q*.

Coresp(Z_f, *Q*) is the function that lets to test the correspondence between the summary Z_f and the query *Q* that we have seen previously.

The idea of our approach is to search using the summary concept; which summary responds to the query and calculate their satisfaction degree. Then we will insert them in a list order by satisfaction degree. We will repeat these steps until ensure that there is not a branch in the summary concept that satisfies the query.

Figure 4. The flexible querying approach of fuzzy summaries

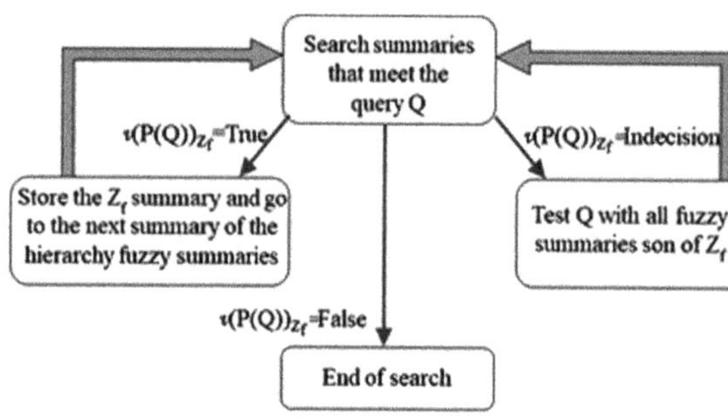

Algorithm 1. SearchResult (Z_f, level, Q)

Require: Z_f the fuzzy concept summary, *level* the current level of summary, *Q* the query.
Ensure: L_{result} list of all fuzzy summary responding to the query *Q*.
$L_{result} \leftarrow \phi$
If (*Coresp*(Z_f, *Q*) = *Exact*) **then**
Insert($Z_{f,}$ L_{result})
Else
If (*Coresp*(Z_f, *Q*) = *indecision*) **then**
For all Fuzzy summary z_f son of Z_f **do**
level← *level*+1
L_{result}← L_{result}+ *SearchResult* (Z_f, *level*, *Q*)
End for
End if
End if
Return L_{result}

Finally we can display the top k α*.summary* from the list of results. Figure 5 shows the principal of our approach.

The satisfaction degree SD of summary Z_f is calculated by a function that returns the best route that lets us to find Z_f. It is determined as follows:

$$SD = \max(\sum_{j=1}^{p} Fuzzy\_score(Zj, Zj-1)). \tag{6}$$

with p the current level of Z_f

$$\text{Fuzzy\_score (Zp, Zp-1)} = \frac{|\ \phi(\text{Zp}) \cap \phi(\text{Zp}-1)|}{|\ \phi(\text{Zp}) \cup \phi(\text{Zp}-1)|} \tag{7}$$

The equation of Fuzzy-score is proposed by (T.T. Quan, 2004).

Generation of Results

Recall that the hierarchy of summary concepts in our Employee test table is presented in table 3.

The result of applying the proposed approach on the concept lattice for queries Q_1, Q_2 and Q_3 for k = 3 is given in Table 5.

REPAIRING QUERY

Our objective is to found a result even in the case of the absence of summaries corresponding strictly to the query.

Figure 5. The steps of the flexible querying approach of the top-k fuzzy data summaries

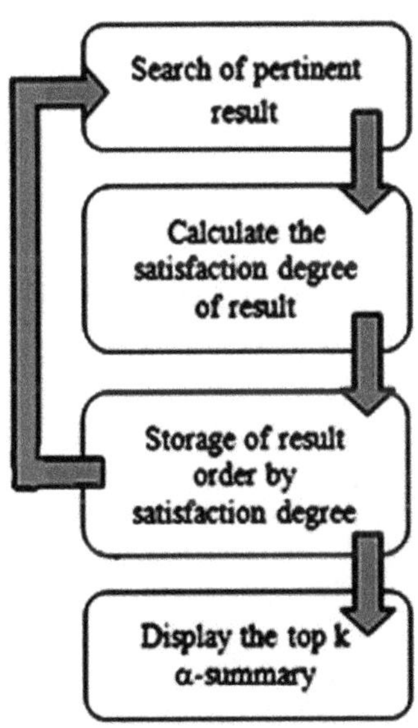

Table 5. Top k α-summary result

Query	α-Summary
Q1	$\alpha - z_{13}=\{t_1^{0.5}, t_3^{0.7}, t_5^{0.6}, t_6^{0.5}\}$
Q2	$\alpha - z_{42} = \{t0.41\}$, $\alpha - z34=\{t0.41, t0.86\}$
Q3	$\alpha - z_{21}=\{t_1^{0.5}, t_2^{0.6}, t_4^{0.4}\}$, $\alpha - z_{22}=\{t_1^{0.5}, t_2^{0.4}, t_3^{0.7}\}$, $\alpha - z_{23}=\{t_2^{0.4}, t_4^{0.6}, t_5^{0.7}\}$

It is thus possible, in an interactive mode, to present to the user the reasons for the failure of his query and to submit alternative queries. To meet this goal, we will present our approach to extracting and querying fuzzy summaries with repairing unanswered query (BenAli-Sougui, I., Sassi-Hidri, M. & Grissa-Touzi, A., 2014) described in Figure 6.

To accommodate this target we will present a new repaired query. Reparation is based on the idea that there could be a result semantically close to the query. To find these results, the query is modified using the best fuzzy summaries which is the near value answering the query.

Repairing queries has already been implemented in the context of a mediator by Bidault et al. (Bidault, A., B. S. & Froidevaux, C., 2000), as one of the cooperative aspects reviewed by Gaasterland et al. in (Gaasterland, T., J. M., & Godfrey, P., 1992).

Basic Principal

This process happens each time a tree exploration that is conducted to find answers for a query Q fails at a specific summary Z_f (i.e., when some required characters are absent from Z_f). Z_f is then called a

Figure 6. The process of extracting and querying fuzzy summaries with unanswered query repair

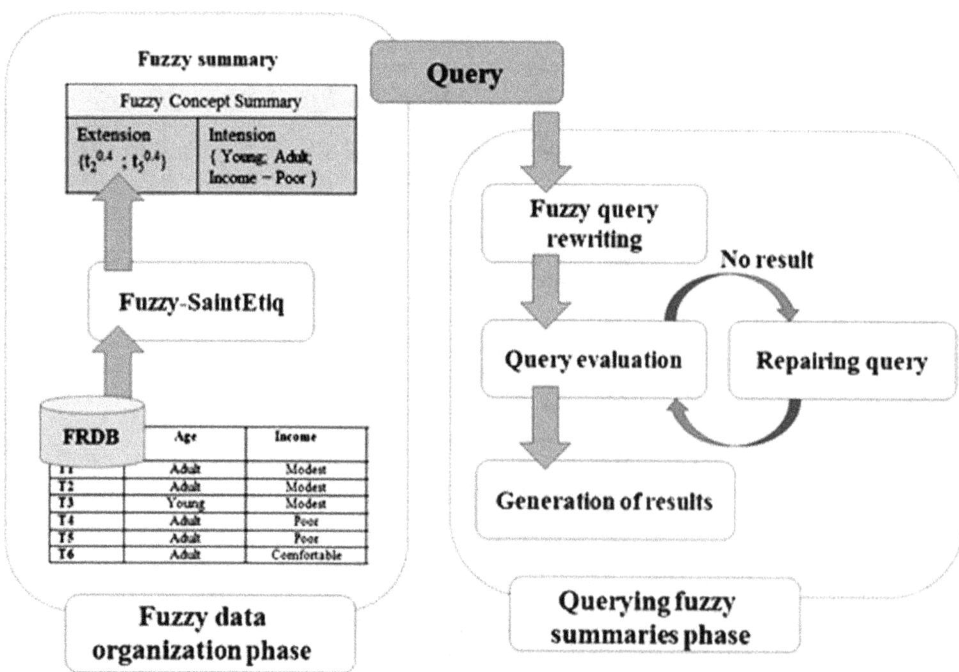

failure node. As the exploration may fail for more than one summary, several modifications of the same query can be proposed to the user.

The strategy is guided by the fuzzy summary hierarchy and it guarantees that results will be found for the new query.

For short, repairing query principe follows these steps:

Step 1: Evaluate the initial query Q;
Step 2: If no result, execute the substitution of query;
Step 3: Determine the best substitution query Q_{sub}, if the son of Z_f have other values for the attributes who has failed then calculate the distance measure between Q and sons of Z_f and use the attributes of the son of Z_f which has the greatest distance value, else use the attributes values of Z_f;
Step 4: Evaluate Q_{sub};
Step 5: Express the results of Q_{sub}.

More distance value is greater, more the summary and query have common attributes.

The distance measure is calculated as follows:

Definition 8: The distance between a query Q and summary Z_f, denoted as $D_Q(Z_f)$, is the sum of the number of common attributes appears in the query Q and *summary* Z_f.

L_{Ak} is the set of linguistic terms of attribute A_k appears in the query Q and I_{Zf} is the sub set of linguistic terms which appears in summary Z_f.

$$D_Q(Z_f) = \sum_{Ak \in Q} \left| L_{Ak} \cap I_{Zf} \right| \tag{8}$$

Query Substitution

Generally, our method produces several alternative answers, more or less close to the original. These queries can be ordered according to the distance to the original one.

To evaluate our query repair method, we will use the food consumption table presented in Table 6, since the Employee database does not contain null queries.

Let consider a relation R = (*Id, Age, Candy, Dairy-product, Lipid*) from an FDB Food consumption table.

The summary concept hierarchy presented in table 7 (APPENDIX 1) is the result generated after applying our Fuzzy-SaintEtiq approach to the Food Consumption table in Table 6.

Let be consider for instance query Q_4:

Q_4: How much is consumed of dairy product and lipid by older people who consume a large amount of candy.
Select 3 0.25 *Dairy-product, Lipid*
From *Food-consumption*
Where *Age FEQ* (*$Old*) *THOLD 0.25*
AND *Candy FEQ* (*$Excessive*) *THOLD 0.25*;

Table 6. Food consumption table

Id	Age	Foods		
		Candy	*Dairy product*	*Lipid*
T1	Young	Great	Average	Great
T2	Child	Great	Excessive	Average
T3	Adult	Low	Average	Great
T4	Old	Low	Excessive	Low
T5	Young	Great	Low	Great
T6	Adult	Low	Great	Great
T7	Old	Low	Great	Low
T8	Child	Excessive	Excessive	Average
T9	Adult	Low	Average	Great
T10	Child	Excessive	Excessive	Great

This query fails $v_f(P_f\,(Q_2)_{Zf}) =$ FALSE and the failure node is z_{42}. From z_{42} two sons' nodes are proposed z_{51} and z_{52}.

$z_{42} = \{\{t_5^{0,2}, t_8^{0,1}, t_{10}^{0,4}\}, \{ YA, CA, GC, EC, AL \}\}$

$z_{51} = \{\{t_5^{0,1}, t_{10}^{0,4}\}, \{ YA, CA, GL, GC, EC, AL \}\}$

$z_{52} = \{\{t_8^{0,1}, t_{10}^{0,4}\}, \{ YA, CA, GC, EC, GD, ED, AL \}\}$

Both '*Young*' and '*Child*' are the labels of attribute *Age* which appear in z_{42}, z_{51} and z_{52}.

So, we notice that we can change the 'old' labels of attribute *Age* by one or both of the 'Young' and 'Child' linguistic labels.

Then from z_{42} the alternative query proposed is Q_{sub}:

Q_{sub}: How much is consumed of dairy product and lipid by children or young people who consume a large amount of candy.

Select 3 0.25 *Dairy-product, Lipid*
From *Food-consumption*
Where *Age FEQ* (*$Child, $Young*)
AND *Candy FEQ* (*$Excessive*);

FUTURE RESEARCH DIRECTIONS

As futures perspectives of this work, we mention to:

- Implement the extracting and querying fuzzy summaries approach.
- Extend the FSQL language to support the top-k best answers.

- Identify the hierarchy generated by our Fuzzy-SaintEtiq model as an index, the querying of summary hierarchies in a DBMS is difficult but still an attractive goal. Indexes are used to accelerate operations on the data, allowing relatively fast access to the data to be processed.

CONCLUSION

Diversification of DB applications highlighted the limitations of Relational Database Management System RDBMS particularly on the modeling plan. In fact, in the real world, we are increasingly faced with the situation where applications need to handle imprecise data and to offer a flexible querying to their users. Several theoretical solutions have been proposed. However, the impact of this work in practice, remained negligible with the exception of a few research prototypes based on the formal model GEFRED.

In this paper, we proposed a new approach of extracting and querying fuzzy summaries. The architecture of our proposed approach is divided in two major phases: fuzzy data organization phase and querying fuzzy summaries phase.

In the fuzzy data organization phase, we proposed a new approach to extract linguistic summarization from fuzzy databases, called Fuzzy-SaintEtiq.

Fuzzy-SaintEtiq takes the FDB records and provides knowledge. The summarization act considered like a process of knowledge discovery from database.

In the querying fuzzy summaries phase, our proposed approach allows users to efficiently querying summaries and exploits the conceptual hierarchical structure, and performs a flexible matching between the summaries and the query to return the top k result.

From an algorithmic point of view, the result of the comparison between the fuzzy summary and the query on the basis of linguistic labels from a user-defined vocabulary determines whether the fuzzy summary is a result but also whether a part of the hierarchy will be explored.

An extension to this method is also proposed. It allows answering queries that have empty result sets with semantically close data.

Our strategy is guided by the fuzzy summary hierarchy and it guarantees that results will be found for the new repaired query.

REFERENCES

Akbarinia, R. P. V., & Pacitti, E. (2007). Processing top-k queries in distributed hash tables. *Euro-Par Conf.* 10.1007/978-3-540-74466-5_53

Azar, A. T. (2012). Overview of Type-2 Fuzzy logic systems. *International Journal of Fuzzy System Applications*, *2*(4), 1–28. doi:10.4018/ijfsa.2012100101

Babcock, B. & C. O. (2003). Distributed top-k monitoring. *SIGMOD Conference*.

Ben Hassine, M. A., Grissa-Touzi, A., Galindo, J., & Ounelli, H. (2008). *How to Achieve Fuzzy Relational Databases. In Handbook of Research on Fuzzy Information Processing in Databases* (pp. 351–380). Information Science Reference. doi:10.4018/978-1-59904-853-6.ch014

BenAli-Sougui, I. (2016). *Vers une nouvelle approche d'exploitation des Bases de Données Relationnelles Floues* (Unpublished doctoral dissertation). National Engineers School of Tunis, El Manar University, Tunisia.

BenAli-Sougui, I., Sassi-Hidri, M., & Grissa-Touzi, A. (2013). About Summarization in Large Fuzzy Databases, *The Fifth International Conference on Advances in Databases, Knowledge, and Data Applications*, 87-94.

BenAli-Sougui, I., Sassi-Hidri, M., & Grissa-Touzi, A. (2014). From User Requirements to Flexible Querying of Fuzzy Summaries. *International Journal of Service Science, Management, Engineering, and Technology*, *5*(1), 84–102. doi:10.4018/ijssmet.2014010105

Bidault, A. B. S., & Froidevaux, C. (2000). Repairing queries in a mediator approach. *Proc. of ECAI00.*

Bosc, P., O., & Litard. L. (1998). Bases de donnes et flexibilité: Les requêtes graduelles. *Techniques et Sciences Informatiques*, *7*, 355-378.

Buckles, B. P., & Petry, F. E. (1982). A fuzzy representation of data for relational databases. *Journal of Fuzzy Sets and Systems*, *7*(3), 213–226. doi:10.1016/0165-0114(82)90052-5

Chaudhuri, S. A. M., & Gravano, L. (2004). Optimizing top-k selection queries over multimedia repositories. *IEEE Transactions on Knowledge and Data Engineering*, 16.

Ciaccia, P., & M. P. (2002). Searching in metric spaces with userdefined and approximate distances. *ACM Transactions on Database Systems*, 27.

Fagin, R. M. N., & Lotem, J. (2003). Optimal aggregation algorithms for middleware. *Journal of Computer and System Sciences*, 66.

Gaasterland, T. J. M., & Godfrey, P. (1992). An overview of cooperative answering. *Journal of Intelligent Systems*, *1*, 123157.

Galindo, J. (2005). New characteristics in fsql, a fuzzy sql for fuzzy databases. *WSEAS Transactions on Information Science and Applications*, *2*(2), 161–169.

Galindo, J. M. P., & Urrutia, A. (2006). *Fuzzy databases: modeling, design and implementation.* Idea Group Publishing Hershey. doi:10.4018/978-1-59140-324-1

Grissa-Touzi, A. (2010). An Alternative Extension of the FCM Algorithm for Clustering Fuzzy Databases. *The Second International Conference on Advances in Databases, Knowledge, and Data Applications*, 135-142. 10.1109/DBKDA.2010.35

Grissa-Touzi, A. & H.O. (2012). A new approach for top-k flexible queries in large database using the knowledge discovered. *The Fourth International Conference on Advances in Databases, Knowledge, and Data Applications.*

Hjaltason, G.R., & H. S. (2003). Index-driven similarity search in metric spaces. *ACM Transactions on Database Systems*, 28.

Kimelfeld, B. & Y. S. (2006). Finding and approximating top-k answers in keyword proximity search. *PODS Conference*. 10.1145/1142351.1142377

Larsen, H. L. (1999). An approach to flexible information access systems using soft computing. *32nd Annual Hawaii International Conference on System Sciences*, 6042.

Medina, J. M. M., & Pons, O. (1994). Gefred. a generalized model of fuzzy relational databases. *Information Sciences*, 76.

Mouratidis, K. D. P., & Bakiras, S. (2006). Continuous monitoring of top-k queries over sliding windows. *SIGMOD Conference*. 10.1145/1142473.1142544

Prade, H., & Testemale, C. (1987). Fuzzy relational databases: Representational issues and reduction using similarity measures. *Journal of the American Society for Information Science*, *38*(2), 118–126. doi:

Quan, T. T., Hui, S. C., & Cao, T. H. (2004). A Fuzzy FCA-based Approach to Conceptual Clustering for Automatic Generation of Concept Hierarchy on Uncertainty Data. *Concept Lattices and Their Applications Workshop*, 1-12.

Raschia, G., & Mouaddib, N. (2002). SaintEtiq: A fuzzy set-based approach to database summarization. *Journal of Fuzzy Sets and Systems*, *129*(2), 137–162. doi:10.1016/S0165-0114(01)00197-X

Sassi, M., Touzi, A. G., Ounelli, H., & Aissa, I. (2010). About database summarization. *International Journal of Uncertainty, Fuzziness and Knowledge-based Systems*, *18*(02), 133–151. doi:10.1142/S0218488510006453

Silberstein, A., & Braynard, R. (2006). A sampling-based approach to optimizing top-k queries in sensor networks. *IEEE International Conference on Data Engineering*. 10.1109/ICDE.2006.10

Umano, M., & Fukami, S. (1980). Retrieval processing from fuzzy databases. *Technical Reports of IECE of Japan*, *80*, 45-54.

Vanisri, D. & D. L. (2010). Survey on fuzzy clustering and rule mining. *International Journal of Computer Science and Information Security*, *8*, 183–187.

Wu, M., & Xu, J. (2006). Monitoring top-k query in wireless sensor networks. *IEEE International Conference on Data Engineering*.

Zadeh, L. A. (1965). Fuzzy Sets. *Journal of Information and Control*, *8*(3), 338–353. doi:10.1016/S0019-9958(65)90241-X

Zemankova-Leech, M. (1985). Implementing imprecision in information systems. *Information Sciences*, *37*(1-3), 107–141. doi:10.1016/0020-0255(85)90008-8

APPENDIX

The following abbreviations are used in the summary concept hierarchy of the Food Consumption table:

- For *Age* attribute: *CA* (Child Age), *YA* (Young Age), *AA* (Adult Age), *OA* (Old Age).
- For *Candy* attribute: *LC* (Low Candy), *AC* (Average Candy) and *GC* (Great Candy) and *EC* (Excessive Candy).
- For *Dairy-product* attribute: *LD* (Low Dairy-product), *AD* (Average Dairy-product), *GD* (Great Dairy-product) and *ED* (Excessive Dairy-product).
- For *Lipid* attribute: *LL* (Low Lipid), *AL* (Average Lipid), *GL* (Great Lipid) and *EL* (Excessive Lipid).

Table 7. Fuzzy Concept summaries' generation

Level	Fuzzy Summaries
0	$Z_1 = \{\{t_1^{0,7}, t_2^{0,2}, t_3^{0,4}, t_4^{0,1}, t_5^{0,1}, t_6^{0,2}, t_7^{0,2}, t_8^{0,6}, t_9^{0,3}, t_{10}^{0,4}\}, \{AL\}\}$
1	$Z_{21} = \{\{t_1^{0,7}, t_2^{0,2}, t_5^{0,9}, t_6^{0,3}, t_8^{0,1}, t_9^{0,2}, t_{10}^{0,4}\}, \{YA, AL\}\}$ $Z_{22} = \{\{t_2^{0,8}, t_3^{0,6}, t_5^{0,9}, t_6^{0,8}, t_9^{0,7}, t_{10}^{0,4}\}, \{GL, AL\}\}$ $Z_{23} = \{\{t_2^{0,2}, t_3^{0,2}, t_4^{0,1}, t_6^{0,9}, t_7^{0,8}, t_8^{0,1}, t_9^{0,4}, t_{10}^{0,4}\}, \{GD, AL\}\}$ $Z_{24} = \{\{t_1^{0,2}, t_2^{0,1}, t_3^{0,2}, t_4^{0,1}, t_6^{0,3}, t_7^{0,2}, t_9^{0,4}\}, \{AC, LC\}\}$ $Z_{25} = \{\{t_1^{0,8}, t_3^{0,8}, t_5^{0,2}, t_6^{0,1}, t_7^{0,2}, t_9^{0,6}\}, \{AD, AL\}\}$ $Z_{26} = \{\{t_1^{0,3}, t_4^{0,9}, t_7^{0,8}, t_8^{0,4}\}, \{LL, AL\}\}$
2	$Z_{31} = \{\{t_1^{0,3}, t_2^{0,2}, t_5^{0,1}, t_8^{0,1}, t_{10}^{0,4}\}, \{YA, CA, GC, AL\}\}$ $Z_{32} = \{\{t_2^{0,2}, t_5^{0,9}, t_6^{0,3}, t_9^{0,2}, t_{10}^{0,4}\}, \{YA, GL, AL\}\}$ $Z_{33} = \{\{t_2^{0,2}, t_6^{0,3}, t_8^{0,1}, t_9^{0,2}, t_{10}^{0,4}\}, \{YA, GD, AL\}\}$ $Z_{34} = \{\{t_1^{0,2}, t_2^{0,1}, t_6^{0,3}, t_9^{0,2}\}, \{YA, AC, AL\}\}$ $Z_{35} = \{\{t_2^{0,2}, t_3^{0,2}, t_6^{0,8}, t_9^{0,4}, t_{10}^{0,4}\}, \{GL, GD, AL\}\}$ $Z_{36} = \{\{t_2^{0,2}, t_4^{0,1}, t_8^{0,1}, t_{10}^{0,4}\}, \{ED, GD, AL\}\}$ $Z_{37} = \{\{t_1^{0,7}, t_5^{0,2}, t_6^{0,1}, t_9^{0,2}\}, \{YA, AD, AL\}\}$ $Z_{38} = \{\{t_2^{0,1}, t_3^{0,2}, t_4^{0,1}, t_6^{0,3}, t_7^{0,2}, t_9^{0,4}\}, \{GD, AC, AL\}\}$ $Z_{39} = \{\{t_3^{0,6}, t_5^{0,2}, t_6^{0,1}, t_9^{0,6}\}, \{GL, AD, AL\}\}$ $Z_{310} = \{\{t_1^{0,2}, t_3^{0,2}, t_6^{0,1}, t_7^{0,2}, t_9^{0,4}\}, \{AC, AD, AL\}\}$ $Z_{311} = \{\{t_1^{0,2}, t_4^{0,1}, t_7^{0,2}\}, \{AC, LL, AL\}\}$ $Z_{312} = \{\{t_4^{0,1}, t_7^{0,8}, t_8^{0,1}\}, \{GD, LL, AL\}\}$
3	$Z_{41} = \{\{t_2^{0,2}, t_5^{0,1}, t_{10}^{0,4}\}, \{YA, CA, GL, GC, AL\}\}$ $Z_{42} = \{\{t_5^{0,2}, t_8^{0,1}, t_{10}^{0,4}\}, \{YA, CA, GC, EC, AL\}\}$ $Z_{43} = \{\{t_1^{0,2}, t_2^{0,1}\}, \{YA, CA, GC, AC, AL\}\}$ $Z_{44} = \{\{t_1^{0,3}, t_8^{0,1}\}, \{YA, CA, GC, LL, AL\}\}$ $Z_{45} = \{\{t_1^{0,2}, t_5^{0,1}\}, \{YA, CA, GC, AD, LD, AL\}\}$ $Z_{46} = \{\{t_2^{0,2}, t_8^{0,1}, t_{10}^{0,4}\}, \{YA, CA, GC, GD, ED, AL\}\}$ $Z_{47} = \{\{t_2^{0,2}, t_6^{0,3}, t_9^{0,2}, t_{10}^{0,4}\}, \{YA, GL, GD, AL\}\}$ $Z_{48} = \{\{t_5^{0,2}, t_6^{0,1}, t_9^{0,2}\}, \{YA, GL, AD, AL\}\}$ $Z_{49} = \{\{t_1^{0,2}, t_6^{0,1}, t_9^{0,2}\}, \{YA, AC, AD, AL\}\}$ $Z_{410} = \{\{t_2^{0,1}, t_3^{0,2}, t_6^{0,3}, t_9^{0,4}\}, \{GL, GD, AC, AL\}\}$ $Z_{411} = \{\{t_2^{0,1}, t_4^{0,1}\}, \{ED, GD, AC, AL\}\}$ $Z_{412} = \{\{t_3^{0,2}, t_4^{0,1}, t_6^{0,3}, t_7^{0,4}, t_9^{0,4}\}, \{GD, AC, LC, AL\}\}$ $Z_{413} = \{\{t_1^{0,2}, t_7^{0,2}\}, \{AC, AD, LL, AL\}\}$ $Z_{414} = \{\{t_4^{0,1}, t_7^{0,2}, t_8^{0,1}\}, \{ED, GD, LL, AL\}\}$
4	$Z_{51} = \{\{t_5^{0,1}, t_{10}^{0,4}\}, \{YA, CA, GL, GC, EC, AL\}\}$ $Z_{52} = \{\{t_8^{0,1}, t_{10}^{0,4}\}, \{YA, CA, GC, EC, GD, ED, AL\}\}$ $Z_{53} = \{\{t_2^{0,2}, t_{10}^{0,4}\}, \{YA, CA, GC, GL, ED, GD, AL\}\}$ $Z_{54} = \{\{t_2^{0,1}, t_6^{0,3}, t_9^{0,2}\}, \{YA, GL, GD, AC, AL\}\}$ $Z_{55} = \{\{t_3^{0,2}, t_6^{0,1}, t_7^{0,2}, t_9^{0,4}\}, \{AD, GD, AC, LC, AL\}\}$ $Z_{56} = \{\{t_3^{0,1}, t_4^{0,1}, t_7^{0,4}\}, \{GD, AC, LC, OA, AL\}\}$
6	$Z_{61} = \{\{t_3^{0,2}, t_6^{0,1}, t_9^{0,4}\}, \{AD, GD, AC, LC, GL, AL\}\}$ $Z_{62} = \{\{t_4^{0,1}, t_7^{0,2}\}, \{OA, LC, AC, LL, GD, AL\}\}$ $Z_{63} = \{\{t_3^{0,1}, t_7^{0,4}\}, \{AD, GD, AC, LC, OA, AL\}\}$
7	$Z_{71} = \{\{t_{10}^{0,4}\}, \{YA, CA, GC, EC, GD, ED, GL, AL\}\}$ $Z_{72} = \{\{t_2^{0,1}\}, \{YA, CA, AC, GC, GL, GD, ED, AL\}\}$ $Z_{73} = \{\{t_1^{0,2}\}, \{YA, CA, GC, AC, LL, AD, LD, AL\}\}$ $Z_{74} = \{\{t_8^{0,1}\}, \{YA, CA, GC, EC, GD, ED, LL, AL\}\}$ $Z_{75} = \{\{t_5^{0,1}\}, \{CA, YA, GC, EC, AD, LD, GL, AL\}\}$ $Z_{76} = \{\{t_6^{0,1}, t_9^{0,2}\}, \{YA, GL, AD, AC, GD, LC, AL\}\}$ $Z_{77} = \{\{t_4^{0,1}\}, \{OA, LC, AC, LL, GD, ED, AL\}\}$ $Z_{78} = \{\{t_3^{0,1}\}, \{OA, AD, GD, AC, LC, GL, AL\}\}$ $Z_{79} = \{\{t_7^{0,2}\}, \{OA, LC, AC, LL, GD, AD, AL\}\}$
8	$Z_8 = \{\{ \varnothing ., \{EL\}\}$

Compilation of References

451 Group. (2010). *Report on Cloud Computing 'As-a-service' market sizing - Report II*. 451 Research.

Aaqib Iqbal, F. M., Khan, S. & Khan, A. (2009). A Critical Analysis of Techniques for Requirement Prioritization and Open Research Issues. *International Journal of Reviews in Computing, 1*, 8-18.

Aasem, M. (2010). *Analysis and optimization of software requirements prioritization techniques*. IEEE.

Abel, F., Gao, Q., Houben, G., & Tao, K. (2011). Semantic enrichment of twitter posts for user profile construction on the social web. In G. Antoniou, M. Grobelnik, E. Simperl, B. Parsia, D. Plexousakis, P. D. Leenheer, & J. Pan (Eds.), *The Semanic Web: Research and Applications* (pp. 375–389). Berlin: Springer. doi:10.1007/978-3-642-21064-8_26

Acharjee, S., Chakraborty, S., Karaa, W. B. A., Azar, A. T., & Dey, N. (2014). Performance evaluation of different cost functions in motion vector estimation. *International Journal of Service Science, Management, Engineering, and Technology*, *5*(1), 45–65. doi:10.4018/ijssmet.2014010103

Act, F. C. R. (2009). *Fair Credit Reporting Act*. Flood Disaster Protection Act and Financial Institute.

Aderaldo, C. M., Mendonca, N. C., Pahl, C., & Jamshidi, P. (2017). Benchmark requirements for microservices architecture research. In *1st International Workshop on Establishing the Community-Wide Infrastructure for Architecture-Based Software Engineering*. IEEE.

Afuah, A. (2000). How much do your "co-opetitors" capabilities matter in the face of technological change? *Strategic Management Journal*, *21*(3), 387–404. doi:

Aggarwal & Kaur. (2004b). Code Coverage Based Technique For Prioritizing Test Cases For Regression Testing. *ACM SIGSOFT Software Engineering Notes, 25*, 1-4.

Aggarwal, Y. S. (2004a). *Software Engineering*. New Age International.

Aghabozorgi, S. R., & Wah, T. Y. (2009). Dynamic modeling by usage data for personalization systems. *Proceedings of 13th International Conference on Information Visualisation*, 450-455, 10.1109/IV.2009.111

Agrawal, H. (1994). Dominators, Super Blocks, and Program Coverage. *Proceedings of the 21st ACM SIGPLAN-SIGACT Symposium on Principles of Programming Languages*, 25-34. 10.1145/174675.175935

Agrawal, H., & Horgan, J. (1990). Dynamic program slicing. *In Proceedings of SIGPLAN '90 Conference on Programming Language Design and Implementation* (pp.246-256). ACM.

Agrawal, R., & Srikant, R. (1994). *Fast algorithms for mining association rules in large databases*. VLDB.

Ahl, V. (2005). An experimental comparison of five prioritization methods - Investigating ease of use, accuracy and scalability. Department of Systems and Software Engineering, School of Engineering, Sweden, Blekinge Institute of Technology.

Akbarinia, R. P. V., & Pacitti, E. (2007). Processing top-k queries in distributed hash tables. *Euro-Par Conf.* 10.1007/978-3-540-74466-5_53

Akhigbe, A., & Whyte, A. M. (2004). The Gramm-Leach-Bliley Act of 1999: Risk implications for the financial services industry. *Journal of Financial Research*, *27*(3), 435–446. doi:10.1111/j.1475-6803.2004.00093.x

Akoglu, L., Melo, P. O. S. V. D., & Faloutsos, C. (2012). *Quantifying Reciprocity in Large Weighted Communication Networks* (Vol. 7302). Advances in Knowledge Discovery and Data Mining Lecture Notes in Computer Science; doi:10.1007/978-3-642-30220-6_8

Alkhanak, E. N., Lee, S. P., & Khan, S. U. R. (2015). Cost-Aware Challenges for Workflow Scheduling Approaches in Cloud Computing Environments: Taxonomy and Opportunities. *Future Generation Computer Systems*, *50*, 3–21. doi:10.1016/j.future.2015.01.007

Alliance, C. (2011b). *Security as a Service*. Retrieved from https://cloudsecurityalliance.org/wpcontent/uploads/2011/09/SecaaS_V1_0.pdf

Alliance, C. (2011a). *Security guidance for critical areas of focus in cloud computing v3. 0*. Cloud Security Alliance.

Almarri, K., & Gardiner, P. (2014). Application of resource-based view to project management research: Supporters and opponents. *Procedia: Social and Behavioral Sciences*, *119*, 437–445. doi:10.1016/j.sbspro.2014.03.049

Al-Nazer, A., Helmy, T., & Al-Mulhem, M. (2014). User's profile ontology-based semantic framework for personalized food and nutrition recommendation. *Procedia Computer Science*, *32*(1), 101–108. doi:10.1016/j.procs.2014.05.403

Amazon Web Services. (2014, Novemeber). *Overview of Security Processes*. Retrieved from http://d0.awsstatic.com/whitepapers/Security/AWS%20Security%20Whitepaper.pdf

Aminia, B., Ibrahima, R., Othmana, M. S., & Nematbakhshb, M. A. (2015). A reference ontology for profiling scholar's background knowledge in recommender systems. *Expert Systems with Applications*, *42*(2), 913–928. doi:10.1016/j.eswa.2014.08.031

Anand, S. S., Kearney, P., & Shapcott, M. (2007). Generating semantically enriched user profiles for web personalization. *ACM Transactions on Internet Technology*, *7*(4), 22, es. doi:10.1145/1278366.1278371

Anand, S. S., & Mobasher, B. (2005). Intelligent Techniques for Web Personalization. In B. Mobasher & S. S. Anand (Eds.), Lecture Notes on Computer Science: Vol. 3169. *Intelligent Techniques for Web Personalization* (pp. 1–36). Berlin: Springer-Verlag. doi:10.1007/11577935_1

Andrea, A., & Xin, Y. (2008). A Novel Co-Evolutionary Approach to Automatic Software Bug Fixing. In *IEEE World Congress on Computational Intelligence*. IEEE Press.

Anker, T. B., Sparks, L., Moutinho, L., & Grönroos, C. (2015). Consumer Dominant Value Creation: A Theoretical Response to the Recent Call for a Consumer Dominant Logic for Marketing. *European Journal of Marketing*, *49*(3/4), 532–560. doi:10.1108/EJM-09-2013-0518

Antonik, A., & Huth, M. (2006). Effcient patterns for model checking partial state spaces in CTL intersection ltl. *Electronic Notes in Theoretical Computer Science*, *158*, 41–57. doi:10.1016/j.entcs.2006.04.004

Apiwattanapong, T., Orso, A., & Harrold, M. J. (2005). Efficient and Precise Dynamic Impact analysis using Execute-After Sequences. In *Proceeding of ACM- International Conference on Software Engineering* (pp. 432-441). ACM.

Apiwattanapong, O. T., & Harrold, M. J. (2003). Leveraging field data for impact analysis and regression testing. In *Proceedings of the ACM SIGSOFT Symposium on Foundations of Software Engineering* (pp.128-137). ACM.

Apostolico, A., Barbares, M., & Pizzi, C. (2011). Speedup for a periodic subgraph miner. *Information Processing Letters*, *111*(11), 521–523. doi:10.1016/j.ipl.2011.02.016

Arabnejad, H., Jamshidi, P., El Ioini, N., Estrada, G., & Pahl, C. (2016). An auto-scaling cloud controller using fuzzy q-learning-implementation in openstack. *European Conference on Service-Oriented and Cloud Computing ESOCC*. 10.1007/978-3-319-44482-6_10

Archibald, B. (2009). Expect More from Your SaaS Vendor. *Occupational Health & Safety (Waco, Tex.)*, *78*(10), 83–84. PMID:19856771

Argyres, N., & Bercovitz, J. (2015). Franchisee associations as sources of bargaining power? Some evidence. *Journal of Economics & Management Strategy*, *24*(4), 811–832. doi:10.1111/jems.12111

Armbrust, M., Fox, A., Griffith, R., Joseph, A. D., Katz, R., Konwinski, A., ... Zaharia, M. (2009). A view of cloud computing. *Communications of the ACM*, *53*(4), 50–58. doi:10.1145/1721654.1721672

Armstrong, C., & Shimizu, K. (2007). A review of approaches to empirical research on the resource-based view of the firm. *Journal of Management*, *33*(6), 959–986. doi:10.1177/0149206307307645

Arnold, R. S., & Bohner, S. A. (1993). Impact analysis - towards a framework for comparison. In *Proceedings of IEEE International Conference on Software Maintenance*, (pp.292-301). IEEE. 10.1109/ICSM.1993.366933

Arthur, W. B. (1994). *Increasing returns and path dependence in the economy*. University of Michigan Press. doi:10.3998/mpub.10029

Artikis, A., Sergot, J. M., & Paliouras, G. (n.d.). Run-time composite event recognition. In *DEBS* (pp. 69-80). ACM.

Austin, T., Iliffe, S., Leaning, M., & Modell, M. (1996). A prototype computer decision support system for the management of asthma. *Journal of Medical Systems*, *20*(1), 45–55. doi:10.1007/BF02260873 PMID:8708491

Ayers, J., & Odegaard, M.-A. (2017). *Retail supply chain management* (2nd ed.). Boca Raton, FL: CRC Press.

Azad, P., & Navimipour, N. J. (2017). An Energy-Aware Task Scheduling in the Cloud Computing Using a Hybrid Cultural and Ant Colony Optimization Algorithm. *International Journal of Cloud Applications and Computing*, *7*(4), 20–40. doi:10.4018/IJCAC.2017100102

Azar, A. T. (2012). Overview of Type-2 Fuzzy logic systems. *International Journal of Fuzzy System Applications*, *2*(4), 1–28. doi:10.4018/ijfsa.2012100101

Babcock, B. & C. O. (2003). Distributed top-k monitoring. *SIGMOD Conference*.

Badger, L., Grance, T., Patt-Corner, R., & Voas, J. (2012, May). *Cloud Computing Synopsis and Recommendations: Recommendations of the National Institute of Standards and Technology*. CreateSpace Independent Publishing Platform.

Baghdadi, Y. (2015). Service-Oriented Software Engineering. *International Journal of Systems and Service-Oriented Engineering*, *5*(2), 1–19. doi:10.4018/IJSSOE.2015040101

Bahari, N., Maniak, R., & Fernandez, V. (2015). *Ecosystem Business Model Design*. Paper presented at the 16th Conférence Internationale de Management Stratégique, Paris, France.

Bahga, A., & Madisetti, V. (2013). Performance evaluation approach for multi-tier cloud applications. *Journal of Software Engineering and Applications*.

Baihaqi, I., & Sohal, A. (2013). The impact of information sharing in supply chains on organisational performance: An empirical study. *Production Planning and Control*, *24*(8-9), 743–758. doi:10.1080/09537287.2012.666865

Bandara, K. Y., Wang, M. X., & Pahl, C. (2009). Context modeling and constraints binding in web service business processes. In *Proceedings of the Workshop on Context-Aware Software Technology and Applications CASTA* (pp. 29-32). ACM Press. 10.1145/1595768.1595780

Banerjee, S., & Jain, S. (2014). A survey on Software as a Service (SaaS) using quality model in cloud computing. *International Journal of Engineering and Computer Science*, *3*(1), 3598–3602.

Banirostam, H., Hedayati, A., Zadeh, A. K., & Shamsinezhad, E. (2013, April). A Trust Based Approach for Increasing Security in Cloud Computing Infrastructure. In *Computer Modelling and Simulation (UKSim), 2013 UKSim 15th International Conference* (pp. 717-721). IEEE. 10.1109/UKSim.2013.39

Baresi, L., & Guinea, S. (2005). Towards Dynamic Monitoring of WS-BPEL Processes. In *Proceedings International Conference on Service-Oriented Computing ICSOC'05*. Springer-Verlag.

Barkaoui, K., Eslamichalandar, M., & Kaabachi, M. (2010). A structural verification of web services composition compatibility. *Proceedings of the 6th International Workshop on Enterprise & Organizational Modeling and Simulation, EOMAS '10*, 30-41.

Barney, J. (1991). Firm resources and sustained competitive advantage. *Journal of Management*, *17*(1), 99–120. doi:10.1177/014920639101700108

Barrett, M., Davidson, E., Prabhu, J., & Vargo, S. L. (2015). Service Innovation in the Digital Age: Key Contributions and Future Directions. *Management Information Systems Quarterly*, *39*(1), 135–154. doi:10.25300/MISQ/2015/39:1.03

Baryannis, G., Garefalakis, P., Kritikos, K., Magoutis, K., Papaioannou, A., Plexousakis, D., & Zeginis, C. (2013). *Lifecycle Management of Service-based Applicaitons on Multi-Clouds: A research roadmap. In Multi-Cloud* (pp. 13–20). ACM.

Basili V. R., Briand L. C., & Melo W. L. (1996). A validation of object-oriented design metrics as quality indicators. *IEEE Transaction on Software Engineering, 22*(10), 751–761.

Bass, F. (1969). A New Product Growth Model for Consumer Durables. Management Sciences, 15(5).

Basturk & Karaboga. (2006). An Artificial Bee Colony (ABC) Algorithm for Numeric function Optimization. *IEEE Swarm Intelligence Symposium*.

Batet, M., Moreno, A., Sánchez, D., Isern, D., & Valls, A. (2012). Turist@: Agent-based personalised recommendation of tourist activities. *Expert Systems with Applications*, *39*(8), 7319–7329. doi:10.1016/j.eswa.2012.01.086

Batoulis, K., Meyer, A., Bazhenova, E., Decker, G., Weske, M., & 1. (2015). Extracting Decision Logic from Process Models. *Lecture Notes in Computer Science, 9097*, 520–521. 10.1007/978-3-319-19069-3

Baykasolu, Özbakır, & Tapkan. (2009). *Artificial Bee Colony Algorithm and Its Application to Generalized Assignment Problem*. Academic Press.

Beere, R. (2017). *The role of franchising on industry evolution: assessing the emergence of franchising and its impact on structural change.* Basingstoke, UK: Palgrave Macmillan. doi:10.1007/978-3-319-49064-9

Bella, G., Giunchiglia, F., & McNeill, F. (2017). Language and domain aware lightweight ontology matching. *Journal of Web Semantics*, *43*, 1–17. doi:10.1016/j.websem.2017.03.003

Ben Azaiez, A., & Sbaï, Z. (2015). Model Checking Web Services Choreography. *Lecture Notes in Business Information Processing*, *231*, 171-186.

Ben Hassine, M. A., Grissa-Touzi, A., Galindo, J., & Ounelli, H. (2008). *How to Achieve Fuzzy Relational Databases. In Handbook of Research on Fuzzy Information Processing in Databases* (pp. 351–380). Information Science Reference. doi:10.4018/978-1-59904-853-6.ch014

BenAli-Sougui, I. (2016). *Vers une nouvelle approche d'exploitation des Bases de Données Relationnelles Floues* (Unpublished doctoral dissertation). National Engineers School of Tunis, El Manar University, Tunisia.

BenAli-Sougui, I., Sassi-Hidri, M., & Grissa-Touzi, A. (2013). About Summarization in Large Fuzzy Databases, *The Fifth International Conference on Advances in Databases, Knowledge, and Data Applications*, 87-94.

BenAli-Sougui, I., Sassi-Hidri, M., & Grissa-Touzi, A. (2014). From User Requirements to Flexible Querying of Fuzzy Summaries. *International Journal of Service Science, Management, Engineering, and Technology*, *5*(1), 84–102. doi:10.4018/ijssmet.2014010105

Berander, P., & Andrews, A. (2005). *Requirements prioritization. Engineering and Managing Software Requirements* (A. Aurum & C. Wohlin, Eds.). Berlin, Germany: Springer Verlag.

Berberidis, C., Aref, G. W., Atallah, M., Ioannis, V. I., & Elmagarmid, K. A. (2002). Multiple and Partial Periodicity Mining in Time Series Databases. *Proceedings of the 15th European. Conference on Artificial Intelligence*, 370-374.

Berkovsky, S., Eytani, Y., Kuflik, T., & Ricci, F. (2005). Privacy-enhanced collaborative filtering. *Proceedings of PEP05 UM05 Workshop on Privacy-Enhanced Personalization*, 75-83.

Bernardi, S., Campos, J., & Merseguer, J. (2011, February). Timing-Failure Risk Assessment of UML Design Using Time Petri Net Bound Techniques. *IEEE Transactions on Industrial Informatics*, *7*(1), 90–104. doi:10.1109/TII.2010.2098415

Berry, M., & Linoff, G. (1997). *Data Mining Technique for Marketing, Sales and Customer Support.* New York: John Wiley.

Bettini, C., Wang, X. S., Jajodia, S., & Lin, J. L. (1998, March/April). Discovering frequent event patterns with multiple granularities in time sequencies. *Transactions on Knowledge and Data Engineering*, *10*(2), 222–237. doi:10.1109/69.683754

Beydeda, S., & Gruhn, V. (2003). Test Case Generation According To The Binary Search Strategy. *Lecture Notes in Computer Science*, *2869*, 1000–1007. doi:10.1007/978-3-540-39737-3_124

Bharadwaj, A. (2000). A Resource-Based Perspective on Information Technology Capability and Firm Performance: An Empirical Investigation. *Management Information Systems Quarterly*, *24*(1), 169–196. doi:10.2307/3250983

Bhowmick, P. K., Sarkar, S., & Basu, A. (2010). Ontology based user modeling for personalized information access. *International Journal of Computer Science and Applications*, *7*(1), 1–22.

Biard, T., Mauff, A. L., Bigand, M., & Bourey, J.-P. (2015). Separation of Decision Modeling from Business Process Modeling Using New "Decision Model and Notation" (DMN) for Automating Operational Decision-Making. In IFIP Advances in Information and Communication Technology (Vol. 463, pp. 489–496). Academic Press. doi:10.1007/978-3-319-24141-8

Bidault, A. B. S., & Froidevaux, C. (2000). Repairing queries in a mediator approach. *Proc. of ECAI00.*

Biswas, Satpathy, & Sukumaran. (2011). Regression Test Selection Techniques: A Survey. *Informatica, 35*, 289–321.

Black, J., & Boal, K. (1994). Strategic resources: traits, configurations and paths to sustainable competitive advantage. *Strategic Management Journal, 15*(SI), 131-148. doi:10.1002/smj.4250151009

Blair, G. S., Bencomo, N., & France, R. B. (2009). Models@Runtime. *IEEE Computer, 10*(10), 22–27. doi:10.1109/MC.2009.326

Blair, R., & Lafontaine, F. (2010). *The economics of franchising* (2nd ed.). New York: Cambridge University Press.

Blanco-Fernández, Y., López-Nores, M., Gil-Solla, A., Ramos-Cabrer, M., & Pazos-Arias, J. J. (2011). Exploring synergies between content-based filtering and spreading activation techniques in knowledge-based recommender systems. *Information Sciences, 181*(21), 4823–4846. doi:10.1016/j.ins.2011.06.016

Blome, C., Schoenherr, T., & Eckstein, D. (2014). The impact of knowledge transfer and complexity on supply chain flexibility: A knowledge-based view. *International Journal of Production Economics, 147*, 307–316. doi:10.1016/j.ijpe.2013.02.028

Boehm, B. W. (1988). A Spiral Model of Software Development and Enhancement. *Computer, 21*(5), 61–72. doi:10.1109/2.59

Bogdan, R., Taylor, S., & DeVault, M. (2015). *Introduction to qualitative research methods: a guidebook and resource* (4th ed.). New York: John Wiley & Sons.

Böhm, M., Koleva, G., Leimeister, S., Riedl, C., & Krcmar, H. (2010). *Towards a Generic Value Network for Cloud Computing*. Paper presented at the 17th Workshop on Economics of Grids, Clouds, Systems, and Services, Ischia, Italy. 10.1007/978-3-642-15681-6_10

Böhm, M., Leimeister, S., Riedl, C., & Krcmar, H. (2009). Cloud Computing: Outsourcing 2.0 oder ein neues Geschäftsmodell zur Bereitstellung von IT-Ressourcen?. *IM – Die Fachzeitschrift für Information Management und Consulting, 24*(2), 6-14.

Bohner, S. A. (2002). Software change impacts - an evolving perspective. In *Proceedings of IEEE International Conference on Software Maintenance* (pp. 263-272). IEEE. 10.1109/ICSM.2002.1167777

Bohner, S. A., & Arnold, R. S. (1996). *Software Change Impact Analysis*. IEEE Computer Society Press.

Bordeaux, L., Salaun, G., Berardi, D., & Mecella, M. (2005). When are two web services compatible? *Proceedings of the 5th International Conference on Technologies for E-Services, TES'04,* 15-28. 10.1007/978-3-540-31811-8_2

Borg, M., De la Vara, J., & Wnuk, K. (2016). Practitioners' perspectives on change impact analysis for safety-critical software—a preliminary analysis. *Proceedings of 5th International Workshop Next Generation Syst. Assurance Approaches Safety-Critical Syst.*, 346–358.

Borg, M., Gotel, O., & Wnuk, K. (2013). Enabling traceability reuse for impact analyses: A feasibility study in a safety context. *Proc. 7th International Workshop Traceability Emerging Forms Software Engineering*, 72–78. 10.1109/TEFSE.2013.6620158

Borgwardt, K. M., Kriegel, H. P., & Wackersreuther, P. (2006). Pattern Mining in Frequent Dynamic Subgraphs. *Proceedings of the Sixth International Conference on Data Mining (ICDM'06)*, 818 – 822. 10.1109/ICDM.2006.124

Bosc, P., O., & Litard. L. (1998). Bases de donnes et flexibilité: Les requêtes graduelles. *Techniques et Sciences Informatiques, 7*, 355-378.

Boukadi, K., Ghedira, C., Chaari, S., Vincent, L., & Bataineh, E. (2008). CWSC4EC: how to employ context, web service, and community in enterprise collaboration. In *Proceedings of the 8th International Conference on New Technologies in Distributed Systems (NOTERE '08)*. ACM. 10.1145/1416729.1416737

Boumahdi, F., Boulefrakh, H. E., & Chalal, R. (2015). Decision Making and Service Oriented Architecture for Recruitment Process. Using the New Standard Decision Model and Notation (DMN). *ICSEA 2015: The Tenth International Conference on Software Engineering Advances Decision,* 489–493.

Boumahdi, F., & Chalal, R. (2014). SOAda: A New Architecture to Enrich SOA with a Decisional Aspect. *International Journal of Systems and Service-Oriented Engineering*, *4*(2), 13–27. doi:10.4018/ijssoe.2014040102

Boumahdi, F., Chalal, R., Guendouz, A., & Gasmia, K. (2016). SOA (Formula presented.) : A new way to design the decision in SOA—based on the new standard Decision Model and Notation (DMN). *Service Oriented Computing and Applications*, *10*(1), 35–53. doi:10.100711761-014-0162-x

Brandon-Jones, E., Squire, B., Autry, C., & Petersen, K. (2014). A contingent resource-based perspective of supply chain resilience and robustness. *The Journal of Supply Chain Management*, *50*(3), 55–73. doi:10.1111/jscm.12050

Breech, B., Danalis, A., Shindo, S., & Pollock, L. (2004). Online impact analysis via dynamic compilation Technology. In *Proceedings of the International Conference of Software Maintenance* (pp. 453-457). Academic Press. 10.1109/ICSM.2004.1357834

Breech, B., Tegtmeyer, M., & Pollock, L. (2005). A Comparison of Online and Dynamic Impact Analysis Algorithms. In *Proceedings of Ninth European Conference on Software Maintenance and Reengineering* (pp. 143-152). Academic Press. 10.1109/CSMR.2005.1

Briand, L. C., Labiche, Y., & O'Sullivan, L. (2003). Impact Analysis and Change Management of UML Models. In *Proceedings of the International Conference on Software Maintenance* (pp. 256-265). Academic Press. 10.1109/ICSM.2003.1235428

Brumec, S., & Vrček, N. (2013). Cost Effectiveness of Commercial Computing Clouds. *Information Systems*, *38*(4), 495–508. doi:10.1016/j.is.2012.11.002

Brunette, G., & Mogull, R. (2009, December). Security guidance for critical areas of focus in cloud computing. *Cloud Security Alliance*. Retrived from http://www.cloudsecurityalliance.org/guidance/csaguide.v2.1.pdf

Buchan, J. (2013). *Franchisees as consumers: benchmarks, perspectives and consequences*. New York: Springer. doi:10.1007/978-1-4614-5614-8

Buckles, B. P., & Petry, F. E. (1982). A fuzzy representation of data for relational databases. *Journal of Fuzzy Sets and Systems*, *7*(3), 213–226. doi:10.1016/0165-0114(82)90052-5

Burnside, R. S. (1987). Electronic Communications Privacy Act of 1986: The Challenge of Applying Ambiguous Statutory Language to Intricate Telecommunication Technologies, The. *Rutgers Computer & Technology Law Journal*, *13*, 451.

Burt, R. S. (2000). The network structure of social capital. *Research in Organizational Behavior*, *22*, 345–423. doi:10.1016/S0191-3085(00)22009-1

Burt, R. S. (2001). Structural Holes versus Network Closure as Social Capital. In N. Lin, K. S. Cook, & R. S. Burt (Eds.), *Social Capital: Theory and Research* (pp. 31–56). Transaction Publishers.

Buyya, R., Broberg, J., & Goscinski, A. (2011). *Cloud Computing - Principles and Paradigms*. New York: Wiley. doi:10.1002/9780470940105

Buyya, R., Yeo, C. S., Venugopal, S., Broberg, J., & Brandic, I. (2009). Cloud Computing and Emerging IT Platforms: Vision, Hype, and Reality for Delivering Computing as the 5th Utility. *Future Generation Computer Systems, 25*(6), 599–616. doi:10.1016/j.future.2008.12.001

Cai, H., & Santelices, R. (2015). A framework for cost-effective dependence-based dynamic impact analysis. *IEEE 22nd International Conference on Software Analysis, Evolution and Reengineering (SANER),* 231-240. 10.1109/SANER.2015.7081833

Cao, F., Huang, H. K., & Zhou, X. Q. (2003). Medical image security in a HIPAA mandated PACS environment. *Computerized Medical Imaging and Graphics, 27*(2), 185–196. doi:10.1016/S0895-6111(02)00073-3 PMID:12620309

Carmagnola, F., Cena, F., & Gena, C. (2011). User model interoperability: A survey. *User Modeling and User-Adapted Interaction, 21*(3), 285–331. doi:10.100711257-011-9097-5

Castro-Herrera, C., Duan, C., Cleland-Huang, J., & Mobasher, B. (2009). A recommender system for requirements elicitation in large-scale software projects. *Proceedings of the 2009 ACM symposium on Applied Computing,* 1419-1426, 10.1145/1529282.1529601

Catteddu, D. (2010). Cloud Computing: benefits, risks and recommendations for information security. In Web Application Security (pp. 17-17). Springer Berlin Heidelberg. doi:10.1007/978-3-642-16120-9_9

Ceccarelli, M., Cerulo, L., Canfora, G., & Penta, M. D. (2010). An Eclectic Approach for Change Impact Analysis. In *Proceedings of International Conference on Software Engineering* (pp. 163-166). Academic Press. 10.1145/1810295.1810320

Cesarotti, V., Giuiusa, A., Kwan, S. K., Introna, V., & Spohrer, J. (2014). Designing Multichannel Value Propositions to Enhance Value-Cocreation Phenomenon. *International Journal of Service Science, Management, Engineering, and Technology, 5*(1), 14–44. doi:10.4018/ijssmet.2014010102

Chabowski, B., Hult, T., & Mena, J. (2011). The retailing literature as a basis for franchising research: Using intellectual structure to advance theory. *Journal of Retailing, 87*(3), 269–284. doi:10.1016/j.jretai.2011.04.001

Chaisiri, S., Bu-Sung, L., & Niyato, D. (2012). Optimization of Resource Provisioning Cost in Cloud Computing. *IEEE Transactions on Services Computing, 5*(2), 164–177. doi:10.1109/TSC.2011.7

Chamiel, G., & Pagnucco, M. (2009).Ontology Guided Dynamic Preference Elicitation, *Proceedings of 3rd ACM Conference on Recommender Systems & the Social Web,* 41-48.

Chancey, E., Flores, M., Luis, J., Báez Palma, M., Badillo Valenzuela, O., Cabrera, S., & Arlethe, M. (2016). Redesign of the supply chain of a restaurant franchise in the food industry. *Global Journal of Business Research, 10*(2), 103–111.

Chaudhuri, S. A. M., & Gravano, L. (2004). Optimizing top-k selection queries over multimedia repositories. *IEEE Transactions on Knowledge and Data Engineering,* 16.

Chellappa, R. K., & Sin, R. G. (2005). Personalization versus privacy: An empirical examination of the online consumer's dilemma. *Information Technology Management, 6*(2-3), 181–202. doi:10.100710799-005-5879-y

Chellatamilan, T., & Suresh, R. M. (2011). Intelligent agents for the semantic adaptive e-learning system. *International Journal of Computer Applications ICACT, 1*(1), 1–5.

Chen, D., & Zhao, H. (2012, March). Data security and privacy protection issues in cloud computing. In *Computer Science and Electronics Engineering (ICCSEE), International Conference on* (Vol. 1, pp. 647-651). IEEE. 10.1109/ICCSEE.2012.193

Cheng, J. H., Chen, M. C., & Huang, C. M. (2014). Assessing Inter-Organizational Innovation Performance through Relational Governance and Dynamic Capabilities in Supply Chains. *Supply Chain Management, 19*(2), 173–186. doi:10.1108/SCM-05-2013-0162

Chen, H., Wu, Z., & Cudré-Mauroux, P. (2012). Semantic web meets computational intelligence: State of the art and perspectives. *IEEE Computational Intelligence Magazine, 7*(2), 67–74. doi:10.1109/MCI.2012.2188592

Chen, P. M., Lee, E. K., Gibson, G. A., Katz, R. H., & Patterson, D. A. (1994). RAID: High-performance, reliable secondary storage. *ACM Computing Surveys, 26*(2), 145–185. doi:10.1145/176979.176981

Chen, Y.-S. (2018). Utilizing information science and technology in franchise organizations. In M. Khosrow-Pour (Ed.), *Encyclopedia of information science and technology* (4th ed.; pp. 4822–4835). Hershey, PA: IGI Global. doi:10.4018/978-1-5225-2255-3.ch418

Chesbrough, H. (2003). *Open Innovation: The New Imperative for Creating and Profiting from Technology*. Harvard Business Press.

Chesbrough, H. (2010). *Open services innovation: Rethinking your business to grow and compete in a new era*. John Wiley & Sons.

Chetna Gupta & Chauhan. (2010a). DU-Regs: Online Dynamic Approach to Visualize Impact Analysis for Regression Testing. *International Journal of Computer Applications, 1*, 8-11.

Chetna Gupta & Chauhan. (2010b). A Dynamic Approach to Estimate Change Impact using Type of Change Propagation. *Journal of Information Processing Systems, 6*, 597-608.

Chou, I. (2011). Secure software configuration management processes for nuclear safety software development environment. *Ann. Nuclear Energy, 38*(10), 2174–2179.

Chowdhuri, S., Chakraborty, S., Dey, N., Azar, A. T., Salem, M. A. M. M., Chaudhury, S. S., & Banerjee, P. (2014). Recent Research on Multi Input Multi Output (MIMO) based Mobile ad hoc Network: A Review. *International Journal of Service Science, Management, Engineering, and Technology, 5*(3), 54–65. doi:10.4018/ijssmet.2014070104

Christopher, M. (2016). *Logistics and supply chain management* (5th ed.). Harlow: Pearson Education.

Ciaccia, P., & M. P. (2002). Searching in metric spaces with userdefined and approximate distances. *ACM Transactions on Database Systems*, 27.

Cimatti, A., Clarke, E. M., Giunchiglia, F., & Roveri, M. (1999). Nusmv: A new symbolic model verifier. *Proceedings of the 11th International Conference on Computer Aided Verification, CAV '99*, 495-499.

Cimitile, A., Fasolino, A. R., & Visaggio, G. (1999). A software model for impact analysis: a validation experiment. In *Proceedings of the IEEE Sixth Working Conference on Reverse Engineering* (pp.212-222). IEEE. 10.1109/WCRE.1999.806962

Cindy Marling, J. S., & Schwartz, F. (2008). Case-Based Decision Support for Patients with Type 1 Diabetes on Insulin Pump Therapy. *Advances in Case-Based Reasoning: 9th European Conference, ECCBR 2008*, 325. 10.1007/978-3-540-85502-6_22

Ciss'ee, R., & Albayrak, S. (2007). An agent-based approach for privacy-preserving recommender systems. *Proceedings of the 6th international joint conference on autonomous agents and multi-agent systems ACM New York USA*. 10.1145/1329125.1329345

Clohessy, T., Acton, T., Morgan, L., & Conboy, K. (2016). *The Times They are A-Chaning for ICT Service Provision: A Cloud Computing Business Model Perspective*. Paper presented at the 23rd European Conference on Information Systems, Istanbul, Turkey.

Cohen, W. M., & Levinthal, D. A. (1990). Absorptive Capacity: A New Perspective on Learning and Innovation. *Administrative Science Quarterly*, *35*(1), 128–152. doi:10.2307/2393553

Coleman, J. S. (1988). Social capital in the creation of human capital. *American Journal of Sociology*, *94*, S95–S120. doi:10.1086/228943

Combs, J., Ketchen, D. Jr, Shook, C., & Short, J. (2011). Antecedents and consequences of franchising: Past accomplishments and futures challenges. *Journal of Management*, *37*(1), 99–126. doi:10.1177/0149206310386963

Conchúir, Holmström, & Fitzgerald. (2009). Global Software Development: Never Mind the Problems – Where are the Benefits. *Communications of the ACM*, 52.

Conchúir. Ågerfalk, & Fitzgerald. (2006). Global Software Development: Never Mind the Problems – Are There Really Any Benefits? *29th Information Systems Research Seminar in Scandinavia (IRIS)*.

Cordobés Madueño, M., & Soldevila García, P. (2015). Management control in inter-organizational relationships: The case of franchises. *Rivista Innovar*, *25*(58), 23–36. doi:10.15446/innovar.v25n58.52357

Costa, V., & Monteiro, S. (2016). Key knowledge management processes for innovation: A systematic literature review. *Journal of Information & Knowledge Management Systems*, *46*(3), 386–410. doi:10.1108/VJIKMS-02-2015-0017

Cox, A. (1996). Relational competence and strategic procurement management: Towards an entrepreneurial and contractual theory of the firm. *European Journal of Purchasing & Supply Management*, *2*(1), 57–70. doi:10.1016/0969-7012(95)00019-4

Croonen, E., & Brand, M. (2013). What makes franchisees trust their franchisors? In T. Ehrmannn, J. Windsperger, G. Cliquet, & G. Hendrikse (Eds.), *Network governance: alliances, cooperatives and franchise chains* (pp. 207–229). Heidelberg, Germany: Physica-Verlag. doi:10.1007/978-3-7908-2867-2_12

Dai, H., & Mobasher, B. (2003). A road map to more effective web personalization: integrating domain knowledge with web usage mining. *Proceedings of the International Conference on Internet Computing*, 1-8.

Damian, D. Z. (2003). Requirements Engineering challenges in multi-site software development organizations. *Requirements Engineering Journal*, *8*, 149–160. doi:10.100700766-003-0173-1

Damian, D. (2006). Requirements Engineering in Distributed Projects. *IEEE International Conference on Global Software Engineering (ICGSE'06)*. 10.1109/ICGSE.2006.261218

Danesh. (2009). Requirements prioritization in on-line banking systems: using value-oriented framework. *2009 International Conference on Computer Technology and Development*. 10.1109/ICCTD.2009.41

Dapp, T. F. (2014). *Fintech - The digital (r)evolution in the financial sector: Algorithm-based banking with the human touch*. Retrieved 2015/1/6 from https://www.dbresearch.com/PROD/DBR_INTERNET_EN-PROD/PROD0000000000345837.pdf

David, B., & Leake. (1996). *Case-Based Reasoning: Experiences, lessons and future directions* (AAAI Press). MIT Press. Retrieved from http://dl.acm.org/citation.cfm?id=524680

David, C., & Yang, J. (2005). Software Assurance by Bounded Exhaustive Testing. *IEEE Transactions on Software Engineering*, *31*(4), 328–339. doi:10.1109/TSE.2005.52

Davis, E., & Serna Molina, J. (2011). *Great Harvest Bread Company: leveraging the supply chain for competitive advantage (Darden Case No. UVA-OM-1437)*. Charlottesville, VA: Darden Graduate School of Business Administration, University of Virginia.

De Castro, L., Mota, J., & Marnoto, S. (2009). Toward a relational perspective of franchising chains. *Service Business*, *3*(1), 15–30. doi:10.100711628-008-0050-6

de la Iglesia, J. L. M., & Gayo, J. E. L. (2009). *Doing business by selling free services. In Web 2.0* (pp. 1–14). Springer.

De la Vara, J., Borg, M., Wnuk, K., & Moonen, L. (2016). An industrial survey of safety evidence change impact analysis practice. *IEEE Transactions on Software Engineering*, *42*(12), 1095–1117. doi:10.1109/TSE.2016.2553032

Decker, G., Overdick, H., & Zaha, J.M. (2006). On the suitability of ws-cdl for choreography modeling. In *Proceedings of Methoden, Konzepte Und Technologien Fur Die Entwicklung Von Dienstebasierten Informationssystemen.*

Dennis, F. S., & Chang, D. J. (2006). Modular Verification of Code with Sat. *ISSTA-2006*, 1-11.

Desikan, P., & Srivastava, J. (2006). *Mining Temporally Changing Web Usage Graphs* (Vol. 3932). Advances in Web Mining and Web Usage Analysis Lecture Notes in Computer Science; doi:10.1007/11899402_1

Devamalar, P. M. B., Bai, V. T., & Srivatsa, S. (2007). An Architecture for a Fully Automated RealTime Web-Centric Expert System. *World Academy of Science, Engineering and Technology*, *13*. Retrieved from https://static.aminer.org/pdf/PDF/000/262/917/a_framework_for_the_delivery_of_web_centric_knowledge_management.pdf

Dhavachelvan, P., & Uma, G. V. (2003). Multi-agent based integrated framework for intra-class testing of object-oriented software. *LNCS, 2869*.

Di Martino, B., Cretella, G., & Esposito, A. (2017). Legislation-Aware Cloud Computing: An Overview. In Reshaping Accounting and Management Control Systems (pp. 77-87). Springer.

Di Martino, B., Cretella, G., & Esposito, A. (2017). Cloud services composition through cloud patterns: A semantic-based approach. *Soft Computing*, *21*(16), 4557–4570. doi:10.100700500-016-2264-1

Diaz, G., Cambronero, M. E., Pardo, J. J., Valero, V., & Cuartero, O. (2007). *Model Checking Techniques applied to the design of Web Services*. Academic Press.

Dorigo, M., Maniezzo, V., & Colorni, A. (1996). The Ant System: Optimization by a Colony of Cooperating Agents. *IEEE Transactions on Systems, Man, and Cybernetics. Part B, Cybernetics*, *26*(1), 29–41. doi:10.1109/3477.484436 PMID:18263004

Doulkeridis, C., Loutas, N., & Vazirgiannis, M. (2006, January). A system architecture for context aware service discovery. *Electronic Notes in Theoretical Computer Science*, *146*(1), 101–116. doi:10.1016/j.entcs.2005.11.010

Duan, D., Li, Y., Jin, Y., & Lu, Z. (2009). Community Mining on Dynamic Weighted Directed Graphs. *Proceeding of the 1st ACM international workshop on Complex networks meet, information & knowledge management*, 11-18. 10.1145/1651274.1651278

Duncan, R. B. (1976). The ambidextrous organization: Designing dual structures for innovation. *The Management of Organization, 1*, 167-188.

Dupuis, M. (1986). *Distribution, la nouvelle donne*. Paris: Les Editions d'Organisation.

Dyer, J., & Nobeoka, K. (2000). Creating and managing a high-performance knowledge-sharing network: The Toyota case. *Strategic Management Journal*, *21*(3), 345–367. doi:

Ebert, C. (2001). Metrics for identifying critical components in software projects. In *Handbook of Software Engineering and Knowledge Engineering* (Vol. 1). World Scientific Publishers. doi:10.1142/9789812389718_0017

Ehlert, P. (2003). *Intelligent user interfaces: introduction and survey*. Research Report DKS03-01/ICE 01. Faculty of information technology and systems, Delft University of Technology.

Eirinaki, M., & Vazirgiannis, M. (2003). Web mining for web personalization. *ACM T Internet Tech.*, *3*(1), 1–27. doi:10.1145/643477.643478

Eirinaki, M., Vazirgiannis, M., & Varlamis, I. (2003). SEWeP: Using Site Semantics and a Taxonomy to Enhance the Web Personalization Process. *Proceeding of SIGKDD '03*, 99-108. 10.1145/956750.956765

Eisenhardt, K. M. (1989). Building theories from case study research. *Academy of Management Review*, *14*(4), 532–550.

Eisenhardt, K. M., & Graebner, M. E. (2007). Theory building from cases: Opportunities and challenges. *Academy of Management Journal*, *50*(1), 25–32. doi:10.5465/AMJ.2007.24160888

Eisenmann, T., Parker, G., & Van Alstyne, M. W. (2006). Strategies for two-sided markets. *Harvard Business Review*, *84*(10), 92. PMID:16649701

El Akremi, A., Herrbach, O., Mignonac, K., & Perrigot, R. (2009). *Quels sont les savoir-faire sources de l'avantage concurrentiel en franchise?* Paris: Fédération Française de la Franchise.

Elango, D. M., Fowley, F., & Pahl, C. (2017). Pattern-driven Architecting of an Adaptable Ontology-driven Cloud Storage Broker. University of Oslo.

Elango, D. M., Fowley, F., & Pahl, C. (2018 To appear). An Ontology-based Architecture for an Adaptable Cloud Storage Broker. In *Advances in Service-Oriented and Cloud Computing*. Springer CCIS.

Elango, D. M., Fowley, F., & Pahl, C. (2018 To appear). Testing and Comparing the Performance of Cloud Service Providers using a Service Broker Architecture. In *Advances in Service-Oriented and Cloud Computing*. Springer CCIS.

Emeakaroha, V., Brandic, I., Maurer, M., & Dustdar, S. (2010). *Low level Metrics to High level SLAs - LoM2HiS framework: Bridging the gap between monitored metrics and SLA parameters in cloud environments. HPCS*. IEEE.

Emeakaroha, V., Netto, M., Calheiros, R., Brandic, I., Rajkumar, B., & De Rose, C. (2012). Towards autonomic detection of SLA violations in Cloud infrastructures. *Future Generation Computer Systems*, *28*(7), 1017–1029. doi:10.1016/j.future.2011.08.018

EU Commission (2010). Report on The Future of Cloud Computing - Opportunities for European Cloud Computing Beyond 2010. European Commission.

Fagin, R. M. N., & Lotem, J. (2003). Optimal aggregation algorithms for middleware. *Journal of Computer and System Sciences*, 66.

Fathian, M., Amiri, B., & Maroosi, A. (2007). Babak Amiri and Ali Maroosi, "Application of honey-bee mating optimization algorithm on clustering. *Applied Mathematics and Computation*, *190*(2), 1502–1513. doi:10.1016/j.amc.2007.02.029

Faust, K. (2006). Comparing Social Networks: Size, Density, and Local Structure. *Metodološki Zvezki*, *3*(2), 185–216.

Favre-Bonte, V., Elodie G. & Thevenard-Puthod, C. (2013). Identifying different forms of innovation in retail banking. *International Journal of Service Science, Management, Engineering, and Technology, 4*(4), 43-57, doi: 10.4018/ijssmet.2013100103

Fernández-Tobías, I., Kaminskas, M., Cantador, I., & Ricci, F. (2011). A generic semantic-based framework for cross-domain recommendation. *Proceedings of HetRec '11 Chicago USA*, 25-32. 10.1145/2039320.2039324

Fernie, J., & Sparks, L. (Eds.). (2014). *Logistics and retail management: emerging issues and new challenges in the retail supply chain* (4th ed.). London: Kogan Page.

Fingar, P. (2009). Cloud computing and the promise of on-demand business innovation. *Intelligent Enterprise.*

Fischer, R., Janiesch, C., Strach, J., Bieber, N., Zink, W., & Tai, S. (2013). *Eine Bestandsaufnahme von Standardisierungspotentialen und -lücken im Cloud Computing.* Paper presented at the 11. Internationale Tagung Wirtschaftsinformatik, Leipzig, Germany.

Floerecke, S., & Lehner, F. (2016). *Cloud Computing Ecosystem Model: Refinement and Evaluation.* Paper presented at the 23rd European Conference on Information Systems, Istanbul, Turkey.

Floerecke, S., Herzfeldt, A., & Krcmar, H. (2012). Risiken bei IT-Lösungen – Ein Risikokatalog für Praktiker aus Anbietersicht. *IM – Die Fachzeitschrift für Information Management und Consulting, 27*(4), 22-30.

Floerecke, S., Wolfenstetter, T., & Krcmar, H. (2015). Hybride Produkte – Stand der Literatur und Umsetzung in der Praxis. *IM+io – Magazin für Innovation. Organisation und Management, 30*(2), 61–66.

Foltz, P. W., & Dumais, S. T. (1992). Personalized information delivery: An analysis of information filtering methods. *Communications of the ACM, 35*(12), 51–60. doi:10.1145/138859.138866

Fong, A. C. M., Zhou, B., Hui Siu, C., Tang, J., & Hong, G. Y. (2012). Generation of personalized ontology based on consumer emotion and behavior analysis. *IEEE Transactions on Effective Computing, 3*(1), 152–164. doi:10.1109/T-AFFC.2011.22

Foster, H., Uchitel, S., Magee, J., & Kramer, J. (2004). Compatibility verification for web service choreography. *Proceedings of the IEEE International Conference on Web Services, ICWS '04*, 738. 10.1109/ICWS.2004.1314806

Fournier-Viger, P., Lin, J. C.-W., Kiran, R. U., Koh, Y. S., & Thomas, R. (2017). A Survey of Sequential Pattern Mining. *Data Science and Pattern Recognition, 1*(1), 54–77.

Fowley, F., Pahl, C., Jamshidi, P., Fang, D., & Liu, X. (2017). A Classification and Comparison Framework for Cloud Service Brokerage Architectures. IEEE Transactions on Cloud Computing.

Fowley, F., Pahl, C., & Zhang, L. (2013). A Comparison Framework and Review of Service Brokerage Solutions for Cloud Architectures. *1st International Workshop on Cloud Service Brokerage.*

Fox, J., Johns, N., Lyons, C., Rahmanzadeh, A., Thomson, R., & Wilson, P. (1997). *PROforma: A general technology for clinical decision support systems, 54 Computer Methods and Programs in Biomedicine.* Elsevier. doi:10.1016/S0169-2607(97)00034-5

Frikha, M., Mhiri, M., & Gargouri, F. (2014). Toward a User Interest Ontology to Improve Social Network-Based Recommender System. In J. Sobecki, V. Boonjing, & S. Chittayasothorn (Eds.), *Advanced Approaches to Intelligent Information and Database Systems* (Vol. 551, pp. 255–264). Springer International Publishing. doi:10.1007/978-3-319-05503-9_25

Fristedt, M., Hansson, A., Huge-Brodin, M., Rehme, J., & Sandberg, E. (2012). *Supply chain management in practice. A case study of McDonald's Sweden* (Report LIU-IEI-R-12/0006). Linköping: Department of Management and Engineering, Linköping University.

Fu, B., Brennan, R., & O'Sullivan, D. (2009). Multilingual Ontology Mapping: Challenges and a Proposed Framework. *Proceedings of the 23rd Convention of the Society for the Study of Artifitial Intelligence and Simulation of Behaviour - Adaptive & Emergent Behaviour & Complex Systems*, 32 - 35.

Fujii, K., & Suda, T. (2009). Semantics-based context-aware dynamic service composition. *ACM Transactions on Autonomous and Adaptive Systems*, *4*(2), 1–31. doi:10.1145/1516533.1516536

Fulconis, F., Paché, G., & Roveillo, G. (2011). *La prestation logistique: origines, enjeux et perspectives*. Caen: Editions Management & Société.

Gaasterland, T. J. M., & Godfrey, P. (1992). An overview of cooperative answering. *Journal of Intelligent Systems*, *1*, 123157.

Gacitua-Decar, V., & Pahl, C. (2017). Structural Process Pattern Matching Based on Graph Morphism Detection. *International Journal of Software Engineering and Knowledge Engineering*, *27*(2), 153–189. doi:10.1142/S0218194017500073

Gagnon, Y.-C. (2012). *L'étude de cas comme méthode de recherche* (2nd ed.). Québec: Presses de l'Université du Québec.

Galindo, J. (2005). New characteristics in fsql, a fuzzy sql for fuzzy databases. *WSEAS Transactions on Information Science and Applications*, *2*(2), 161–169.

Galindo, J. M. P., & Urrutia, A. (2006). *Fuzzy databases: modeling, design and implementation*. Idea Group Publishing Hershey. doi:10.4018/978-1-59140-324 1

García-Valls, M., Cucinotta, T., & Lu, C. (2014). Challenges in real-time virtualization and predictable cloud computing. *Journal of Systems Architecture*, *60*(9), 726–740. doi:10.1016/j.sysarc.2014.07.004

Garfinkel, T., & Rosenblum, M. (2005, May). When Virtual Is Harder than Real: Security Challenges in Virtual Machine Based Computing Environments. In *Procedding of the 10th Conference on Hot Topics in Operating System* (Vol. 10, pp. 20-20). ACM.

Garfinkel, T., Pfaff, B., Chow, J., Rosenblum, M., & Boneh, D. (2003, October). Terra: A virtual machine-based platform for trusted computing. *Operating Systems Review*, *37*(5), 193–206. doi:10.1145/1165389.945464

Gavalas, D., & Kenteris, M. (2011). A web-based pervasive recommendation system for mobile tourist guides. *Personal and Ubiquitous Computing*, *15*(7), 759–770. doi:10.100700779-011-0389-x

Gawer, A. (2014). Bridging differing perspectives on technological platforms: Toward an integrative framework. *Research Policy*, *43*(7), 1239–1249. doi:10.1016/j.respol.2014.03.006

Gawer, A., & Cusumano, M. A. (2002). *Platform leadership: How Intel, Microsoft, and Cisco drive industry innovation*. Harvard Business School Press Boston.

Gawer, A., & Cusumano, M. A. (2008). How companies become platform leaders. *MIT Sloan Management Review*, *49*(2), 28.

Geer, D., Bustard, D., & Sunazuka, T. (1999). Prioritization of system changes using cost-benefit and risks assessments. *Proceedings of Fourth IEEE International Symposium on Requirements Engineering*. 10.1109/ISRE.1999.777998

Gegick, M., Rotella, P., & Williams, L. (2009). Predicting Attack-prone Components. *2009 International Conference on Software Testing Verification and Validation*, 181-190. 10.1109/ICST.2009.36

Geguang, P., Jianqi, S., Zheng, W., Lu, J., Jing, L., & Jifeng, H. (2007). The Validation and Verification of WSCDL. *Asia-Pacific Software Engineering Conference*, 81-88.

Genero, L. M., Piattini, M., & Calero, C. (2000). Early measures for UML class diagrams. *L'Objet*, *6*, 2000.

Genoves, S., & Akamai. (2009). *Cloud basedfire wall.* Retrived from http://cloudcomputing.sys-con.com/node/1219023

Ghorab, M. R., Zhou, D., O'Connor, A., & Wade, V. (2013). Personalised information retrieval: Survey and classification. *User Modeling and User-Adapted Interaction*, *23*(4), 381–443. doi:10.100711257-012-9124-1

Gibson, C. B., & Birkinshaw, J. (2004). The antecedents, consequences, and mediating role of organizational ambidexterity. *Academy of Management Journal*, *47*(2), 209–226. doi:10.2307/20159573

Gim, J., Jung, H., & Jeong, D. (2015). XOnto-Apriori: An Effective Association Rule Mining Algorithm for Personalized Recommendation Systems. In Computer Science and its Applications(pp 1131-1138). Berlin: Springer-Verlag.

Gonzalez, N., Miers, C., Redígolo, F., Simplicio, M., Carvalho, T., Naslund, M., & Pourzandi, M. (2012). A quantitative analysis of current security concerns and solutions for cloud computing. *Journal of Cloud Computing*, *1*(1), 1–18.

Google Apps Service Level Agreement. (2010, November). Retrieved from http://www.google.com/apps/intl/en/terms/sla.html

Gopalachari, M. V., & Sammulal, P. (2014). Personalized collaborative filtering recommender system using domain knowledge. *Proceedings of International Conference on Computer and Communications Technologies (ICCCT)*, 1066-1071. 10.1109/ICCCT2.2014.7066693

Grace, A., Frazer, L., Weaven, S., & Dant, R. (2016). Building franchisee trust in their franchisor: Insights from the franchise sector. *Qualitative Market Research*, *19*(1), 65–83. doi:10.1108/QMR-09-2014-0085

Granovetter, M. (1983). The strength of weak ties: A network theory revisited. *Sociological Theory*, *1*(1), 201–233. doi:10.2307/202051

Grant, R. (1991). The resource-based theory of competitive advantage: Implications for strategy formulation. *California Management Review*, *33*(3), 114–135. doi:10.2307/41166664

Grimm, K. (2003). Software technology in an automotive company: Major challenges. *Proceedings of 25th International Conference Software. Engineering*, 498–503.

Grissa-Touzi, A. & H.O. (2012). A new approach for top-k flexible queries in large database using the knowledge discovered. *The Fourth International Conference on Advances in Databases, Knowledge, and Data Applications.*

Grissa-Touzi, A. (2010). An Alternative Extension of the FCM Algorithm for Clustering Fuzzy Databases. *The Second International Conference on Advances in Databases, Knowledge, and Data Applications*, 135-142. 10.1109/DBKDA.2010.35

Grönroos, C. (1994). From scientific management to service management: A management perspective for the age of service competition. *International Journal of Service Industry Management*, *5*(1), 5–20. doi:10.1108/09564239410051885

Grönroos, C. (2011). Value Co-Creation in Service Logic: A Critical Analysis. *Marketing Theory*, *11*(3), 279–301. doi:10.1177/1470593111408177

Grönroos, C., & Ravald, A. (2011). Service as Business Logic: Implications for Value Creation and Marketing. *Journal of Service Management*, *22*(1), 5–22. doi:10.1108/09564231111106893

Grönroos, C., & Voima, P. (2013). Critical Service Logic: Making Sense of Value Creation and Co-Creation. *Journal of the Academy of Marketing Science*, *41*(2), 133–150. doi:10.100711747-012-0308-3

Grozev, N., & Buyya, R. (2012). InterCloud architectures and application brokering: taxonomy and survey. Software: Practice and Experience.

Gruber, R. T. (1993). Toward principles for the design of ontologies used for knowledge sharing. *International Journal of Human-Computer Studies*, *43*(5-6), 907–928. doi:10.1006/ijhc.1995.1081

Gruman. (2008). *Early experiments in cloud computing*. Retrieved from http://www.infoworld.com/article/08/04/07/15FE-cloud-computing-utility_1.html

Guermouche, N., Perrin, O., & Ringeissen, C. (2008). Timed Specification For Web Services Compatibility Analysis. *Electronic Notes in Theoretical Computer Science*, *200*(3), 155–170. doi:10.1016/j.entcs.2008.04.098

Gupta, A., Thakur, K.H., Garg, A., & Garg, D. (2016). Mining and Analysis of Periodic Patterns in Weighted Directed Dynamic Network. *International Journal of Service Science, Management, Engineering, and Technology*, *7*. doi: 10.4018/IJSSMET.2016010101

Gupta, C., Singh, Y., & Chauhan, D.S. (2009). An Efficient Dynamic Impact Analysis using Definition and Usage Information. *International Journal of Digital Content Technology and its Applications*, *3*(4), 112-115.

Gupta, V., & Chauhan, D. S. (2011c). Hybrid Regression Testing Technique: A Multi Layered Approach. In *IEEE annual Conference "INDICON"*. Hyderabad, India: IEEE. 10.1109/INDCON.2011.6139363

Gupta, A., Thakur, K. H., & Kishore, P. (2014). Mining Regular Patterns in Weighted Directed Networks. *Proceedings of 13th International Conference Of Information Technology (ICIT '14)*, 215-220. 10.1109/ICIT.2014.44

Gupta, C., Singh, Y., & Chauhan, D. S. (2010). A Dynamic Approach to Estimate Change Impact using Type of Change Propagation. *International Journal of Information Processing Systems*, *6*(4), 597–608. doi:10.3745/JIPS.2010.6.4.597

Gupta, C., Singh, Y., & Chauhan, D. S. (2010). Dependency based Process Model for Impact Analysis: A Requirement Engineering Perspective. *International Journal of Computers and Applications*, *6*(6), 28–33. doi:10.5120/1079-1410

Gupta, C., Singh, Y., Chauhan, D. S., & Srivastav, M. (2010). DU-Regs: Online Dynamic Approach to Visualize Impact Analysis for Regression Testing. *International Journal of Computers and Applications*, *1*(19), 8–11. doi:10.5120/411-609

Gupta, C., Srivastav, M., & Gupta, V. (2015). Software Change Impact Analysis: An Approach to Differentiate Type of Change to Minimize Regression Test Selection. *Special Issue on Software Architecture, Evaluation and Testing for Emerging Paradigms. International Journal of Computer Applications in Technology*, *51*(4), 1–10. doi:10.1504/IJCAT.2015.070498

Gupta, V., Chauhan, D. S., & Dutta, K. (2012a). *Decision Aspect Prioritization Technique for Globally Distributed Developments: A Hybrid Approach*. Elsevier.

Gupta, V., Chauhan, D. S., & Dutta, K. (2012b). Impact Analysis of Requirement Prioritization on Regression Testing. *AWER Procedia Information Technology & Computer Science*, *2*, 379–383.

Gupta, V., & Srivastav, M. (2011a). Web based tool supported requirement prioritization: Based on multiple stakeholder preferences. *International Journal on Computer Engineering and Information Technology*, *25*(1), 12–19.

Gupta, V., & Srivastav, M. (2011b). *Overcoming Global Distance in Globally Distributed Developments by Web Based Framework*. CiiT International Journal of Software Engineering and Technology.

Gyimothy, T., Ferenc, R., & Siket, I. (2005, October). TiborGyimo´thy, Rudolf Ferenc, and Istva'nSiket (2005), "Empirical Validation of Object-Oriented Metrics on Open Source Software for Fault Prediction. *IEEE Transactions on Software Engineering*, *31*(10), 897–910. doi:10.1109/TSE.2005.112

Haag, S., & Eckhardt, A. (2014). Organizational Cloud Service Adoption: A Scientometric and Content-Based Literature Analysis. *Journal of Business Economics*, *84*(3), 407–444. doi:10.100711573-014-0716-6

Hagiu, A. (2009). *Multi-sided platforms: From microfoundations to design and expansion strategies*. Harvard Business School Strategy Unit Working Paper (09-115).

Hagiu, A. (2007). Merchant or two-sided platform? *Review of Network Economics*, *6*(2). doi:10.2202/1446-9022.1113

Halder, S., Han, Y., & Lee, Y. (2013). Discovering Periodic Patterns using Supergraph in Dynamic Networks. *Research Notes in Information Science*, *14*, 148–151.

Han, J., Dong, G., & Yin, Y. (1999). Efficient mining of partial periodic patterns in time series database. *Proceedings 15th International Conference on Data Engineering*, 106-115.

Hariri, N., Mobasher, B., & Burke, R. (2013). Personalized text-based music retrieval. *Intelligent Techniques for Web Personalization and Recommendation: Papers from the AAAI 2013 Workshop*, 24-30.

Hasan, M., Mahmood, A., Alam, M., Hasan, M., & Rahman, F. (2010). An Evaluation of Software Requirement Prioritization Techniques. *International Journal of Computer Science and Information Security*, *8*(9), 83–94.

Hashizume, K., Rosado, D. G., Fernández-Medina, E., & Fernandez, E. B. (2013). An analysis of security issues for cloud computing. *Journal of Internet Services and Applications*, *4*(1), 1–13. doi:10.1186/1869-0238-4-5

Hassanien, A. E., Azar, A. T., Snasel, V., Kacprzyk, J., & Abawajy, J. H. (2015). *Big Data in Complex Systems: Challenges and Opportunities, Studies in Big Data* (Vol. 9). Springer-Verlag GmbH Berlin/Heidelberg.

Hassanien, A. E., Azar, A. T., Snasel, V., Kacprzyk, J., & Abawajy, J. H. (2015). Big Data in Complex Systems: Challenges and Opportunities, Studies in Big Data. Springer-Verlag GmbH. doi:10.1007/978-3-319-11056-1

Hawalah, A., & Fasli, M. (2011). A multi-agent system using ontological user profiles for dynamic user modeling. *IEEE/WIC/ACM International Conferences on Web Intelligence and Intelligent Agent Technology*, 430-437. doi:10.1109/WI-IAT.2011

Hawalah, A., & Fasli, M. (2015). Dynamic user profiles for web personalisation. *Expert Systems with Applications*, *42*(5), 2547–2569. doi:10.1016/j.eswa.2014.10.032

Heinrich, R., van Hoorn, A., Knoche, H., Li, F., Lwakatare, L. E., Pahl, C., . . . Wettinger, J. (2017). Performance Engineering for Microservices: Research Challenges and Directions. *Proceedings of the 8th ACM/SPEC on International Conference on Performance Engineering Companion.*

Hellerstein, J. L., Ma, S., & Perng, C. S. (2002). Discovering actionable patterns in event data. *IBM Systems Journal*, *41*(3), 475–493. doi:10.1147j.413.0475

Helou, M. A., Palmonari, M., & Jarrar, M. (2016). Effectiveness of Automatic Translations for Cross-Lingual Ontology Mapping. [JAIR]. *Journal of Artificial Intelligence Research*, *55*, 165–208.

Henzinger, T.A., Nicollin, X., Sifakis, J., & Yovine, S. (1994). Symbolic Model Checking for Real-Time Systems. *Information and Computation, 111*(2), 193–244.

Herzfeldt, A., Schermann, M., & Krcmar, H. (2010). *Towards A Set of Requirements for A Holistic IT Solution Engineering Approach.* Paper presented at the 21st Australasian Conference on Information Systems, Brisbane, Australia.

Higashino, W. (2016). *Complex Event Processing as a Service in MultiCloud Environments.* The University of Western Ontario.

Historical Data for S&P 500 Stocks. (n.d.). Retrieved from http://pages.swcp.com/stocks/#historical%20data

Hjaltason, G.R., & H. S. (2003). Index-driven similarity search in metric spaces. *ACM Transactions on Database Systems*, 28.

Hoberg, P., Wollersheim, J., Böhm, M., & Krcmar, H. (2012). Cloud Computing – Überblick und Herausforderungen für das Controlling. *Controlling – Zeitschrift für erfolgsorientierte Unternehmensführung*, *24*(6), 294-300.

Hong, J., Suh, E., Kim, J., & Kim, S. (2009). Context-aware system for proactive personalized service based on context history. *Expert Systems with Applications*, *36*(1), 7448–7457. doi:10.1016/j.eswa.2008.09.002

Hongli, Y., Xiangpeng, Z., Zongyan, Q., Geguang, P., & Shuling, W. (2006). A formal model of web service choreography description language (WS-CDL). *International Conference on Web Services.*

Hoppe, A., Nicolle, C., & Roxin, A. (2013). Automatic ontology based user profile learning from heterogeneous web resources in a big data context. *Proceedings of the VLDB Endowment International Conference on Very Large Data Bases*, *6*(12), 1428–1433. doi:10.14778/2536274.2536330

Horgan, J., London, S., & Lyu, M. (1994). Achieving Software Quality with Testing Coverage Measures. *IEEE Computer*, *27*(9), 60–69. doi:10.1109/2.312032

Horwitz, S., Reps, T., & Binkley, D. (1990). Interprocedural Slicing Using Dependence Graphs. *ACM Transactions on Programming Languages and Systems*, *12*(1), 27–60. doi:10.1145/77606.77608

Huang, L., & Song, Y. T. (2006). Dynamic Impact Analysis Using Execution Profile Tracing. In *Proceedings of the Fourth International Conference on Software Engineering Research, Management and Applications* (pp. 237-243). Academic Press. 10.1109/SERA.2006.30

Huber, G. P. (1991). Organizational Learning: The Contributing Processes and the Literatures. *Organization Science*, 2(1), 88–115. doi:10.1287/orsc.2.1.88

Hulme, G. V. (2011). *Cloud passsge aims to ease cloud server security management.* Available from http://www.csoonline.com/artical/658121/cloudpassage-aims-to-ease-cloud-server-security-management

Hult, G., Ketchen, D. Jr, & Nichols, E. Jr. (2003). Organizational learning as a strategic resource in supply management. *Journal of Operations Management*, *21*(5), 541–556. doi:10.1016/j.jom.2003.02.001

Hunt, S., & Davis, D. (2008). Grounding supply chain management in resource-advantage theory. *The Journal of Supply Chain Management*, *44*(1), 10–21. doi:10.1111/j.1745-493X.2008.00042.x

Hunt, S., & Davis, D. (2012). Grounding supply chain management in resource-advantage theory: In defense of a resource-based of the firm. *The Journal of Supply Chain Management*, *48*(2), 14–20. doi:10.1111/j.1745-493X.2012.03266.x

Huo, B., Han, Z., & Prajogo, D. (2016). Antecedents and consequences of supply chain information integration: A resource-based view. *Supply Chain Management*, *21*(6), 661–677. doi:10.1108/SCM-08-2015-0336

Iansiti, M., & Levien, R. (2004). *The keystone advantage: what the new dynamics of business ecosystems mean for strategy, innovation, and sustainability*. Harvard Business School Press.

Iivari, M. M., Ahokangas, P., Komi, M., Tihinen, M., & Valtanen, K. (2016). Toward Ecosystemic Business Models in the Context of Industrial Internet. *Journal of Business Models*, *4*(2), 42–59.

IJntema, W., Frasincar, F., Goossen, F., & Hogenboom, F. (2010). Ontology-Based News Recommendation. *Proceedings of EDBT '10, New York, USA*.

In. (1998). *Conflict Identification and Resolution For Software Attribute Requirements*. Faculty of The Graduate School, University of Southern California.

Inokuchi, A., Washio, T., & Motoda, H. (2000). An Apriori-based Algorithm for Mining Frequent Substructures from Graph Data. Principles of Data Mining and Knowledge Discovery. *Lecture Notes in Computer Science*, *1910*, 13–23. doi:10.1007/3-540-45372-5_2

Inzinger, C., Satzger, B., Leitner, P., Hummer, W., & Dustdar, S. (2013). *Model-based adaptation of cloud computing applications*. MODELSWARD.

Iqbal, M., Zaidi, A., & Murtaza, S. (2010). A New Requirement Prioritization Model for Market Driven Products Using Analytical Hierarchical Process. *2010 International Conference on Data Storage and Data Engineering*. 10.1109/DSDE.2010.49

ISO/IEC. (2006). *International Standard - ISO/IEC 14764, IEEE Std 14764-2006, IEEE Standard for Software Maintenance*. IEEE Computer Society.

Iyer, B., & Henderson, J. C. (2012). Business Value from Clouds: Learning from Users. *MIS Quarterly Executive*, *11*(1), 51–61.

Jacobides, M., & Winter, S. (2005). The co-evolution of capabilities and transaction costs: Explaining the institutional structure of production. *Strategic Management Journal*, *26*(5), 395–413. doi:10.1002mj.460

Jaison, G. V., & Varghese, C. M. (2015). Privacy protection in personalized web search using homomorphic encryption, *Int. J. of Scientific and Research Publications*, *5*(9), 1–5.

Jamshidi, P., Ahmad, A., & Pahl, C. (2013). Cloud Migration Research: A Systematic Review. *IEEE Transactions Cloud Computing*.

Jamshidi, P., Pahl, C., & Mendonca, N. C. (2017). Pattern-based multi-cloud architecture migration. *Software, Practice & Experience*, *47*(9), 1159–1184. doi:10.1002pe.2442

Jamshidi, P., Sharifloo, A., Pahl, C., Arabnejad, H., Metzger, A., & Estrada, G. (2016). Fuzzy self-learning controllers for elasticity management in dynamic cloud architectures. *12th International ACM SIGSOFT Conference on Quality of Software Architectures QoSA*. 10.1109/QoSA.2016.13

Jansen, W. A. (2011, January). Cloud hooks: Security and privacy issues in cloud computing. In *System Sciences (HICSS), 44th Hawaii International Conference* (pp. 1-10). IEEE.

Jansen, W., & Grance, T. (2011). Guidelines on security and privacy in public cloud computing. *NIST special publication, 800*, 144.

Jasti, A., Shah, P., Nagaraj, R., & Pendse, R. (2010, October). Security in multi-tenancy cloud. In *Security Technology (ICCST), IEEE International Carnahan Conference* (pp. 35-41). IEEE. 10.1109/CCST.2010.5678682

Javed, M., Abgaz, Y. M., & Pahl, C. (2013). Ontology change management and identification of change patterns. *Journal on Data Semantics*, *2*(2-3), 119–143. doi:10.100713740-013-0024-2

Jbeli, N., Sbaï, Z., & Ben Ayed, R. (2016). On expressiveness of TCTLhdelta for Model Checking Distributed Systems. *International Conference on Computational Collective Intelligence.*

Jeckmans, A.J.P., Beye, M.R.T., Erkin, Z., Hartel, P.H., Lagendijk, R.L. & Tang, Q. (2013). Privacy in recommender systems. *Proceedings of Social Media Retrieval. Computer Communications and Networks*, 263-281.

Jeon, H.-J., Dant, R., & Baker, B. (2016). A knowledge-based explanation of franchise system resources and performance. *Journal of Marketing Channels*, *23*(3), 97–113. doi:10.1080/1046669X.2016.1186470

Jeya Mala, D., & Iswarya, R. (2014). Multi Agents Based Approach for Critical Components Identification and Verification. *International Journal of Systems and Service-Oriented Engineering*, *5*(1), 21–39. doi:10.4018/ijssoe.2014010102

Jeya Mala, D., Kamalapriya, S., & Mohan, V. (2010). Automated Software Test Optimization Framework – an Artificial Bee Colony Optimization based Approach. *IET Software*, *4*(5), 334–348. doi:10.1049/iet-sen.2009.0079

Jeya Mala, D., & Mohan, V. (2009). On the Use of Intelligent Agents in Test Sequence Selection and Optimization. *International Journal of Computational Intelligence and Applications*, *8*(2), 155–179. doi:10.1142/S1469026809002515

Jim, A., Randy, S., & David, C. (2007). Value-Oriented requirements Prioritization in a Small Development Organization. *IEEE Software*, 32–37.

Jin, C., Jin, S.-W., & Ye, J.-M. (2012). Artificial neural network-based metric selection for software fault-prone prediction model. *IET Software, Vol*, *6*(6), 479–487. doi:10.1049/iet-sen.2011.0138

Jones, S., Irani, Z., Sivarajah, U., & Love, P. E. (2017). Risks and Rewards of Cloud Computing in the UK Public Sector: A Reflection on Three Organisational Case Studies. *Information Systems Frontiers*, 1–24.

Kalfoglou, Y., Domingue, J., Vargas-Vera, E. M. M., & Shum, S. B. (2001). *myPlanet: an ontology-driven Web-based personalized news service*. Retrieved from www.kmi.open.ac.uk/tr/papers/kmi-tr-102.pdf

Kamkar, M. (1995). An Overview and Comparative Classification of Program Slicing Techniques. *Journal of Systems and Software*, *31*(3), 197–214. doi:10.1016/0164-1212(94)00099-9

Kandukuri, B. R., Paturi, V. R., & Rakshit, A. (2009, September). Cloud security issues. In *Services Computing, 2009. SCC'09. IEEE International Conference* (pp. 517-520). IEEE.

Kaner, M., & Karni, R. (2010). An ideation framework for service process improvement. *International Journal of Service Science, Management, Engineering, and Technology*, *1*(3), 17–49. doi:10.4018/jssmet.2010070102

Kapitsaki, G., Kateros, D., Prezerakos, G., & Venierris, I. (2009). Model-driven development of composite context-aware web applications. *Information and Software Technology*, *51*(8), 1244–1260. doi:10.1016/j.infsof.2009.03.002

Karaboga, D., & Basturk, B. (2008). On The Performance Of Artificial Bee Colony (ABC) Algorithm. Applied Soft Computing, 8(1), 687-697.

Karaboga, N. (2009). A new design method based on artificial bee colony algorithm for digital IIR filters. *Journal of the Franklin Institute*, *346*(4), 328–348. doi:10.1016/j.jfranklin.2008.11.003

Karastoyanova, D., & Leymann, F. (2009). BPEL'n'Aspects: Adapting service orchestration logic. In *International Conference on Web Services* (pp. 229-229). IEEE Press.

Karlsson, J., & Ryan, K. (1997). A Cost-Value Approach for Prioritizing Requirements. *IEEE Software*, *14*(Sept), 67–74. doi:10.1109/52.605933

Karlsson, J., Wohlin, C., & Regnell, B. (1998). An Evaluation of Methods for Prioritizing Software Requirements. *Journal of Information and Software Technology*, *39*(14-15), 939–947. doi:10.1016/S0950-5849(97)00053-0

Karlsson, L., Berander, P., Regnell, B., & Wohlin, C. (2004). Requirements Prioritisation: An Experiment on Exhaustive Pair-Wise Comparisons versus Planning Game Partitioning. *Proceedings of the 8th International Conference on Empirical Assessment in Software Engineering (EASE 2004)*. 10.1049/ic:20040407

Karunagaran, S., Mathew, S., & Lehner, F. (2016). *Differential Adoption of Cloud Technology: A Multiple Case Study of Large Firms and SMEs*. Paper presented at 36th International Conference on Information Systems, Dublin, Ireland.

KaterinaGoseva, P. (2003, October). Architectural-Level Risk Analysis Using UML. *IEEE Transactions on Software Engineering*, *29*(10), 946–959.

Kaur, I. (2015). A Compound Metric for Identification of Fault Prone Modules. *IOSR Journal of Computer Engineering, 17*(6), 31-35.

Kaur, R., & Pateriya, P. K. (2013, March). A Study on Security Requirements in Different Cloud Frameworks. International Journal of Soft Computing and Engineering, 3(1), 133-136.

Keller, A., & Ludwig, H. (2002). *The WSLA Framework: Specifying and monitoring Service Level Agreements for Web Services*. IBM.

Keller, A., & Kar, G. (2001). Determining service dependencies in distributed systems. *ICC*, *7*, 2084–2088.

Kellner, A. (2017). Human resource management standardisation and adaptation in franchises. *Service Industries Journal*, *37*(9-10), 545–566. doi:10.1080/02642069.2017.1339794

Kemmerer, R. A. (1985, January). Testing Formal Specifications to Detect Design Errors. *IEEE Transactions on Software Engineering*, *1*(1), 32–44.

Kharbili, M. E., & Keil, T. (2010). Bringing agility to business process management: Rules deployment in an SOA. Emerging Web Services Technology: Vol. 3. *Whitestein Series in Software Agent Technologies and Autonomic Computing*. Birkhaeuser Verlag.

Kilpinen, M., Eckert, C., & Clarkson, P. (2009). Assessing impact analysis practice to improve change management capability. *Proceeding of 17th International Conference Engineering Discipline*, 205–216.

Kimelfeld, B. & Y. S. (2006). Finding and approximating top-k answers in keyword proximity search. *PODS Conference*. 10.1145/1142351.1142377

Kim, G., Shin, B., Kim, K. K., & Lee, H. G. (2011). IT Capabilities, Process-Oriented Dynamic Capabilities, and Firm Financial Performance. *Journal of the Association for Information Systems*, *12*(7), 487–517.

Kim, H., & Boldyreff, C. (2002). Developing Software Metrics Applicable to UML Models. *ECOOP Workshop on Quantitative Approaches in Object-Oriented Software Engineering (QAOOSE)*.

Kim, H.-J. (2017). The complementary effects of transaction cost economics and resource-based view: A technological alliance perspective. *International Journal of Business Excellence*, *13*(3), 355–376. doi:10.1504/IJBEX.2017.087230

Kim, P., & Aldrich, H. (2005). *Social capital and entrepreneurship*. Now Publishers Inc.

Kobsa, A. (2001). Generic User Modeling Systems. *User Modeling and User-Adapted Interaction*, *11*(1/2), 49–63. doi:10.1023/A:1011187500863

Kobsa, A., Knijnenburg, B. P., & Livshits, B. (2014). Let's do it at my place instead?: attitudinal and behavioural study of privacy in client-side personalization. *Proceedings of the SIGCHI Conference on Human Factors in Computing Systems (CHI '14) ACM New York USA*, 81–90. 10.1145/2556288.2557102

Kogut, B., & Zander, U. (1993). Knowledge of the firm and the evolutionary theory of the multinational corporation. *Journal of International Business Studies*, *24*(4), 625–645. doi:10.1057/palgrave.jibs.8490248

Korel, B., & Laski, J. (1990). Dynamic slicing in computer programs. *Journal of Systems and Software*, *13*(3), 187–195. doi:10.1016/0164-1212(90)90094-3

Kor, Y., & Mahoney, J. (2000). Penrose resource-based approach: The process and product of research creativity. *Journal of Management Studies*, *37*(1), 109–139. doi:10.1111/1467-6486.00174

Ko, S. R., & Lee, S. S. (2013). *Cloud computing vulnerability incidents: A statistical overview*. Cloud Security Alliance.

Kosala, R., & Blockee, H. (2000). Web mining research: A survey. *SIGKDD Explorations*, *2*(1), 1–15. doi:10.1145/360402.360406

Kraaijenbrink, J., Spender, J., & Groen, A. (2010). The resource-based view: A review and assessment of its critiques. *Journal of Management*, *36*(1), 349–372. doi:10.1177/0149206309350775

Kumar Jha, S., & Singh, D. K. (2012). Development of knowledge Base Expert System for Natural treatment of Diabetes disease. *International Journal of Advanced Computer Science and Applications*, *3*(3). Retrieved from www.ijacsa.thesai.org

Kumar, P., & Kumar, L. (2013). Security Threats to Cloud Computing. *International Journal of IT, Engineering and Applied Sciences Research*, *2*(1).

Kumar, V. (2014). Making" freemium" work. *Harvard Business Review*, *92*(5), 27–29.

Labes, S., Hanner, N., & Zarnekow, R. (2017). Successfull Business Model Types of Cloud Providers. *Business & Information Systems Engineering*, *59*(4), 223–233. doi:10.100712599-016-0455-z

Lahiri, M., & Berger-Wolf, Y. T. (2010). Periodic subgraph mining in dynamic networks. *Journal of Knowledge and Information Systems*, *24*(3), 467–497.

Langheinrich, M. (2001, January). Privacy by design—principles of privacy-aware ubiquitous systems. In Ubicomp: Ubiquitous Computing (pp. 273-291). Springer Berlin Heidelberg.

Laplante, P. A., Zhang, J., & Voas, J. (2008). What's in a Name? Distinguishing between SaaS and SOA. *IT Professional*, *10*(3), 46–50. doi:10.1109/MITP.2008.60

Larsen, H. L. (1999). An approach to flexible information access systems using soft computing. *32nd Annual Hawaii International Conference on System Sciences*, 6042.

Larus, J. (1999). Whole Program Paths. In Proceedings of Programming Language Design and Implementation SIGPLAN (pp. 1-11). ACM.

Lassen, K. B., & van der Aalst, W. M. P. (2006). Workownet2bpel4ws: A tool for translating unstructured workflow processes to readable bpel. *Proceedings of the 2006 Confederated International Conference on On the Move to Meaningful Internet Systems: CoopIS, DOA, GADA, and ODBASE*, 127-144.

Lavie, D. (2006). The competitive advantage of interconnected firms: An extension of the resource-based view. *Academy of Management Review*, *31*(3), 638–658. doi:10.5465/AMR.2006.21318922

Law, J., & Rothermel, G. (2003). Incremental dynamic impact analysis for evolving software systems. In *Proceedings of the International Symposium on Software Reliability Engineering* (pp. 430-441). Academic Press. 10.1109/ISSRE.2003.1251064

Law, J., & Rothermel, G. (2003). Whole program path-based dynamic impact analysis. In *Proceedings of the International Conference on Software Engineering* (pp. 308-318). Academic Press.

Lee, M., Offutt, A. J., & Alexander, R. T. (2000). Algorithmic Analysis of the Impacts of Changes to Object oriented Software. In Proceedings of The Technology of Object-Oriented Languages and Systems (pp.61-70). Academic Press.

Lee, T. B., Hendler, J., & Lassila, O. (2000). The semantic web. *The Scientific Americans, 15*(1), 36. Retrieved from http://www.scientificamerican.com/2001/0501issue/0501berbers-lee.html

Lee, W., Liu, C., & Lu, C. (2002). Intelligent agent-based systems for personalized recommendations in Internet commerce. *Expert Systems with Applications*, *22*(4), 275–284. doi:10.1016/S0957-4174(02)00015-5

Lehtola, Kauppinen, & Kujala. (2004). Requirements prioritization challenges in practice. *Product focused software process improvement*, 497-508.

Leitner, P., Inzinger, C., Hummer, W., Satzger, B., & Dustdar, S. (2012). *Application-level performance monitoring of cloud services based on the complex event processing paradigm. In SOCA* (pp. 1–8). IEEE.

Leuschner, R., Rogers, D., & Charvet, F. (2013). A meta-analysis of supply chain integration and firm performance. *The Journal of Supply Chain Management*, *49*(2), 34–57. doi:10.1111/jscm.12013

Leusse, P. D., Dimitrakos, T., & Brossard, D. (2009). A governance model for SOA. In *IEEE International Conference on Web Services* (pp. 1020-1027). IEEE Press.

Lewis, D., Curran, S., Feeney, K., Etzioni, Z., Keeney, J., Way, A., & Schäler, R. (2009). Web service integration for next generation localisation. *Workshop on Software Engineering, Testing, and Quality Assurance for Natural Language Processing*, 47-55. 10.3115/1621947.1621956

Li, H. F., Lee, S. Y., & Shan, M. K. (n.d.). An efficient algorithm for mining frequent itemsets over the entire history of data streams. In *ECML/PKDD* (p. 2004). Springer.

Li, H., & Peng Lam, C. (2004). Software Test Data Generation using Ant Colony Optimization. *Transactions on Engineering, Computing and Technology*.

Li, C., & Li, L. Y. (2012). Optimal Resource Provisioning for Cloud Computing Environment. *The Journal of Supercomputing*, *62*(2), 989–1022. doi:10.100711227-012-0775-9

Li, J., & Sun, J. (2007). Automated Test Data Generation Algorithm Based on Reversed Binary Tree. *SNPD*, *3*, 1124–1128.

Li, L., & Offutt, A. J. (1996). Algorithmic analysis of the impact of changes to object-oriented software. In *Proceedings of IEEE International Conference on Software Maintenance* (pp.171-184). IEEE. 10.1109/ICSM.1996.565004

Lindblom, A., & Tikkanen, H. (2010). Knowledge creation and business format franchising. *Management Decision*, *48*(2), 179–188. doi:10.1108/00251741011022563

Linthicum. (2009). *Approaching SaaS Integration with Data Integration Best Practices and Technology*. White Paper.

Li-Pei, W., Puan, C. Y., Malcolm, Y. H. L., & Chong, C. S. (2008). Bee colony optimization algorithm with big valley landscape exploitation for job shop scheduling problems. *Applied Soft Computing*, *8*(1), 687–697.

Liu, G., Wong, L., & Chua, H. N. (2009). Complex discovery from weighted PPI networks. *Journal of Bioinformatics*, *25*(15), 1891–1897. doi:10.1093/bioinformatics/btp311 PMID:19435747

Liu, H., Ke, W., Wei, K. K., & Hua, Z. (2013). The Impact of IT Capabilities on Firm Performance: The Mediating Roles of Absorptive Capacity and Supply Chain Agility. *Decision Support Systems*, *54*(3), 1452–1462. doi:10.1016/j.dss.2012.12.016

Lockhart, H., Andersen, S., Bohren, J., Sverdlov, Y., Hondo, M., Maruyama, H., . . . Prafullchandra, H. (2006). Web services federation language (WS-federation) version 1.1. *International Business Machines Corp.* Retrived from http://www-128. ibm. com/developerworks/library/specification/ws-fed/(17.08. 2007)

Loebbecke, C., van Fenema, P., & Powell, P. (2016). Managing inter-organizational knowledge sharing. *The Journal of Strategic Information Systems*, *25*(1), 4–14. doi:10.1016/j.jsis.2015.12.002

Lohmann, N. (2008). *oWFN2BPEL: translates a Petri net model into an abstract BPEL process.* GNU Free Documentation License.

Lohmann, N., & Kleine, J. (2008). Fully-automatic translation of open workflow net models into simple abstract bpel processes. Modellierung 2008, 57-72.

López-Fernández, B., & López-Bayón, S. (2018). (forthcoming). Antecedents of early terminations in franchising: Franchisor versus franchisee cancelations. *Small Business Economics*. doi:10.10071118

Loyall, J. P., Mathisen, S. A., & Satterthwaite, C. P. (1997). Impact analysis and change management for avionics software. In *Proceedings of IEEE National Aerospace and Electronics Conference* (pp.740-747). IEEE. 10.1109/NAECON.1997.622723

Ludwig, H., Stamou, K., Mohamed, M., Mandagere, N., Langston, B., Alatorre, G., . . . Keller, A. (2015). rSLA: Monitoring SLAs in Dynamic Service Environments. In ICSOC (pp. 139-153). Springer.

Luo, S., Lin, Z., Chen, X., Yang, Z., & Chen, J. (2011, December). Virtualization security for cloud computing service. In *Cloud and Service Computing (CSC), International Conference* (pp. 174-179). IEEE.

Lusch, R. F., & Nambisan, S. (2015). Service innovation: A Service-Dominant Logic Perspective. *Management Information Systems Quarterly*, *39*(1), 155–175. doi:10.25300/MISQ/2015/39.1.07

Ma, C., Xu, Q., & Sanders, J. W. (2009). *A survey of business process execution language (bpel)*. UNU-IIST Report No. 425.

Ma, Q. (2009). The Effectiveness of Requirements Prioritization Techniques for a Medium to Large Number of Requirements: A Systematic Literature Review. School of Computing and Mathematical Sciences, Auckland University of Technology.

Ma, D. (2012, March). *Cryptographic Approach for Delegation and Authorization in Cloud Computing*. Presentation at *NSF Workshop on Security for Cloud Computing*.

Magnusson, M. (2000). Discovering hidden time patterns in behavior: T-patterns and their detection. *Behavior Research Methods, Instruments, & Computers*, 93-110.

Magoutis, K., Devarakonda, M. V., Joukov, N., & Vogl, N. G. (2008). Galapagos: Model-driven discovery of end-to-end application - storage relationships in distributed systems. *Journal of Research and Development (Srinagar)*, *52*(4-5), 367–378. doi:10.1147/rd.524.0367

Mahmood, Z. (2011, September). Data location and security issues in cloud computing. In *Emerging Intelligent Data and Web Technologies (EIDWT), International Conference* (pp. 49-54). IEEE. 10.1109/EIDWT.2011.16

Mahmudul Hasan, M. (2016). Regulatory Requirements Compliance in Requirements Engineering: A Systematic Classification and Analysis. *International Journal of Systems and Service-Oriented Engineering*, *6*(4). doi:10.4018/IJSSOE.2016100102

Mahoney, J., & Pandian, J. (1992). The resource-based view within the conversation of strategic management. *Strategic Management Journal*, *13*(5), 363–380. doi:10.1002mj.4250130505

Malhotra & Singh. (2010). A Regression Test Selection and Prioritization Technique. *Journal of Information Processing Systems, 6*, 235-252.

Malik, Z. K., & Fyfe, C. (2012). Review of web personalization. *Journal of Emerging Technologies in Web Intelligence*, *4*(3), 285–296. doi:10.4304/jetwi.4.3.285-296

Manku, G. S., & Motwani, R. (2002). *Approximate frequency counts over data streams. In VLDB* (pp. 346–357). ACM.

Marc, F., Benoit, T., Romain, C., Raphael, C. Y. F., & Mohammed, O. (2010). Comparison of Model Checking Tools for Information Systems. *Proceedings of the 12th International Conference on Formal Engineering Methods and Software Engineering*, 581-596.

March, J. G. (1991). Exploration and exploitation in organizational learning. *Organization Science*, *2*(1), 71–87. doi:10.1287/orsc.2.1.71

Marston, S., Li, Z., Bandyopadhyay, S., Zhang, J., & Ghalsasi, A. (2011). Cloud Computing – The Business Perspective. *Decision Support Systems*, *5*(1), 176–189. doi:10.1016/j.dss.2010.12.006

Martens, A. (2003). On compatibility of web services. *Petri Net Newsletter,* (65), 12-20.

Martina, M., & Antonia, B. (2003). Using Spanning Sets for Coverage Testing. *IEEE Transactions on Software Engineering*, *29*(11), 974–984. doi:10.1109/TSE.2003.1245299

Ma, S., & Hellerstein, L. J. (2001). Mining partially periodic event patterns with unknown periods. *Proceedings of 17th International Conference on Data Engineering*, 205 – 214.

Mayring, P. (2010). *Qualitative Inhaltsanalyse – Grundlagen und Techniken*. Weinheim, Germany: Beltz Verlag.

McGregor, J. D. (1997). Component Testing. *Journal of Object-Oriented Programming*, *10*(1), 6–9.

Medina, J. M. M., & Pons, O. (1994). Gefred. a generalized model of fuzzy relational databases. *Information Sciences*, 76.

Meiklejohn, A. (Ed.). (2014). *Franchising: cases, materials, and problems*. Chicago, IL: American Bar Association.

Meister, A., Breß, S., & Saake, G. (2014). *Cost-Aware Query Optimization during Cloud-Based Complex Event Processing*. GI-Jahrestagung.

Mentzer, J., Min, S., & Bobbitt, M. (2004). Toward a unified theory of logistics. *International Journal of Physical Distribution & Logistics Management*, *34*(8), 606–627. doi:10.1108/09600030410557758

Mertens, S., Gailly, F., & Poels, G. (2015). Enhancing Declarative Process Models with DMN Decision Logic Steven. In Lecture Notes in Business Information Processing (Vol. 214, pp. 151–165). Springer. doi:10.1007/978-3-319-19237-6

Metzger, A., Pohl, K., Papazoglou, M., Di Nitto, E., Marconi, A., & Karastoyanova, D. (2012). Research challenges on adaptive software and services in the future internet: towards an S-Cube research roadmap. *Workshop on European Software Services and Systems Research-Results and Challenges*, 1-7. 10.1109/S-Cube.2012.6225501

Miao, C., Yang, Q., Fang, H., & Goh, A. (2007). A cognitive approach for agent-based personalized recommendation. *Knowledge-Based Systems*, *20*(4), 397–405. doi:10.1016/j.knosys.2006.06.006

Middleton, S. E., Shadbolt, N. R., & De Roure, D. C. (2004). Ontological User Profiling in Recommender Systems. *ACM Transactions on Information Systems*, *22*(1), 54–88. doi:10.1145/963770.963773

Mietzner, R., Leymann, F., & Papazoglou, M. P. (2008). Defining composite configurable SaaS application packages using SCA, variability descriptors and multi-tenancy patterns. In *International Conference on Internet and Web Applications and Services* (pp. 156-161). IEEE Press. 10.1109/ICIW.2008.68

Mietzner, R., Unger, T., Titze, R., & Leymann, F. (2009). Combining different multi-tenancy patterns in service-oriented applications. In *IEEE International Enterprise Distributed Object Computing Conference* (pp. 131-140). IEEE Press. 10.1109/EDOC.2009.13

Min, S., Roath, A., Daugherty, P., Genchev, S., Chen, H., Arndt, A., & Richey, G. (2005). Supply chain collaboration: What's happening? *International Journal of Logistics Management*, *16*(2), 237–256. doi:10.1108/09574090510634539

Mishra, P., Singh, D. B. V., Rana, N., & Shailendra, S. (2014). Clinical Decision Support System for Diabetes Disease Diagnosis. *International Conference On Emerging Trends in Mechanical and Electrical Engineering (ICETMEE- 13th-14th March 2014)*, 105–110.

Mizuno, O., & Hata, H. (2010). An integrated approach to detect fault-prone modules using complexity and text feature metrics. *Proceedings of the 2010 international conference on Advances in computer science and information technology*, 457-468. 10.1007/978-3-642-13577-4_41

Mizuno, Y., & Odake, N. (2016b). A Study of Diffusion of Cloud Service with Two-sided Markets Structure and Freemium Business Model. *Production management: journal of Japan Society for Production Management, 23*(2), 7-15. (in Japanese)

Mizuno, Y., & Odake, N. (2015). A Case Study of Progressive Formation of Accounting Cloud Services in JAPAN. *International Journal of Service Science, Management, Engineering, and Technology*, *6*(1), 1–21. doi:10.4018/ijssmet.2015010101

Mizuno, Y., & Odake, N. (2016a). A Case Study of Progressive Formation of Accounting Cloud Service in Japan: The Second Stage. *International Journal of Service Science, Management, Engineering, and Technology*, *7*(3), 19–37. doi:10.4018/IJSSMET.2016070102

Mizuno, Y., & Odake, N. (2017). A Study of Development and Formation of Personal Information Trust Service in Japan. *International Journal of Service Science, Management, Engineering, and Technology*, *8*(3), 108–129. doi:10.4018/IJSSMET.2017070107

Mobasher, B., Honghua, D., Tao, L., & Miki, N. (2001). Effective personalization based on association rule discovery from web usage data. *Proceeding of 3rd International Workshop on Web Information and Data Management*, 9–15. 10.1145/502932.502935

Modi, B., & Mabert, A. (2007). Supplier development: Improving supplier performance through knowledge transfer. *Journal of Operations Management*, *25*(1), 42–64. doi:10.1016/j.jom.2006.02.001

Moore, J. (1996). *The death of competition: leadership and strategy in the age of business ecosystems*. New York: HarperBusiness.

Moore, J. F. (1993). Predators and Prey. *Harvard Business Review*, *71*(3), 75–83. PMID:10126156

Moore, J. F. (1993). Predators and Prey: A new Ecology of Competition. *Harvard Business Review*, *71*(3), 75–83. PMID:10126156

Moreton, R. (1990). A process model for software maintenance. *Journal of Information Technology*, *5*(2), 100–104. doi:10.1057/jit.1990.19

Mouratidis, K. D. P., & Bakiras, S. (2006). Continuous monitoring of top-k queries over sliding windows. *SIGMOD Conference*. 10.1145/1142473.1142544

Munday, J. (2016). *Introducing translation studies: Theories and applications*. Routledge.

Myers. (1979). The Art of Software Testing. John Wiley & Sons, Inc.

Myers, M. D., & Newman, M. (2007). The Qualitative Interview in IS Research: Examining the Craft. *Information and Organization*, *17*(1), 2–16. doi:10.1016/j.infoandorg.2006.11.001

Myres, G. J. (1979). *Art of Software Testing*. New York: John Wiley & Sons.

Nagappan, N., Ball, T., & Zeller, A. (2006). Mining Metrics to Predict Component Failures. *28th International Conference on Software Engineering (ICSE)*.

Nahrstedt, K., & Campbell, R. (2012). *Security for Cloud Computing. In A Report: Directorate for Computer and Information Science and Engineering* (pp. 1–19). CISE.

Naruchitparames, J., & Giine, M. H. (2011, July). Enhancing data privacy and integrity in the cloud. In *High Performance Computing and Simulation (HPCS), 2011 International Conference* (pp. 427-434). IEEE. 10.1109/HPCSim.2011.5999856

Neagu, G., Preda, Ş., Stanciu, A., & Florian, V. (2017). *A Cloud-IoT Based Sensing Service for Health Monitoring*. Paper presented at the 6th IEEE International Conference on E-Health and Bioengineering Conference (EHB), Sinaia, Romania. 10.1109/EHB.2017.7995359

Neves, A. R. M., Carvalho, Á. M. G., & Ralha, C. G. (2014). Agent-based architecture for context-aware and personalized event recommendation. *Expert Systems with Applications*, *41*(2), 563–573. doi:10.1016/j.eswa.2013.07.081

Nielsen, B. (2005). The role of knowledge embeddedness in the creation of synergies in strategic alliances. *Journal of Business Research*, *58*(9), 1194–1204. doi:10.1016/j.jbusres.2004.05.001

Object Management Group Inc. (2012). Service oriented architecture Modeling Language (SoaML) Specification. *OMG*,. Retrieved from http://www.omg.org/spec/SoaML/

Object Management Group. (2015). *Decision Model and Notation, 1.0(September), 172*. Retrieved from http://www.omg.org/spec/DMN/

Obulesu, O., Reddy, M. R., & Reddy, T. M. (2014). Finding Frequent and Maximal Periodic Patterns in Spatiotemporal Databases towards Big Data. *International Journal of Emerging Technology and Advanced Engineering*, *4*(1), 265–266.

Ogawa, K. (2014). *An Open & Close strategy: Condition of the Japanese company revival.* Tokyo: Shoeisha Publisher. (in Japanese)

Oke, A., Prajogo, D., & Jayaram, J. (2013). Strengthening the innovation chain: The role of internal innovation climate and strategic relationships with supply chain partners. *The Journal of Supply Chain Management, 49*(4), 43–58. doi:10.1111/jscm.12031

Olavarrieta, S., & Ellinger, A. (1997). Resource-based theory and strategic logistics research. *International Journal of Physical Distribution & Logistics Management, 27*(9-10), 559–587. doi:10.1108/09600039710188594

Oleshchuk, V. A., & Koien, G. M. (2011, February). Security and Privacy in the Cloud-A Long-Term View. In *Security and Privacy in the Cloud-A Long-Term View. In: 2nd International Conference on Wireless Communications, Vehicular Technology, Information Theory and Aerospace and Electronic Systems Technology (Wireless VITAE), WIRELESS VITAE'11* (pp. 1-5). 10.1109/WIRELESSVITAE.2011.5940876

Oliveira, T., Thomas, M., & Espadanal, M. (2014). Assessing the Determinants of Cloud Computing Adoption: An Analysis of the Manufacturing and Services Sectors. *Information & Management, 51*(5), 497–510. doi:10.1016/j.im.2014.03.006

Orso, A., Apiwattanapong, T., Law, J., Rothermel, G., & Harrold, M. J. (2004). An empirical comparison of dynamic impact analysis algorithms. In *Proceedings of the International Conference on Software Engineering* (pp. 491-500). Academic Press. 10.1109/ICSE.2004.1317471

Osterwalder, A., Pigneur, Y., & Tucci, C. L. (2005). Clarifying Business Models: Origins, Present, and Future of the Concept. *Communications of the Association for Information Systems, 16*(1), 1–43.

Ozden, B., Ramaswamy, S., & Silberschatz, A. (1998). Cyclic Association Rules. *Proceedings of 14th international conference on Data Engineering*, 412-421. 10.1109/ICDE.1998.655804

Pahl, C., & Collins, L. (2015). Software Service Adaptation Based on Interface Localisation. *International Journal of Systems and Service-Oriented Engineering, 5*(1). DOI: 10.4018/IJSSOE.2015010102

Pahl, C., & Xiong, H. (2013). Migration to PaaS Clouds – Migration Process and Architectural Concerns. *IEEE 7th International Symposium on the Maintenance and Evolution of Service-Oriented and Cloud-Based Systems MESOCA.*

Pahl, C., & Zhu, Y. (2005). A Semantical Framework for the Orchestration and Choreography of Web Services. *International Workshop on Web Languages and Formal Methods WLFM'05. ENTCS 151*(2), 3-18.

Pahl, C., Brogi, A., Soldani, J., & Jamshidi, P. (2017). Cloud Container Technologies: a State-of-the-Art Review. IEEE Transactions on Cloud Computing. doi:10.1109/TCC.2017.2702586

Pahl, C. (2005). Layered Ontological Modelling for Web Service-oriented model-Driven Architecture. *European Conference on Model-Driven Architecture ECMDA'05.*

Pahl, C. (2010). Dynamic adaptive service architecture - towards coordinated service composition. In *European Conference on Software Architecture ECSA'2010* (pp. 472-475). Springer-Verlag. 10.1007/978-3-642-15114-9_43

Pahl, C. (2012). Cloud Service Localisation. In *European Conference on Service-Oriented and Cloud Computing ESOCC 2012* (pp. 138-153). Springer-Verlag.

Pahl, C., Giesecke, S., & Hasselbring, W. (2009). Ontology-based Modelling of Architectural Styles. *Information and Software Technology, 1*(12), 1739–1749. doi:10.1016/j.infsof.2009.06.001

Pahl, C., & Jamshidi, P. (2015). Software architecture for the cloud–a roadmap towards control-theoretic, model-based cloud architecture. *European Conference on Software Architecture*. 10.1007/978-3-319-23727-5_17

Pahl, C., & Jamshidi, P. (2016). Microservices: A Systematic Mapping Study. *Proceedings CLOSER Conference*, 137-146.

Pahl, C., Jamshidi, P., & Weyns, D. (2017). Cloud architecture continuity: Change models and change rules for sustainable cloud software architectures. *Journal of Software: Evolution and Process*, *29*(2).

Pahl, C., Jamshidi, P., & Zimmermann, O. (2018). . Architectural Principles for Cloud Software. *ACM Transactions on Internet Technology*.

Pahl, C., Xiong, H., & Walshe, R. (2013). A Comparison of On-premise to Cloud Migration Approaches. *European Conference on Service-Oriented and Cloud Computing ESOCC*. 10.1007/978-3-642-40651-5_18

Pakdeetrakulwong, U., & Wongthongtham, P. (2013). State of the art of a multi-agent based recommender system for active software engineering ontology. *International Journal of Digital Information and Wireless Communications*, *3*(4), 29–42.

Panayides, M. (2007). Effects of organizational learning in third-party logistics. *Journal of Business Logistics*, *28*(2), 133–159. doi:10.1002/j.2158-1592.2007.tb00061.x

Pan, P., Wang, C., Horng, G., & Cheng, S. (2010). The development of an ontology-based adaptive personalized recommender system. *Proceedings of International Conference on Electronics and Information Engineering (ICEIE 2010)*, *1*, V1-76-V1-80. 10.1109/ICEIE.2010.5559842

Parra-Arnau, J., Rebollo-Monedero, D., & Forné, J. (2014). Measuring the privacy of user profiles in personalized information systems. *Future Generation Computer Systems*, *33*, 53–63. doi:10.1016/j.future.2013.01.001

Paswan, A., D'Souza, D., & Rajamma, R. (2014). Value co-creation through knowledge exchange in franchising. *Journal of Services Marketing*, *28*(2), 116–125. doi:10.1108/JSM-09-2013-0254

Patnaik, D., Ramakrishnan, N., Laxman, S., & Chandramouli, B. (2012). *Streaming Algorithms for Pattern Discovery over Dynamically Changing Event Sequences*. CoRR, abs/1205.4477

Paul Meinl. (2014, February). Software as a Service – *Correct Conclusion of Contracts 2nd enhanced edition*. Retrived from http://saas.clusterwien.at/fsDownload/SaaS_Guide_EN_07.pdf?forumid=342&v=1&id=5562745

Paulraj, A., Lado, A., & Chen, I. (2008). Inter-organizational communication as a relational competency: Antecedents and performance outcomes in collaborative buyer-supplier relationships. *Journal of Operations Management*, *26*(1), 45–64. doi:10.1016/j.jom.2007.04.001

Pearson, S. (2013). Privacy, security and trust in cloud computing. In *Privacy and Security for Cloud Computing* (pp. 3–42). Springer London. doi:10.1007/978-1-4471-4189-1_1

Peltoniemi, M. (2006). Preliminary Theoretical Framework for the Study of Business Ecosystems. *Emergence*, *8*(1), 10–19.

Penrose, E. (2013). *The theory of the growth of the firm*. Eastford, CT: Martino Fine Books. (Original work published 1959)

Perdreau, F., Le Nadant, A.-L., & Cliquet, G. (2015). Human capital intangibles and performance of franchise networks: A complementary view between agency and critical resource perspectives. *Managerial and Decision Economics*, *36*(2), 121–138. doi:10.1002/mde.2656

Perini, Ricca, & Bazzanella. (2007). An empirical study to compare the accuracy of AHP and CBRanking techniques for requirements prioritization. *Fifth International Workshop on Comparative Evaluation in Requirements Engineering*, 23-35.

Perini, A., Ricca, F. F., & Susi, A. (2009). Tool Supported requirement prioritization: Comparing the AHP and CBRank methods. *Information and Software Technology*, *51*(6), 1021–1032. doi:10.1016/j.infsof.2008.12.001

Perrigot, R., & Martinez-Ribes, L. (2008). Innovation in franchising: some illustrations from the French franchising chains. *Proceedings of the 7th International Congress on Marketing Trends*, 1-24.

Persais, E. (2004). Les compétences relationnelles peuvent-elles s'avérer stratégiques? *Revue Française de Gestion*, *158*(148), 119–145. doi:10.3166/rfg.148.119-145

Petcu, D. (2014). *SLA-Based Cloud Security Monitoring: Challenges, Barriers, Models and Methods. In Euro-Par Workshops* (pp. 359–370). Springer.

Peteraf, M. (1993). The cornerstones of competitive advantage: A resource-based view. *Strategic Management Journal*, *14*(3), 179–191. doi:10.1002mj.4250140303

Peyravi, F., & Latif, A. (2014). Secure multi agent information filtering. *International Journal of Computer Theory and Engineering*, *6*(3), 240–246. doi:10.7763/IJCTE.2014.V6.869

Pfleeger, S. L. (1998). *Software Engineering: Theory and Practice*. Englewood Cliffs, NJ: Prentice Hall.

Pham, D. T., Ghanbarzadeh, A., Koc, E., Otri, S., Rahim, S., & Zaidi, M. (2006). The Bees Algorithm, (2006), "A Novel Tool for Complex Optimization Problems. In *Proc 2nd International Virtual Conf on Intelligent Production Machines and Systems (IPROMS 2006)*. Oxford, UK: Elsevier.

Phillips, A. P. (2005). *Web Services and Internationalization*. Whitepaper. Retrieved October, 27, 2017 from http://www.inter-locale.com/whitepaper/multilingual/ml73-ws-20050524.xml

Pierrakos, D., Paliouras, G., Papatheodorou, C., & Spyropoulos, C. D. (2003). Web Usage Mining as a Tool for Personalization: A Survey. *User Modeling and User-Adapted Interaction*, *13*(4), 311–372. doi:10.1023/A:1026238916441

Podgurski, A., & Clarke, L. A. (1990). A formal model of program dependences and its implications for software testing, debugging, and maintenance. *IEEE Transactions on Software Engineering*, *16*(9), 965–979. doi:10.1109/32.58784

Polanyi, M. (1967/2009). *The tacit dimension*. Chicago, IL: University of Chicago Press.

Popescu, R., Staikopoulos, A., Liu, P., Brogi, A., & Clarke, S. (2010). *Taxonomy-driven adaptation of multi-tier applications using templates. In SASO* (pp. 213–222). IEEE.

Popovic, K., & Hocenski, Z. (2010, May). Cloud computing security issues and challenges. In *MIPRO, 2010 proceedings of the 33rd international convention* (pp. 344-349). IEEE.

Porter, B. W., & Bareiss, E. R. (1986). *PROTOS: An Experiment in Knowledge Acquisition for Heuristic Classification Tasks*. University of Texas at Austin. Retrieved from https://dl.acm.org/citation.cfm?id=899623

Porter, M. (1980). *Competitive advantage: creating and sustaining superior performance*. New York: The Free Press.

Prade, H., & Testemale, C. (1987). Fuzzy relational databases: Representational issues and reduction using similarity measures. *Journal of the American Society for Information Science*, *38*(2), 118–126. doi:

Prahalad, C.-K., & Hamel, G. (1990). The core competence of the corporation. *Harvard Business Review*, *68*(3), 79–91.

Prakash, K., & Sivakumar, R. (2013). Ontology-based intelligent multi-agent for diet food recommendation. *International Journal of Scientific & Engineering Research*, *4*(1), 1–8.

Pressman, R. S. (2007). *Software Engineering – A Practitioners Approach.* McGraw Hill Publishers.

Priem, R., & Butler, J. (2001). Is the resource-based theory a useful perspective for strategic management research? *Academy of Management Review, 26*(1), 22–40. doi:10.2307/259392

prikladnicki & Evaristo. (2003). Global Software Development in Practice Lessons Learned. *Software Process Improvement and Practice, 8*, 267–281.

Prince, B. (2009). *IBM Discovers Encryption Scheme That Could Improve Cloud Security, Spam Filtering.* E-Week.com.

Privacy, and Compliance in Microsoft Azure. (2014, February). Retrieved from http://azure.microsoft.com/en-in/support/trust-center/security/

Purushothama, B. R., & Amberker, B. B. (2012, December). Efficient Query Processing on Outsourced Encrypted Data in Cloud with Privacy Preservation. In *Cloud and Services Computing (ISCOS), 2012 International Symposium* (pp. 88-95). IEEE. 10.1109/ISCOS.2012.16

Qin, G., Yang, J., Gao, L., & Li, J. (2011). Evolution pattern discovery in dynamic networks. *Proceedings of IEEE International Conference on Signal Processing, Communications and Computing (ICSPCC)*, 1-6. doi:10.1109/ICSPCC.2011.6061813

Quan, T. T., Hui, S. C., & Cao, T. H. (2004). A Fuzzy FCA-based Approach to Conceptual Clustering for Automatic Generation of Concept Hierarchy on Uncertainty Data. *Concept Lattices and Their Applications Workshop*, 1-12.

Quélin, B. (1996). Appropriability and the creation of new capabilities through strategic alliances. In R. Sanchez & A. Heene (Eds.), *Strategic learning and knowledge management* (pp. 139–160). New York: John Wiley & Sons.

Racheva & Buglione. (2008). *Supporting the Dynamic Reprioritization of Requirements in Agile Development of Software Products. 2nd International Workshop on Software Product Management (IWSPM)*, Barcelona, Spain.

Rai, R., Sahoo, G., & Mehfuz, S. (2013). Securing software as a service model of cloud computing: Issues and solutions. International Journal on Cloud Computing: Services and Architecture, 3(4).

Ralston, P., Richey, R., & Grawe, S. (2017). The past and future of supply chain collaboration: A literature synthesis and call for research. *International Journal of Logistics Management, 28*(2), 508–530. doi:10.1108/IJLM-09-2015-0175

Ramayah, T., & Omar, R. (2010). Information exchange and supply chain performance. *International Journal of Information Technology & Decision Making, 9*(1), 35–52. doi:10.1142/S0219622010003658

Ramon, S., & Jose, A. L. (2006). Scatter Search in Software Testing, Comparison and Collaboration with Estimation of Distribution Algorithms. *European Journal of Operational Research, 169*(2), 392–412. doi:10.1016/j.ejor.2004.08.006

Rane, P. (2010). Securing SaaS Applications: A Cloud Security Perspective for Application Providers. *Information Security Management Handbook, 5.*

Rao, J., & Su, X. (2005). A survey of automated web service composition methods. *Proceedings of the First International Conference on Semantic Web Services and Web Process Composition, SWSWPC'04*, 43-54. 10.1007/978-3-540-30581-1_5

Raschia, G., & Mouaddib, N. (2002). SaintEtiq: A fuzzy set-based approach to database summarization. *Journal of Fuzzy Sets and Systems, 129*(2), 137–162. doi:10.1016/S0165-0114(01)00197-X

Rasheed, F., Alshalalfa, M., & Alhajj, R. (2011). Efficient periodic pattern mining in time series & sequence databases. *Journal of IEEE Transactions on Knowledge and Data Engineering, 23*(1), 79-94.

Razmerita, L. (2011). An ontology-based framework for modeling user behavior—A case study in knowledge management. *IEEE Transactions on Systems, Man, and Cybernetics. Part A, Systems and Humans*, *41*(4), 772–782. doi:10.1109/TSMCA.2011.2132712

Ren, S., Hu, C., Ngai, E., & Zhou, M. (2015). An empirical analysis of inter-organisational value co-creation in a supply chain: A process perspective. *Production Planning and Control*, *26*(12), 969–980. doi:10.1080/09537287.2014.1002022

Ribeiro, R., Moreira, A., Broek, P., & Pimentel, A. (2011). Hybrid assessment method for software engineering decisions. *Decision Support Systems*, *51*(1), 208–219. doi:10.1016/j.dss.2010.12.009

Ried, S. (2011). *Cloud Broker - A New Business Model Paradigm*. Forrester.

Rimal, B. P., Choi, E., & Lumb, I. (2009, August). A taxonomy and survey of cloud computing systems. In *INC, IMS and IDC, 2009. NCM'09. Fifth International Joint Conference* (pp. 44-51). IEEE. 10.1109/NCM.2009.218

Rivera, C. (2016). *The Decision. Mouth* (Vol. 1). 10.5281/ZENODO.154280

Rochet, J. C., & Tirole, J. (2003). Platform competition in two-sided markets. *Journal of the European Economic Association*, *1*(4), 990–1029. doi:10.1162/154247603322493212

Rogers, E. M. (2003). *Diffusion of innovations*. Simon and Schuster.

Romer, K. (2006). Distributed Mining of Spatio-Temporal Event Patterns in Sensor Networks. *EAWMS Workshop at DCOSS*, 103-116.

Rosaci, D., & Sarnè, G. M. L. (2014). Multi-agent technology and ontologies to support personalization in B2C E-Commerce. *Electronic Commerce Research and Applications*, *13*(1), 13–23. doi:10.1016/j.elerap.2013.07.003

Rosenberg, F., & Dustdar, S. (2005). Business rules integration in BPEL - a service-oriented approach. In *7th IEEE Intl Conference on E-Commerce Technology* (pp. 476 – 479). IEEE Press.

Rosenthal, S. S., & Strange, W. C. (2004). Evidence on the nature and sources of agglomeration economies. Handbook of regional and urban economics, 4, 2119-2171.

Rothermel, Chu, & Harrold. (1999). Test Case Prioritization: An Empirical Study. *Proceedings of the International Conference on Software Maintenance*.

Roy, P., Goswami, S., Chakraborty, S., Azar, A. T., & Dey, N. (2014). Image Segmentation Using Rough Set Theory: A Review. *International Journal of Rough Sets and Data Analysis*, *1*(2), 62–74. doi:10.4018/ijrsda.2014070105

Ruchika Malhotra, M. G. (2011). An Adequacy Based Test Data Generation Technique Using Genetic Algorithms. *Journal of Information Processing Systems*, *7*(2), 363–384. doi:10.3745/JIPS.2011.7.2.363

Ruchton, A., Croucher, P., & Baker, P. (Eds.). (2014). *The handbook of logistics and distribution management* (5th ed.). London: Kogan Page.

Ruhe, G., Eberlein, A., & Pfahl, D. (2002). Quantitative WinWin – A New Method for Decision Support in Requirements Negotiation. *Proceedings of the 14th International Conference on Software Engineering and Knowledge Engineering*.

Ruhe, G., Eberlein, A., & Pfahl, D. (2003). Trade-off Analysis for Requirements Selection. *International Journal of Software Engineering and Knowledge Engineering*, *13*(4), 345–366. doi:10.1142/S0218194003001378

Rungtusanatham, M., Salvador, F., Forza, C., & Choi, T.-Y. (2003). Supply chain linkages and operational performance: A resource-based view perspective. *International Journal of Operations & Production Management*, *23*(9), 1084–1099. doi:10.1108/01443570310491783

Russel, S., & Norvig, P. (1995). *Artificial Intelligence: A modern approach.* Prentice-Hall Inc.

Ryder, B. G., & Tip, F. (2001). Change impact analysis for object oriented programs. In *Proceedings of the ACM Workshop on Program Analysis for Software Tools and Engineering* (pp. 46-53). ACM. 10.1145/379605.379661

Saaty, T. (1980). *The Analytic Hierarchy Process: Planning, Priority Setting, Resource Allocation.* McGraw-Hill.

Sabahi, F. (2011, May). Cloud computing security threats and responses. In *Communication Software and Networks (ICCSN), 2011 IEEE 3rd International Conference* (pp. 245-249). IEEE. 10.1109/ICCSN.2011.6014715

Safarkhani, B., Mohsenzadeh, T. M., & Mojde, M. M. R. (2009). Deriving semantic sessions from semantic clusters. *International Conference on Information Management and Engineering*, 523-528. doi:10.1109/ICIME.2009.131

Sahai, A., & Waters, B. (2005). Fuzzy identity-based encryption. In *Advances in Cryptology–EUROCRYPT* (pp. 457–473). Springer Berlin Heidelberg.

Sahoo, G., & Mehfuz, S. (2013). Securing Software as a Service Model of Cloud Computing: Issues and Solutions. *International Journal on Cloud Computing: Services & Architecture, 3*(4).

Sanchez, R. (2000). Une comparaison des approches de la ressource, des capacités dynamiques, et de la compétence: une contribution à la théorie du management stratégique. In B. Quélin & J.-L. Arrègle (Eds.), *Le management stratégique des compétences* (pp. 55–81). Paris: Ellipses.

Sanchez, R., Heene, A., & Thomas, H. (1996). Towards the theory and practice of competence-based competition. In R. Sanchez, A. Heene, & H. Thomas (Eds.), *Dynamics of competence-based competition: theory and practice in the new strategic management* (pp. 1–35). New York: John Wiley & Sons.

Sanfelix, G., & Puig, F. (2018). New challenges in franchisor-franchisee relationship: An analysis from agency theory perspective. *Cuadernos de Gestión, 18*(1), 85–102. doi:10.5295/cdg.150610gn

Sassi, M., Touzi, A. G., Ounelli, H., & Aissa, I. (2010). About database summarization. *International Journal of Uncertainty, Fuzziness and Knowledge-based Systems, 18*(02), 133–151. doi:10.1142/S0218488510006453

Sbaï, Z., & Barkaoui, K. (2013). Vérification formelle des processus workflow - Extension aux workflows inter-organisationnels. *Revue Ingénierie des Systèmes d'Information, 18*(5), 33-57.

Sbaï, Z., & Barkaoui, K. (2015). On quantitative Analysis of Time Open Workflow Nets and Parametric Extension. *Proceedings of the 9th Workshop on Verification and Evaluation of Computer and Communication Systems, VECoS 2015*, 97-108.

Sbaï, Z., Barkaoui, K., & Boucheneb, H. (2014). Compatibility Analysis of Time Open Workflow Nets. *Proceedings of the International Workshop on Petri Nets and Software Engineering (PNSE 2014), in conjunction with Petri nets and ACSD*, 249-268.

Sbaï, Z., & Guerfel, R. (2016). CTL Model Checking of Web Services Composition based on Open Workflow Nets Modeling *International Journal of Service Science, Management, Engineering, and Technology, 7*(1), 27–42. doi:10.4018/IJSSMET.2016010102

Schneidewind, N. F. (1979, August). Application of Program Graphs and Complexity Analysis to Software Development and Testing. *IEEE Transactions on Reliability, R-28*(3), 192–198. doi:10.1109/TR.1979.5220563

Seacord, R. C., Plakosh, D., & Lewis, G. A. (2003). *Modernizing Legacy Systems: Modernizing Legacy Systems: Software Technologies, Engineering Processes, and Business Practices*. Addison-Wesley.

Securty and Privecy Overview. (2014, January). Retrived from https://trust.salesforce.com/

Sefion, I., Ennaji, A., Gailhardou, M., & Canu, S. (2003). ADEMA: A decision support system for asthma health care. *Studies in Health Technology and Informatics*, *95*, 623–628. Retrieved from http://www.ncbi.nlm.nih.gov/pubmed/14664057 PMID:14664057

Selznick, P. (1957/1984). *Leadership in administration: a sociological interpretation.* Berkeley, CA: University of California Press.

Senkul, P., & Salin, S. (2011). Improving pattern quality in web usage mining by using semantic information. *Knowledge and Information Systems*, *30*(3), 527–541. doi:10.100710115-011-0386-4

Sezen, B. (2008). Relative effects of design, integration and information sharing on supply chain performance. *Supply Chain Management*, *13*(3), 233–240. doi:10.1108/13598540810871271

Shah, A. (2008). Kernel-based Virtualization with KVM. *Linux Magazine*, *86*, 37–39.

Sharma, A., Al-khwaldeh, A. T., & Singh, A. (2016). ABSEP3S- An Agent Based Security Engine for Privacy Preserving in Personalized Search. *International Journal of Computing Academic Research*, *5*(3), 170–176.

Sharma, D. H., Dhote, C. A., & Potey, M. M. (2013). Security-as-a-Service from Clouds: A Comprehensive Analysis. *International Journal of Computers and Applications*, *67*(3), 15–18. doi:10.5120/11374-6642

Shen, X., Tan, B., & Zhai, C. (2007). Privacy protection in personalized search. *Proceedings of ACM SIGIR Forum*, *41*(1), 4-17.

Sherriff, M., & Williams, L. (2008). Empirical Software Change Impact Analysis using Singular Value Decomposition. In *Proceedings of 1st IEEE International Conference on Software Testing, Verification, and Validation* (pp. 268-277). IEEE. 10.1109/ICST.2008.25

Shin, Y., Meneely, A., & Williams, L. (2011). Evaluating Complexity, Code Churn, and Developer Activity Metrics as Indicators of Software Vulnerabilities. *IEEE Transactions on Software Engineering*, *37*(99), 1–16.

Shou, L., Bai, H., Chen, K., & Chen, G. (2014). Supporting privacy protection in personalized web search. *IEEE Transactions on Knowledge and Data Engineering*, *26*(2), 453–467. doi:10.1109/TKDE.2012.201

Sieg, A., Mobasher, B., & Burke, R. (2007). Ontological user profiles for personalized web search. American Association for Artificial Intelligence, 84-91.

Silberstein, A., & Braynard, R. (2006). A sampling-based approach to optimizing top-k queries in sensor networks. *IEEE International Conference on Data Engineering.* 10.1109/ICDE.2006.10

Silva, D., Oliveira, P., Rabelo, R., Campanhã, M., Neto, P., & Britto, R. (2016). A Hybrid Approach for Test Case Prioritization and Selection. *IEEE Congress on Evolutionary Computation.* 10.1109/CEC.2016.7744363

Sim, A. T., Idrawan, M., Zutshi, S., & Srinivasan, B. (2010). Logic-based pattern discovery. *Transasctions on Knowledge Data Engineering*, *22*(6), 798–811. doi:10.1109/TKDE.2010.49

Simons, A. J., Bratanis, K., Kourtesis, D., Paraskakis, I., Veloudis, S., Verginadis, Y., . . . Rossini, A. (2014). *Advanced Service Brokerage Capabilities as the Catalyst for Future Cloud Service Ecosystems.* Paper presented at the 2nd International Workshop on CrossCloud Systems, Bordeaux, France. 10.1145/2676662.2676675

Singh, A. (2009). An artificial bee colony algorithm for the leaf-constrained minimum spanning tree problem. *Applied Soft Computing*, *9*(2), 625–631. doi:10.1016/j.asoc.2008.09.001

Singh, A. (2012). Wisdom Web: The WWW generation next. *International Journal of Advancements in Technology, 3*(3), 123–126.

Singh, A., & Sharma, A. (2017). MAICBR: A Multi-agent Intelligent Content-Based Recommendation System. *Springer Lecture Notes in Networks and Systems, 10*, 399–411. doi:10.1007/978-981-10-3920-1_41

Singh, A., & Sharma, A. (2017). *Web Semantics for Personalized Information Retrieval. In Web Semantics for Textual and Visual Information Retrieval* (pp. 166–186). IGI Global.

Singh, A., Sharma, A., & Dey, N. (2015). Semantics and agents oriented web personalization: State of the art. *International Journal of Service Science, Management, Engineering, and Technology, 6*(2), 35–49. doi:10.4018/ijssmet.2015040103

Singh, A., & Teng, J. (2016). Enhancing supply chain outcomes through information technology and trust. *Computers in Human Behavior, 54*, 290–300. doi:10.1016/j.chb.2015.07.051

Slimani, T. (2013). Approaches for semantic web service: An overview. *International Journal of Service Science, Management, Engineering, and Technology, 4*(3), 18–36. doi:10.4018/ijssmet.2013070102

Somorovsky, J., Heiderich, M., Jensen, M., Schwenk, J., Gruschka, N., & Lo Iacono, L. (2011, October). All your clouds are belong to us: security analysis of cloud management interfaces. In *Proceedings of the 3rd ACM workshop on Cloud computing security workshop* (pp. 3-14). ACM. 10.1145/2046660.2046664

Soosay, C., Hyland, P., & Ferrer, M. (2008). Supply chain collaboration: Capabilities for continuous innovation. *Supply Chain Management, 13*(2), 160–169. doi:10.1108/13598540810860994

Sosnovsky, S., & Dicheva, D. (2010). Ontological technologies for user modeling. *International Journal of Metadata, Semantics and Ontologies, 5*(1), 32–71. doi:10.1504/IJMSO.2010.032649

Spohrer, J. (2007). Service Science, Management, and Engineering and Its Relation to Academic Disciplines. In B. Stauss, K. Engelmann, A. Kremer, & A. Luhn (Eds.), *Services science: Fundamentals, challenges and future developments* (pp. 11–40). Springer.

Srikant, R., & Agrawal, R. (1996). Mining Sequential Patterns: Generalizations and Performance Improvements. *Proceedings of the 5th International Conference on Extending Database Technology: Advances in Database Technology*, 3-17. 10.1007/BFb0014140

Srinivasa Rao, R., Narasimham, S. V. L., & Ramalingaraju, M. (2008). Optimization of Distribution Network Configuration for Loss Reduction Using Artificial Bee Colony Algorithm. *Proceedings Of World Academy Of Science, Engineering And Technology, 35*, 709–715.

Stare, M., & Rubalcaba, L. B. (2007). Research on Services: From Exploring the "Residual" to Services Science. In B. Stauss, K. Engelmann, A. Kremer, & A. Luhn (Eds.), *Services science: Fundamentals, challenges and future developments* (pp. 71–78). Springer.

Stauss, B., Engelmann, K., Kremer, A., & Luhn, A. (2007). *Services science: Fundamentals, challenges and future developments*. Springer.

Streed, O., & Cliquet, G. (2017). Autonomy in franchising. In F. Hoy, R. Perrigot, & A. Terry (Eds.), *Handbook of research on franchising* (pp. 50–64). Cheltenham, UK: Edward Elgar. doi:10.4337/9781785364181.00011

Subashini, S., & Kavitha, V. (2011). A survey on security issues in service delivery models of cloud computing. *Journal of Network and Computer Applications, 34*(1), 1–11. doi:10.1016/j.jnca.2010.07.006

Subramanyam, R., & Krishnan, M. S. (2003, April). Empirical Analysis of CK Metrics for Object-Oriented Design Complexity: Implications for Software Defects. *IEEE Transactions on Software Engineering*, *29*(4), 297–310. doi:10.1109/TSE.2003.1191795

Sultan, N. (2014). Servitization of the IT Industry: The Cloud Phenomenon. *Strategic Change*, *23*(5-6), 375–388. doi:10.1002/jsc.1983

Sun, W., Zhang, K., Chen, S. K., Zhang, X., & Liang, H. (2007). Software as a service: An integration perspective. *Service-oriented computing–ICSOC*, 558-569.

Sun, W., Zhang, X., Guo, C. J., Sun, P., & Su, H. (2008, September). Software as a service: Configuration and customization perspectives. In Congress on Services Part II, 2008. SERVICES-2 (pp. 18-25). IEEE.

Surl, P. K., & Kumar, S. (2010, June). Simulator for Identifying Critical Components for Testing in a Component Based Software System. *IJCSNS International Journal of Computer Science and Network Security, VOL.*, *10*(6), 250–257.

Sweeny, L. (2002). k-anonymity: A model for protecting privacy. *Int. J. Uncertain. Fuzz.*, *10*(5), 557–570. doi:10.1142/S0218488502001648

Szajnar, W., & Setlak, G. (2011). A concept of design process of intelligent system supporting diabetes diagnostics. *Methods and Instruments of Artificial Intelligence*, 168–178. Retrieved from http://www.foibg.com/ibs_isc/ibs-20/ibs-20-p21.pdf

Szudzik, M. (2006). *An elegant Pairing Function.* Retrieved from http://szudzik.com/ElegantPairing.pdf

Tabibi, S. T., Zaki, T. S., & Ataeepoor, Y. (2013). Developing an Expert System for Diabetics Treatment Advices. *International Journal of Hospital Research*, *2*(3), 155–162.

Taibi, D., Lenarduzzi, V., & Pahl, C. (2018). Processes, Motivations and Issues for Migrating to Microservices Architectures: An Empirical Investigation. IEEE Cloud Computing.

Takabi, H., Joshi, J. B., & Ahn, G. J. (2010, July). Securecloud: Towards a comprehensive security framework for cloud computing environments. In Computer Software and Applications Conference Workshops (COMPSACW), 34th Annual (pp. 393-398). IEEE.

Tan, W., Fan, Y., & Zhou, M. C. (2009). A petri net-based method for compatibility analysis and composition of web services in business process execution language. *IEEE T. Automation Science and Engineering*, 94-106.

Tang, Y., Lee, P. P., Lui, J. C., & Perlman, R. (2010). FADE: Secure overlay cloud storage with file assured deletion. In Security and Privacy in Communication Networks (pp. 380-397). Springer Berlin Heidelberg.

Tan, K.-H., Wong, W.-P., & Chung, L. (2016). Information and knowledge leakage in supply chain. *Information Systems Frontiers*, *18*(3), 621–638. doi:10.100710796-015-9553-6

Tao, Y., Lee, L.-H., & Chew, E.-P. (2016). Quantifying the effect of sharing information in a supply chain facing supply disruptions. *Asia-Pacific Journal of Operational Research*, *33*(4), 1–28. doi:10.1142/S0217595916500299

Tassey, G. (2002). *The Economic Impacts of Inadequate Infrastructure for Software Testing. Final Report.* National Institute of Standards And Technology.

Tatemoto, H., & Takanashi, T. (2010). Competitive strategies on standardization: Establishing and exploiting consensus standardization [in Japanese]. *Journal of Japan Association for Management Systems*, *26*(2), 67–81.

Teodorović, D., & Dell, M. (2006). *Bee Colony Optimization – A Cooperative Learning Approach To Complex Transportation Problems.* Advanced OR and AI Methods in Transportation.

Teshome, A., Rilling, L., & Morin, C. (2017). *Verification for Security Monitoring SLAs in IaaS Clouds: the Example of a Network IDS*. Inria Rennes Bretagne Atlantique.

Tiwari, P. K., & Joshi, S. (2015). Data Security for Software as a Service. *International Journal of Service Science, Management, Engineering, and Technology*, *6*(3), 47–63. doi:10.4018/IJSSMET.2015070104

Tiwari, P. K., & Mishra, B. (2012). Cloud Computing Security Issues, Challenges and Solution. *International Journal of Emerging Technology and Advanced Engineering*, *2*(8), 306–310.

Tiwari, R. G., Husain, M., Srivastava, V., & Agrawal, A. (2011). Web personalization by assimilating usage data and semantics expressed in ontology terms. *Proceedings of International Conference and Workshop on Emerging Trends in Technology (ICWET 2011) – TCET 13-15 December Mumbai India*, 516-521. 10.1145/1980022.1980133

Toch, E., Wang, Y., & Cranor, L. F. (2012). Personalization and privacy: A survey of privacy risks and remedies in personalization-based systems. *User Modeling and User-Adapted Interaction*, *22*(1), 203–220. doi:10.100711257-011-9110-z

Trajkova, J., & Gauch, S. (2004). Improving ontology-based user profiles. *Proceedings of RIAO 2004, University of Avignon (Vaucluse), France, April 26-28*, 380-389.

Tran, H. V. (2013, June). Data Management Challenges in Cloud Computing. In *Computational Science and Its Applications (ICCSA), 2013. 13th International Conference on* (pp. 19-27). IEEE. 10.1109/ICCSA.2013.14

Turver, R. J., & Munro, M. (1994). Early impact analysis technique for software maintenance. *Journal of Software Maintenance: Research and Practice*, *6*(1), 35–52. doi:10.1002mr.4360060104

Umano, M., & Fukami, S. (1980). Retrieval processing from fuzzy databases. *Technical Reports of IECE of Japan, 80*, 45-54.

Vanisri, D. & D. L. (2010). Survey on fuzzy clustering and rule mining. *International Journal of Computer Science and Information Security*, *8*, 183–187.

Vargo, S. L., & Lusch, R. F. (2004). Evolving to a new dominant logic for marketing. *Journal of Marketing*, *68*(1), 1–17. doi:10.1509/jmkg.68.1.1.24036

Vargo, S. L., & Lusch, R. F. (2008). Service-Dominant Logic: Continuing the Evolution. *Journal of the Academy of Marketing Science*, *36*(1), 1–10. doi:10.100711747-007-0069-6

Vargo, S. L., & Lusch, R. F. (2016). Institutions and Axioms: An Extension and Update of Service-Dominant Logic. *Journal of the Academy of Marketing Science*, *44*(1), 5–23. doi:10.100711747-015-0456-3

Vargo, S. L., Maglio, P. P., & Akaka, M. A. (2008). On Value and Value Co-Creation: A Service Systems and Service Logic Perspective. *European Management Journal*, *26*(3), 145–152. doi:10.1016/j.emj.2008.04.003

Varun Gupta & Dutta. (2012a). Decision Aspect Prioritization Technique for Globally Distributed Developments: A Hybrid Approach. Elsevier.

Varun Gupta & Dutta. (2012b). Impact Analysis of Requirement Prioritization on Regression Testing. *2nd World Conference on Innovation and Computer sciences (INSODE).*

Varun Gupta. (2011a). Hybrid Regression Testing Technique: A Multi Layered Approach. I*EEE annual Conference "INDICON"*.

Velten, C., & Janata, S. (2012). *Cloud Vendor Benchmark 2012*. München, Germany: Experton Group AG.

Vigneshwari, S., & Aramudhan, M. (2012). A novel approach for personalizing the web using user profiling ontologies. *IEEE- Fourth International Conference on Advanced Computing (ICoAC), 13-15 Dec., Chennai*, 1-4. doi:10.1109/ICoAC.2012.6416815

Vipindeep, V. (2005). *Efficient Static Analysis with Path Pruning using Coverage Data. International Worskhop on Dynamic Analysis*. WODA.

Vuljanic, D., Rovan, L., & Baranovic, M. (2010). Semantically enhanced web personalization approaches and techniques. *Proceedings of the ITI 2010 32nd Int. Conf. on Information Technology Interfaces Croatia*, 217-222.

W3C. (2005). *Web Services Internationalization Usage Scenarios*. Retrieved October, 27, 2017 from http://www.w3.org/TR/ws-i18n-scenarios

Wade, M., & Hulland, J. (2004). The Resource-Based View and Information Systems Research: Review, Extension, and Suggestions for Future Research. *Management Information Systems Quarterly*, *28*(1), 107–142. doi:10.2307/25148626

Wang, C. L., & Ahmed, P. K. (2007). Dynamic Capabilities: A Review and Research Agenda. *International Journal of Management Reviews*, *9*(1), 31–51. doi:10.1111/j.1468-2370.2007.00201.x

Wang, D., Rundensteiner, E., & Ellison, R. (2011). Active Complex Event Processing over Event Streams. *PVLDB*, *4*(10), 634–645.

Wang, M. X., Bandara, K. Y., & Pahl, C. (2010). Process as a Service - Distributed Multi-tenant Policy-based Process Runtime Governance. In *IEEE International Conference on Services Computing SCC 2010* (pp. 578-585). IEEE Press. 10.1109/SCC.2010.33

Webber, R. (2013). *An introduction to franchising*. New York: Palgrave Macmillan. doi:10.1007/978-1-137-29610-8

Webb, G. I., & Zhang, S. (2005). K-optimal rule discovery. *Data Mining and Knowledge Discovery*, *10*(1), 39–79. doi:10.100710618-005-0255-4

Weigand, H., Heuvel, W.-J. d., & Hiel, M. (2008). Rule-based service composition and service-oriented business rule management. *Interdisciplinary Workshop Regulations Modelling and Deployment*, 1-12. Retrieved October, 27, 2017 from http://ceur-ws.org/Vol-342/

Weiser, M. (1981). Program slicing. In *Proceedings of 5th IEEE International Conference on Software Engineering* (pp.439-49). IEEE.

Wernerfelt, B. (1984). A resource-based view of the firm. *Strategic Management Journal*, *5*(2), 171–180. doi:10.1002mj.4250050207

Wes, M., Andy, P., & David, L. (2007). An Empirical Study of Test Case Filtering Techniques Based on Exercising Information Flows. *IEEE Transactions on Software Engineering*, *33*(7), 454–477. doi:10.1109/TSE.2007.1020

Whipple, J., Wiedmer, R., & Boyer, K. (2015). A dyadic investigation of collaborative competence, social capital, and performance in buyer-supplier relationships. *The Journal of Supply Chain Management*, *51*(2), 3–21. doi:10.1111/jscm.12071

Wiegers. (1999b). First Things First: Prioritizing Requirements. *Software Development, 7*(9).

Wiegers, K. (1999a). *Software Requirements*. Redmond, WA: Microsoft Press.

Williamson, O. (1975). *Markets and hierarchies: analysis and antitrust implications. A study in the economics of internal organization*. New York: The Free Press.

Williamson, O. (1999). Strategy research: Governance and competence perspectives. *Strategic Management Journal, 20*(12), 1087–1108. doi:

Wong, Y.-C., & Karia, N. (2010). Explaining the competitive advantage of logistics service providers: A resource-based view approach. *International Journal of Production Economics, 128*(1), 51–67. doi:10.1016/j.ijpe.2009.08.026

Wooldridge, M., & Jennings, N. (1995). Intelligent agents: Theory and practice. *The Knowledge Engineering Review, 10*(2), 115–152. doi:10.1017/S0269888900008122

Wu, C., Zhu, Y., & Pan, S. (2013). *The SLA Evaluation Model for Cloud Computing. ICCNCE.* Atlantis Press.

Wu, E., Diao, Y., & Rizvi, S. (2006). High-performance complex event processing over streams. In *SIGMOD Conference* (pp. 407-418). ACM.

Wu, F., Yeniyurt, S., Kim, D., & Cavusgil, T. (2006). The impact of information technology on supply chain capabilities and firm performance: A resource-based view. *Industrial Marketing Management, 35*(4), 493–504. doi:10.1016/j.indmarman.2005.05.003

Wu, L., Chuang, C.-H., & Hsu, C.-H. (2014). Information sharing and collaborative behaviors in enabling supply chain performance: A social exchange perspective. *International Journal of Production Economics, 148*, 122–132. doi:10.1016/j.ijpe.2013.09.016

Wu, M., & Xu, J. (2006). Monitoring top-k query in wireless sensor networks. *IEEE International Conference on Data Engineering.*

Xen Architcture Overview. (2008, February). Xen Wiki Whitepaper.

Xiao, M., Mohamed, E. A., Marek, R., & James, M. (2007). Empirical Evaluation Of Optimization Algorithms When Used in Goal-Oriented Automated Test Data Generation Techniques. *Empirical Software Engineering, 12*(2), 183–239. doi:10.100710664-006-9026-0

Xu, Y., Zhang, B., Chen, Z., & Wang, K. (2007). Privacy-enhancing personalized web search. *Proceeding of WWW 2007 Banff Canada*, 591-600.

Yang, J., Wang, W., & Yu, P. (2003). Mining asynchronous periodic patterns in time series data. *Journal of IEEE Transactions on Knowledge and Data Engineering, 15*(3), 613–628. doi:10.1109/TKDE.2003.1198394

Yang, K., & Jia, X. (2014). Expressive, efficient, and revocable data access control for multi-authority cloud storage. *Parallel and Distributed Systems. IEEE Transactions, 25*(7), 1735–1744.

Yang, K., Jia, X., Ren, K., Zhang, B., & Xie, R. (2013). Dac-macs: Effective data access control for multiauthority cloud storage systems. *Information Forensics and Security. IEEE Transactions, 8*(11), 1790–1801.

Yang, Y., Yu, J. X., Gao, H., Pei, J., & Li, J. (2014). Mining most frequently changing component in evolving graphs. *Journal of World Wide Web., 17*(3), 351–376. doi:10.100711280-013-0204-x

Yano Research Institute. (2017). *Forecasts the domestic market scale of Fintech Venture Sales volume in Japan.* Retrieved 2017/11/2 from http://www.yano.co.jp/market_reports/C58122500 (in Japanese)

Yan, R., & Wang, K.-Y. (2012). Franchisor-franchisee supply chain cooperation: Sharing of demand forecast information in high-tech industries. *Industrial Marketing Management, 41*(7), 1164–1173. doi:10.1016/j.indmarman.2012.06.013

Yogesh Singh & Suri. (2010). A Hybrid Approach for Regression Testing in Interprocedural Programs. *Journal of Information Processing Systems, 6*, 21-32.

Yong, P. E. N. G., Wei, Z. H. A. O., & Feng, X. I. E. (2012). Secure cloud storage based on cryptographic techniques. *Journal of China Universities of Posts and Telecommunications*, *19*, 182–189. doi:10.1016/S1005-8885(11)60424-X

Yoo. (2007). Regression Testing Minimization, Selection and Prioritization: A Survey. *Software Testing, Verification & Reliability*.

Yu, Y., Ni, J., Au, M. H., Liu, H., Wang, H., & Xu, C. (2014). Improved security of a dynamic remote data possession checking protocol for cloud storage. *Expert Systems with Applications*, *41*(17), 7789–7796. doi:10.1016/j.eswa.2014.06.027

Zadeh, L. A. (1965). Fuzzy Sets. *Journal of Information and Control*, *8*(3), 338–353. doi:10.1016/S0019-9958(65)90241-X

Zaki, M. J., & Hsiao, C. J. (n.d.). *CHARM: An efficient algorithm for closed itemset mining. SDM.* SIAM.

Zeginis, C., Konsolaki, K., Kritikos, K., & Plexousakis, D. (2011). *Towards practive cross-layer service adaptation. In WISE* (pp. 704–711). Springer.

Zeginis, C., Kritikos, K., Garefalakis, P., Konsolaki, K., Magoutis, K., & Plexousakis, D. (2012). *Towards cross-layer monitoring of multi-Cloud service-based applications. In ESOCC* (pp. 188–195). Springer.

Zemankova-Leech, M. (1985). Implementing imprecision in information systems. *Information Sciences*, *37*(1-3), 107–141. doi:10.1016/0020-0255(85)90008-8

Zengin, A., Marconi, A., & Pistore, M. (2011). *CLAM: Cross-layer adaptation manager for service-based applications. In QASBA* (pp. 21–27). ACM. doi:10.1145/2031746.2031754

Zetta. (2008). *Zetta: Enterprise cloud storage on demand.* Retrieved from http://www.zetta.net/

Zhang, Q., Cheng, L., & Boutaba, R. (2010). Cloud Computing: State-of-the-Art and Research Challenges. *Journal of Internet Services and Applications*, *1*(1), 7–18. doi:10.100713174-010-0007-6

Zhao, J. L., Tanniru, M., & Zhand, L. (2007). Services Computing as the Foundation of Enterprise Agility: Overview of Recent Advances and Introduction to the Special Issue. *Frontiers of Information Systems*, *9*(1), 1–8. doi:10.100710796-007-9023-x

Zhou, M., Zhang, R., Xie, W., Qian, W., & Zhou, A. (2010, November). Security and privacy in cloud computing: A survey. In *Semantics Knowledge and Grid (SKG), 2010 Sixth International Conference* (pp. 105-112). IEEE. 10.1109/SKG.2010.19

Zhou, K., Zhang, Q., Sheng, S., Xie, E., & Bao, Y. (2014). Are relational ties always good for knowledge acquisition? Buyer-supplier exchanges in China. *Journal of Operations Management*, *32*(3), 88–98. doi:10.1016/j.jom.2014.01.001

Zhou, Y., & Leung, H. (2006, October). Empirical Analysis of Object-Oriented Design Metrics for Predicting High and Low Severity Faults. *IEEE Transactions on Software Engineering*, *32*(10), 771–789. doi:10.1109/TSE.2006.102

Zimmerman, T., Nagappan, N., Herzig, K., Premraj, R., & Williams, L. R. (2011). An Empirical Study on the Relation between Dependency Neighborhoods and Failures. *IEEE fourth international conference on Software Testing (ICST)*, 347-356.

Zissis, D., & Lekkas, D. (2012). Addressing cloud computing security issues. *Future Generation Computer Systems*, *28*(3), 583–592. doi:10.1016/j.future.2010.12.006

Zott, C., Amit, R., & Massa, L. (2011). The Business Model: Recent Developments and Future Research. *Journal of Management*, *37*(4), 1019–1042. doi:10.1177/0149206311406265

About the Contributors

Mehdi Khosrow-Pour, D.B.A., received his Doctorate in Business Administration from the Nova Southeastern University (Florida, USA). Dr. Khosrow-Pour taught undergraduate and graduate information system courses at the Pennsylvania State University – Harrisburg for almost 20 years. He is currently Executive Editor at IGI Global (www.igi-global.com). He also serves as Executive Director of the Information Resources Management Association (IRMA) (www.irma-international.org), and Executive Director of the World Forgotten Children's Foundation (www.world-forgotten-children.org). He is the author/editor of more than 100 books in information technology management. He is also the Editor-in-Chief of the *International Journal of Open Source Software and Processes*, *International Journal of Green Computing*, *International Journal of Digital Library Systems*, *International Journal of E-Entrepreneurship and Innovation*, *International Journal of Art, Culture, and Design Technologies*, *International Journal of Signs and Semiotic Systems*, and *International Journal of Disease Control and Containment for Sustainability*, and has authored more than 50 articles published in various conference proceedings and scholarly journals.

* * *

Thierry Allègre is working for a European logistics service provider. He holds a Ph.D. in Management from Aix-Marseille University, France, in November 2017. Member of the Centre de Recherche sur le Transport et la Logistique (CRET-LOG) in Aix-en-Provence, and CEO of the Club Demeter in Paris, his research fields include inter-firm relationships and more precisely the management of franchising system.

Ines Benali-Sougui (Tunisia) obtained the Engineering Master Degree in 2009 and the Ph.D. degree in electric genius in 2016, both from ENIT (Ecole Nationale d'Ingenieurs de Tunis). Her researches interest includes many aspects of fuzzy databases and flexible querying. She has authored three conference articles: A Quantitative Algorithm for Extracting Generic Basis of Fuzzy Association Rules (9th International Conference on Fuzzy Systems and Knowledge Discovery (FSKD 2012)), About Summarization in Large Fuzzy Databases (The Fifth International Conference on Advances in Databases, Knowledge, and Data Applications (DBKDA 2013)) and Flexible SQLf query based on fuzzy linguistic summaries (International Conference on Control, Engineering and Information Technology (CEIT'2013)) and two journal articles: From User Requirements to Flexible Querying of Fuzzy Summaries (International Journal of Service Science, Management, Engineering, and Technology (IJSSMET 2014)) and No-FSQL: a graph-based fuzzy NoSQL querying model (International Journal of Fuzzy System Applications (IJFSA 2016)).

Luke Collins received his B.Sc. degree from National University of Ireland, Maynooth in computer science and software engineering and his M.Sc. degree from Dublin City University in software engineering. His research interests include computer architecture, virtualization technology, cluster computing and grid computing, peer-to-peer computing, and network security. He joined IBM on graduating from Dublin City University where he holds the role of Senior Software Engineer.

Nasrin Dastranj received her B.Sc. in Computer Engineering from Shahid Beheshti University in 2004. She received her M.Sc. in IT Management from Alzahra University in 2010 and her Ph.D. in Science and Technology Policy from Tarbiat Modares Univeristy in 2016. She is currently assistant professor with the department of IT at ICT Research Institute (ITRC), Tehran, Iran. She is the author or a coauthor of more than 30 papers that have appeared in various journals and proceedings. Her main research interests include innovation systems, technology roadmapping and technology strategic management.

Jeya Mala Dharmalingam has a Ph.D. in Software Engineering with Specialization on Software Testing and is Associate Professor in Thiagarajar College of Engineering, –a leading educational and philanthropic institution in Tamil Nadu, India. She had been in the industry for about 4 years. She has a profound teaching and research experience of more than 12 years. She has published a book on "Object Oriented Analysis and Design using UML" for Tata McGraw Hill Publishers. She has published more than forty (40) papers about her research works at leading international journals and conferences such as IET, ACM, Springer, World Scientific, Computing and Informatics etc. As a researcher, Dr. Jeya Mala had investigated practical aspects of software engineering and object oriented paradigms for effective software development. Her work on Software Testing has fetched grants from UGC under Major Research Project scheme. Her dissertation has been listed as one of the best Ph.D. thesis in the CSIR – Indian Science Abstracts. She has successfully guided numerous Software Development based projects for the IBM- The Great Mind Challenge (TGMC) contest. The project she has mentored during 2007, has received national level Best Top 10 Project Award – 2007, from IBM. Currently she is guiding Ph.D. and M.Phil research scholars under the areas of Software Engineering and optimization techniques. She is a life member of Computer Society of India and an invited member of ACEEE. She forms the reviewer board in Journals like IEEE Transactions on Software Engineering, Elsevier – Information Sciences, Springer, World Scientific, International Journal of Metaheuristics etc. She has organized several sponsored national level conferences and workshops, notably she is one of the organizers of "Research Ideas in Software Engineering and Security (RISES'13) – A run-up event of ICSE 2014 sponsored by Computer Society of India". She has been listed in Marquis Who's Who list in 2011. She has completed certification on Software Testing Fundamentals, Brain bench certification on Java 1.1 programming, IBM certification as Associate Developer Websphere Application Studio. She is a proud recipient of several laurels from industries like Honeywell, IBM and Microsoft for her remarkable contributions in the field of Software Development and Object Orientation.

Christoph Ertl graduated with a Diploma in Business Information Systems at the Cooperative State University Stuttgart in 2007 and received his Master's degree in Information Systems at the Technical University of Munich (TUM) in 2010. Since 2011 he is working as an external research assistant at the Chair for Information Systems, Department of Informatics at TUM. His research focus encompasses state-owned enterprises and their challenges in management accounting. He is currently working as Vice President Financial Management in the service division Engineering and Facilities at Munich Airport International.

Sebastian Floerecke graduated with both a Bachelor's and a Master's degree in Information Systems at Technical University of Munich (Germany) in 2011 and 2014, respectively. During his studies, he worked as a student assistant at the Chair for Information Systems (Prof. Dr. Helmut Krcmar) in the research fields of product service systems and enterprise architecture management. In late 2014, he joined the Chair of Information Systems (with focus on Information and IT Service Management) of University of Passau held by Professor Franz Lehner, where he is currently research associate and doctoral candidate. His academic work focuses on success factors of cloud computing business models based on a business ecosystem perspective.

François Fulconis is Assistant Professor of Strategic and Logistics Management at the Avignon Institute of Technology, Avignon University (UAPV), France. Member of the Centre de Recherche sur le Transport et la Logistique (CRET-LOG) in Aix-en-Provence, his research interests include strategic partnerships, organisational theories, network structures, supply chain management, closed-loop logistics system and global competitiveness. He has published several reports, book chapters and papers in academic journals and conference proceedings.

Anshul Garg is working as a Software Engineer at Infoedge India Pvt. Ltd. His areas of interest include data mining and pattern analysis.

Amel Grissa-Touzi (Tunisia) received the diploma of engineering in computer science, Ph.D. and HDR in computer science from the Faculty of Sciences of Tunis, (Tunisia) in 1989, 1994 and 2010, respectively. Actually, Dr. Amel Grissa Touzi is a Professor at Technologies of Information and Communications Department in ENIT (Ecole Nationale d'Ingénieurs de Tunis), Tunisia. She is also a member of the LIPAH (Laboratoire d'Informatique, Programmation, Algorithmique et Heuristiques) laboratory in Computer Science Department in FST (Faculté des Sciences de Tunis), Tunisia. Her researches interest includes many aspects of deductive databases, fuzzy databases, flexible querying, logic programming and artificial intelligence.

Anand Gupta is an Associate Professor in Netaji Subhas Institute of Technology (NSIT). NSIT is affiliated to Delhi University, India. His research interests include Data Mining, Database System, Image Processing and Information Retrieval.

Chetna Gupta is an Associate Professor at Jaypee Institute of Information Technology, India. She obtained her Doctorate in the area of Software Testing. She also holds a Masters of Technology and a Bachelor of Engineering degree in Computer Science and Engineering. Her areas of interest are software engineering, requirement engineering, software testing, software project management, data structures and web applications. She has many publications in international journals and conferences to her credit.

Varun Gupta received his Ph.D & Master of Technology (By Research) in Computer Science & Engineering and Bachelor of Technology (Hon's) respectively. He had qualified State Eligibility Test (SET) conducted by Himachal Pradesh Public Service Commission (HPPSC) Shimla (Accredited by University Grants Commission as equivalent to UGC-NET). Presently he is working as Assistant Professor in the Department of Computer Science & Engineering at Amity University, Noida and had worked with National Institute of Technology, Hamirpur, PEC University of Technology, Chandigarh (Formerly Punjab Engineering College), Jawaharlal Nehru Govt. Engineering College (JNGEC), Sundernagar and Indian Institute of Technology- Mandi, M.G Institute of Engineering & Technology. He had also worked as Scientist-C in Chronicler Solutions, IT Park, Chandigarh. He was Coordinator of Remote Center for National Mission on Education through ICT (NMEICT) project of MHRD for IIT Bombay, Resource Center for Spoken Tutorial project and National Knowledge Network (NKN) coordinator at Jawaharlal Nehru Govt. Engineering College, Sundernagar. His area of interest is Software Engineering (Requirement Engineering, Requirement prioritization, Global Software Development, Software Testing). He has authored numerous research papers in peer reviewed, referred International Journals and various International and National Conferences that were held in India and Abroad. He is Associate Editor of IEEE Access (IEEE Journal, SCIE Indexed), Associate Editor of International Journal of Computer Aided Engineering & Technology (Inderscience Publishers, Scopus indexed), Editorial Team Member (Wiley, SCIE Indexed), Deputy Editor-In Chief of International Journal of Software Engineering And Computing (Serial Publishers, Scopus Indexed), Guest editor and review panel board member of many renowned international journals published by Inderscience, Springer, Wiley, IGI Global etc. He also holds lifetime memberships in various technical societies like IEEE, IEEE Industry Applications Society, IAENG etc. He is also on reviewer panel of Tata McGraw-Hill publisher and is subject matter expert of Software Engineering.

Alexander Herzfeldt earned his PhD in Management Information Systems from Munich University of Technology in 2014. He researched profitability of cloud service providers in his dissertation project. After finishing his PhD, Alexander started as consultant for a German-based private equity company. Since 2015 is working as management consultant at Siemens Management Consulting focusing on strategy and M&A projects in both the oil and gas (service) and digital factory industries. His research interests include service engineering and finance/accounting topics in service businesses.

Sandeep Joshi is a professor, Manipal University Jaipur, India. Dr. Joshi is the PhD director of the "Computer Science and Engineering Department at "Manipal University. He has published more than 30 research article in refereed international journals and conferences. He has given keynote/invited talks at conferences. His research interest covers Cloud Computing, Grid Computing, optical network, MANET and Network security.

Helmut Krcmar (born December 16, 1954 in Hanau, Germany) holds the Chair for Information Systems, Department of Informatics at the Technische Universität München (TUM), Germany since 2002 with a joint appointment to TUM Business School. Helmut is Vice-Dean and served as Dean, Faculty of Informatics from 10/2010 until 09/2013. He is Deputy Director of the Center for Doctoral Studies in Informatics and its Applications and Academic Director of the SAP University Competence Center @

TUM. Since inception of the Association of Information Systems (AIS) in 1995 he has been actively involved in numerous AIS activities. Helmut Krcmar has served in various capacities in the organization of AIS and its conferences. Since 6/2015 he is Immediate Past President of the AIS.

Kyriakos Kritikos received his BSc, MSc, and PhD degrees in Computer Science from the University of Crete. He was a Post-Doc Researcher at Politecnico di Milano and CNR in Italy as well as at CRP Henri Tudor in Luxembourg. He is currently a Researcher at the Information Systems Laboratory of the Institute of Computer Science, FORTH in Greece. His research interests span the following areas: Service-Oriented Computing with focus on quality-based service description, discovery, composition and adaptation and on automated service negotiation; Ontology modeling and reasoning; Constraint and Mathematical Programming; Distributed (Information) Systems; Cloud Computing.

J. Miguel Martins is an Economist and Research Assistant at the CRMC (Centre for Research in Marketing and Consumer Psychology) at Reykjavik University, Iceland. His academic affiliations are the European University of Lisbon (bachelor in Management), the London School of Economics and Political Science (Global Dimensions Programme) and the Reykjavik University (MBA class of 2019). His professional background focused on consulting in Portugal, and banking in the United Kingdom. Furthermore, his research interests include Management, Economics and Consumer Behavior.

Yutaka Mizuno received his Ph.D of Engineering in Civil Engineering & Master of Engineering in Management of Technology in 2017 and 2014 from Nagoya Institute of Technology, and Bachelor of Engineering in 1984 from Keio University, respectively. He joined Mitsubishi Electric Corporation on graduating from Keio University where he held the role of Senior System Engineer. His professional background focused on consulting in enterprise information system in Japan. His research interests include service innovation, platform development and ecosystem formation in social system, and management of technology. He is a member of the Japan Society for Production Management, the Japan Academy of Small Business Studies, the Records Management Society of Japan, Japan Industrial Management Association, and PICMET.

Nobutaka Odake is Director of Academy Industry Cooperative Engineering Center co.Ltd (2017-), Former Professor, Department of Techno-business Administration (2011- 2017). Ph.D., Nagoya Institute of Technology (2002). MD (Eng.), University of Tokyo (1974). Primarily responsible for Education and Research, both Japan and international in scope, in the fields such as academy-industry cooperation, regional development, innovation system and technology management. Concurrently responsible for 'Workforce development for shop manager program' in Nagoya Institute of Technology. Membership in Academic Societies: the Academic Association for Organizational Science; the Japan Society for Science Policy and Research Management; the Japan Society for Production Management; the Japan Academy of Small Business Studies; the Society of Chemical Engineers, Japan; Japan Society for Intellectual Production; Japan MOT Society; Japan Association for Management Systems; International Society for Standardization Studies; the Chubu Society for the Industrial Heritage; PICMET.

Gilles Paché is Professor of Retailing and Logistics Management at Aix-Marseille University, France. He has more than 360 publications in the forms of journal papers, books, edited books, edited proceedings, edited special issues, book chapters, conference papers, and reports. Member of the Centre de Recherche sur le Transport et la Logistique (CRET-LOG) and Director of the University Press of Aix-Marseille in Aix-en-Provence, his major interests are network organisations, supply chain management, distribution channels and retail operations management.

Claus Pahl is a Professor of Software Engineering at Free University of Bozen-Bolzano, Italy. He received this Ph.D. from the University of Dortmund and his M.Sc. from the Technical University of Braunschweig, Germany. His research focuses on software architecture and cloud service engineering. His specific interests include dynamic architectures (dynamic context-dependent composition, service models at runtime; constraints monitoring), business process and service architecture integration (layered architectures for process and architecture configuration); and mediation and integration in cloud computing (interoperability for on-demand architectures; multi-tenancy SOA and policy-based governance). He has published more than 250 papers. He is a reviewer for more than 30 journals and helped organizing more than 100 conferences and workshops. He is on the editorial board of four journals and he has acted as a panel member and an evaluator for various research schemes worldwide.

Dimitris Plexousakis is a Professor at the Department of Computer Science, University of Crete and Director of the Institute of Computer Science, FORTH in Greece. He was Chair of the Department of Computer Science from 2006 to 2010 and Head of the University of Crete's Data Processing Laboratory and Computing Center from 2009 to 2013. He received his BSc degree in Computer Science from the Department of Computer Science, University of Crete in 1988 and MSc and PhD degrees in Computer Science from the Department of Computer Science, University of Toronto in 1990 and 1996, respectively. He was a visiting Professor at the Universities of Vienna, CNAM, Orsay and Paris-Est. His research interests span the following areas: Knowledge Representation and Knowledge Base Design with emphasis on the representation and management of temporal knowledge; Distributed Database Systems and Databases on the Web with focus on the study of formal knowledge representation models and query languages for the Semantic Web; Formal reasoning systems and applications of artificial intelligence in database systems and Ambient Intelligence with focus on dynamic action theories and belief revision; Business process and e-service modeling over wide-area distributed information systems with focus on automated service composition and service adaptation. He has published over 160 articles in international conferences and journals and has served on the program committees of numerous international conferences. He has extensive experience in the scientific coordination of research projects at the national and European levels. He is a member of the ACM, AAAI and IEEE.

Zohra Sbai received a Ph.D. degree in Computer Science in 2010 from the Conservatoire National des Arts et Métiers (CNAM Paris - France) and jointly from the Ecole Nationale d'Ingénieurs de Tunis (ENIT-Tunisia). She also obtained an Engineer Diploma in Computer Science from ENIT in 2003 and a Master Diploma in ATS from ENIT in 2004 . Since 2004, Dr. Sbai is a faculty member (assistant then assistant professor) in the Information and Communication Technologies department of ENIT. Actually (since September 2016), she is attached to the College of Computer Engineering and Sciences, PSAU University, KSA where she is an assistant professor and the deputy head of the Computer Science department. Her research interests include Formal methods, Model checking, Petri nets, Timed and Concurrent

Systems, Application to the Modelling and Formal Verification of Workflow Processes and Web Services Composition. Dr. Sbai has authored or co-authored more than 30 journals and conference papers and has served as a reviewer for some journals and a PC member for many international conferences and workshops. She is a member of the Association for Computing Machinery (ACM).

Anu Sharma is working as a Scientist at ICAR-IASRI. She is Ph.D. in Computer Science from Maharishi Markandeshwar University, Mullana, Haryana. She has completed MCA from Maharishi Dyanand University, Rohtak, India. She has eighteen years of experience in research, teaching and training in computer application in agriculture and bioinformatics. She has published many research papers in various national and international journals. Her areas of research are web personalization using ontology and semantic web, bioinformatics, data mining and statistical software development. She is the member of editorial board of many good journals.

Aarti Singh is an Assistant Professor in Guru Nanak Girls College, Yamuna Nagar, India. She is Ph.D. in Computer Science from Maharishi Markandeshwar University, Mullana, Haryana. She also has M.Phil. (Computer Science), MCA, M.Sc. (Computer Science) and B.Sc. (Computer Science & App.) degrees to her credit. She has 15 years experience of teaching and research. She owes credit of 78 published research papers in various national & International Journals of repute, with one paper awarded as the Best Paper in an IEEE Conference. She has participated in many International conferences within India and abroad. She is member of many professional bodies such as: Life Member of Institution of Doctors, Engineers and Scientists (IDES),India, Member of Scientific Technical Committee & Associate Editor for International Journal of Computing Academic Research (IJCAR), Jordan. Editorial Member of International Journal of Advancements in Technology (IJoAT), Reviewer in Springer Journal of Institution of Engineers (India):Series B, Member of International Association of Engineers (IAENG). Her research interests include Semantic Web, Agent Technology, Web Mining and Web Personalization.

Maria José Sousa (PhD in Management) is an University Professor and a research fellow of CIEO (Algarve University) and BRU-ISCTE/IUL. Her research interests currently are information science, innovation and business management issues, with a particular focus on the learning and knowledge management processes. She is a best seller author in ICT and People Management and has co-authored over 50 articles and book chapters and published in several scientific journals (e.g. I. J. of Knowledge, Culture and Change Management, Knowledge Management, I.J. of Systems and Service-Oriented Engineering, Wseas Transactions on Business and Economics and Information Systems Frontiers and others), she has also organized and peer-reviewed international conferences, and is the guest-editor of three Special Issues. She has participated in several European projects of innovation transfer and is also External Expert of COST Association - European Cooperation in Science and Technology and President of the ISO/TC 260 – Human Resources Management, representing Portugal in the International Organization for Standardization.

Minyar Sassi Hidri (Tunisia) obtained the Engineering Master Degree in 2003 from ENIT (Ecole Nationale d'Ingénieurs de Tunis), the Ph.D. degree in electric genius in 2007 from ENIT, and the Habilitation Universitaire in computer science from FST (Faculté des Sciences de Tunis). From 2007 to 2008, she was an Assistant with department of Multimedia in the ISLAIB (Institut Supérieur des Langues

Appliquées et Informatique de Béja), Tunisia. From 2008 to 2011, she was an Assistant Professor with the department of Networks and Multimedia from ISITCom (Institut supérieur d'informatique et des techniques de communication de Hammam Sousse), Tunisia. From 2011 to 2017, she was an Assistant Professor with the department of computer science and communication technologies from the ENIT. From 2017 to present, she was an assistant professor with Imam Abdulrahman Bin Faisal University, Deanship of Preparatory Year and Supporting Studies, Dammam, Kingdom of Saudi Arabia. She has authored or coauthored of more than 30 refereed journal and conference articles in the field of fuzzy sets systems and applications. She has served on PCs for numerous international conferences. She is a guest editor of International Journal of Intelligent Engineering Informatics.

Hardeo Kumar Thakur is an Associate Professor in Manav Rachna University, India. His area of interest include Dynamic graph mining, Data Mining, Pattern Mining.

Pradeep Kumar Tiwari is a research scholar in Computer Science & Engineering Department at Manipal University Jaipur, India. He received the M.Phil degree in Computer Science from the Mahatma Gandhi Chitrakoot Gramodaya Vishwavidyalaya, M.P., India, 2012. The M.C.A degree in Computer Application from the Rajiv Gandhi Proudyogiki Vishwavidyalaya, M.P., India, in 2008. He has held Asst. Professor Positions at The Vindhya Institute of Technology, and the Aditya College of Technology, Satna, M.P, India. His research interests cover the Distributed Computing, Cloud Computing, Virtualization, Grid Computing, Cluster Computing and Network Security. He has published 6 article in refereed international journal and 10 publications in international conferences (i.e. IEEE, Springer). He is reviewer of 5 international conferences and at present, he is reviewer and editorial member of 4 journals.

Chrysostomos Zeginis received his BSc degree in Applied Informatics from the University of Macedonia and his MSc and PhD degrees in Computer Science from the University of Crete. He is currently a Post-doc Researcher at the Information Systems Laboratory of the Institute of Computer Science, FORTH in Greece. His research interests span the following areas: Service-Oriented Computing with focus on Monitoring and Adaptation; Pattern Discovery; Meta-modeling; Distributed (Information) Systems; Cloud Computing; Multi-cloud systems.

Index

L

M

N

O

P

R

S

T

W

www.ingramcontent.com/pod-product-compliance
Lightning Source LLC
LaVergne TN
LVHW060405300926
849162LV00006B/1352